D1175660

Agriculture and Environmental Change

For J.C.M., M.M. and A.N.C: they are sadly missed. And for M.D.T. who helps to make life worthwhile.

Naturam expellas furca
tamen usque recurret

(Though you drive out nature with a pitchfork, yet will she always return)

Horace *Epistles* I.x.24

Agriculture and Environmental Change

Temporal and Spatial Dimensions

A.M. Mannion
Department of Geography,
University of Reading, UK

JOHN WILEY & SONS
Chichester • New York • Brisbane • Toronto • Singapore

Other Wiley Editorial Offices

John Wiley & Sons, Inc., 605 Third Avenue,
New York, NY 10158-0012, USA

Jacaranda Wiley Ltd, 33 Park Road, Milton,
Queensland 4064, Australia

John Wiley & Sons (Canada) Ltd, 22 Worcester Road,
Rexdale, Ontario M9W 1L1, Canada

John Wiley & Sons (SEA) Pte Ltd, 37 Jalan Pemimpin #05-04,
Block B, Union Industrial Building, Singapore 2057

British Library Cataloguing in Publication Data

A catalogue record for this book is available from the British Library

ISBN 0-471-95478-0

Typeset in 10/12pt Times by Mayhew Typesetting, Rhayader, Powys
Printed and bound in Great Britain by Redwood Books Ltd, Trowbridge, Wiltshire

This book is printed on acid-free paper responsibly manufactured from sustainable forestation,
for which at least two trees are planted for each one used for paper production.

Contents

Preface

By the time I had completed the research for the two chapters on agriculture in my earlier book *Global Environmental Change* (Longman, Harlow, 1991), I had become convinced that the role of agriculture as an agent of environmental change had been inadequately documented. This, in combination with the fact that for several years I have been teaching an undergraduate course on biotic resources which focuses on agriculture, led me to design and complete this volume. As far as I am aware, there is no comparable text that examines the temporal and spatial context of agriculture and its changing practice in combination with its impact on the Earth's surface. As with my earlier books, this text arises and benefits considerably from my research interests in palaeoenvironments and people/environment relationships in prehistory as well as my teaching commitments that involve the global biogeography of the present-day, biotic resource use and environmental change. This illustrates and highlights the close relationship that exists between research, scholarship and teaching at a time when that unique and productive relationship, the mainstay of undergraduate degree courses in higher education establishments in the UK, is being trivialised and subverted under the now scurrilous term 'rationalisation'.

The use of another equally discredited phrase 'back to basics' is also unavoidable in relation to this book. Not, of course, because this book is discreditable but because agriculture is basic. It provides what so many of the population of the developed world take for granted: a plentiful and varied food supply. There are few people in this sophisticated, high-technology world who know what photosynthesis is; even fewer people realise how much they, and people at large, are dependent upon it. In shops, bread is now available in several flavours and is sold in plastic bags, whilst water not only emanates from taps but can also be found in plastic bottles in supermarkets. Whilst almost all foods reflect technology to some degree (agriculture itself is a form of technology) they are not only a means of sustenance but also a means of wealth generation and political security. Without adequate food production, secondary and tertiary industries would collapse as would world political order.

To transpose any component of the Earth's surface into a resource requires ingenuity. This unpredictable factor militates against analysis and projection. For this reason, there are considerable difficulties in the establishment of frameworks within which to examine people/environmental relationships. Although such a grave issue is worthy of a text in its own right, it is important to emphasise the pivotal role of agriculture within this relationship. This book attempts to do so by examining the temporal and spatial dimensions of agriculture. It encompasses palaeoenvironmental and archaeological evidence for the initiation and spread of agricultural systems in prehistory. It includes a brief review of the historical aspects of changes in agricultural practice and, mainly through case studies, it describes the transitory and permanent arable and pastoral agricultural systems that characterise the Earth's surface today. Unlike most books that deal with agriculture, several chapters are devoted to the environmental impact of agriculture; notably on natural habitats and biodiversity, soil erosion and water quality. Penultimately, I have provided a chapter on current and likely future developments in the reciprocal relationship that exists between agriculture and environmental change. The factors involved include advances in biotechnology (including genetic engineering), the need for sustainable agricultural systems as a basic requirement of sustainable development policies and the likely impact of global warming. The endpiece considers the impact of two vital factors, i.e. technological development and population pressure, on agricultural systems in first, a temporal context and, secondly, in a spatial context. It also addresses the challenges that the practice and organisation of world agriculture will need to meet as the millennium approaches.

This book is intended for students on higher education courses in geography, environmental sciences, applied biological sciences, agriculture and related disciplines. Each chapter has recommendations for further reading and there is an extensive reference list to help those who wish to read further on a topic that is so vital for the Earth's and its population's health.

ANTOINETTE M. MANNION
University of Reading
November 1994

Acknowledgements

I should like to record my thanks to Sonia Luffrum and Lisa Perkins for typing the manuscript and to Judith Fox for drawing the diagrams. I have had many valuable conversations with my colleague Dr Erlet Cater, for whose time I am most grateful whilst Mr John Creasy of the Rural History Centre, University of Reading, and the staff of the Inter-Library Loans section of the University Library have given invaluable assistance in the search for relevant literature. My appreciation also goes to Dr Michael D. Turnbull who has patiently read chapters, organised the index and assisted with all the chores inherent in the production of a book.

CHAPTER 1

Introduction: Ecosystems, Agricultural Systems and Energy

Agriculture is, without a doubt, unsurpassed as a cultural agent of environmental change. By its very nature, agriculture requires modification of the environment in which it is practised. The degree of modification varies temporally and spatially. In a temporal context, the early development of agriculture is difficult to ascertain. Scavenging, hunting and gathering, and eventually plant and animal domestication represent a continuum of human manipulation of biotic resources; there are few sharp boundaries in the archaeological record that indicate when one activity ceased and another began. Such obfuscation is inevitable in view of the nature of archaeological evidence, its spatial unevenness and the diversity of pathways along which agriculture may have developed, as well as the likelihood of overlap between phases. Throughout prehistory and history agricultural systems have spread and evolved, affecting all the continents except Antarctica. The resulting food resource allowed other changes in society to occur, such as craftwear production and, eventually, industrialisation. Through the ages science and technology have been brought to bear on food production and so intensified the role of agriculture as an agent of environmental change. Currently, such innovations include genetic engineering; this is beginning to facilitate the design of crops for the environment rather than the converse and more traditional approach to food, fuel and fibre production.

Spatially, the variety of agricultural systems is enormous, as is the variety in Earth-surface modification that each system creates. This modification may be comparatively slight, as exemplified by some forms of nomadic pastoralism. Other types of agricultural systems, such as intensive cereal production, have wide-reaching environmental implications. All agricultural systems are characterised by inputs, processes and outputs and thus there are parallels with the natural ecosystems from which agricultural systems derive. In effect, agricultural systems are a form of control system and the degree of control influences the extent of environmental change they may create. Since its inception, some 10 000 years ago, agriculture has contributed to soil erosion, desertification and deforestation. More recent problems include cultural eutrophication which has developed as a response to intensive artificial fertiliser use. These, and possibly other problems that may arise from biotechnology, are likely to increase in the twenty-first century as agricultural systems come under increased pressure to produce more food, a necessary response to a growing world population.

All agricultural systems, no matter how simple or sophisticated, have many features in common. These are the manipulation, through human controls, of trophic energy transfers and nutrient exchanges that are components of biogeochemical cycles. These ecosystem characteristics are controlled, to a lesser or greater degree, by the varying amounts of energy and particularly fossil-fuel energy inputs, to give the variety of agricultural systems that have occurred temporally and/or spatially.

1.1 WHAT IS AGRICULTURE?

Agriculture, or farming, is the production of crop plants through cultivating the soil and the rearing of animals. The world itself derives from the latin word *agar* meaning field and the greek word *agros* which also means field. This simple definition, however, whilst being succinct belies the diversity of agricultural types that occur over the Earth's surface. These, in turn, reflect the degree of control that is exerted by the farmer on the plants, animals and their milieu of growth that comprise an agricultural system. How the control operates, whether it is intensive or extensive, and what it consists of in terms of energy inputs, is a response to social and economic requirements (Figure 1.1). The latter are, thus, forcing factors but they can only operate within certain boundaries which are defined by environmental characteristics; these act as constraints. As social and economic factors change, so too does agriculture which is, consequently, a dynamic activity in a continuous state of flux. This interrelationship reflects the dependency of society on agricultural systems that, in turn, represent a means of tapping the natural, and in principle renewable, resources of the Earth's surface. Agriculture is thus a manifestation of the people–environment relationship. Temporally, it allowed human societies to develop and differentiate, the degree of which is manifest to a large extent in today's spatial distribution of agricultural systems. Many people, especially a substantial part of the developed world, do not recognise this relationship and fail to appreciate its significance. In one respect, this is because agriculture has been 'too successful'; most people in the developed world enjoy a high degree of food security but are divorced from the primary production of agriculture. Such people see only the end products, neatly packaged, in supermarket refrigerators and on shop shelves. It is all too easily forgotten that this food security underpins political stability and pre-eminence, industrial activity, services, comfortable standards of living as well as cultural and educational achievements. The most appropriate analogy is that of a house of cards. Agriculture occupies the basal layer; if it fails then the rest collapses.

Agriculture is clearly influenced by social and economic factors of which population numbers and rates of population growth are very important. These, amongst many other factors, represent inputs into agricultural systems. The systems themselves, however, comprise an array of natural resources, the dominant control on which is climate. Average temperatures and precipitation amounts and their annual distribution constrain agricultural activity more than any other environmental factors. Soil type, itself related to geology, also influences agriculture; factors such as erosivity, nutrient status, structure, drainage and water content all affect the crops that can be grown and the animals that can be reared (Figure 1.2). The plants and animals produced in agricultural systems are all components of the Earth's biota, another resource that provides products and services. Cultivated species have been selected over time, and bred, because of specific attributes such as high protein or high carbohydrate contents. Remarkably few species, in comparison with the Earth's

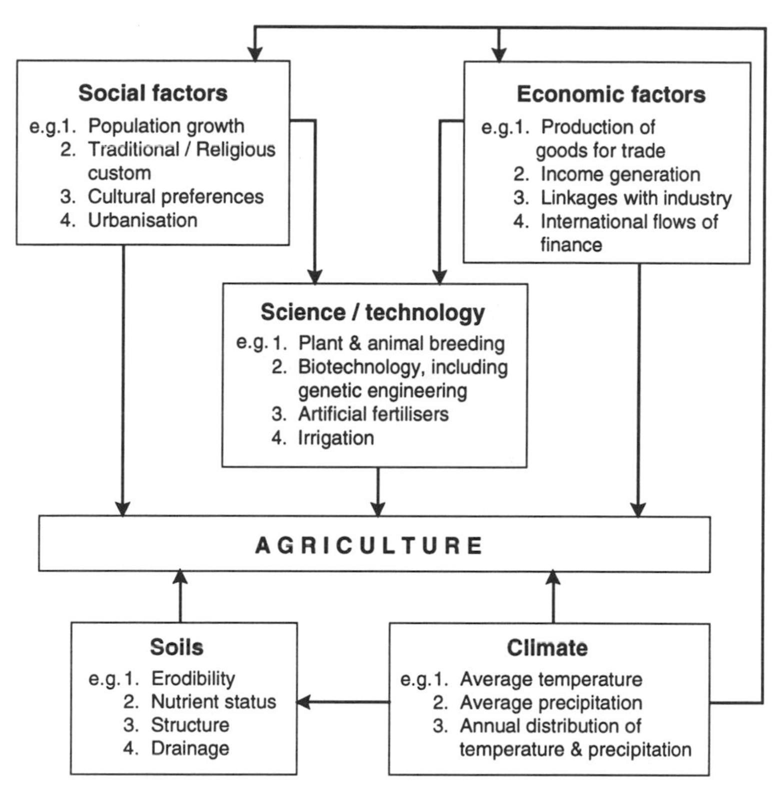

Figure 1.1 The factors involved in agriculture

total biota, have been domesticated and the majority of the world's major crop plants and livestock emanate from prehistoric farming practices in the Near East, Asia and the Americas. The agricultural biota is the recipient via the farmer of the social and economic forcing factors and it is manipulated to react to these within the limitations imposed by environmental factors.

However, there is another facet of agriculture that is produced by society but which acts directly on the biota and/or its environmental resources. This is the input of science and technology that, over the years, has been brought to bear on altering the biota (through breeding programmes and now modern biotechnology) or the environment, e.g. the use of artificial fertilisers. Such endeavours have been aimed at increasing agricultural productivity in two ways. First, plant and animal breeding programmes have been directed at improving species' productivity of food, fuel or fibre. This has taken many forms, including the breeding of plants and animals with disease resistance, drought resistance, etc. Such work is being continued via modern biotechnology that includes genetic engineering and which is opening up a whole new range of possibilities. Secondly, efforts have been, and will continue to be, directed at changing the environment in which agriculture takes place. Examples of this include artificial fertiliser use and irrigation. In the first instance nitrate availability, which is often a limiting factor to growth, is enhanced and in the second instance, another constraint on productivity, that of water availability, is artificially improved. All of these inputs have indeed increased productivity but they are not without

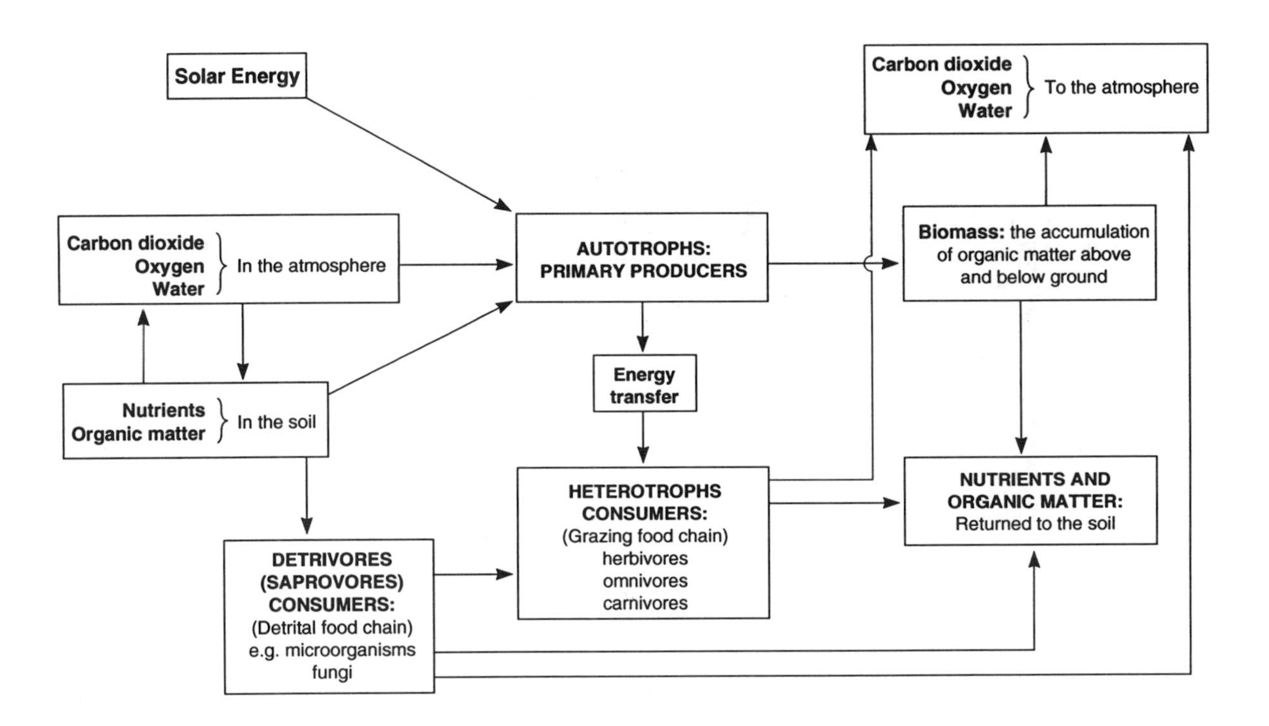

Figure 1.2 The major components and relationships in a terrestrial ecosystem

environmental cost. Occasionally, that cost may result in severe land degradation and a concomitant loss of its total productivity.

Defining agriculture simply in terms of plant and animal production is, consequently, naïve and misleading. As Figures 1.1 and 1.2 show, agriculture is a reaction to a complex set of factors that include social and economic circumstances, usually mediated through political policies that prompt scientific enquiry and which operate under an umbrella of environmental constraints. It is an activity that underpins all other human activities. This highlights the dependence of human endeavour in all its forms on the ability of green plants to produce food and agriculture, directly or indirectly, is a major means of wealth generation. Agriculture also represents one of the main ways in which society attempts not simply to dominate but to subjugate nature, an attitude that has pervaded Western intellectualism since the days of Roman imperialism.

1.2 THE RELATIONSHIP BETWEEN ECOSYSTEMS AND AGRICULTURAL SYSTEMS

All agricultural systems are derived from natural ecosystems and both have many features in common. As in any system, both have inputs and outputs and are characterised by components and processes. The major difference between the two is that the inputs, outputs and processes that operate in agricultural systems are controlled to a large extent by human decision making.

The term 'ecosystem', defined in the 1930s by the British botanist Sir Arthur Tansley, means a unit comprising a group of organisms (the biota) and their environment (the abiotic elements, e.g. soils). It can be defined at any scale and operates at or close to equilibrium. Any ecosystem thus has homeostasis, whereby there is a balance between the inputs, outputs, components and processes. Such a system is not static and its continued operation without major changes in its integrity is achieved only through the operation of self-regulatory controls. These are negative feedbacks which have a stabilising or dampening effect; in many ecosystems such feedbacks comprise the rise and/or fall of populations of organisms. Positive feedback disrupts ecosystem homeostasis as a component, a set of components or an external factor intensifies at the expense of the existing structure and function. The ecosystem will change as a result; it will become a new ecosystem with a set of attributes that differ substantially from those of the original ecosystem. Figure 1.2 illustrates the various characteristics of ecosystems in some detail. Overall, there is a recycling of nutrients and water, whilst solar energy is converted into biomass and/or soil organic matter, which are the energy-storage compartments in the ecosystem, or it is dissipated via the grazing and the detrital food chains.

Virtually all ecosystems operate in this way though there is a significant variation in the relative importance of the grazing *versus* the detrital food chains/webs. Such systems are, in the absence of positive feedback or extrinsic forcing factors such as climatic change, self-sustaining entities. They operate within the biosphere, i.e. that part of the Earth's surface and its atmosphere in which living organisms occur. As the climate of the biosphere is not uniform, the character of the Earth's ecosystems and their distribution reflects the constraints of moisture and temperature regimes. These characteristics are illustrated in Figures 1.3 and 1.4. Other, more localised limitations

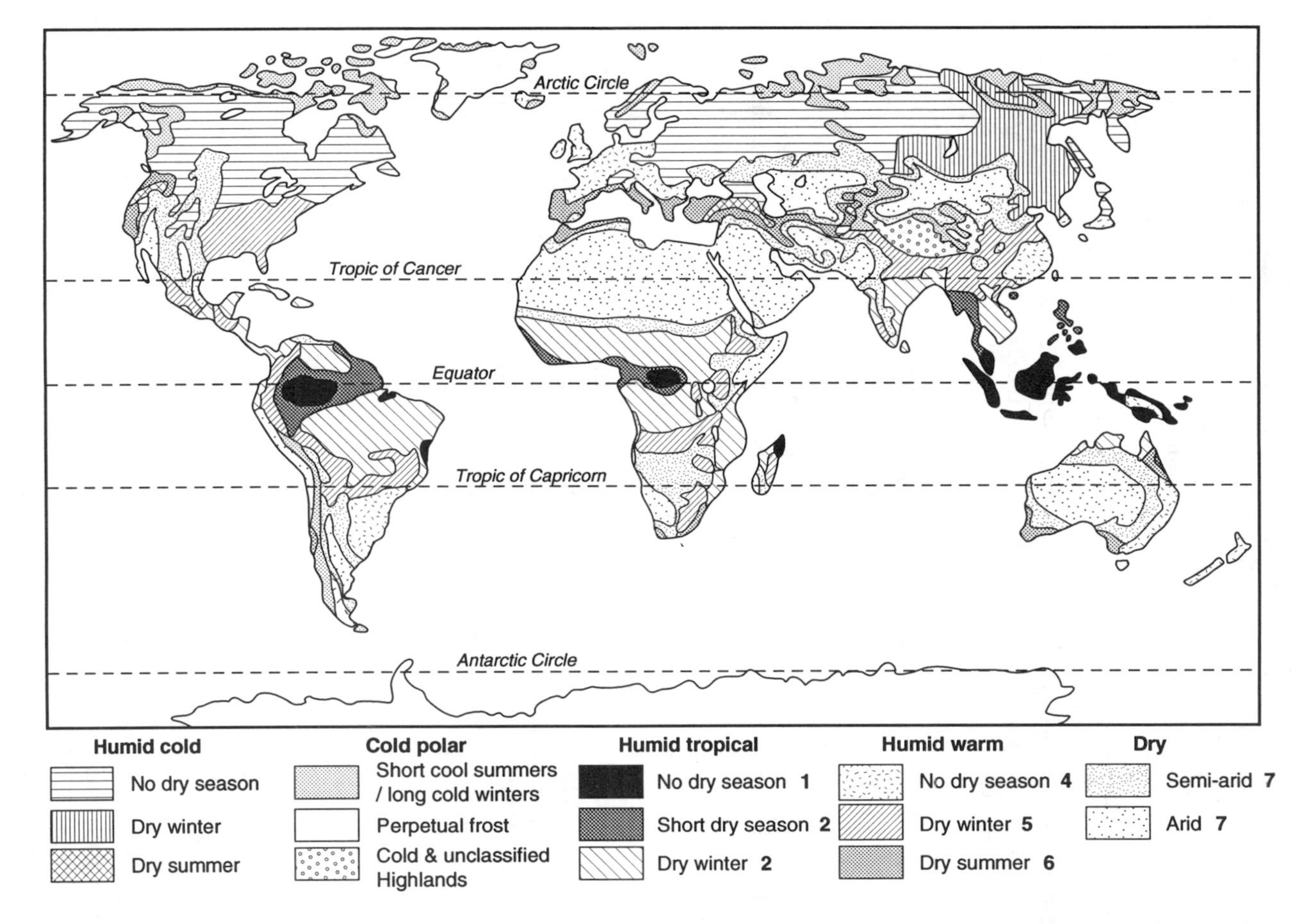

Figure 1.3 World climatic zones (adapted from the *Oxford Hammond Atlas of the World*, 1993) and their relationship with the agroclimatic zones described in Table 1.1: 1, wet tropical; 2, wet–dry tropical; 3, not shown; 4, moist mid-latitude; 5, dry mid-latitude; 6, Mediterranean; 7, arid

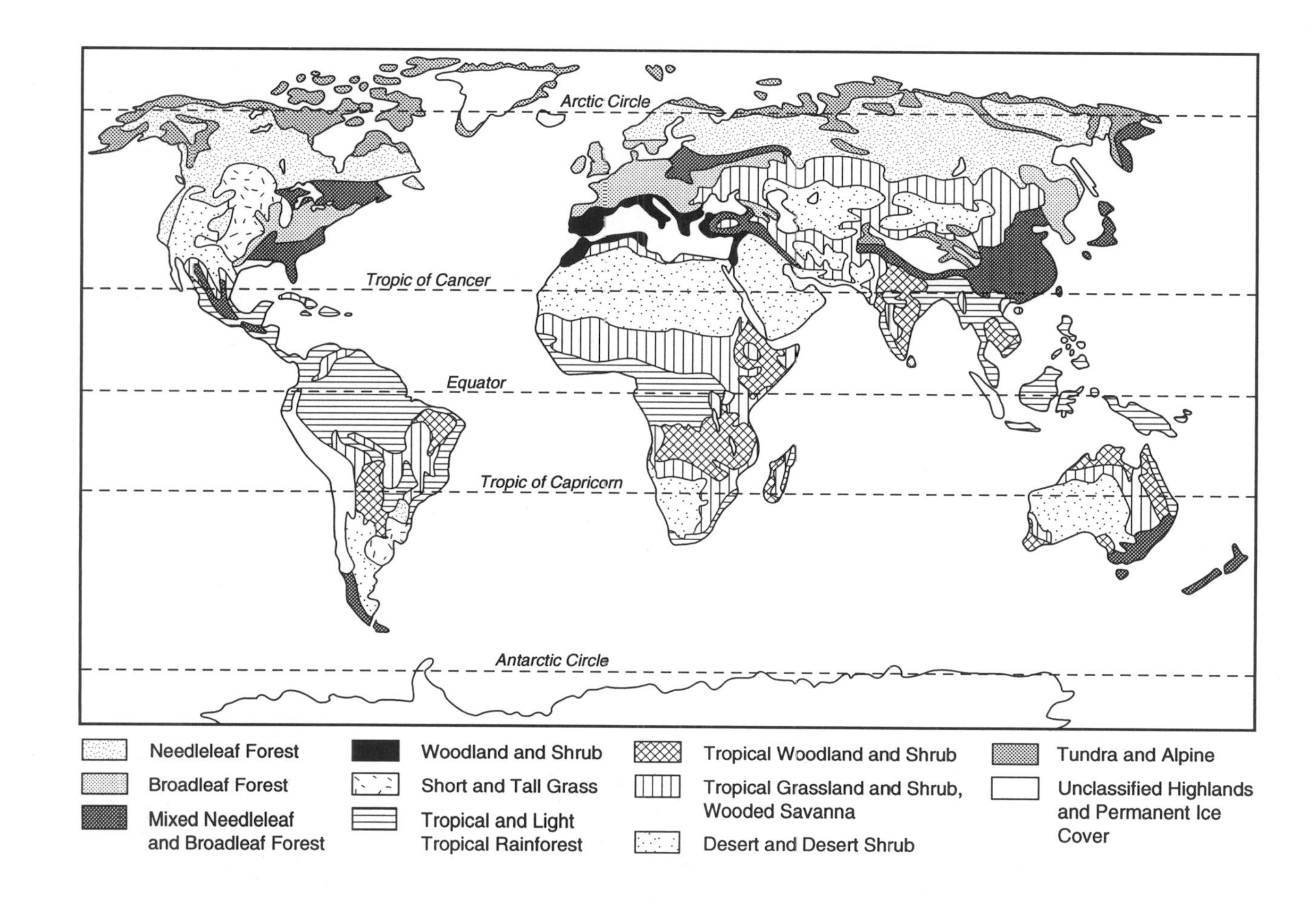

Figure 1.4 World vegetation zones (adapted from the *Oxford Hammond Atlas of the World*, 1993)

on productivity are soil type, nutrient availability, topography, aspect and drainage.

The environmental constraints on agricultural systems are the same as those on ecosystems; climate, in particular, determines the geographical region in which any given crop can be cultivated. Even crops which are versatile in relation to their moisture and temperature requirements still have their limits; climatic determinism thus remains an important factor in agriculture. Most crops also have specific edaphic requirements for optimal yield and so soil type plays also a major role in shaping the geography of agriculture. On the basis of these factors, Rice and Vandermeer (1990) have presented what they describe as a sevenfold classification of the Earth's agroclimatological characteristics. These are referred to in Figure 1.5 and Table 1.1 which illustrate, first, the close relationship between agriculture and climate and, secondly, the dominant cropping systems that exist. This is just one of several ways of classifying agricultural systems, none of which are ideal because of the difficulties of taking into account all the variables involved. The classification presented here relies entirely on physical environmental characteristics, though the output from such systems represents a form of economic order.

The significance of climate to agricultural systems has also been discussed by Tivy (1990) who considers the influence of individual climate parameters on crop growth. In particular, she makes the point about the importance of and connection between temperature and light as determinants of the rate and quantity of crop growth, provided there is adequate water. Indeed, this relationship has allowed crop physiologists to distinguish between crops of tropical, subtropical, temperate and cool-temperate regions of the world. In particular, it is possible to distinguish between three biochemical pathways of carbon dioxide fixation in photosynthesis (Osmond *et al.*, 1982). These represent adaptations of plants, and crops, to different climatic regimes. The three categories are: the carbon-3 or Calvin cycle pathway (C3), the carbon-4 (C4) pathway and the Crassulacean Acid Metabolic pathway (CAM). The majority of crop plants, including the temperate cereals, rice, woody crop plants and the white potato, have the C3 pathway. Some tropical grasses, however, are characterised by the C4 mechanism, typical crops of which are sugar cane, maize and sorghum. Of plants operating on the CAM pathway, only a few are used in agriculture; examples include sisal and pineapple.

In general, in hot and sunny environments C4 plants are more productive, i.e. they assimilate more carbon dioxide per unit leaf area and more biomass per unit of water transpired than C3 plants. There is also the possibility that C4 plants use nitrogen more efficiently than C3 plants. CAM plants are similar to C4 plants in their method of carbon dioxide fixation but they have adaptions for conserving water. Where water is a limiting factor the stomata of CAM plants close during the day, so reducing evaporation, and open at night allowing them to absorb carbon dioxide which is then synthesised the next day. Such species are thus well adapted to arid and drought-prone regions. C3 plants are adapted to relatively temperate environments with reduced sunshine, in comparison to tropical regions, and lower temperatures. Thus, adaptation to climate, another influence of environment on evolution, characterises plants from which the majority of modern crop plants have been domesticated and bred (as noted above). Although there is considerable geographical overlap in terms of photosynthetic efficiency between the groups, these basic physiological differences remain. There are two observations of interest in this context. Evans (1993), for

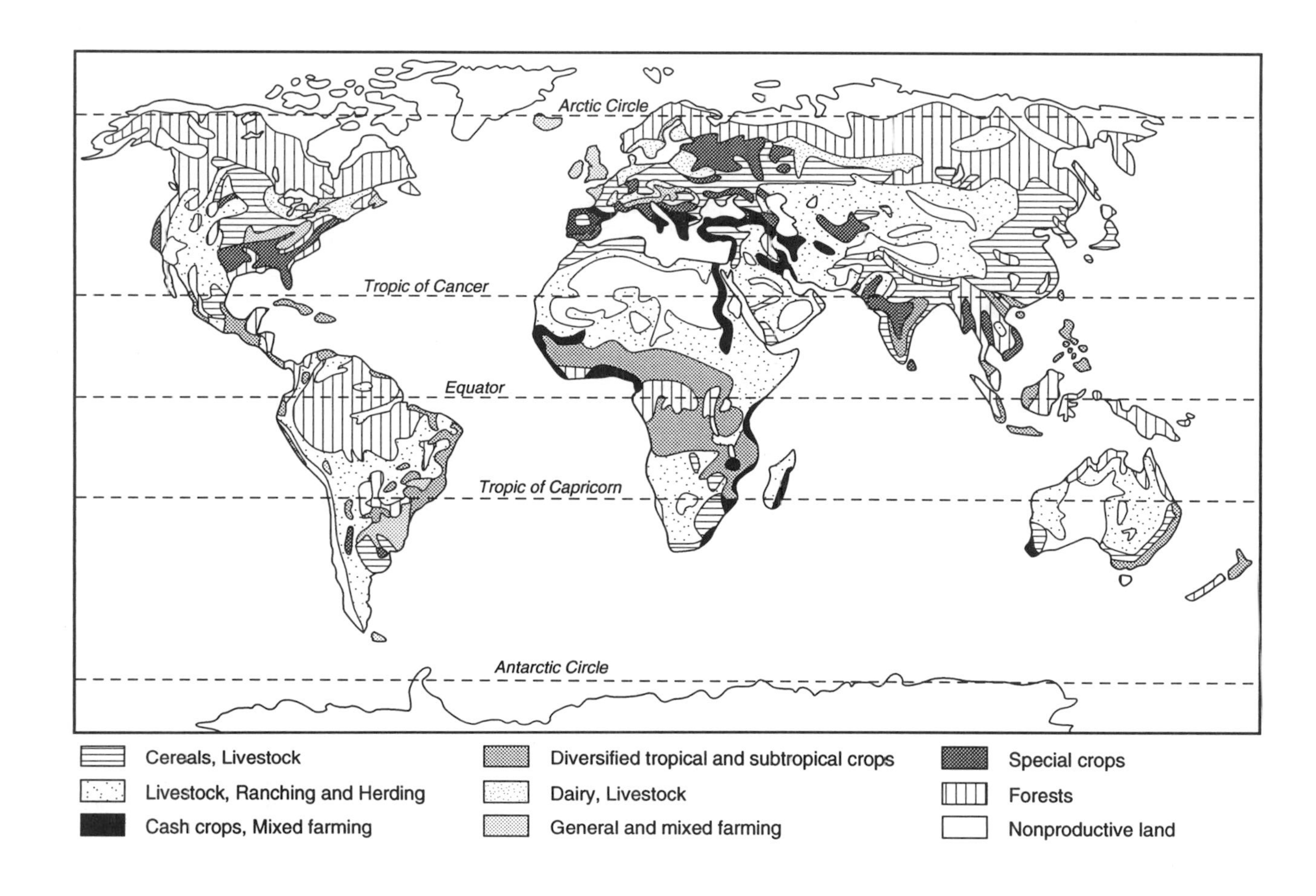

Figure 1.5 World agricultural types (adapted from the *Oxford Hammond Atlas of the World*, 1993)

Table 1.1 A classification of agricultural systems based on climate–soil–crop interrelationships, also known as agroclimatological types (based on Rice and Vandermeer, 1990)

Agroclimatology	Approximate distribution	Soil type*	Cropping systems
1. Wet tropical	Lat. 5°N to 5°S	OXISOLS, ULTISOLS Nutrient poor	Shifting cultivation, plantation cropping
2. Wet–dry tropical	Lat. 5° to 25°N 5° to 25°S	VERTISOLS, ALFISOLS, MOLLISOLS, Water content varies; high clay content; fire is important	Shifting cultivation, rice cultivation, maize production, dryland rainfed agriculture
3. Cool tropical	Mountainous zones of the tropics elevations > 1000 m, e.g. Andes and high regions of southern Asia	Soils are very varied	Diverse agriculture, e.g. tea plantations, coffee plantations, dairy cattle
4. Moist mid-latitude	Lat. 25° to 55°N mostly northern hemisphere Frost threat present	ULTISOLS, MOLLISOLS, ALFISOLS	Cotton, peanuts, tobacco, soy beans, rice, maize, tomatoes, multiple cropping systems, e.g. three main crops per year
5. Dry mid-latitude	Lat. 30° to 50°N mostly northern hemisphere	MOLLISOLS Fire is important	Small grains, e.g. wheat, maize (e.g. US corn belt), oil-seed crops
6. Mediterranean	Lat. 30° to 40°N Wet winter Dry summer	INCEPTISOLS High clay content	Small grain production, grazing, rainfed cereals, viticulture, olives, citrus fruit production
7. Arid	23.5°N, 23.5°S band either side of these two tropics	ARIDISOLS Moisture deficit all year; irrigation	Pastoral nomadic systems, some rainfed agriculture

* Soil classification is the US Seventh Approximation or Comprehensive Soil Classification Scheme.

example, makes the point that such variation, since it is genetically controlled, offers opportunities to manipulate organisms via genetic engineering. Moreover, and rather ironically, Hall (1990) opines that the differences in photosynthetic rates, water-use efficiency, etc., that are obvious at present may disappear in response to increasing concentrations of carbon dioxide that also threaten global warming.

Another adaptation of crops to climate is related to plant response to seasonal changes in weather (Hadley *et al.*, 1982). This is phenology. In particular, changes in day length or photoperiod prompt developmental changes such as seed germination, flowering or tuber production. In relation to flowering, there are four groups. Short-day species will flower with a photoperiod of less than 10 hours; examples include soy beans, sorghum and millet. Long-day species such as wheat, rye and barley require a photoperiod of more than 14 hours. There are species that require an intermediate-length photoperiod and others that are unaffected by day length, e.g. cultivars of tomato, maize, cotton and rice.

The climatic factors discussed above provide the backdrop against which agricultural systems in today's world operate. According to FAO (1990), about 11 per cent (1475×10^6 ha) of the world's total land area of $13\,382 \times 10^6$ ha ($13\,069 \times 10^6$ ha are ice-free) are being cultivated. A further 24 per cent, 3211×10^6 ha, comprise permanent pasture. There is some dissension as to whether these figures represent the limits of agricultural potential or not. For example, Pimentel *et al.* (1987) state that most of the world's fertile land is already under cultivation. However, Tolba (1992), quoting FAO sources, states that the total area of potential arable land is about 3200×10^6 ha. This means that, in principle, the world's cultivated land could double. Indeed, it seems certain that it will increase in order to accommodate the growing world population. However, all of the land presently cultivated or occupied by pasture has its productive potential constrained by the environmental factors discussed above. Some amelioration of environmental constraints may be possible; for example, water availability can be enhanced by irrigation. Such modification may itself create environmental degradation through positive feedback and in any case it too has limits within which it is effective. As suggested above, it would appear that climatic determinism is particularly appropriate in the context of agriculture.

Of the inputs into agricultural systems, solar energy cannot be modified by human endeavour though eventually crop plants that are more efficient energy trappers than are current crop plants may be produced, via biotechnology. Other inputs (Figure 1.6) can be altered artificially. Nutrient availability, especially nitrate, may be enhanced through artificial fertiliser applications; similarly micronutrient, e.g. molybdenum, availability can be improved. Soil tilth and structure can be improved through mechanical means, as can water availability. The biota within an agricultural system can also be manipulated to improve outputs. The early domestication of plants and animals some 10 000 years ago represents just such a modification (Section 2.2). Subsequent plant and animal breeding programmes (Chapter 3) and current biotechnology (Sections 10.1 and 10.2) are all geared to improving the inherent productivity of crop plants and livestock. Productivity in an agricultural system can also be improved by reducing the competition for light and nutrients between the crop and so-called pests. These can be weed plants, insects, viruses that cause disease, bacteria that cause disease and fungi. A whole range of substances, e.g. pesticides (which are also known as crop-protection chemicals) and animal health products, are

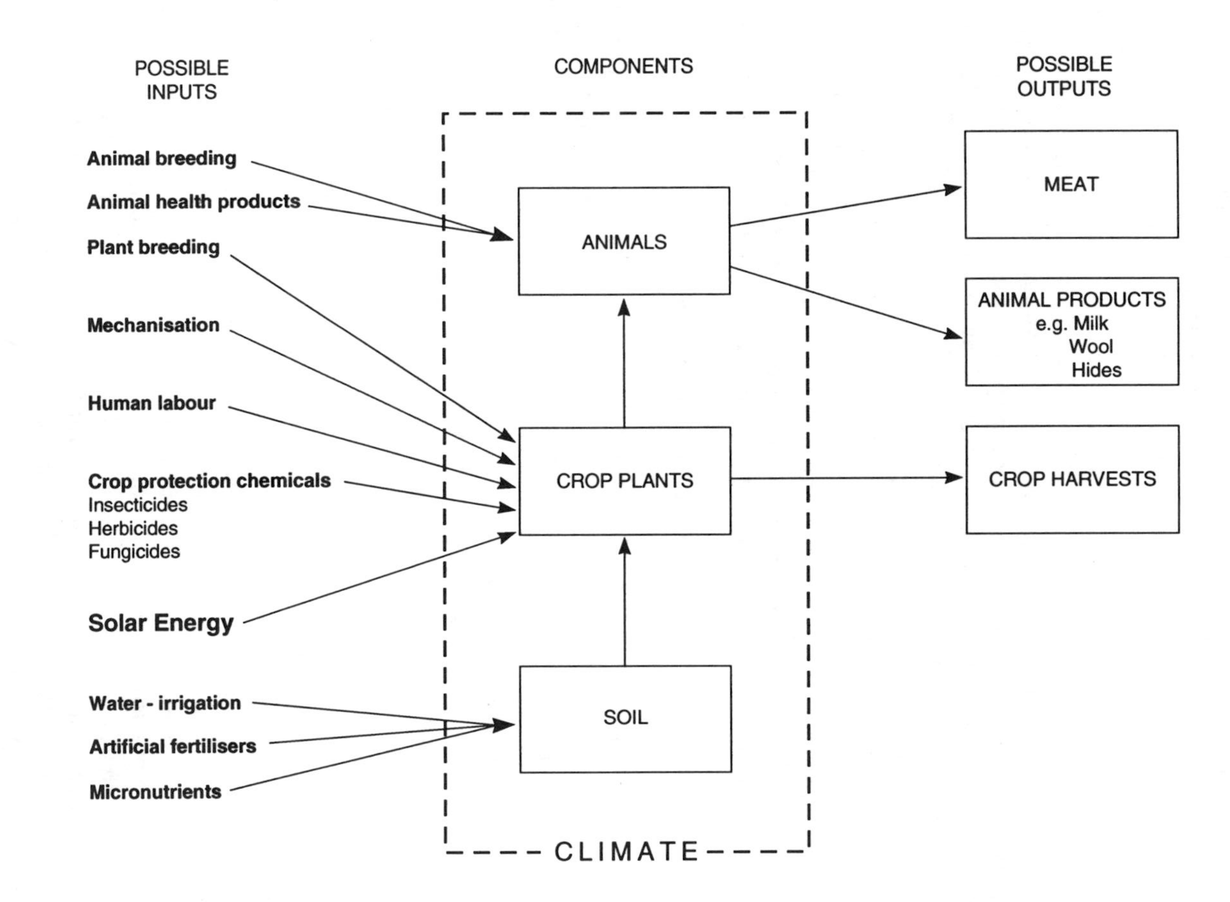

Figure 1.6 The possible range of inputs, components and outputs of an agricultural system

available to reduce this competition and so improve the end product. There are also other chemicals available that enhance animal productivity, not through combatting disease, but through stimulating milk or meat production, for example.

All of these innovations represent increased inputs of energy to the agricultural system. This energy is derived mostly from fossil fuels; agricultural systems that are typified by high mechanical, fertiliser and pesticide usage thus have high fossil-fuel energy subsidies. As discussed in Section 1.4, this is a means of using fossil-fuel energy to enhance the conversion of energy into food energy. The importance of this energy subsidy varies considerably between agricultural systems but even the most unsophisticated system is augmented by human labour at the very least. This energy subsidy is a characteristic that sets agricultural systems apart from natural or semi-natural ecosystems.

Another crucial difference between the two relates to the products of the systems. As Figure 1.2 shows, biomass, i.e. organic productivity, accumulates in an ecosystem and comprises one of its storage elements. A balance between accumulation and storage is accomplished through decomposition: leaf litter and the biomass in organisms that die are consumed by detrivores. Colinvaux (1993) says 'Most of the energy of net primary productivity [in natural ecosystems] goes to rot'. Up to 93 per cent of primary productivity may enter the detrital food chain and the energy released will simply be dissipated as entropy. In an agricultural system the situation is quite different. The biomass, i.e. the harvest, is the object of the exercise and its production on a cyclical basis is vital to the sustenance of human communities and their other enterprises that it supports. The harvest is thus removed from the agricultural system to be processed and/or packaged and certainly consumed outside the system in which it was produced. This reduces, significantly, the role of the detrital pathway of energy transfer in all agricultural systems when compared with their counterpart natural ecosystems. There is also likely to be a further energy subsidy as the harvest is stored, processed and eventually marketed (Section 1.4).

Both types of system rely on the fixation of solar energy in photosynthesis. Green plants, the primary producers, provide the power base for food webs and food chains which dominate ecosystems and agricultural systems respectively. In agricultural systems, however, selected organisms are cultivated to generate the desired plant or animal product. These organisms, in most cases, have been altered through breeding programmes to emphasise particular attributes. Ecosystems, conversely, are populated by organisms that have evolved through the ages to become adapted to the environmental conditions of the present and the competition of equally well-adapted niche sharers. On the whole, ecosystems are much less open than agricultural systems into which there are usually increased inputs and outputs, especially designed to enhance food energy production.

1.3 THE TRANSFORMATION OF ECOSYSTEMS INTO AGRICULTURAL SYSTEMS

The transformation of ecosystems into agricultural systems has been occurring since the end of the last ice age about 10 000 years ago (Chapter 2). Then, as now, this transformation represents a major human impact on the natural environment. Apart from impacts such as soil erosion and desertification, this transformation has also caused a diminution of genetic resources. The transformation process is, however, set to proceed well into the twenty-first century as world population continues to grow. The diverse nature of agricultural systems means that the process of transformation has also varied. In some instances the transformation has been comparatively minimal, e.g. nomadic pastoralism. In others, the changes have been enormous, e.g. the intensive production of wheat in East Anglia, UK, or on the prairies of North America. There are further types of agricultural systems between these two extremes. In all cases, energy flows (Section 1.4) have been altered.

Although the transformation process has been heterogeneous, it is possible to establish three fundamental ways in which ecosystems can be transposed into agricultural systems. First, an ecosystem can be manipulated without any significant change in its overall biotic characteristics. Selected components only are altered but the ecosystem's integrity remains intact. In many pastoral systems, for example, there is little change in the natural ecosystem except for the installation of domesticated or semi-domesticated animals (Section 4.1). Such activities tend to occur in regions that are marginal for agriculture, as shown in Figure 1.5 and Table 1.1. The low, and sometimes poor quality, primary productivity usually means that migration is also an essential component of management. The major change in this system is the selection of suitable animals, e.g. camels and reindeer, which replace wild counterparts and, possibly, the control of predators. Energy flows and nutrient cycles are not substantially altered; primary productivity is much the same as it would be under natural conditions though repeated annual or biennial grazing may influence species composition and reduce diversity. Secondary productivity is mainly in the form of milk products and occasionally blood, e.g. in Masai Mara, Kenya, and also meat. Such systems can be stable over long periods of time, though overgrazing will render them liable to degradation which reduces carrying capacity.

Secondly, and the most common way in which ecosystem transformation is effected, is the replacement, either temporarily or permanently, of the naturally occurring biota by fully domesticated species. Where this occurs on a transitory basis the agricultural system is known as shifting cultivation (Sections 4.3–4.5) which, today, is mainly restricted to tropical regions. Provided the time period, or fallow, between cultivation is sufficiently long to allow the natural vegetation to regenerate and reinstate nutrient cycles, shifting cultivation can be efficient and sustainable. Not all the natural biota may be removed; the most efficient of such systems are those which allow useful trees, such as oil palms and fruit trees, to remain so that they may contribute to the harvest. Where there is permanent replacement of naturally occurring biota there are lasting changes in the inputs, outputs (see Figure 1.6) and components of the system. Crop plants and/or livestock replace the biota, some of which may maintain only an ephemeral presence as weeds and pests; energy flows and nutrient (biogeochemical) cycles are totally altered and controlled to some degree (Figure 1.6). Examples are crop, livestock and mixed farming systems that occur throughout the world. This replacement of the natural biota substantially alters energy flows, emphasising the importance of the grazing food chain and reducing the importance of the detrital food chain (Section 1.6). Where there is little fossil-fuel input, net primary productivity may be less than that of the replaced natural ecosystem. The inception of such systems is also paralleled by a major reduction in species diversity because the wide variety of wild species is substituted by just a few domesticated plants and/or animals. In fact this type of transposition has been responsible for a major reduction in global biodiversity (Sections 8.1 and 9.4).

In most cases, ecosystems with diverse biotas are replaced by agricultural systems that are much less biologically diverse. However, there are some instances which represent the third way in which ecosystems are transformed into agricultural systems. Examples include the inception of polyculture in desert lands which are irrigated, and the extension of arable agriculture into some of the world's grassland biomes. In the latter, mixed farming was introduced in the nineteenth century and in both of these examples, the inception of agriculture has created an increase in species diversity. Primary and/or secondary productivity are increased in comparison with the indigenous ecosystem. Such systems, however, are not always successful and require a high degree of management with major inputs.

The transformation of 35 per cent (FAO, 1990) of the world's land area into agricultural systems has already had major repercussions. Many of these are discussed in Chapters 8 and 9. Statistics for the world's land area and its land use (World Resources Institute, 1992) for the last decade reveal trends that represent considerable regional variations and which together add up to an overall increase in agricultural land use globally. Much of this increase has been at the expense of forests and woodlands. The data given in Table 1.2 show that in many countries of the developed world cropland and permanent pasture have actually decreased. This is a response, through government action, to the production of food surpluses. In many of these countries, e.g. in the UK, the same period has been characterised by an increase in forest cover; some of this now occupies former agricultural land but most of it represents upland moorland and peatland afforestation and thus represents another means whereby semi-natural habitats can be lost. Conversely, in the developing world, most notably in Africa, the amount of cropland has increased. In Uganda, for example, it has increased by 20 per cent, much of which derives from forest and woodland. There have been comparatively few changes in the amount of pasture, the exceptions being Canada, Costa Rica, Paraguay and the Philippines. In many developing countries, there has been much pressure to increase food production because of increasing populations. This is reflected in the data of Table 1.2 though the increase in cropland would probably have been higher than this if it were not for improved varieties of crops that have been developed as part of the so-called green revolution.

Using data (Table 1.3) for the period 1700–1980 (Richards, 1990), it is evident that many of the changes currently taking place in the developed world have counterparts in the events of a century or more ago in Europe. Forests and woodlands declined substantially between 1700 and 1850 and corresponded with a massive increase in cropland whereas in Africa the biggest changes in forests and woodlands began in 1950. Globally, the greatest removal of forest and woodland occurred between 1950 and 1980; there was little change overall, only 0.1 per cent in the amount of pasture. Croplands, however, increased most significantly in the period 1700 to 1850. All of these data reflect the importance and dynamism of agriculture as an agent of environmental change in the past and the present. It will no doubt continue to alter the face of the Earth; possibly even more so than now if global warming ensues in the twenty-first century and the world's food production systems are obliged to adjust rapidly (Section 10.5).

Table 1.2 Data on changes in the amount of agricultural land cover and forest/woodland (abstracted from World Resources Institute, 1992)

Region	Land area (000s of hectares)	Cropland ha 1987–1989	Cropland % change since 1977–1979	Permanent pasture ha 1987–1989	Permanent pasture % change since 1977–1979	Forest/ woodland (% decrease)
AFRICA	2 964 138	186 392	4.4	890 899	–0.5	–3.6
e.g. Burkina Faso	27 380	3 423	27.5	10 000	0.0	–8.2
Cote d'Ivoire	31 800	3 653	20.8	13 000	0.0	–24.1
The Gambia	1 000	174	13.7	90	0.0	–26.2
Uganda	19 955	6 705	20.0	1 800	0.0	–8.1
Zimbabwe	38 667	2 796	10.3	4 856	0.0	–4.0
NORTH & CENTRAL AMERICA	2 137 700	273 816	1.1	368 631	3.1	1.0
e.g. Canada	922 097	45 977	4.4	32 500	26.7	5.8
Costa Rica	5 106	527	5.5	2 310	24.0	–17.0
Mexico	190 869	24 708	1.9	74 499	0.0	–12.0
Nicaragua	11 875	1 270	2.8	5 300	11.5	–23.5
USA	916 660	189 915	0.0	241 467	1.0	–1.1
SOUTH AMERICA	1 752 926	141 578	10.9	477 863	4.1	–4.6
e.g. Argentina	273 669	35 750	1.9	142 400	–0.7	–1.3
Brazil	845 651	78 233	17.1	169 000	6.3	–4.2
Paraguay	39 730	2 203	46.7	20 420	32.6	–27.7
ASIA	2 731 228	454 456	0.8	694 251	–0.3	–5.3
e.g. China	932 641	96 615	–3.9	319 080	0.0	–7.7
India	297 319	169 357	0.5	11 923	–3.4	–0.7
Malaysia	32 855	4 880	2.5	27	0.0	–11.0
Philippines	29 817	7 957	4.4	1 220	23.1	–16.4
EUROPE	472 953	140 409	–1.3	83 177	–4.0	1.1
e.g. Finland	30 461	2 435	–6.4	125	–27.3	–0.4
France	55 010	19 210	1.6	11 757	–9.5	1.2
UK	24 160	6 888	–1.2	11 119	–3.3	13.8
FORMER USSR	2 227 200	231 871	–0.2	371 500	–0.6	1.7
WORLD 1	13 128 841	1 477 877	2.2	3 322 934	0.1	–1.8

Table 1.3 Changes in global land use, 1700–1980 (reproduced by permission of Cambridge University Press from Richards, 1990)

Regions	Vegetation types	Area (million ha)					Percentage changes from:				
		1700	1850	1920	1950	1980	1700 to 1850	1850 to 1920	1920 to 1950	1950 to 1980	1700 to 1980
Tropical Africa	Forests & woodlands	1358	1336	1275	1188	1074	−1.6	−4.6	−6.8	−9.6	−20.9
	Grassland & pasture	1052	1061	1091	1130	1158	0.9	2.8	3.6	2.5	10.1
	Croplands	44	57	88	136	222	29.5	54.4	54.5	63.2	404.5
North Africa/ Middle East	Forests & woodlands	38	34	27	18	14	−10.5	−20.6	−33.3	−22.2	−63.2
	Grassland & pasture	1123	1119	1112	1097	1060	−0.4	−0.6	−1.3	−3.4	−5.6
	Croplands	20	27	43	66	107	35.0	59.3	53.5	62.1	435.0
North America	Forests & Woodlands	1016	971	944	939	942	−4.4	−2.8	−0.5	0.3	−7.3
	Grasslands & pasture	915	914	811	789	790	−0.1	−11.3	−2.7	0.1	−13.7
	Croplands	3	50	179	206	203	1556.7	258.0	15.1	−1.5	6666.7
Latin America	Forests & woodlands	1445	1420	1369	1273	1151	−1.7	−3.6	−7.0	−9.6	−20.3
	Grassland & pasture	608	621	646	700	767	2.1	4.0	8.4	9.6	26.2
	Croplands	7	18	45	87	142	157.1	150.0	93.3	63.2	1928.6
China	Forests & woodlands	135	96	79	69	58	−28.9	−17.7	−12.7	−15.9	−57.0
	Grassland & pasture	951	944	941	938	923	−0.7	−0.3	−0.3	−1.6	−2.9
	Croplands	29	75	95	108	134	158.6	26.7	13.7	24.1	362.1
South Asia	Forests & woodlands	335	317	289	251	180	−5.4	−8.8	−13.1	−28.3	−46.3
	Grassland & pasture	189	189	190	190	187	0.0	0.5	0.0	−1.6	−1.1
	Croplands	53	71	98	136	210	34.0	38.0	38.8	54.4	296.2
South-East Asia	Forests & woodlands	253	252	247	242	235	−0.4	−2.0	−2.0	−2.9	−7.1
	Grassland & pasture	125	123	114	15	92	−1.6	−7.3	−7.9	−12.4	−26.4
	Croplands	4	7	21	35	55	75.0	200.0	66.7	57.1	1275.0
Europe	Forests & woodlands	230	205	200	199	212	−10.9	−2.4	−0.5	6.5	−7.8
	Grassland & pasture	190	150	139	136	138	−21.1	−7.3	−2.2	1.5	−27.4
	Croplands	67	132	147	152	137	97.0	11.4	3.4	−9.9	104.5
USSR	Forests & woodlands	1138	1067	987	952	941	−6.2	−7.5	−3.5	−1.2	−17.3
	Grassland & pasture	1068	1078	1074	1070	1065	0.9	−0.4	−0.4	−0.5	−0.3
	Croplands	33	94	178	216	233	184.8	89.4	21.3	7.9	606.1
Pacific developed countries	Forests & woodlands	267	267	261	258	246	0.0	−2.2	−1.1	−4.7	−7.9
	Grassland & pasture	639	638	630	625	608	−0.2	−1.3	−0.8	−2.7	−4.9
	Croplands	5	6	19	28	58	20.0	216.7	47.4	107.1	1060.0
Total	Forests & woodlands	6215	5965	5678	5389	5053	−4.0	−4.8	−5.1	−6.2	−18.7
	Grasslands & pasture	6860	6837	6748	6780	6788	−0.3	−1.3	−0.5	0.1	−1.0
	Croplands	265	537	913	1170	1501	102.6	70.0	28.1	28.3	466.4

1.4 ENERGY IN AGRICULTURE

Energy is a key issue in agriculture which is essentially an energy processing system. Energy (as food energy) production is the main raison d'être of agriculture, whilst increasing inputs of fossil-fuel energy are rendering many modern agricultural systems unsustainable. Inevitably, any activity that relies on fossil-fuel energy is, on the one hand, susceptible to the caprices of world fuel prices and, on the other hand, contributing to the accumulation of carbon dioxide in the atmosphere and its potential role in global warming. Energy in agriculture is thus not only a key issue but a controversial one.

A consideration of the energy characteristics of agricultural systems allows them to be compared with the other types of system on the Earth's surface. Odum (1975) designated four such basic systems: unsubsidised natural solar-powered ecosystems, naturally subsidised solar-powered ecosystems, human subsidised solar-powered ecosystems and the fuel-powered urban-industrial systems. These and their interrelationships are illustrated in Figure 1.7. The last of these systems consumes the products of the former three systems of which the human subsidised solar-powered ecosystems comprise agricultural systems. The importance of solar power as an energy source is highlighted in Figure 1.7. Indeed, either directly or indirectly, solar power accounts for some 89 per cent of the current world energy consumption (Tolba, 1992). Of this some 77 per cent comprises fossil fuel, itself a product of solar power captured aeons ago (Figure 1.7). Approximately 12 per cent consists of biomass fuel, including fuelwood, agricultural residues and dung, which is used mainly in developing countries. Clearly, this too is a product of solar energy. The remaining 11 per cent of the world's energy derives from hydroelectricity and nuclear fuel.

The behaviour of energy is governed by the laws of thermodynamics. The first law states that energy can be transformed from one form into another but it can never be created or destroyed. For example, light energy can be fixed by green plants to produce food. The second law states that any process requiring a transformation of energy will cause energy to be degraded from a concentrated to a dispersed form. Thus the consumption of food, a concentrated energy form, provides energy for respiration that transposes the energy to a dispersed form, i.e. heat. This is known as 'entropy' and the second law of thermodynamics is often referred to as the entropy law. As a result of this dissipation, the energy can never be reused. This is a characteristic of energy transfer that makes it unique from all other transfers, notably those of water and nutrients, that occur in ecosystems and agricultural systems. Water and nutrients are continually recycled but to effect recycling there is an energy requirement. This is another call on solar energy. These relationships are discussed in detail in Mannion (1986) and apply equally to ecosystems and agricultural systems (Section 1.2).

Solar energy is the most important input into agricultural systems. How well it is used and/or stored depends on many other factors (Section 1.2), particularly the biota, climate, abundance of water and nutrient availability. Although energy is vital a surprisingly small proportion of it is actually used in photosynthesis. Solar radiation is part of the spectrum of electromagnetic radiation that reaches the Earth's surface. It has wavelengths between 0.1 and 10 micrometres and comprises visible light, along

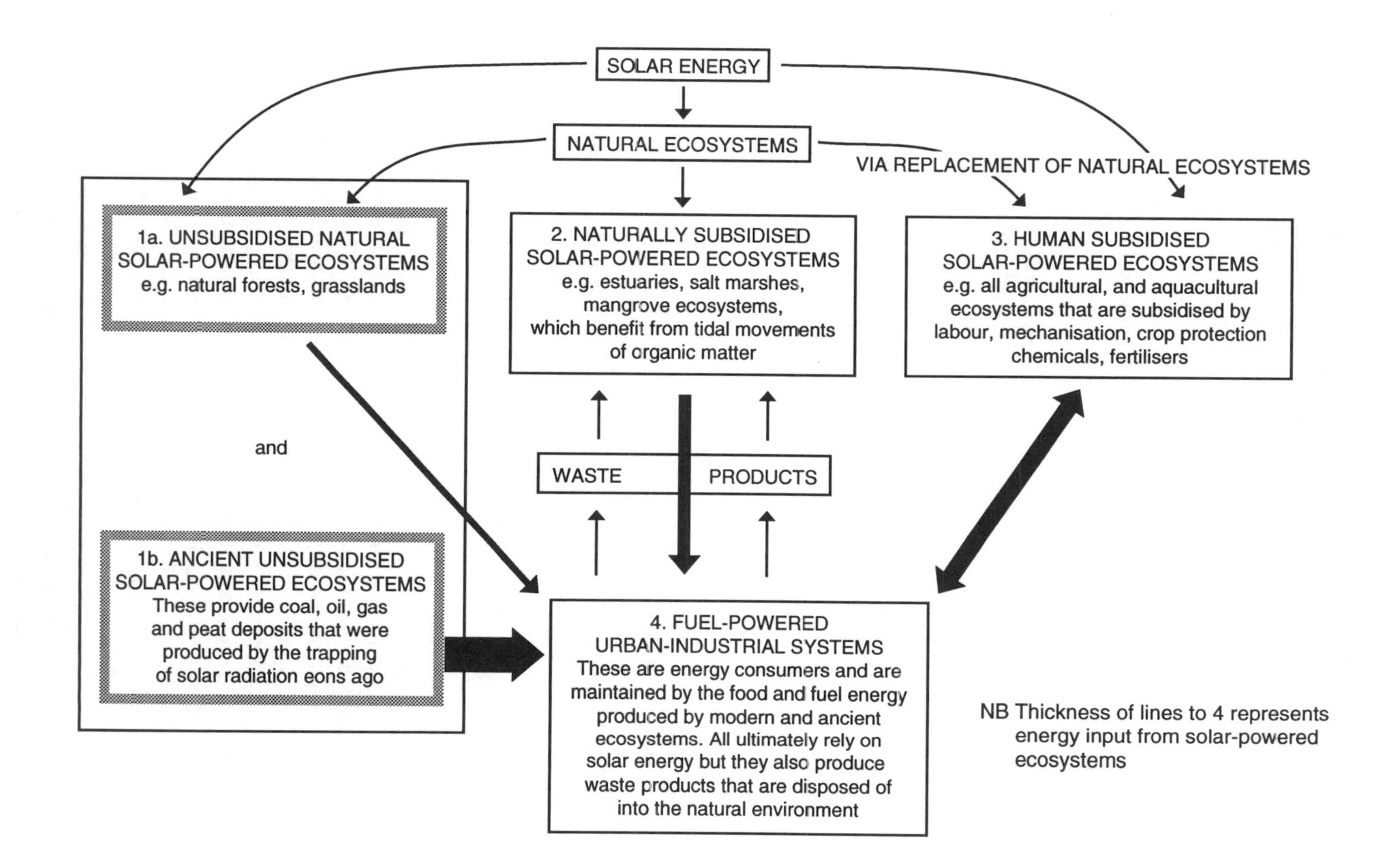

Figure 1.7 An interpretation of E.P. Odum's (1975) energy flows juxtaposing natural ecosystems, agricultural systems and urban-industrial complexes (from Mannion and Bowlby, 1992)

with ultraviolet light and infra-red light which are invisible. These three components have different roles and are subject to different processes. Ultraviolet light, for example, is absorbed by the ozone layer in thc stratosphere; infra-red radiation is responsible for heating the Earth's surface, while the visible light is used in photosynthesis. All three are absorbed to a greater or lesser degree by the atmosphere as solar radiation approaches the Earth's surface. In the biosphere the relationship between the radiation components and the biota, especially the plant species, is complex (as shown in Figure 1.8), but only about 0.8 per cent of the total radiation from the sun is used in photosynthesis. According to Simmons (1989), approximately 170×10^{19} J of energy appear per year as plant matter, i.e. net primary productivity, after taking into account that more energy than this is actually fixed, i.e. gross primary productivity, but that some of it is consumed by the plants themselves for their metabolism. These values are for the whole of the biosphere of which only 35 per cent (Section 1.2) is agricultural and occupied by cropland and permanent pasture. It is rather sobering to acknowledge that virtually all of society, and in particular the densely populated, non-energy-productive and industrially intense fuel-powered systems (Figure 1.7) are totally sustained by this comparatively small amount of solar radiation. In these terms society and its infrastructure appear very fragile and vulnerable. Developing efficient and sustainable ways to use this all important solar energy will be discussed in Chapter 10.

The process of food production begins with photosynthesis which can be summarised in the following equation:

$$\underset{\text{Carbon dioxide}}{6\ CO_2} + \underset{\text{Water}}{6\ H_2O} \xrightarrow[\text{CHLOROPHYLL}]{\text{SOLAR ENERGY}} \underset{\text{Carbohydrate}}{C_6 H_{12} O_6} + \underset{\text{Oxygen}}{6\ O_2}$$

Photosynthesis consists of several stages that can be designated as either light reactions or dark reactions. The former consists of the absorption of light energy (i.e. photons) by chlorophyll. This energy splits a molecule of water by dissociation into hydrogen ions, oxygen which is released to the atmosphere, and electrons. In the dark reactions, energy is used in several complex stages to synthesise carbohydrate from carbon dioxide absorbed from the atmosphere and water. In an ecosystem, the rate of photosynthesis is limited by the architecture of the vegetation and the photosynthetic efficiency of individual species. It is also limited by environmental conditions that are external to the plant community such as varying light intensity, temperature regimes, water availability, nutrient availability, topography and soil structure. The management element in agricultural systems is targeted at removing or reducing these constraints on productivity in order to maximise the output from the useable solar radiation. This is one of the reasons why agricultural systems are so varied, as will be discussed in Chapters 4–7.

A comparison of net primary productivity (or biological productivity) of a natural ecosystem with that of an agricultural system growing under the same climate conditions is not always very meaningful. This is because it is rare in agricultural systems to use for human and/or animal consumption all the organs of any crop

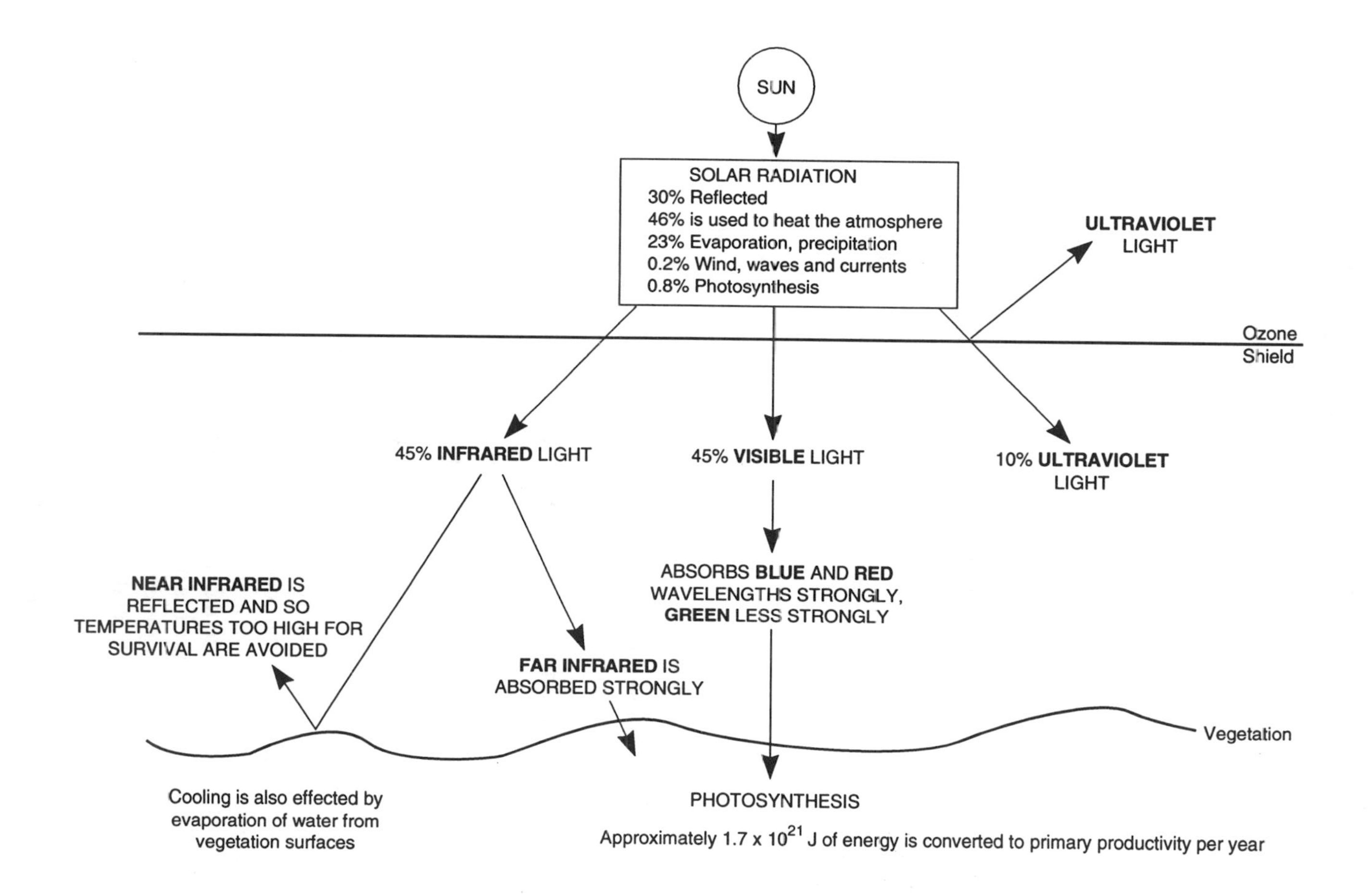

Figure 1.8 The characteristics and distribution of solar radiation (data from various sources quoted in the text)

Table 1.4 The components of biological yield in three cereals and improvements in harvest index through winter wheat breeding (a) from Donald and Hamblin (1976) and (b) from Austin *et al.* (1989)

(a) The contribution to biological yield (biomass) of various organs in three cereal crops

	Winter wheat (%)	Spring barley (%)	Oats (%)
Grain	40	51	41
Stems	33	28	34
Chaff	10	10	15
Leaves	9	6	7
Vegetative tillers	8	5	3

(b) Changes in the grain, straw, biomass and harvest index of four groups of winter wheat

	$g\ m^{-2}$			
	Grain	Straw	Biomass	Harvest
1. Very old group grown in 19th century before era of modern plant breeding	505	995	1500	34
2. Old group grown in early 20th century	557	983	1541	36
3. Intermediate group 1955–1975	669	814	1484	45
4. Modern group introduced since 1980	805	784	1588	51

plants. In the case of cereals, for example, only the grains have a food value for human/animal consumption. The leaves, stems, tillers and chaff are discarded. As Table 1.4 shows, one-half, or slightly less in the case of spring barley, of the biological yield is not consumable. Thus yield in an agricultural system is always less than the total productivity and, obviously, always less than net primary productivity. The overall productivity of agricultural ecosystems has been discussed by Colinvaux (1993) who points out that many forms of agriculture have a lower productivity (total or biological productivity, not yield) than the natural ecosystems they replaced because the crop does not totally cover the land surface and because the land may be left bare for part of the year between reaping and sowing. Whittaker and Likens (1973) estimated mean net primary productivity for the world's major ecosystems. Colinvaux (1993) has converted these values for biomass into values for carbon production, with forests, marshes and reefs highest at >350 g of carbon $m^{-2}\ yr^{-1}$ and tundra, deserts and oceans lowest at <70 g carbon $m^{-2}\ yr^{-1}$. Agricultural systems, along with grasslands, lakes and upwelling zones of oceans, lie between these two extremes in the range 70–350 g carbon $m^{-2}\ yr^{-1}$, though some may be just as productive as the forests if there is a large energy subsidy. These values, however, mask the yield values which are considerably lower but which are nevertheless the prime objectives of any agricultural system.

When comparing the productivity between agricultural systems it is possible to use a variety of approaches. For example, the energy or calorie content of the crop component may be determined. This is particularly important for ascertaining energy ratios which represent the proportion of energy input in relation to the energy output. Er values are very important aids in the management of agricultural systems as will be referred to in Chapters 4–7. Such data have, for example, been compiled by Stanhill (1981) for eighteenth-century Egypt. On the basis of animal and human workdays, i.e. the labour input, comprising 9.61 GJ ha^{-1} yr^{-1} some 19.25 GJ ha^{-1} yr^{-1} of edible energy suitable for human consumption were produced. In addition, sufficient clover to provide animal feed for 62 days along with enough straw for 122 days was generated. This high level of productivity was made possible through irrigation facilitating crop growth on a year-round basis. Stanhill suggests that this level of food production was at least four times that of English wheat during the same period, and possibly as much as ten times, whilst it was more than twice that of France. Moreover, if productivity was as high during the centuries around the birth of Christ it is little wonder that the Nile Valley became the centre of a major civilisation. To switch some of the available labour to public building works would not have jeopardised food production, including the production of sufficient wheat for export to the Roman Empire.

Another measure of crop productivity is the harvest index or crop index, the details of which have been discussed by Tivy (1990). This is the ratio of commercial (or economic) to recoverable yield for any given crop. In wheat production, for example, the harvest index can be expressed as the ratio of grain to straw. It relates to the proportion of the crop that is useable/consumable, i.e. the yield. The harvest index varies considerably between crops; it is highest in crops like potatoes at 85 per cent, is about 50 per cent in cereals and lowest in oil-seed crops at *c.* 30 per cent (Harris, 1992). The harvest index is another measure that can be manipulated by management. The introduction of low-stemmed or small-leaved cultivars of cereals reduces the above-ground parts in relation to the grain. Such attributes may be the product of plant breeding or the application of growth-regulating chemicals, e.g. chlormequat (Harris, 1992). Austin *et al.* (1989), for example, have demonstrated such improvements in winter wheat. Their data (given in Table 1.4) show that improved strains now have a higher harvest index, due to increased grain production, whilst biomass for old and new strains has remained roughly constant.

Potential crop yield is another measure that reflects the energy characteristics of an agricultural system and is a useful tool for management purposes. It allows comparisons between actual yields and potential yields and may offer ways to determine if there is scope for improvement. Potential crop yield is subject to the same limitations as the productivity of natural ecosystems, e.g. the environmental conditions, especially climatic conditions, under which the crop is grown. Other limitations include the yield potential of the cultivar being grown and the efficiency of management in attenuating the limiting factors to growth, e.g. nitrate deficiency. The measure relates to the efficiency with which light energy is converted into chemical energy within the crop. As discussed above, only a small proportion of solar radiation is actually used for photosynthesis (Figure 1.8) so this represents a major limiting factor that cannot be much altered by management. In north-west Europe, for example, the short-term maximum growth rates of many cereal, root and legume crops are approximately 200 kg ha^{-1} day^{-1} (data in Harris, 1992). The potential crop yield, calculated using the following equation, is 243 kg ha^{-1}.

$$\frac{\text{Amount of radiation per day (June)} = 17 \text{ MJ m}^{-2} \times 2.5\% \; \text{(\% of radiation converted into crop dry matter)}}{\text{Energy content of crop} = 17.5 \text{ kJ g}^{-1}}$$

Thus the actual yields are fairly close to the potential yields.

The efficiency with which plants convert solar energy to chemical energy is also a function of carbon dioxide concentrations in the atmosphere, and their diurnal changes, as well as the architecture of the crop canopy. The latter, and especially the leaf area index (LAI), which is the ratio of total leaf surface area to ground surface area, determines the amount of light intercepted. The development of the maximum LAI depends on the growth pattern, or phenology, of the crop plant. Factors such as the rate of germination, the establishment of new shoots and rate of leaf growth all affect LAI. Some variations between LAI in a range of crops are given in Figure 1.9 which also shows how the method of propagation can influence the amount of light interception in potato crops. According to Parsons (1994), for example, the control of LAI in grasslands is an important management technique to optimise yield. Another factor in biological productivity is the length of time that a closed canopy can be sustained during the growth period. Again, this varies between crops. Some, such as the cereals, are characterised by leaf growth that ceases prior to flowering; others, such as potato and sugar beet, continue to produce leaves throughout the growing season. The final economic yield thus relies on how much of the food product, or harvest, is produced in, and/or accumulated by the yield organ. According to Bunting (1975), there are three categories of crop plants that can be designated depending on when during the plant's life-cycle and where in the plant the yield is located. These are given in Table 1.5 which shows that for some harvest indices, the accumulation occurs for almost all of the growing season, e.g. in potatoes, but for others the harvest index reflects a yield which is produced towards the end of the plant's life, e.g. cereals, or on an annual/biennial cycle. Thus, whilst LAI is a vital ingredient in productivity rates, the length of the yield-forming phase is also important, as is the availability and uptake efficiency of all the resources necessary for crop plant growth, e.g. water, nutrients, etc. As Scott *et al.* (1994) point out, maximising crop yield is becoming increasingly reliant on an holistic appraisal of crop resource capture. Ensuring that this is balanced, i.e. to maximise carbon uptake as well as curtail pollution from nitrate fertilisers and, where appropriate, to limit water loss, is an essential ingredient of sustainable agricultural systems (see Section 10.4).

All of the foregoing relates to primary production within agricultural systems. Many such systems are, however, designed as producers of animals or animal products and are thus concerned with secondary productivity. In a natural ecosystem a proportion of the net primary productivity (see Section 1.2) is consumed directly by herbivores and/or omnivores. These, in turn, may be consumed by carnivores. This is, in simplistic terms, a grazing food chain. At each stage in the producers → herbivore/omnivore → carnivore chain, energy is transferred. To conform with the second law of thermodynamics (see above), each exchange of energy results in the dispersal of some of that energy to a degraded or entropic form. Thus the amount of energy passed on to the herbivore or omnivore is less than that available in the producers. For every further exchange there is less energy available. This means that in natural

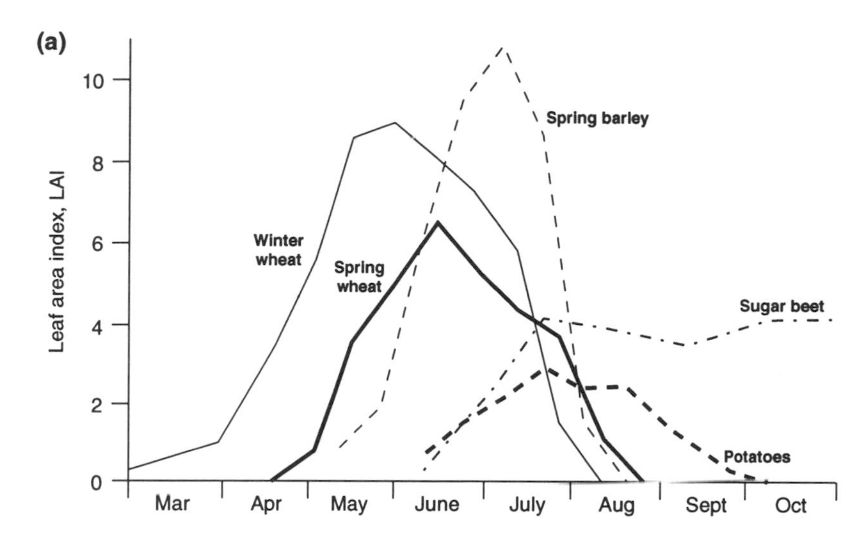

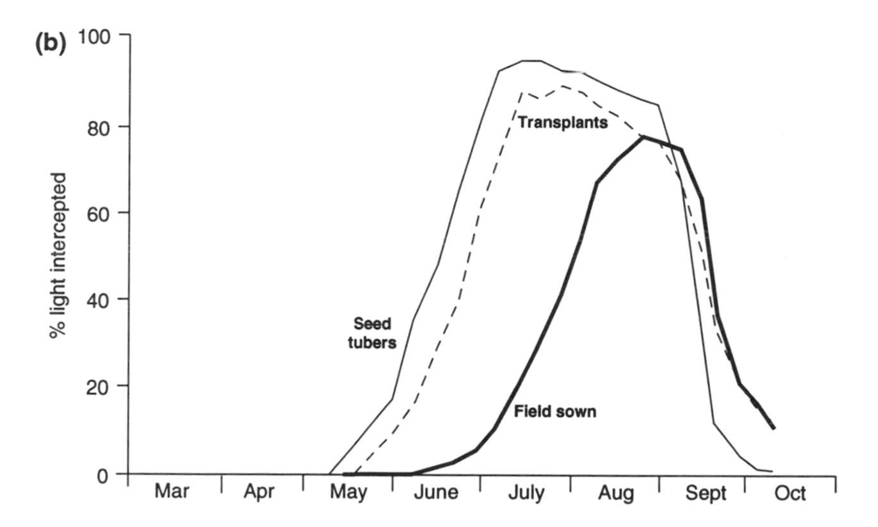

Figure 1.9 (a) Changes over time in the leaf area index (LAI) of various crop plants in the UK (adapted from Watson, 1971). (b) The effect of methods of propagation on light interception in potato crops (adapted from Shakya, 1985)

ecosystems there are fewer carnivores than there are herbivores. The corollary of this grazing food chain in an agricultural system is that of grass (producers) → sheep (herbivores) → humans (carnivores).

Even at this simple level it is immediately obvious that large amounts of energy are lost where meat is the desired end product. It would be much more energy efficient if humans were all vegetarian! However, for most of the history of modern humans (*Homo sapiens sapiens*) they have been omnivores and meat has featured significantly in their diets. Many animals, particularly domesticated species, have the ability to convert poor-quality plant protein into high quality animal protein. A herbivore can

Table 1.5 The economic yield of various categories of crop plants in relation to plant phenology (adapted from Bunting, 1975)

Category A	The economic yield is produced throughout nearly all of the growing season. It comprises the vegetative parts of a perennial or biennial crop. Examples include fodder grasses, forage crops, silage crops, sugar cane and many root and tuber crops, tea, cocoa, rubber. In all of these the yield accumulates in the organs responsible for its acquisition, i.e. leaves and stems or in storage organs such as roots and tubers. The accumulation occurs as long as suitable growing conditions prevail.
Category B	The economic yield is produced during a proportion (small or large) of the life of the crop, usually in fruits or seeds that begin to be formed at an early stage in the plant's life. Examples include grain-legumes, oil-seeds, tomatoes, cotton, fruit trees and fruit bushes.
Category C	The economic yield is produced in the later stages of a crop's lifespan or growing cycle. In annual crops such as the cereals inflorescences are formed late. In some perennial crops, such as bananas and plantains, the economic yield is the annual shoot.

produce lysine, an amino acid, from plant material in which lysine is deficient. The lysine is then used to produce meat protein. One important consequence of this ability, coupled with the aptitude of cud-chewing ruminants to digest cellulose-rich plants, is that large areas of land unsuitable for cultivation can be managed to produce food in the form of meat or animal products. In addition, many of the waste products from arable agricultural systems, i.e. the biological yield minus the harvest (see above) which can be a considerable volume, can also be used to produce food energy by using it to feed livestock. Many such animals, notably the horse, camel and water buffalo, also return some of the energy they produce to agricultural systems in which they are used as draught animals.

The energetics of animal production systems are just as important management tools as are the energetics of arable production systems. The emphasis is on maximising returns whilst maintaining environmental quality; this is the essence of sustainable agriculture though, as will become evident in Chapters 8 and 9, few agricultural systems achieve this status. Like arable systems, animal production systems are subject to similar limitations in relation to environmental constraints. They are also, though not exclusively, in receipt of fossil-fuel energy subsidies. In general, there are three basic types of animal production systems which vary, primarily, in the degree of management intensity, which, in turn, reflects the degree of fossil-fuel energy subsidy. These are: extensive grazing on natural and semi-natural vegetation communities, animal production from cultivated crops (usually grasses) which are grazed (foraged), and animal production that relies on the production of fodder crops, such as cereals, root crops (e.g. turnips), legumes and grasses, that are grown solely for the purpose. In the latter two, management is intensive as is the fossil-fuel energy subsidy.

Much of the management effort is designed to maintain animals in a healthy condition. Providing the correct diet is only one aspect of this; animal health products are widely available and include growth hormones and antibiotics. In the developed world much effort is put into animal breeding, now being further manipulated by genetic engineering, to produce optimum livestock for specific purposes (reviewed in Greenhalgh, 1992). In relation to efficiency within an animal producing system, one of the most important characteristics is the conversion of plant biomass to animal biomass. A simple measure of such efficiency is to calculate the livestock conversion efficiency. This is the ratio of food intake, i.e. the resources used to produce either the energy or protein over a specified time period by an animal. The quality of the feed is a major variable in calculating this ratio which often gives rise to apparently inconsistent values for any given animal. Any impairment of animal health will also produce variable values (see discussion in Tivy, 1990).

There are other ways to measure efficiency. For example, edible energy and edible protein can be calculated as percentages of total energy and total protein produced per animal. This is a valuable approach because it provides information on the consumable portion of the secondary productivity. In this respect it is a measure that resembles the harvest index. Some examples are given in Table 1.6 which shows that there are not only variations between animal types but also for the same animal type at different ages. Rao and Singh (1977) have used other but related measures: energy yield, the percentage gross energy recovery from animal products and the conversion efficiencies of vegetable to animal protein. As Table 1.5 illustrates, the respective values for meat production are all much lower than those for milk production. The overall energy yield is also much lower for all animal products than it is for crops; the energy yields are also higher for category A crops when compared with category C crops (Table 1.5). The energy characteristics of different animal production systems will be discussed in more detail in Chapters 4 and 6.

Finally, there are additional components of both ecosystems and agricultural systems that deserve some consideration. These are the production of detritus and its storage which relate to the role of the detrital system of energy transfer (Section 1.3). This is particularly important because it has implications for global carbon cycling. It has already been stated (Section 1.3) that the transformation of ecosystems into agricultural systems reduces the importance of detrital energy flow (Figure 1.2) in relation to the grazing pathway. This essentially means that more carbon is harvested than is stored in the detritus of the soil. Thus the storage compartment of the global cycle of carbon is reduced and carbon, via the harvest, is actively circulated as it is consumed by livestock and/or humans. It would be interesting to know what contribution this is making to the ever increasing concentrations of carbon dioxide in the atmosphere. Clearly, the spread of agricultural systems and their intensification over time must have contributed to this to some degree. This is in addition to the removal of natural biomass, especially forest, that is necessary before an agricultural system can be initiated.

Reiners (1973), calculating data similar to that of Whittaker and Likens (1973; see above), has ascertained that globally agricultural land produces 9.1×10^9 t yr^{-1} of world net primary productivity of which 50 per cent is converted to detritus. This may seem a relatively high value until it is compared with the 90 per cent or more of primary productivity that is converted to detritus in tropical rain forest, temperate forest, boreal forest, swamp and marsh, tundra, alpine environments and hot deserts. Indeed, the only natural ecosystem type that equates with agriculture is grassland. Moreover, agricultural land produces detritus at a rate of 4.6×10^9 t yr^{-1} of dry weight. Although this figure is higher than those for temperate grassland, tundra,

Table 1.6 Food values and energy yields for specific animals and food types. (a) From Holmes (1977); (b) from Rao and Singh (1977)

(a) Food production: efficiency of specific animals

	Edible protein (%)*	Edible energy (%)†	Edible energy (%)‡	g protein MJ Total ME	g protein MJ Total GE
Broiler	19.0	16.0	11.0	2.9	2.1
Turkey					
stag	20.0	9.0	6.0	3.1	2.2
hen	20.0	9.0	6.0	3.1	2.2
Rabbit	17.0	13.0	8.0	3.2	1.9
Pig					
pork	27.0	31.0	20.0	3.4	2.2
bacon	22.0	25.0	17.0	2.6	1.8
heavy hog	15.0	22.0	15.0	1.7	1.1
Lamb					
early	28.0	28.0	25.0	3.3	3.0
late	10.0	15.0	9.0	1.3	0.8
Cow	25.0	12.0	10.0	2.9	2.2
cereal-beef	12.0	10.0	6.0	1.3	0.8
18-month beef	11.0	10.0	6.0	1.3	0.7
24-month beef	9.0	10.0	6.0	1.1	0.6
Milk					
low conc.	20.0	21.0	11.0	2.5	1.4
high conc.	21.0	23.0	13.0	2.8	1.5
Chicken	25.0	21.0	14.0	3.2	2.1

* Percentage total crude protein eaten.
† Percentage total metabolizable energy (ME) eaten.
‡ Percentage total gross energy (GE) eaten.

(b) Energy yield, recovery and conversion for major food items

Food	Energy yield (kcal×10^6/A)	Average gross energy recovery	Average conversion of vegetable to animal protein
Animal products			
Milk	4.5	15	38
Butter	2.0	–	–
Eggs	1.25	7	31
Beef	1.0	4	6
Crops			
Sugar	62.5	–	–
Potatoes	29.0	–	–
Rice	16.5	–	–
Grain	13.25	–	–

alpine and hot desert environments, it is considerably lower than those for the world's forests types. For example, tropical forest has the highest value at 38×10^9 t yr^{-1} of dry weight and temperate forest has the next highest value at 22.2×10^9 t yr^{-1} of dry weight. These values represent 19.0 and 11.1×10^9 t of carbon per year. The value for agricultural land is only 2.3×10^9 t of carbon per year. Since many of the world's agricultural systems have replaced forest it must have had a substantial impact on the amount of carbon dioxide in the atmosphere. This is yet another reason why agriculture must be considered as a major agent of environmental change.

1.5 CONCLUSION

Agriculture is a major component of the management of planet Earth. In all its forms it both relies on and alters the environment in which it takes place. It is limited by environmental factors, such as climate, but innovations through the ages have facilitated the mitigation of some environmental constraints. Structurally and functionally, agricultural systems have much in common with the natural ecosystems they replaced. They are both characterised by inputs, processes, outputs and feedback mechanisms, all of which involve biotic elements. Ecosystems are usually more biologically diverse than agricultural systems which are thus simplified ecosystems and which have, through their spread, caused a reduction in global biodiversity. The major difference between the two is that the productivity of an ecosystem is recycled within it but in an agricultural system much of the productivity is exported, for human consumption. The inputs and processes in the latter are thus controlled by management in order to ensure as large a harvest as possible. In many agricultural systems the management requires a substantial input of fossil-fuel energy and so represents a means of using such energy to enhance the production of food energy.

In changing ecosystems into agricultural systems all of the basic processes are altered to a greater or lesser degree. Energy flows, for example, in agricultural systems are dominated by the grazing food chain and the storage element is much reduced; detrital energy flows dominate in most natural ecosystems. Nutrient cycles are altered in agricultural systems; there is a net export of nutrients in the harvest but in an ecosystem nutrients are recycled mainly within the system itself.

Various measures are available to determine the productivity of agricultural systems: calorie or joule content, potential crop yield, harvest index, edible energy and edible protein contents. The energy characteristics of agricultural systems are important for management since the basic objective of such systems is to enhance the trapping of solar energy and so increase food productivity. Crop production systems are more energy efficient at food production than are livestock producing systems but the latter convert poor quality protein into high quality protein.

Agriculture is thus a complex process that is reliant on the physical environment and biota but which is also responsive to economic and social processes, particularly population change. It underpins all other forms of human activity; its success or failure can confer political pre-eminence through food security on the one hand, or famine and poverty on the other hand.

1.6 FURTHER READING

Grigg, D. (1995) *An Introduction to Agricultural Geography*, 2nd edn. Routledge, London.
Mannion, A.M. (1991) *Global Environmental Change*. Longman, Harlow.
Monteith, J.L., Scott, R.K. and Unsworth, M.H. (eds) (1994) *Resource Capture by Crops*. Nottingham University Press.
Odum, E.P. (1993) *Ecology and Our Endangered Life-Support Systems*, 2nd edn. Sinauer Associates, Inc., Sunderland, Massachusetts.

Open Universiteit, The Netherlands and University of Greenwich, London (1994) *Crop Productivity*. Butterworth Heinemann, Oxford.
Peart, R.M. and Brook, R.C. (eds) (1992) *Analysis of Agricultural Systems*. Elsevier, Amsterdam.
Spedding, C.R.W. (ed.) (1992) *Fream's Principles of Food and Agriculture*, 17th edn. Blackwell Scientific Publications, Oxford.

CHAPTER 2

The Beginnings of Agriculture

The domestication of plants and animals that accompanied the inception of agricultural systems is considered to be a major turning point in the history of modern humans. More than a century ago Darwin (1859) commented on the significance of domestication as 'that which enables the agriculturalist, not only to modify the character of his flock but to change it altogether. It is the magician's wand, by means of which he may summon into life whatever form and mould he pleases.' More recently MacNeish (1992) stated that 'One of humanity's most important inventions is agriculture. This decisive step freed people from the quest for food and released energy for other pursuits. No civilisation has existed without an agricultural base, either in the past or today. Truly, agriculture was the first great leap forward by human beings'. There is no doubt that these developments set in motion major changes in the social and economic fabrics of human society. The moments of these changes led the British archaeologist Vere Gordon Childe (1936) to coin the term 'Neolithic revolution', since it is during the archaeological period known as the Neolithic, or new stone age, that the foundations of agriculture were laid. Inevitably, the material culture reflected in the archaeological record on which such observations depend more than likely reflects the stage at which agriculture was sufficiently well established to create such changes. This is why it is often referred to as a revolution even though the term is misleading if interpreted conventionally in relation to timescales experienced by individuals. Agriculture did not arise spontaneously but has its origins in its precursor hunter–gatherer societies. The term 'Neolithic' is also time transgressive since cultures so characterised arose at different times and in different places. On the Eurasian landmass, for example, the earliest Neolithic societies developed in what is today the Near East about 12 000 years ago. It was not until *c.* 5000 years ago that such a culture arose in the British Isles. In addition, the term 'Neolithic' is not applied outside Eurasia; in new world archaeology, for example, local terms are used, often denominating the geographic field site from which a specific culture was first identified.

According to MacNeish (1992), both Vere Gordon Childe (see above) and Ivan Vavilov, a renowned Russian botanist, were instrumental in developing modern ideas on plant domestication and the origins of agriculture. Both were also Marxists and believed that cultural change was wrought via economic revolutions fuelled by improved production. They nevertheless held very different views as to what precipitated the switch from foraging to farming. Childe believed that the climatic oscillations that occurred as the last ice sheet waned created just the right combination of conditions for humans to begin to select, more positively than their hunter–gatherer ancestors, specific plant and animal species and to manage them. There is an element of climatic determinism in this theory and it gives rise to what MacNeish describes as an environmental cause for the origin of agriculture. The resulting food security then facilitated population increase. This contrasts with Vavilov's (reviewed in Vavilov, 1992) theories, which like Childe's were formulated in the 1920–1950 period, and assume that the motive for the inception of agriculture was primarily the need to generate a food surplus to support increasing populations. This perspective MacNeish describes as materialism. The input of the Vavilov model, i.e. population increase, is the output of the Childe model. Despite this dichotomy, both scholars made considerable contributions to elucidating the origins of agriculture, the motives or reasons for which still remain obscure despite the fact that many additional research data are now available, particularly for the Near East and the Americas.

Although there is an environmental component in Childe's hypothesis, virtually all accounts of the beginnings of agriculture refer to its significance for society, as is epitomised by MacNeish's quotation (above). However, it must also be acknowledged that the commencement of agriculture was a major turning point in environmental as well as cultural history. In terms of geological time, the age of the Earth being 5000×10^6 years, the onset of agriculture was indeed a revolution. As discussed in Sections 1.2 and 1.3, it resulted in the partial control of energy flows, transformed biogeochemical cycles and biotas, set in train a massive extinction event that continues to impoverish the Earth's genetic resources and altered all Earth surface processes, e.g. weathering, soil erosion, the hydrological cycle, etc. The repercussions for the environment were, and continue to be, just as important as those for society (see Figure 1.1).

The initial domestication of plants and animals represented a crude form of genetic selection and began a close link between humans and certain components of the biota. Genetic selection intensified with the establishment of breeding programmes in the nineteenth century. Today, the acme of genetic selection is genetic engineering (Chapter 10). The oldest known domesticated animal and plant remains derive from archaeological sites in the Near East. There is no doubt that this was a major centre of domestication but there were also others. According to MacNeish (1992), the other centres are the Andes, Mesoamerica and the Far East. Here there was *in situ* domestication. In a number of other areas, described by MacNeish as non-centres, e.g. India, South-East Asia, Africa and north-western tropical South America, there was an importation of domesticated species from the centres and the domestication of other plants occurred *in situ*. As will be discussed below, there are other such schemes that have been devised to indicate foci of plant domestication. Furthermore, animal domestication occurred initially in the Near East but other centres can also be identified. In these primary and secondary areas plants and/or animals were harnessed as natural ecosystems were transformed into agricultural systems.

Gradually, and often improving and changing in the process, agricultural systems spread beyond their centres of origin at the expense of natural ecosystems. Thus began forest clearance, the extent of which spatially was on a par with modern tropical rain forest demise but which occurred over several thousand years rather than in just a century or so. These systems underpinned some of the most influential civilisations that have ever existed. Without food security, and a food surplus to facilitate trade, the social and environmental legacies of the ancient Egyptians, Greeks and Romans would have been much impoverished. The course of history may have been very different.

2.1 HUNTING AND GATHERING

A major difficulty of assessing the impact of, and the success or otherwise of, hunter–gatherer societies in pre-agricultural prehistory is a dearth of evidence and, often, an inherent bias in available evidence. This is because plant remains are not readily preserved until they are, serendipitously for the archaeologist and palaeoecologist, deposited in an anaerobic and/or acidic environment. Both lack of oxygen and low pH conspire to inhibit microbial decomposition and so provide ideal conditions for plant preservation. Moreover, the fruits of many plants are completely edible so no remains would occur, except possibly in human faeces. These too would oxidise unless encapsulated in an anaerobic environment. Even in the wetter environments of the world, conditions rarely occur that provide preserving milieu. Bones, however, are not so susceptible to oxidation and the chances of preservation are greater than those for plant remains. As a result more is known about the animal component of hunter–gatherer diets than about the plant constituents. There is an obvious parallel here between wood and stone implements that were also used by hunter–gatherers. The significance of the former is probably understated because it does not preserve as readily as stone axe heads. In addition, hunter–gatherer societies were generally nomadic. This probably meant that they had meagre belongings in order to remain unencumbered as they travelled. Although such people made temporary, or even seasonally occupied, camp sites, the chances of finding an abundance of food remains and/or implements is slim. Overall then, comparatively little is known about the hunter–gatherer societies of pre-agricultural *Homo sapiens sapiens* or of the earlier hominids (reviewed in Mannion, 1991).

The dental characteristics of hominid skeletons imply that plant resources have always constituted a major source of food. What evidence there is for plant consumption in prehistoric hunter–gatherer societies derives mainly from the last 20 000 years. For example, Wendorf *et al.* (1988) and Hillman (1989) have reported on plant remains and human palaeofaeces from a palaeolithic site at Wadi Kubbaniya in Egypt. These remains are between 18 000 and 17 000 years old and their discovery is particularly important because they constitute one of the oldest-known examples of pre-agricultural plant resource use in the Old World (Eurasia). Wadi Kubbaniya is located 12 km north of Aswan and is a palaeo-drainage system of the south-western desert. Four of the 16 excavated sites contained food remains and/or faeces which had been preserved by charring. As shown in Table 2.1, some 25 different types of seeds, fruits and soft vegetable tissues have been recognised; of these 13 have been ascribed to species or genus. Hillman (1989) states that all 13 are edible, that their modern counterparts grow wild in the Nile Valley and that most have been used as sources of food by contemporary or near-contemporary hunter–gatherers. The most prolific species is the wild nut-grass (*Cyperus rotundus*). This is not the only sedge as club rush (*Scirpus* sp.) is also found, as is the fruit of the dóm palm (*Hyphaene thebica*). The sedges (and some others) are typical of wet environments so it is likely that swamps existed within this arid zone; this occurred at the same time as the

Table 2.1 The charred remains of food plants from Wadi Kubbaniya, Egypt (adapted from Hillman, 1989)

(a) Charred remains of vegetative tissues and organs	
Cyperus rotundus (wild nut-grass)	Whole tubers and fragments
C. rotundus or *Scirpus maritimus*	Whole tubers and fragments
Scirpus maritimus/tuberosus type	Whole tubers and fragments
Pteridophyte (fern)	Rhizome fragments
Indeterminate monocotyledon stem fragments	
Indeterminate monocotyledon leaf fragments	
Indeterminate parenchyma (soft, thin-walled tissue) of five types of plant	

(b) Charred remains of fruits and seeds	
Hyphaene thebica (dóm palm)	Fruit mesocarp fragments
Scirpus maritimus/tuberosus (club rush)	Nutlets
Compositae—tribe A themidae (chamomile tribe)	Achenes
Compositae embedded in coprolites	Three types of achenes
Nymphaceae (waterlily family)	Immature bud receptacles
Schizocarpous fruit (*Tribulus* type)	Green and mature when charred
Balanites type	Cotyledon fragment
Lilaceae type a (like *Asparagus*)	Seed
Lilaceae type b	Seed
Umbelliferae type (aniseed family)	Kernal of mericarp
Indeterminate seed/fruits of five types of plant	
Indeterminate fruit-skin fragment (*Capparis* type)	

maximum extent of global ice volume. There are also several grinding stones and mortars which may have been used for grinding the tubers of wild nut-grass and club rush. Additional evidence from the same site has been presented for the exploitation of fish and possibly large mammals. From this work, Hillman (1989) concludes that the diet of these Palaeolithic hunter–gatherers was probably as diverse as hunter–gatherers in the arid zone of Africa today. Another site, 19 000 years old, has been reported by Nadel and Hershkovitz (1991) and Kislev *et al.* (1992) from Ohalo II in the Jordan Valley. Faunal remains indicate a broad-based diet which included birds, reptiles, rodents, turtles, hares, foxes, gazelles, deer and fish; floral remains include the oldest grains of wild barley and wild wheat ever found in an archaeological site and 30 other plant species including fruits.

N.F. Miller (1991) has reviewed the evidence for plant consumption in the late Epipalaeolithic (*c.* 13 000 to 11 000 years ago) groups of the Near East. The plant remains indicate a reliance on a range of species including legumes, chenopods, sedges and large- and small-seeded grasses. These were the precursors of the domesticated species discussed in Section 2.2. Similar work at Abu Hureyra, Syria, is reported by Hillman *et al.* (1989). There are also records of hazel nuts in many Mesolithic (pre-Neolithic and pre-agricultural) sites in Britain (Greig, 1991) which are 8000 or 9000 years old. However, Loy *et al.* (1992) have recently reported the presence of starch residues, notably starch grains and crystalline raphides (needle-like crystals), on stone artifacts from the Solomon Islands. Remarkably, these artifacts are 28 000 years old, as revealed by radiocarbon dating. The species to which the starch remains belong have been identified as taros (*Colocasia* spp.) which are currently cultivated for their edible roots in many Pacific islands. The potential of this residue-detection technique for increasing the understanding of plant resource use in prehistoric hunter–gatherer groups is considerable.

The evidence for animal consumption is hardly more prolific than that for plant use. For much of their early history (Figure 2.1) hominids were scavengers of dead or trapped animals (see discussion in Lewin, 1993). According to Davis (1987), a shift from scavenging to actual hunting may have occurred between one and two million years ago. It coincided with the emergence of *Homo erectus* from its Australopithecine ancestors who were mainly vegetarian. There is also the possibility that hominids had by this time learnt to manipulate fire which would have been useful as a means of herding animals. Shipman *et al.* (1981) have examined the bone assemblages from Olorgesailie, a site in Kenya dated to between 0.7 and 0.4 million years. The bones are predominantly those of a giant gelada baboon *Theropithecus oswaldi.* The size of this animal, about 656 kg, and its abundance suggests deliberate hunting rather than scavenging, as does the abundance of juveniles. Marks in the bones also indicate dismembering using hand axes. There are, however, no clues as to how the animals might have been hunted but it is likely that it required co-operation and organisation. Moreover, the sharing of a dead animal, especially one as large as a gelada baboon, would have represented efficient resource use. The onset of organised hunting could well have had just as profound social repercussions as the later domestication of plants and animals (Sections 2.2 and 2.3).

From about 0.2 million years ago, the time when *Homo sapiens sapiens* emerged, there is a good continuous and unique record of animal remains in South Africa. It has facilitated a comparison between the Middle Palaeolithic (old stone age) *c.* 0.2 million years ago, and the Late Palaeolithic about 35 000 years ago (summarised in Klein, 1989, 1992). Both the species composition of the bone assemblages and the artifacts vary considerably reflecting a change in resource exploitation. Some of these data are summarised in Table 2.2. They indicate that hunting and gathering were more considered and possibly more organised during the Late Palaeolithic than in the early Middle Palaeolithic. The reduced size of limpet shells, for example, in the Late Palaeolithic indicates that they were heavily exploited so that old and thus large specimens were rare. The presence of the remains of particularly ferocious animals in the younger sites indicates that by then improved methods of trapping had been developed, as is suggested by artifacts. Seal culling was most selective in the Late Palaeolithic, also indicating an understanding of population cycles. The even-aged distribution of remains of eland, a type of antelope, reflects what Klein describes as a catastrophic pattern. This animal, and the bastard hartebeest, were probably hunted by driving them over cliffs as is also indicated in a Transkei rock painting of the Late Palaeolithic. This study also indicates that marine resources played a role in Palaeolithic diets.

The apparently more sophisticated food-procurement strategies of the Late Palaeolithic compared with the Middle Palaeolithic represent an improved human understanding of environmental resources and their annual patterns. Although far from conclusive, the evidence may also indicate increased communal activity to facilitate at least the procurement of meat. This, more directed, hunting strategy is

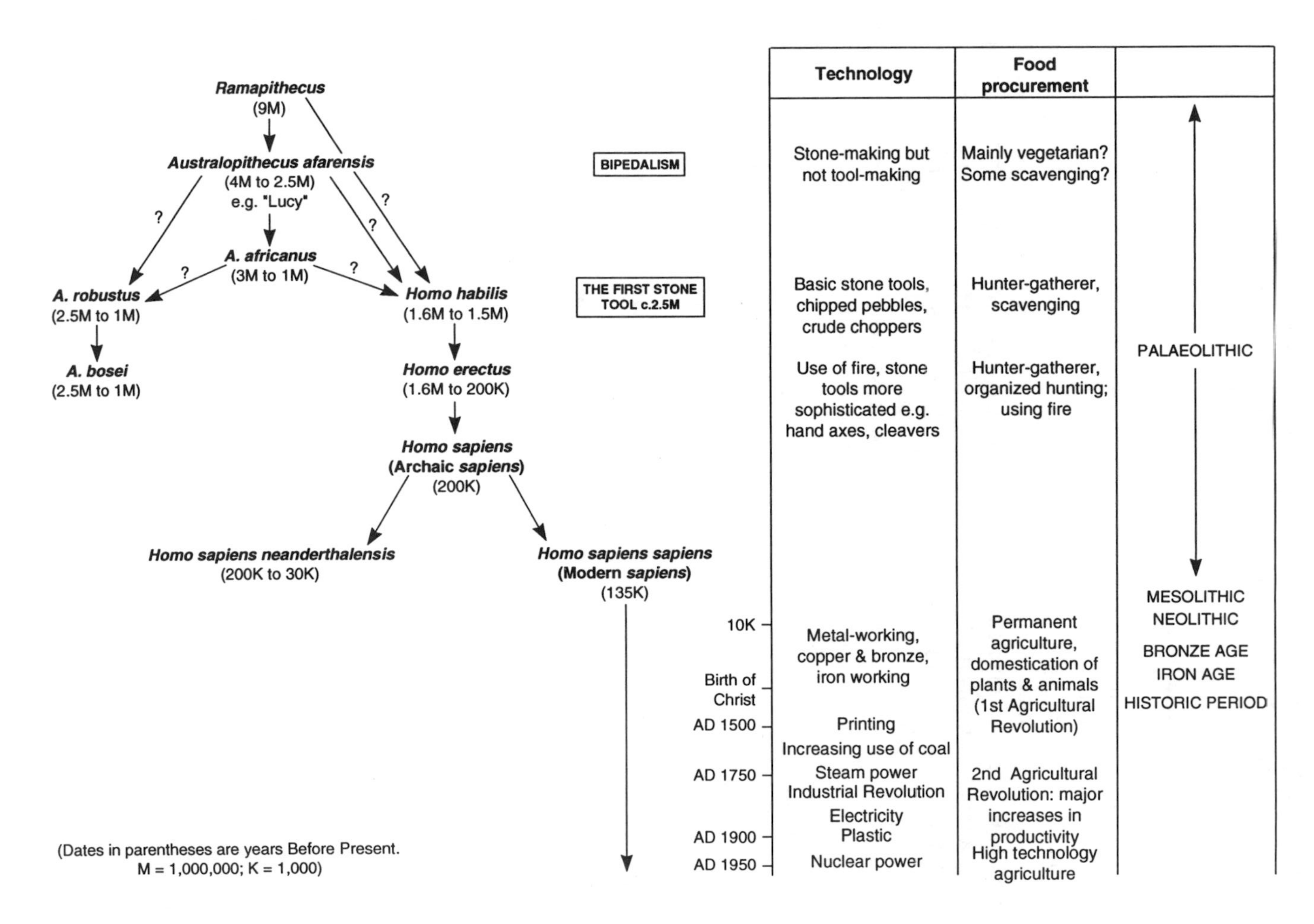

Figure 2.1 Human evolution in relation to technological and food-procurement strategies (adapted from Mannion, 1991)

Table 2.2 Variations in animal and artifact assemblages of Middle and Late Palaeolithic Age in South Africa (based on Klein, 1989, 1992)

Cultural period	Species exploited	Artifacts
Late Palaeolithic (began 30 000 to 40 000 years ago) modern humans	Abundant fish Abundant flying birds Warthog, Bushpig, Cape buffalo — ferocious species Seals—most killed near age of weaning Limpets, Angulate tortoises — smaller shells due to more intensive exploitation Some eland	Indicate fishing and fowling; probably bow and arrow, spears
Middle Palaeolithic (began about 0.2 million years ago)	Few or no fish Few or no flying birds Few or no ferocious species Seals of all ages killed Young and old eland Limpets, Angulate tortoises — large shells due to a less intensive exploitation Abundant eland of all ages	No indication of bow and arrow, or hunting-related artifacts

accompanied by a change in artifact assemblages, notably the inclusion of bone, antler and ivory tools. Generally, Late Palaeolithic groups are considered to be more efficient hunters than their ancestors. However, Chase (1989), reviewing evidence for the European Middle to Late Palaeolithic transition, states that 'It is possible that, if Palaeolithic hunters were indeed more efficient than their predecessors, it is due less to an increase in foresight and planning, and more to the existence of language and the transfer of intelligence.' This and other attributes of modern humans are discussed by Lewin (1993). Whatever the reason for the changes in hunting strategy, the infrastructure—i.e. societal organisation, knowledge of plant phenology (seasonal growth patterns), animal life-cycles and selection—for the establishment of permanent agriculture was being inaugurated. Undoubtedly, the development of language would have greatly assisted this process.

The next 10 000 years and the events thereof are not well represented in the fossil or archaeological record. In Europe the last ice age was at a maximum 18 000 years ago; by this time modern humans had become well established in Australia (*c.* 50 000 years ago) though they had yet to enter the Americas. This latter is a subject of considerable controversy (see Mannion, 1991, for a review). In Europe itself, cave painting was already being practised, providing considerable insight into the activities of the time and indicating that human groups were skilled hunters. Examples include the famous cave paintings of Lascaux in the Dordogne and of Altamira in the Pyrenees. Here animals such as aurochs, horse, deer and wolf were depicted for posterity by human groups who occupied these areas about 17 000 years ago. Judging by the recent evidence from a Greenland ice core (Dansgaard *et al.*, 1993), these frescoes were painted during a lengthy cold period of the last ice age. By this time it is clear, from pictorial and archaeological evidence, that skill and selection in hunting were established.

As the last ice age waned a major change occurred in faunal assemblages in many parts of the world. This episode has precipitated a controversy that is equalled in importance only by that which surrounds the advent of agriculture itself (Section 2.4). In Australia and North America especially, but elsewhere as well, there was an extinction episode of large herbivorous mammals. As is well established (see Mannion, 1991, for a review), the end of the last ice age was a period of rapid climatic and associated environmental change, the very *raison d'être* that Childe in 1936 opined was essential for the initiation of agriculture (see above). Temperatures were changing rapidly; Atkinson *et al.* (1987) indicate that changes of as much as 2.6°C per century were usual. By this time natural change was particularly variable, possibly even at the scale of the human lifetime. Cultural change was also underway. The controversy relates to whether the herbivore demise was due to natural or cultural factors. In view of the abundant evidence for climatic instability it would be surprising if this did not have some impact on animal populations. Whether it could have caused extinctions at such a rate is another matter; there appear to be no precedents for similar extinctions in earlier Quaternary times that were just as climatically variable. There is some evidence to suggest that human populations may have been the cause of the extinctions. At High Furlong near Blackpool (Hallam *et al.*, 1973) barbed bone points have been discovered in association with a skeleton of the elk *Alces alces* which is between 11 000 and 12 000 years old. In addition, Martin (1984) has suggested that the extinction event in North America may have been associated with the arrival of the first humans. There is also the possibility that climatic change, possibly through its impact on human groups, and cultural factors were both contributory factors.

Whatever the case, there is abundant evidence for substantial organisation within human groups at this time. Scavenging, the indiscriminate exploitation of animals in an opportunistic fashion, had largely been replaced by barbarism which required deliberate manipulation of animal herds. By the end of the last ice age the dog (*Canis* sp.) had already been domesticated, probably from the wolf (*Canis lupus*). According to Davis (1987), the domestication of the dog occurred in the Near East at least 12 000 years ago. There is little evidence from the fossil or archaeological record to indicate that the species was butchered so it is most likely that dogs were used, as they are today, for herding, hunting and possibly guarding. The association between people and dogs is well illustrated by the discovery, in a tomb at Ein Mallaha in the Hula basin of northern Israel, of a skeleton of an elderly human with the remains of a wolf or dog puppy beneath the left hand (Davis and Valla, 1978). Dogs were probably valuable assets that assisted in the intensified management of hunting in the Late Palaeolithic. In the Near East, there is evidence for a reliance on gazelle herds prior to agriculture (Henry,1989) indicating that hunting had become specialised. In the Dordogne region of France, there is evidence that reindeer (*Rangifer tarandus*) was the chosen animal (Mellars, 1973). Some two to three thousand years later the domestication of sheep, goats and cattle occurred and herding became a predominant food-procurement activity. Hunting, however, remained an important activity to supplement diets.

This phase of climatic change and highly organized hunting and gathering provided the backdrop to a different though related set of activities in the Near East where plant domestication was in progress and cultivation was beginning. What was the stimulus, or stimuli, that prompted such a radical change of existence? This is one of the most debated aspects of human history. It was also a climacteric. Although there is much evidence for domestication there is little in the archaeological or palaeoecological record that indicates why it was occurring. Harlan (1992a) states 'The question must be raised. Why farm? Why give up the 20-h work week and the fun of hunting in order to toil in the sun? Why work harder for food less nutritious and a supply more capricious? Why invite famine, plague, pestilence and crowded living conditions? Why abandon the Golden Age, and take up the burden?' As discussed in the introduction to this chapter, there are essentially two schools of thought: environmentalism and materialism, which originated from Vere Gordon Childe and Ivan Vavilov respectively. These are illustrated in Figure 2.2. Boserup (1965), for example, presents a convincing case for population pressure as the driving force behind plant and animal domestication. Her ideas, based on analogies with processes operative in modern hunter–gatherer communities, focus on the impact of sedentary life in the Late Palaeolithic/Mesolithic period on hunted and gathered resources. The latter would have become depleted as human populations increased, causing people to turn to cultivation and animal husbandry. The question of why agriculture began may never be resolved and it is likely to have been a response to a complex set of factors; it may even have been a proactive response to planning for the future. It becomes even more enigmatic when ethnographic data and palaeopathological evidence are considered, as discussed in Harlan (1992a). Ethnographic studies show that hunter–gatherers engage in most of the activities that cultivators practise but they apparently do not work as hard. Their level of knowledge about plant and animal life-cycles, etc., is just as great as that of cultivators. Palaeopathological evidence also reveals some interesting facts. Roosevelt (1984), for example, states that Palaeolithic skeletons do not show evidence of severe or chronic stress, nor evidence of frequent or severe malnutrition. She suggests that this indicates an adequate diet and contrasts with the characteristics of younger, Neolithic skeletons which show evidence of chronic malnutrition despite the fact that they derive from a sedentary agricultural society. This leads Roosevelt to the conclusion that population pressure was unlikely to have been the primary motivating force behind the inception of agriculture. A similar conclusion is reached by Layton *et al.* (1991) who have examined both archaeological and ethnographical evidence for hunter–gatherers and cultivators. It is not possible, however, to rule out materialism entirely because there may have been other non-essential reasons for the establishment of agriculture, such as the desire to produce surplus food with which to trade and barter. The motives behind the onset of one of the most significant human activities are indeed enigmatic and may always remain obscure.

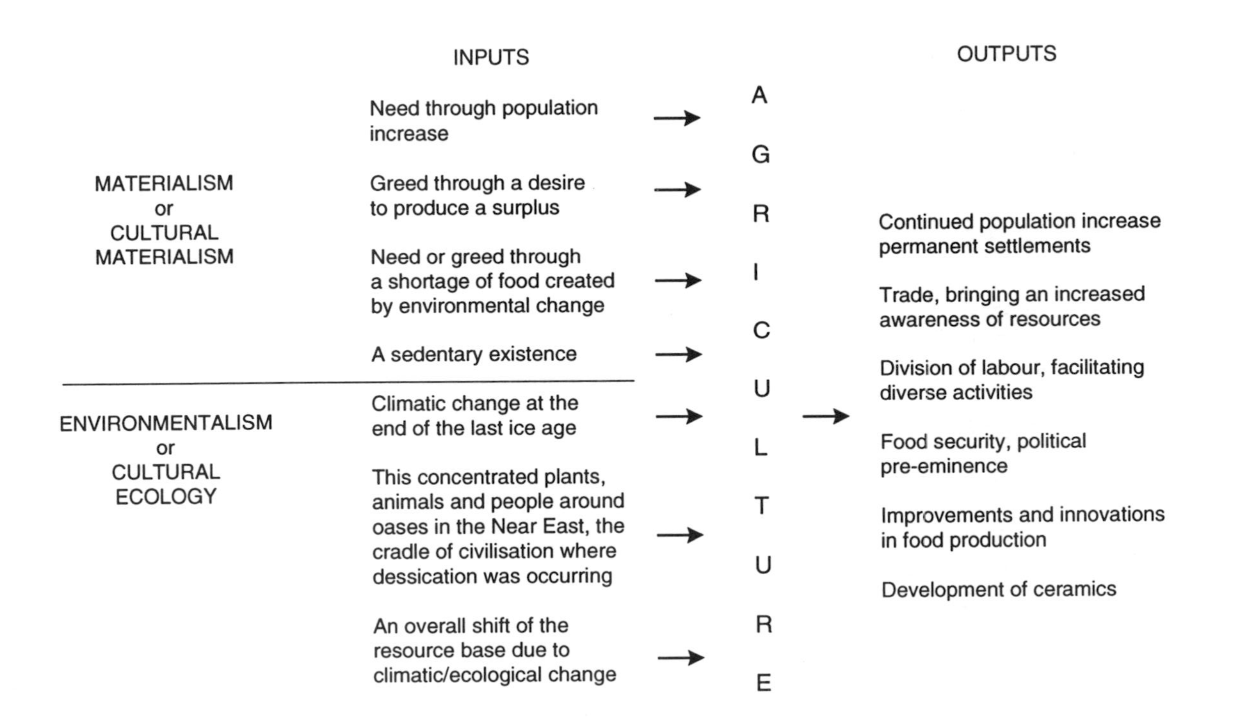

Figure 2.2 Models for the underpinning causes of the beginnings of agriculture

2.2 THE DOMESTICATION OF PLANTS

Although there is no evidence to confirm or refute either the environmentalism or materialism hypotheses for the origins of agriculture it has not prevented researchers from embellishing and/or modifying the hypotheses of Childe and Vavilov (see Section 2.1). In general, Vavilov's ideas on materialism prevailed through the work of the botanist Edgar Anderson and the geographer Carl Sauer, both in the USA. Sauer (1952), like Vavilov, hypothesised that the likely centres of plant domestication were in those areas characterised by high biological diversity so providing an abundance of many different sources of food. The limited amount of foraging required to survive may have led people to develop a semi-sedentary lifestyle with a considerable amount of leisure time. This, in turn, could have provided opportunities for experimentation and innovation with plants and technical skills. Ultimately, control was exerted via planting and as population increased agriculture developed; alternatively, population may have increased as the result of a proximal reliable food supply (Figure 2.2). Anderson (1956) presented a similar hypothesis known as the 'dump heap theory'. This hypothesis contains elements of environmentalism, in so far as the ecological preferences of specific plants are important, and materialism. Anderson, concurring with Vavilov, recognised that many domesticated species have wild relatives which thrive best in open and/or distributed habitats. Examples of such habitats include the nutrient-enriched rubbish heaps of campsites in lightly- or non-forested regions or around lakes. Such areas would also have contained a rich seed bank from which stock seeds, including mutants with particularly favourable characteristics, could be collected by humans. Population increase could also have encouraged the exploitation of such resources. Many early domesticates and village agriculturists thus developed a mutual dependence or symbiosis. Many later researchers, such as Braidwood, Mangelsdorf and MacNeish (reviewed in MacNeish, 1992) became supporters of environmentalism, alias cultural ecology (Figure 2.2), because of the belief that ecological/environmental factors as well as, or instead of, population and/or cultural factors were involved. This attitude was also an outcome, at least partially, of growing liaisons between different disciplines, notably archaeology, geography and botany, which brought a variety of methodologies to this fundamental problem. This multidisciplinary approach has provided an enormous body of data on plant (and animal) domestication but it has still not facilitated the identification of stimuli.

Recently, both MacNeish (1992) and Harlan (1992a) have summarised the available data on plant domestication. The former describes himself as a proponent of cultural ecology and his schemes for the initiation of agriculture are given in Figure 2.3. For all of the routes described, the changes were gradual and evolutionary over several generations and not revolutionary, i.e. occurring within one or two human lifetimes. Harlan (1992a), in contrast, though described as a cultural materialist by MacNeish (1992), offers a synthesis of available research and opts for what he describes as a 'no-model model'. This he believes emphasises the individuality of agricultural development in different regions. Such an approach has parallels with the individualistic concept of the plant association proposed by Gleason in 1926 (Gleason, 1926).

Regardless of how or why plant domestication transpired there is a general consensus that it occurred in different regions at different times though, as Blumler

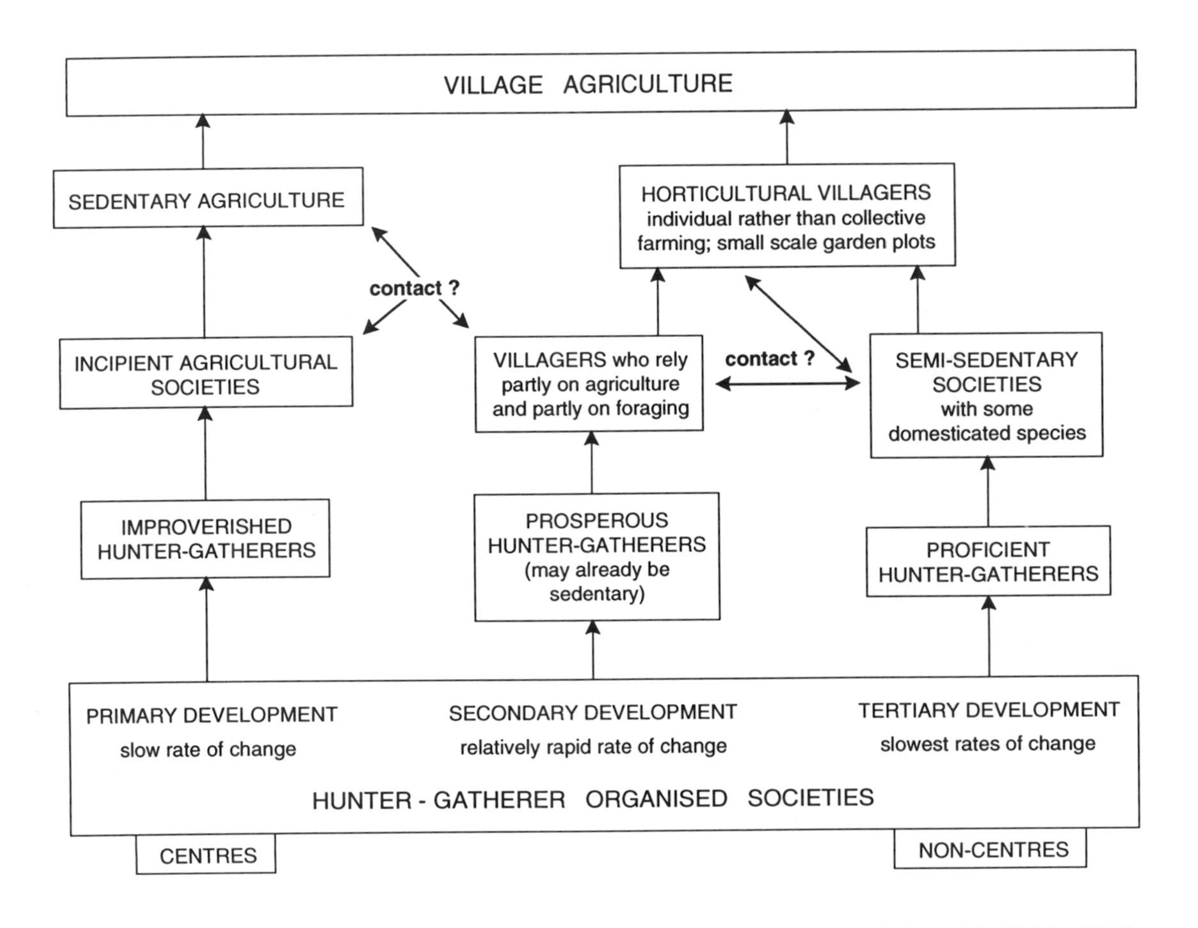

Figure 2.3 MacNeish's (1992) schemes for the initiation of agriculture (adapted from MacNeish, 1992). See text for definition and Figures 2.5 and 2.6

(1992) has discussed, there were likely to have been independent domestications outside the centres and probably several points in space and time when domestication occurred within the same centre. In an attempt to identify potential regions, Vavilov (reviewed in Vavilov, 1992) proposed that there were eight centres of origin; these are given in Figure 2.4. Vavilov suggested that these centres of origin should be referred to as the hearths of domestication. Vavilov's ideas were based on the geographical distribution of varieties of domesticated plants; the areas that contained the greatest concentration of domesticated plant varieties, i.e. contained the greatest genetic diversity, were the hearths. The concept of a centre of domestication is still widely used, though the large body of evidence now available has resulted in much modification of the Vavilov model. In 1971 Harlan proposed three independent units, each with a centre and a non-centre. The former is a relatively small definable geographical area in which a number of plants were domesticated and then diffused outward into neighbouring regions. An example is the Near East (Figure 2.5). A non-centre, conversely, is a relatively large geographical region in which plant domestication occurred widely, e.g. Africa. Harlan (1992a) has recently reiterated his scheme which is given in Figure 2.5. MacNeish's (1992) scheme on the other hand, consists of four centres and six non-centres (Figure 2.6). The latter, however, are defined as areas that received their first domesticates from the centres and then domesticated their own species, areas in which there was no primary development of agriculture and areas in which there were secondary and/or tertiary domestications. Yet another scheme is given by Cowan and Watson (1992). They refer to domestication in pristine and secondary settings. The former relate to the centres of Harlan and MacNeish. Secondary settings are of two types: a suite of domesticated plants is introduced into a hunting–gathering economy which is transformed into an agricultural economy, or a new domesticated crop that will ultimately dominate may be introduced into an economy already reliant to some degree on domesticated species.

There is also some debate about when, in relation to cultivation, actual domestication occurred. The term 'domestication' means to tame or, metaphorically, to bring into the household or human domain. A domesticated plant (or animal) is, according to MacNeish (1992), one 'that differs from its wild ancestor because it has been changed genetically through human selection, either consciously or unconsciously'. Such species may lose their ability to survive and reproduce in the wild. Blumler and Byrne (1991) have questioned the assumption that domestication could only have occurred after cultivation. In fact the 'dump heap hypothesis' (see above) may well have provided the setting for at least partial domestication prior to cultivation. In addition, Blumler and Byrne suggest that the harvesting of small wild cereal stands with seed beaters could have begun the process of domestication by increasing the frequency of plants with indehiscent rachises (flowers that do not open when mature). The first stages of domestication may well have preceded cultivation, as may a sedentary lifestyle.

A question related to this concerns the rate at which a species may become domesticated. Hillman and Davies (1990) have examined the possibilities for wild wheats and barley under primitive cultivation, mainly in the Near East. First, they dispute the 'dump heap hypothesis' (see above) and suggest that the first cereal crops were sown from seed collected from wild stands. Thereafter these crops were domesticated; this precludes the possibility of pre-cultivation domestication suggested

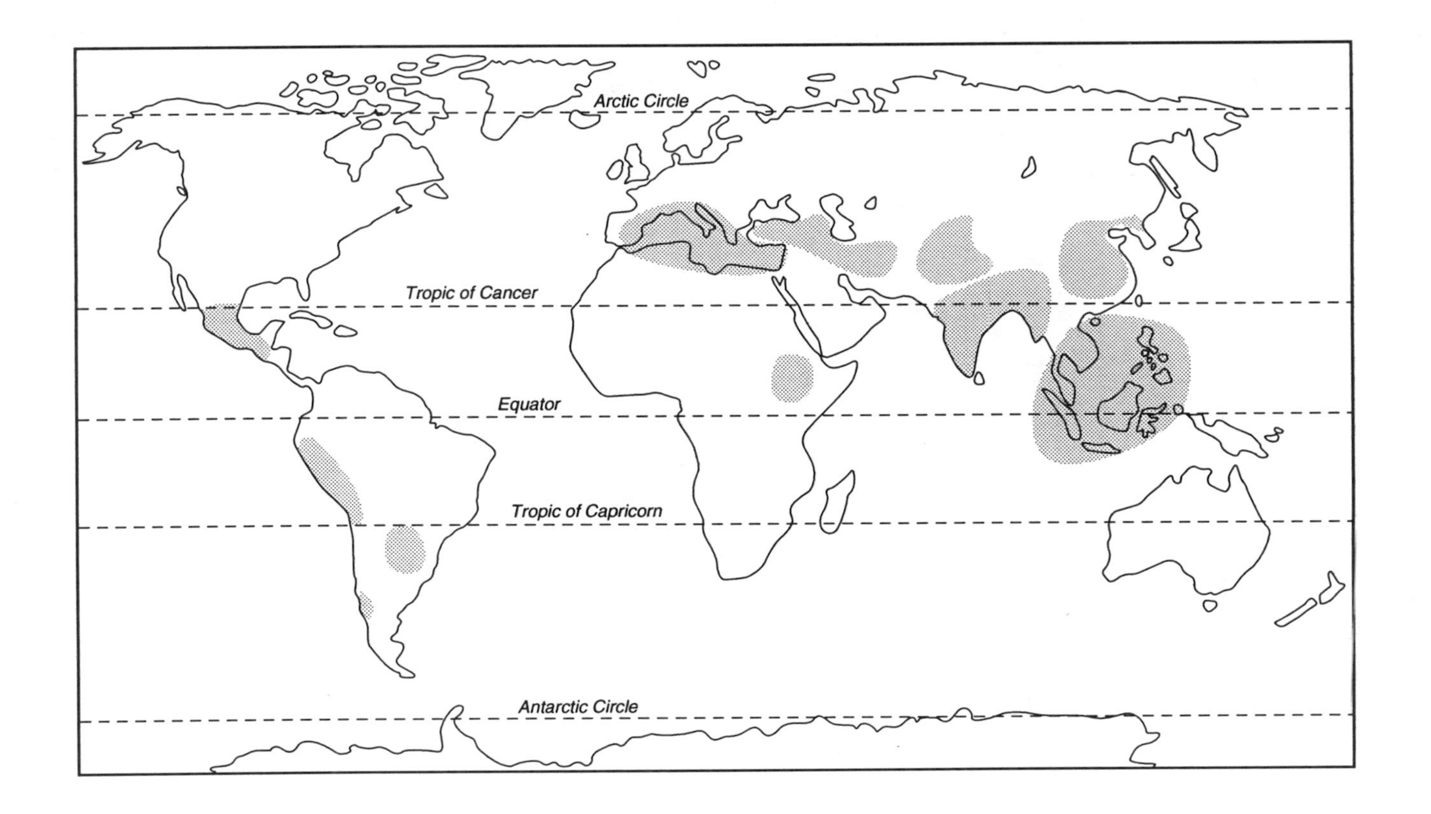

Figure 2.4 The centres of crop origin suggested by Vavilov in 1926 (based on Harlan, 1992a)

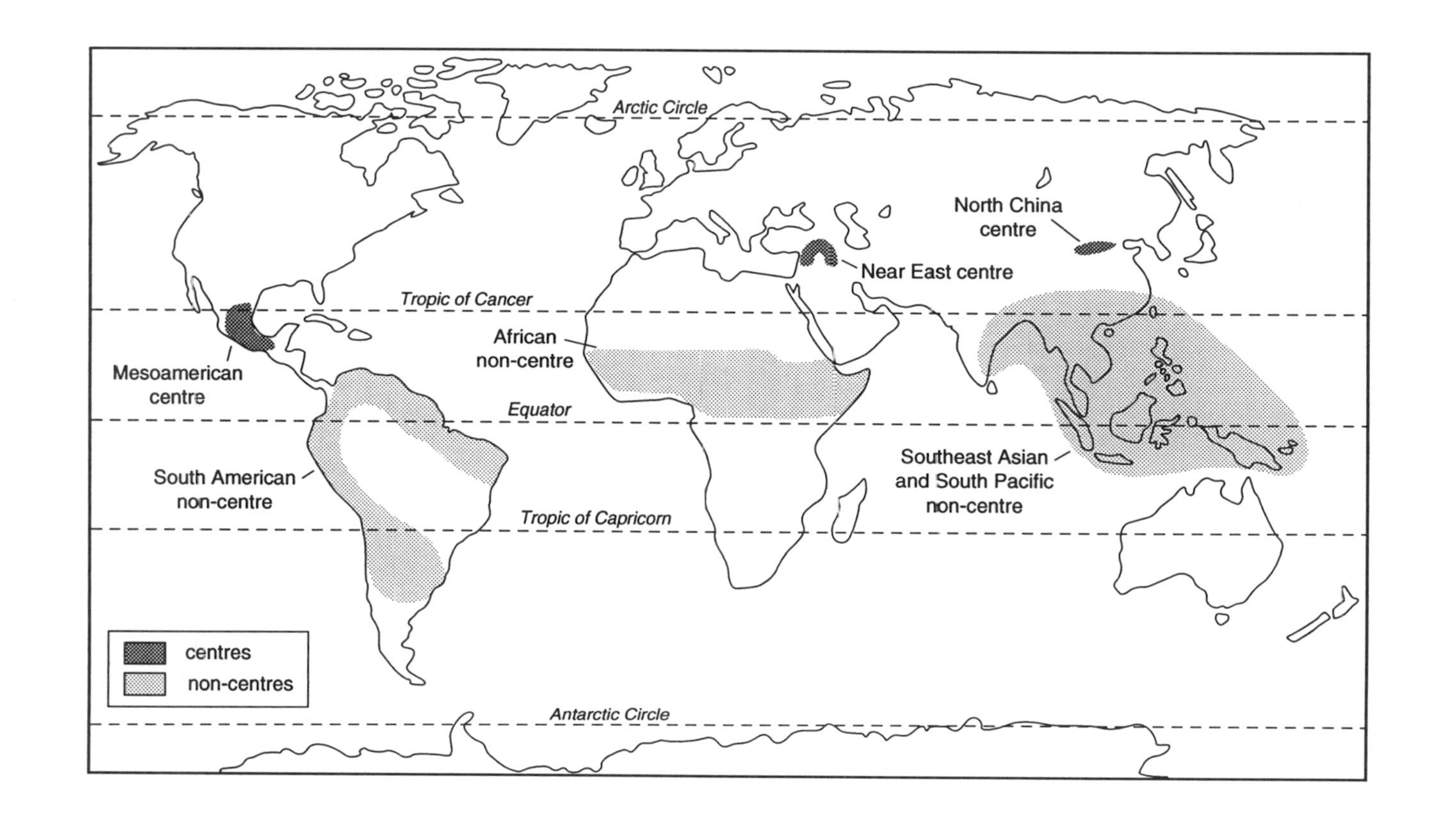

Figure 2.5 Harlan's (1971, 1992a) centres and non-centres of plant domestication (based on Harlan, 1992a)

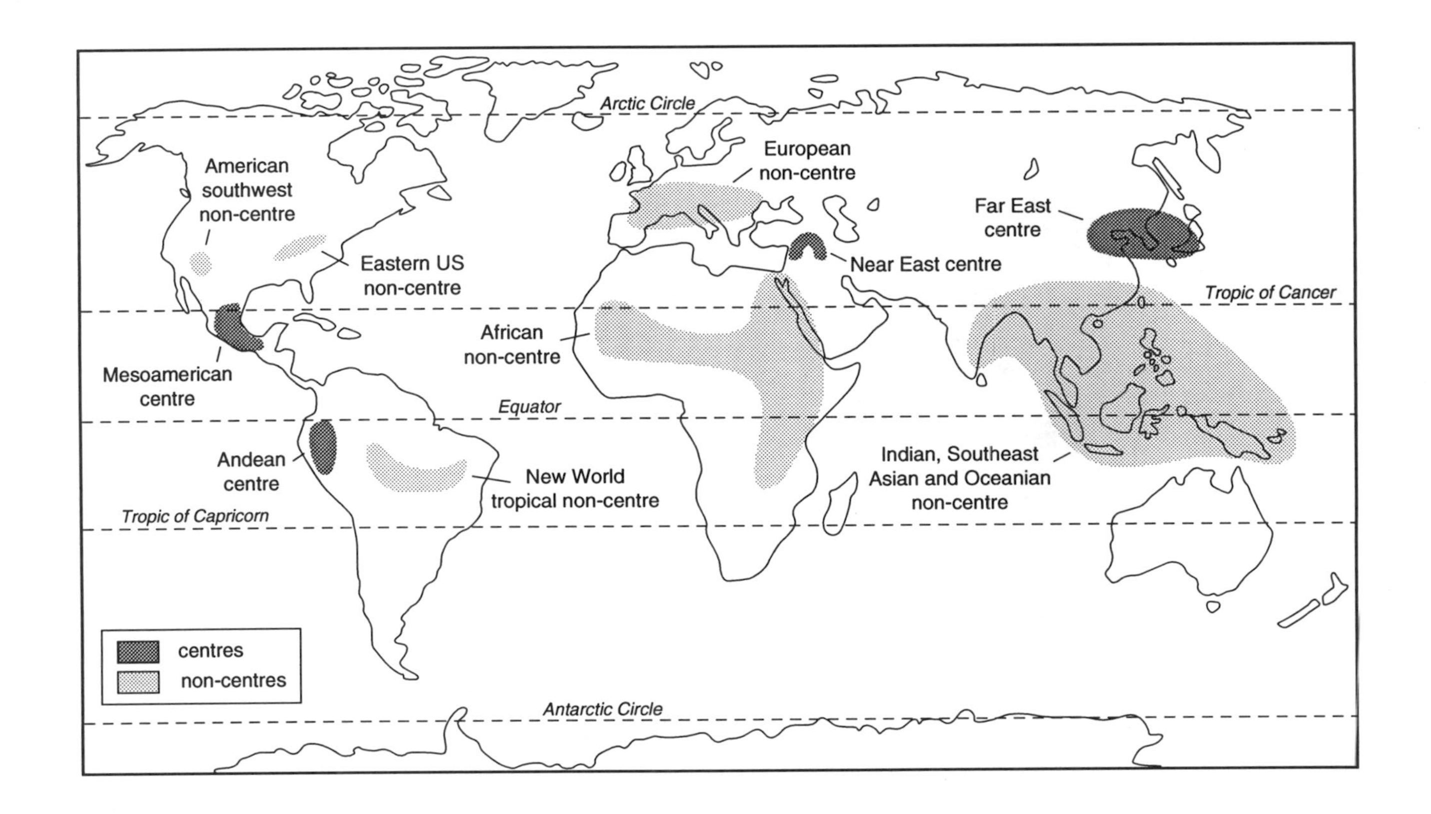

Figure 2.6 MacNeish's (1992) centres and non-centres of plant domestication (based on MacNeish, 1992)

by Blumler and Byrne (1991, see above). Hillman and Davies (1990) show that domestication could be achieved within 20–30 years, given ideal conditions, and certainly within 200 years. Similarly, Ladizinsky (1987) suggests that lentils could have been domesticated in approximately 25 years, with some domestication occurring prior to cultivation. In the case of maize, Galinat (1992) advocates 100–400 years. Accordingly, it would not have taken that many generations of transitional farmers to establish a strong bond between plants and people. This gives a little more credence to the idea of a Neolithic revolution.

Although plant domestication occurred in spatially and temporally disparate loci there are some common features. In the centres of Harlan (1971, 1992a) and MacNeish (1992), which are shown in Figures 2.5 and 2.6, agriculture began in the period between 12 000 and 8000 years ago. Some of the evidence for this is given in Table 2.3. Such a clustering of dates, albeit in a two-millennium period, may be more than coincidence. It has led to some suggestions that materialism is an unacceptable stimulus for the origins of agriculture and that environmentalism is more likely. Flannery (1986) makes this inference when dealing primarily with Mesoamerican data. Blumler and Byrne (1991) as well as McCorriston and Hole (1991), all of whom are concerned with studies in the Near East, also conclude that climatic change played a major role in the instigation of agriculture. The latter review the evidence for environmental and cultural change in the region in the light of information on domestication (see also the review by Byrd, 1994). First, the abundance of information in the Jordan valley and Southern Levant indicates that by 10 000 years ago a number of stimuli were in operation to prompt cultivation. Culturally, these changes occurred at the end of the Natufian period which was characterised by what is known as an incipient agricultural economy. There is evidence for the harvesting of wild cereals and associated technology. This means that opportunity, i.e. sufficient populations of the required plants, was there and that adaptations to improve collecting were underway. There is also evidence (Henry, 1989) that at least a proportion of the population were sedentary in villages with food storage facilities and cereal-grinding equipment. In terms of the environment, most village sites occur in the Mediterranean zone, utilising the resources of forest, steppe and permanent water. In addition, there were other habitat types within or near the core region of the Jordan Valley that provided a further array of resources. Climatic change was also underway. A strongly seasonal Mediterranean climate was developing with hyper-arid summers. Not only are Mediterranean regions biologically diverse, and so contain many potential species for domestication, but strongly seasonal climates also tend to favour annuals, notably grasses and legumes. In particular, the long dry season encouraged people already used to collecting wild grasses, etc., to store food. Thus both environmental and cultural factors were operating to select for some species that were subsequently domesticated. This is a particularly plausible argument in view of polar ice core and other evidence for rapid climatic change during this period (Dansgaard *et al.*, 1993).

Supporting commentaries and data have been presented by several researchers, each dealing with a different aspect of archaeology and/or palaeoecology. Moore and Hillman (1992) refer to existing palynological data from the Near East and a new pollen diagram from Lake Huleh (in present-day Israel) covering the period 13 000 to 10 000 years BP. In relation to plant remains from the archaeological site of Abu

Table 2.3 Evidence for the comparative synchroneity of plant domestication in the centres

Centre	Date in 000s years ago	Site location	Plant remains	Reference
Mesoamerica	10.7 to 9.8	Oaxaca, Mexico	Squash-pumpkin (*Cucurbita pepo*)	Flannery (1986)
	9.0	Ocampo Caves, Mexico	Bottle gourd (*Lagenaria*)	Quoted in Harlan (1992a)
Near East	10 to 9.3	Jericho	Emmer wheat (*Triticum dicoccum*) Two-rowed barley (*Hordeum vulgare*)	Quoted in Zohary and Hopf (1988)
	9.8 to 8.6	Tell Aswad	Emmer wheat (*Triticum dicoccum*) Einkorn wheat (*Triticum monococcum* spp. *boeoticum*) Pea (*Pisum sativum*) Lentil (*Lens esculenta*)	Quoted in Hopf and Zohary (1988)
Far East	8.0 to 7.0	Peulang, north China	Broomcorn millet (*Panicum miliaceum*)	Quoted in MacNeish (1992) and Chang (1986)
	8.0 to 7.0	Yang-shao, north China	Rice (*Oryza sativa*) Gourd (*Lagenaria*) Water Chestnut (*Trapa natans*)	Quoted in MacNeish (1992) and Chang (1986)
Andes*	9.4 to 9.2	Guitarrero Cave, Western Sierra	Chile pepper (*Capsicum chinense*) Common bean (*Phaseolus vulgaris*)	Quoted in MacNeish (1992)
	9.0 to 8.0	Tres Ventanas Cave, Upper Lurin region	Ullucu (*Ullucus tuberosus*) White potatoes (*Solarium tuberosum*)	Hawkes (1989)

* A non-centre according to Harlan (1992a).

Hureyra, some 500 km north-east of Lake Huleh, they conclude that the vegetation changes caused by rapid climatic change compelled the population, if Abu Hureyra is representative of the region, to alter their plant-gathering habits. The disruption of existing plant communities containing wild foods may well have prompted the establishment of permanent agriculture, especially after the adoption of a sedentary lifestyle. The archaeological record also indicates that this may have been the case. Wright (1994) has examined grinding stones and mortars from the Near East which attest to the intensive labour required for the processing of cereals. She suggests that such plants would only have been adopted as a major food source when other wild foods were scarce, as in cold and dry periods. Such a climate prevailed between 11 500 and 10 500 years ago (late Natufian culture) immediately after which time cultivated forms of wheat and barley appeared at several places in the region and there was a substantial increase in processing equipment. Further evidence from Abu Hureyra, notably the skeletal remains of humans of the same period, reflects this hard work and the effects of a comparatively poor diet (Molleson, 1994). For example, vertebral deformities are attributed to the carrying of heavy loads and periods of prolonged kneeling associated with the preparation of cereals for cooking using the grinding tools referred to above. In addition, the coarsely ground grain caused significant and distinctive wear on the teeth found at Abu Hureyra. It seems unlikely that people would willingly undertake such strenuous and debilitating tasks or eat food that damaged the teeth if there were better and more easily prepared alternatives.

In this section most emphasis has been placed on the beginnings of agriculture. As illustrated by Cowan and Watson's (1992) scheme for the establishment of agriculture (see above), there are many examples of secondary development whereby a domesticated species is introduced and transforms a hunter–gatherer economy in which local plants are then also domesticated. Alternatively, a domesticated crop may be introduced into an economy already reliant on domesticated species. Examples of the former occur in Western Europe, particularly in the British Isles, parts of Scandinavia and in the regions adjacent to the Mediterranean coast (Dennell, 1992). In these regions farming emerged gradually from Mesolithic hunter–gatherer communities. The adoption of cereals, legumes and livestock, notably sheep and goats, occurred by degrees; there is virtually no evidence for *in situ* domestication. Moreover, the adoption of farming practices occurred much later than it did in the Near East. In Britain, for example, the earliest finds of domesticated species date to about 5000 years ago. Why the well-organised hunter–gatherer communities should have turned to agriculture is almost as enigmatic as why agriculture began initially.

An example of an economy in which one domesticated species replaced others is that of pre-Columbian Eastern North America. Events here have been discussed by B. D. Smith (1989, 1992) who demonstrates that three distinct phases in agricultural development and change can be distinguished over the period 4000 to 1000 years ago. These are given in Table 2.4. By 800 years ago maize (*Zea mays*) had come to be a staple and primary field crop. This occurred partly because a new eight-row variety had developed which was adapted to a shorter growing season. This study, as well as many of those referred to above, indicates the emergence of a comparatively few plant species as major crop plants. Some of these are examined in more detail below.

Table 2.4 The stages in the inception and development of agriculture in eastern North America (based on B.D. Smith, 1989, 1992)

Period in years ago	Activity
A post-800	Field agriculture based on maize. Development of maize (*Zea mays*), beans (*Phaseolus vulgaris*), squash (*Cucurbita spp.*) complex developed, though there was much variation.
B 1700–800	The expansion of field agriculture. Plant husbandry systems with plants listed below developed. Maize was introduced early on but did not become a staple until later.
C 3000–1700	The development of farming economies. Knotweed (*Polygonum erectum*), maygrass (*Phalaris caroliniana*) and little barley (*Hordeum pusillum*) join the high-carbohydrate-yielding domesticates. Others are high-oil-yielding.
D 4000–3000	Initial domestication of seed plants. Evidence for the domestication of sumpweed (*Iva annua*), sunflower (*Helianthus annuus*), goosefoot (*Chenopodium berlandieri*).
E 7000–4000	Middle Holocene collectors. *Cucurbita* (gourds), goosefoot, sumpweed and sunflower began to proliferate in disturbed fluvial environments that provided potential campsites. Species like hickory nuts may also have been favoured.
F pre-7000	Early and Middle Holocene foragers. Evidence for the collection of seeds, berries and nuts of 20 plant species, e.g. beech, hazel, oak, knotweed, hickory nuts.

2.3 THE MAJOR CROP PLANTS

According to the FAO (1992), the most important sources of food today are the cereals, notably wheat and rice (Table 2.6). The centres of origin, with approximate dates for domestication, are given in Table 2.5. In fact it is surprising how much the world's population relies on so few staple crops. Some crops, of course, are grown for the purposes of producing fibre or biomass fuels and not for food energy. Apart from wheat and rice, maize, barley and potato are staple crops whilst cotton is the world's major fibre crop. Reference to some of these has already been made in Section 2.2, while Table 2.5 lists some of the other most important cultivated plants and their centres of origin.

2.3.1 Wheat

As Table 2.5 shows, wheat was one of the earliest species to be domesticated, dating back about 10 000 years in the Near East. The earliest domesticated cereals have been recovered from Jericho (Bar-Josef and Kislev, 1989) and from Tell Aswad near Damascus (van Zeist and Bakker-Heeres, 1979). Previously, wild ancestors of emmer (*Triticum dicoccum*) and einkorn (*T. monococcum*) had been collected from wild stands that were widespread in the Near East. Evidence from many archaeological sites suggests that both wild and cultivated species of wheat were harvested (Kislev *et al.*, 1986).

Domesticated species differ from wild species in that the former have a tough rachis (inflorescence). This ensures that the grain is retained in the ear until harvest.

Table 2.5 Some cultivated plants, and their likely places and dates of origin (based on Harlan, 1992a; L.T. Evans, 1993, and sources quoted in the text)

(a) THE NEAR EAST (AND MEDITERRANEAN)

Crop	Common name and additional locus of origin	Approximate date of domestication in years BP
Cereals		
Avena sativa	Oats: secondary crop	
Hordeum vulgare	Barley	10 200
Secale cereale	Rye: secondary crop	
Triticum aestivum	Bread wheat: addition crop	7800
T. dicoccum	Emmer	9500
T. monococcum	Einkorn	9500
Pulses		
Lens esculenta	Lentil	9500
Pisum sativum	Garden pea	
Vicia faba	Broadbean: wild form unknown	8500
Roots and tubers		
Beta vulgaris	Beet, mangel	
Brassica rapa	Turnip: possibly also China	
Daucus carota	Carrot	
Oil crops		
Brassica napus	Rapeseed	
B. nigra	Mustard and mustard oil	
Olea europea	Olive	7000
Fruits and nuts		
Prunus avium	Cherry: Balkans to Caspian	
Pyrus malus	Apples: Balkans–Transcaucasia–Caspian	
Vitis vinifera	Grape	7000
Vegetables and spices		
Allium cepa	Onion	
A. ativum	Garlic	
Brassica oleraceae	Cabbage; cauliflower, kale, etc.	
Fibre plants		
Cannabis sativa	Hemp: widespread, Eurasian	9500
Linum usitatissimum	Flax, also cultivated for oil	

(b) AFRICA

Crop	Common name and additional locus of origin	Approximate date of domestication in years BP
Cereals		
Eleusine coracana	Finger millet: Ethiopia–Uganda highlands	
Oryza glaberrima	African rice: West African savanna	
Pennisecum glaucum	Pearl millet: dry savanna, Sudan to Senegal	
Sorghum bicolor	Sorghum: savanna zones Sudan to Chad	
Pulses		
Vigna linguiculata	Cowpea: West African forest margins	3400
Roots and tubers		
Dioscorea cayenensis	Yam: Ivory Coast to Cameroon	possibly 10 000
Sphenostylis stenocarpa	Yampea: West Africa, forest zone	
Oil crops		
Elaeis guineensis	Oil palm: West Africa, forest margins	
Ricinus communis	Castor bean, castor oil: Ethiopia to Egypt	

continued overleaf

Table 2.5 (*continued*)

Crop	Common name and additional locus of origin	Approximate date of domestication in years BP
Fruit and nuts		
Colocynthis citrullus	Watermelon: Dry savanna: South and East Africa	
Adansonia digitata	Baobab: savannas	
Vegetables and spices		
Abelmoschus esculentus	Okra: West Africa	
Fibre plants		
Gossypium herbaceum	Old World Cotton: Sudan?	
Other		
Coffea arabica	Coffee: Ethiopia, forests	
C. canephora	Robusta coffee: lowland forests	

(c) FAR EAST, CHINA

Crop	Common name and additional locus of origin	Approximate date of domestication in years BP
Cereals		
Oryza sativa	Rice: south China to India	9000
Fagopyrum esculentum	Buckwheat: western China	
Pulses		
Glycine max	Soy bean: north-east China	3000
Vigna angularis	Adzuki beans: south China	
Roots and tubers		
Brassica rapa	Turnip: see Near East	
Nelumbium speciosum	Lotus: seeds and tubers eaten	
Oil crops		
Aleurites fordii	Tung oil: south China	
Fruit and nuts		
Castanea henryi	Chinese chestnut: temperate China	
Juglans regia	Walnut: mountains of south-west China	
Pyrus spp.	Chinese pears: temperate China	
Vegetables and spices		
Benincasa hispida	Winter melon: widespread	
Wasabia japonica	Horseradish: widespread	
Zingiber officinale	Ginger: south China	

(d) SOUTHEAST ASIA AND PACIFIC ISLANDS

Crop	Common name and additional locus of origin	Approximate date of domestication in years BP
Cereals		
Panicum miliare	Slender millet: India to China	
Oryza sativa	Rice: Himalayas and Burma	
Pulses		
Cajanus cajan	Pigeonpea: India	
Vigna spp.	Various beans: widespread	
Roots and tubers		
Colocasia esculenta	Taro: Assam, Upper Burma	possibly 9000
Tacca leontopetaloides	Arrowroot: South Pacific Islands	
Oil crops		
Cocos nucifera	Coconut: South Pacific Islands	5000
Sesamum indicum	Sesame: India	4000

continued

Table 2.5 (*continued*)

Crop	Common name and additional locus of origin	Approximate date of domestication in years BP
Fruit and nuts		
Citrus aurantiifolia	Lime: South-East Asia and south China	
Mangifera indica	Mango: Indo-Malaysia	9200
Musa acuminta	Banana: Malaysia–Thailand–Indonesia	
Vegetables and spices		
Syzygium aromaticum	Clove: Spice Islands	
Piper nigrum	Black Pepper: south-east Asia	
Myristica fragrans	Nutmeg: Spice Islands	

(e) THE AMERICAS

Crop	Common name and additional locus of origin	Approximate date of domestication in years BP
Cereals		
Zea mays	Indian corn	7700
Pulses		
Phaseolus lunatus	Lima bean	7700
Phaseolus vulgaris	Common bean	
Roots and tubers		
Manihot esculenta	Casava	4500
Dioscorea trifida	Yam	
Ipomoea batatus	Sweet potato	4500
Solanum tuberosum	Potato	7000
Oil crops		
Arachis hypogaea	Peanut	
Helianthus annuus	Sunflower	
Fruit and nuts		
Carica papaya	Papaya	
Psidium guajava	Guava	
Vegetables and spices		
Capsicum annuum	Pepper	8500
C. frutescens	Chilli	
Cucurbita spp.	Various squashes	10 700
Fibre plants		
Gossypium hirsutum	Upland cotton	5500
G. barbadense	Sea island cotton	4500

Wild species were thus selected for this characteristic which made harvesting more efficient. According to Harlan (1992b), the reaping and planting cycles favoured and selected for uniform rather than uneven ripening and increased seed retention. The genetics of modern domesticated wheat have been discussed by Lagudah and Appels (1992). Bread wheat (*Triticum aestivum* L.) is a polyploid species (with more than twice the normal haploid number of chromosomes); it originated from the union of a tetraploid wheat (*T. turgidum*) with a diploid species (*T. tanschii*), probably in the region adjacent to the Caspian Sea about 8000 years ago (Harlan, 1992b). Emmer and bread wheat spread from their centres of origin by different routes. The former spread into the Balkans, along the Danube, down the Rhine and around the shores of the Mediterranean Sea. It was associated with other crops, such as pulses and flax, which also spread eastward into India and southward to Ethiopia. In its homeland its domestication was paralleled by the domestication of animals (Section 2.4), notably goats, sheep and cattle. Bread wheat, on the other hand, took a more northerly route than emmer through Europe; it spread through the Caucasus and into Eastern and Northern Europe. Today, it is the most important wheat type produced, having been subject to intense breeding programmes in the twentieth century and is now subject to genetic manipulation (Chapter 10).

2.3.2 Barley

The history of barley (*Hordeum vulgare*) is similar to that of emmer wheat (see above). The domesticated species derives from a group of wild types collectively known as *Hordeum vulgare* L. subsp. *spontaneum*, which are two-rowed and diploid types. These species are distributed in the wild over a wide area in the Near East and Western Asia, even as far as China. They were present in a wide range of vegetation communities, including open herbaceous formations, oak forest formations surrounding the Syrian desert and even the steppe and semi-arid communities to the east of the Jordan rift valley. Some of the oldest remains of barley have been found at the early Neolithic site of Netiv Hagud and are dated at 10 200 years, i.e. 500 or so years older than the earliest domesticated wheat (see above). In fact Bar-Josef and Kislev (1989) suggest that before its inherent food value was recognised *Triticum dioccoides* may initially have grown as a weed in fields of barley. Six-rowed barley appears in the archaeological records at about 9500 years ago. Both two-rowed and six-rowed types are found abundantly in the remains of early farming settlements throughout Europe, as are those of wheat.

Barley was, thus, a founder crop of the early agriculture of the Near East though it was probably not as well regarded as wheat. From the Near East it spread westward into Turkey, Greece and the Mediterranean basin, reaching southern Sweden and Britain by about 5000 years ago (Zohary and Hopf, 1993). Barley also spread to the east into Afghanistan and central Asia. Like wheat, the most obvious difference between wild and domesticated species is the presence of a tough rachis to ensure retention of the grain until harvest.

2.3.3 Rice

As shown in Table 2.5 there are two centres of origin for rice. *Oryza glaberrima* was domesticated in Africa while *O. sativa* probably originated in China. According to Chang (1989), there are also 20 wild species all of which originated in the Gondwana supercontinent which began to break up in the early Cretaceous period. Rice undoubtedly provided pre-agricultural people in the humid tropics and subtropics with a source of wild food. There is, however, no single focus of domestication (cf. wheat; see above) in Asia presently known. Instead, *O. sativa* developed from its wild ancestors that grew in a broad belt from the southern foothills of the Himalayas, across Upper Burma, northern Thailand, into Laos and northern Vietnam and then into southern China (Chang, 1976). Temporally, the origins of cultivated rice also occur over a broad timescale between 10 000 and 15 000 years ago. There is also the possibility that the domestication of rice took place in several different centres at different times. There is some debate as to the nature of *O. sativa*'s wild progenitors; Chang (1976) suggests it was an annual species while Oka (1988) believes it was a perennial.

Chang (1989) has summarised the chronology of the oldest rice remains in Asia. Grains, hulls, straw and stems have been recovered from Ho-mu-tu, Zhejiang in southern China. These are approximately 7000 years old and the same age as carbonised grains and broken tusks from Luo-jia-jiao, Zhejiang. More recently, however, Yan (1991) has reported the remains of cultivated rice from Pengtoushan in the mid-Yangtze basin. This predates the south China sites by some 2000 years. Outside China, the earliest dates for domesticated rice come from India. At Koldihwa, near Allahabad in Uttar Pradesh, rice grains (of the *indica* variety) embedded in earthen potsherds and husks are 6500–8500 years old. Subsequently, cultivated species diffused throughout Asia as dryland, shallow-water and wetland adapted cultivars developed. Zohary and Hopf (1993) suggest that *Oryza sativa* may have been introduced into the Near East during the Hellenistic period. Sallares (1991) notes that it is referred to in Assyrian cuneiform texts *c.* 700 BC and that Strabo mentions its cultivation in Palestine in the first century BC. Today, rice is the world's second most important cereal crop (Table 2.6).

African rice, *Oryza glaberrima*, originated in the savanna zone of West Africa from an annual grass adapted to the seasonality of rainfall (Harlan, 1989). This is *O. glaberrima* subsp. *barthii*. Asian rice was domesticated much earlier than its African counterpart which was domesticated only about 1500 years ago. Today, African rice is grown only in small areas of West Africa where *O. sativa* is gradually replacing the native cultivar as the favoured crop. Rice has been the subject of intensive plant breeding programmes in the last 30 years and crop yields have dramatically increased as a result (coupled with improved crop protection, fertiliser use, etc.). High-yielding varieties now dominate rice production and research is underway to improve productivity still further (Hawkes, 1993) using genetic engineering as well as conventional plant breeding.

2.3.4 Maize

Despite a great deal of research, the origin and domestication of maize is a source of much controversy. There is also some debate as to whether the domestication of this most important crop could have occurred in a number of centres in Central or South America. Whatever the reality of the matter turns out to be, maize, once domesticated, became the most important crop in the pre-Columbian Americas, underpinning the great empires of the Mayas, Incas and Aztecs as well as the native American groups of North America. Europeans recognised its food value and introduced it to other parts of the world where it has become a staple crop. Today, it is one of the world's most important crops (Table 2.6), its annual production being surpassed only by those of wheat and rice.

The domestication of maize has recently been reviewed by Galinat (1992) and Davies and Hillman (1992). The former states that 'It is spectacular in terms of its wide divergence from the wild types regarding ear structure.' This is one of the reasons why it has been difficult to pinpoint the ancestor of maize. Now, however, there are two schools of thought. The first is that the wild ancestor of maize is a grass

Table 2.6 Production figures for the world's major crops, 1990–1992 (data abstracted from FAO, 1993)

	Yield (tonnes × 10^6)		
Crop	1990	1991	1992
Wheat	592.918	546.119	563.649
Barley	177.670	167.992	160.134
Rice	521.140	517.875	525.475
Maize	479.140	491.001	526.410
Potato	267.586	257.929	268.492
Cotton (lint)	18.443	20.668	18.430

known as teosinte (*Zea mays* (L.) subsp. *mexicana* Iltis and supsp. *parviglumis* Iltis and Doebley), a conclusion that is supported by molecular evidence (Doebley, 1990). The second theory involves hybridisation between an early domesticated but now extinct maize and a teosinte type (Wilkes, 1989). Whatever the origins, the reason for the marked differences between the two-rowed teosinte and the eight-rowed domesticated maize is because of a condition known as monoecism. This means that the plant has both male and female flowers. As these can be recognised, either male or female flowers can be removed to ensure interbreeding, rather than self-pollination, between plants. The development of hybrids with advantageous characteristics could thus proceed rapidly.

The oldest known remains of maize occur in the Tehuacán valley of central Mexico. These remains are believed to be nearly 8000 years old (Table 2.5), though there is evidence in the pollen record from nearby Oaxaca that cultivation occurred as long as 10 000 years ago (referred to in MacNeish, 1992). This area contains many locations in which teosinte grows though it is not certain if Tehuacán/Oaxaca really is the region of domestication. Moreover, there is evidence to suggest that maize was domesticated from a wild ancestor quite separately in South America (Bonavia and Grobman, 1989), notably in the Central Andes, at about the same time. Precisely what relationship, if any, there is between these spatially separate but temporally convergent finds remains to be ascertained. Interestingly, however, there is evidence from an intermediate area, that of the central part of the Panama isthmus, for the cultivation of domesticated maize about 7000 years ago (Piperno, 1989) and from the Calima Valley of Colombia, South America (Monsalve, 1985).

From Mexico, maize cultivation, along with other crops such as beans and squash, spread into North America (Section 2.2 and Table 2.3) where it ousted indigenous crops as the staples. Van de Merwe (1982) provides evidence from carbon isotope analyses of human skeletons that reflects the predominance in the diet of C3 or C4 plants (see Section 1.2). C4 plants, of which maize is one, have low values of isotope ^{13}C and so the collagen fibres in the bones of their consumers is also low in ^{13}C. Such data from Tehuacán skeletons reveals that by 6000 years ago maize had become a primary crop; in mid-western bone remains the shift occurred between 500 and 800 years ago. This latter concurs with ^{14}C dates for archaeological discoveries in eastern North America (see Table 2.4) that confirm the shift. However, van de Merwe's dates for the adoption of a maize-dominated diet in the Orinoco River lowlands of

Venezuela, i.e. 1600 years ago, are somewhat curious in the light of Piperno's (1990) research in the Amazon Basin of Ecuador. She has obtained phytoliths (deposits of silica within a plant that are species-specific) of cultivated maize from sediments 5300 years old. This is currently one of the oldest dates for agriculture reported for South America. Clearly, there is still a great deal to be learnt about the early development of maize-based agriculture in the Americas.

In the fifteenth century, soon after Columbus' discovery of the Americas, maize was introduced into Europe from where it spread into the Middle East and beyond. Today, it is widely grown for direct consumption and as an animal feed (Chapter 5).

2.3.5 Potato

Like maize, the potato originated in the Americas and, like maize, it has become a crop that is grown all over the world. It ranks fourth in global food production (FAO, 1993) after wheat, maize and rice, and it is the most abundantly produced root crop. According to Hawkes (1990) there are seven cultivated species of potato, the most common one being *Solanum tuberosum* subsp. *tuberosum*. There are also a great many wild species, probably more than for any other crop plant. This means that the potato has a large genetic pool or genetic resource that is already providing opportunities to improve crops via genetic engineering (Chapter 10).

The potato was probably domesticated in the central Andes of Peru; the oldest remains of domesticated potatoes so far discovered derive from the Chilca Canyon in Peru and are dated to about 7000 years ago (Hawkes, 1991). The site is in the cool temperate high-mountain zone in which there is an abundance of wild potato species today. The wild ancestor of the cultivated potato has not been established with certainty though it is likely to have been *Solanum stenotomum* which currently has a distribution from northern to central Bolivia. This may have interbred with *S. sparsipilum*, a weed species of Peru and Bolivia, to produce *S. tuberosum* subsp. *andigena*. As the name suggests, this species occurs in the Andes. It is also likely that *S. tuberosum* subsp. *andigena* then crossed with another species to produce *S. tuberosum* subsp. *tuberosum* which is characteristic of southern Chile (Grun, 1990). Hawkes (1990) believes that it was subsp. *andigena* that was first introduced into Europe, possibly from Colombia.

Hawkes (1990) has also reviewed the evidence for the introduction of the potato into Europe which was not, as is popularly believed, the work of either Sir Walter Raleigh or Sir Francis Drake. There were two early introductions into Europe: the first was to Spain in 1570 and the second was to England, probably in 1590. Potatoes achieved great popularity in Spain and Italy by the close of the sixteenth century. By the 1770s they were widely grown in France and Ireland but it was not until the mid-eighteenth to early nineteenth century that potatoes became popular in England or Eastern Europe. It also seems likely that England was the source of potatoes introduced into North America in the colonial times; surprisingly, it does not appear to have been introduced directly from South America. Missionaries and explorers introduced the crop to India, China and New Zealand between the seventeenth and nineteenth centuries thus ensuring its wide distribution.

The susceptibility of the potato to the late blight fungus (*Phytophthora infestans*) has given rise to an historical event that is widely documented. This was the Irish potato famine of 1845–1847. During these years entire potato crops, on which whole families depended, were lost; the resulting famine, disease and high mortality triggered one of the greatest diaspora in history as Irish people left in their droves for the New World. Today it is possible to genetically engineer potatoes with resistance to the late blight fungus. Moreover, the International Potato Centre (IPC) is now engaged in the breeding of potatoes for developing countries of the humid tropical zone because of their high food value.

2.3.6 Cotton

As Table 2.5 shows, there are several varieties of domesticated cotton. Two of these, upland cotton (*Gossypium hirsutum*) and sea island cotton (*G. barbadense*), derive from the Americas whilst the old world cottons (*G. herbaceum* and *G. arboreum*) originated in Sudan/Ethiopia. According to Harlan (1992a), *G. herbaceum* now constitutes less than 1 per cent of total world cotton production which mainly comprises the American species, especially *G. barbadense*. All three domesticated species have wild counterparts. Today, cotton is the world's most important fibre crop.

The earliest remains of cotton (*G. hirsutum*) associated with human activity occur in the Tehuacán Valley of Mexico and are *c.* 5500 years old. There are also remains of cotton in an archaeological site, which is *c.* 4500 years old, on the central coast of Peru. Numerous records also exist from the highland regions of Peru and Bolivia (Pearsall, 1992). The earliest archaeological records of cultivation of Old World cottons have been found at Mohenjo-Daro in Pakistan. Here, fragments of cotton textiles and cotton strings have been discovered in association with artifacts of the Harappan civilisation (referred to in Zohary and Hopf, 1993). There is, however, little evidence for indigenous use in Africa itself, the home of domesticated cotton's wild counterparts. The earliest domesticated cotton plants were perennial and evolved into annual types in recent historical time as cotton cultivation was introduced into temperate climatic zones. Harlan (1992a) states 'Cotton was grown mostly for home use until Whitney invented the saw gin in 1793. Mechanization of spinning and weaving followed, and the cotton industry became a major component of the Industrial Revolution'. In the eighteenth century, small volumes of cotton were imported into Britain from the Mediterranean and the West Indies, whence it had spread from South America. The industrialisation process, however, brought cotton into the realm of big business and the resulting demand produced a swift expansion of cotton-growing agriculture, notably in the USA.

Cotton is no less important today. It is the world's leading fibre crop and the major producers are the USA, Egypt, the Sudan and the Central Asian Republics (former USSR). For all of these countries cotton is also a major export crop.

2.4 THE DOMESTICATION OF ANIMALS

Much the same set of questions that is addressed to plant domestication applies to animal domestication. The answers are also similar: the motives that led human communities to change their lifestyles so fundamentally remain unknown. The environmentalism versus materialism debate (see Sections 2.1 and 2.2) cannot easily be resolved though MacNeish (1992) does offer a range of possibilities (Figure 2.3).

Apart from that of the dog (discussed in Section 2.1) which took place about 12 000 years ago, there were several domestications in the period between 10 000 and 8000 years ago (Figure 2.7). These involved sheep, goats, cattle and chickens; these are amongst the most economically important livestock species of today. However, surprisingly few animal species have been domesticated, especially if these are compared with the much wider range of plant species that underpin modern farming systems. Compare, for example, Table 2.5 with Figure 2.7. Some animals, notably the horse and the donkey, were domesticated not for their meat but for their ability to provide transport for people and their goods, and for traction. Yet other animals have been domesticated for secondary products, i.e. wool and milk. Figure 2.7 also gives a few avian species that have been domesticated, notably the chicken and the turkey. Most domesticated animals are large mammals; there are few rodents or birds.

Why this is so is a matter for conjecture and has been addressed by Clutton-Brock (1992). She suggests that the social behavioural patterns of animals are of fundamental significance because certain types of patterns predispose some animals to a close relationship with humans. A tamed animal is not necessarily a domesticated animal. It would appear that domestication is generally only possible with those animals that are social, rather than solitary, species. Clutton-Brock proposes that it is possible to tame virtually any mammal if it is manipulated at a young age. This is how the association between people and wolves, that eventually produced the domesticated dog, may have begun. Many other species of mammal, e.g. foxes, may have enjoyed the similar attentions of ancient human groups but because of their innate solitariness they never become domesticated. One exception to this rule is the domestic cat (*Felis* spp.). For many social animals bonding or imprinting between people and young animals can occur. So strong can this relationship be that young animals will look to humans rather than their natural mother, to provide sustenance. Indeed, this is one of the reasons why attempts to rear, in captivity, wild species threatened with extinction are often thwarted. To overcome this bonding process, wild species like the American condor are fed using glove puppets that are designed to resemble adult birds.

It is also interesting to note that many of the Epipalaeolithic sites of the Near East contain abundant remains of gazelle bones (e.g. Henry, 1989). Horizons and/or sites of later eras when agriculture was widely practised also contain bones of gazelle but there is no evidence at all for domestication. Instead, goats and sheep, which like the gazelle were wild herbivores, were domesticated first, about 9000 years ago, probably because of their social and less territorial behaviour. Cattle and pigs were domesticated slightly later (Figure 2.7) but still in the Near East. About 6000 years ago the horse was domesticated in the Ukraine. In the Americas animal domestication appears generally to have post-dated plant domestication by at least a millennium. The llama, for example, was domesticated by about 6000 years ago by which time at least domesticated species of gourds and squashes (Table 2.5) were in widespread use.

As Figure 2.7 shows, the Near East was a very important centre of animal domestication, though it post-dated the earliest plant domestication. The occurrence of a domesticated species in bone assemblages is, like that of plants, considered to represent the outcome of a fairly long period of transition from the wild. Thus the presence of a domesticated species, which is diagnosed on the basis of morphological variations, size differences, sex- and age-related culling, frequency changes within a

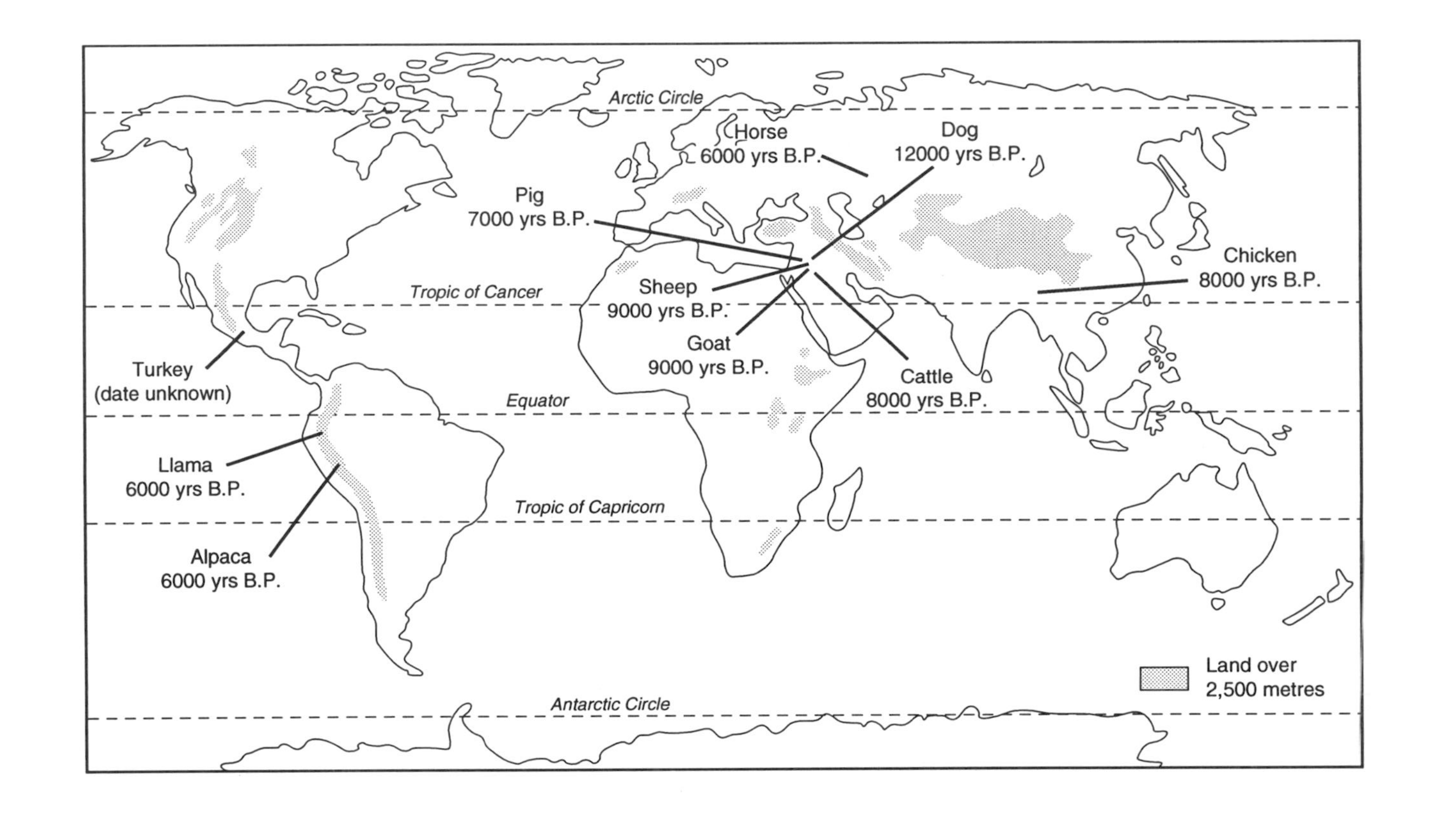

Figure 2.7 The places of origin, with approximate dates for the most common domesticated animals (data from sources quoted in the text)

succession of faunas and the presence of a foreign species (Davis, 1987), represents the culmination of a process begun many, possibly even hundreds of years earlier. It may be that the sedentary lifestyles that became more prevalent and possibly more secure with the advent of a reliable proximal food supply, encouraged animal domestication. Alternatively, the hunting territories of people as population expanded, either before or after plant domestication, may have declined in area so that returns were reduced. It is also possible that the comparatively rapid climatic changes characterising the period between 12 000 and 10 000 years ago reduced the extent of suitable habitats for herbivores. The motives for animal domestication are, like those for plant domestication, enigmatic.

According to Davis (1987), there is firm evidence that the modern domestic sheep originated from the mouflon (*Oris orientalis*) which is indigenous in the Near East (Figure 2.7). Unlike other species of wild sheep, the mouflon has 54 chromosomes which is the same number of chromosomes as domesticated sheep. Of two species of wild goat, the bezoar goat (*Capra aegagrus*) is the most likely ancestor of the domesticated species. It too was widely distributed in the Near East in the wild. The domestication of wild boar (*Sus scrofa*) to produce the pig also occurred initially in the Near East but its wide distribution throughout Europe and into Asia means that it could have been domesticated in many different places at many different times. The wild ancestor of modern cattle was the auroch *Bos primigenius* which also enjoyed a wide distribution throughout Europe and Asia. Interestingly, the distributions in the wild of these four animals all overlap in the Near East. Consequently, the domestication of one species providing meat, milk, wool and/or hide may have stimulated the domestication of others. From the Near East these species were introduced into the rest of Europe and parts of Asia though there is biomolecular evidence, i.e. mitochondrial DNA, to suggest that there was a separate centre of domestication in India (Loftus *et al.*, 1994).

Of the domesticated birds, the chicken is the oldest. According to West and Zhou (1988), there is strong evidence for its earliest domestication in South-East Asia from the red jungle-fowl (*Gallus gallus*). From South-East Asia it was introduced into China where it had become well established by 8000 years ago. There are, for example, abundant archaeological sites with the remains of chicken bones in the lower reaches of the Yellow River. From China the chicken was introduced into Japan and Korea. It is also likely that the species was domesticated in independent centres, such as India about 4000 years ago. There is also the possibility that it spread from China through Russia and into Europe where it was present about 5000 years ago.

There appear to have been few primary domestications in Europe though, as stated above, it is quite possible that some species like the pig were domesticated in independent centres. One species that did originate in Europe, however, is the horse whose domestication occurred in the Ukraine. Anthony *et al.* (1991) state that prior to 6000 years ago wild horses roamed the extensive grasslands that even today characterise the Ukraine. They probably formed large herds and were the dominant grazing animals of the steppes that extended as far east as Mongolia. Archaeological sites in the Ukraine yield evidence for three uses of the horse: as wild game, as a domesticated source of meat and as mounts, the latter providing a source of power or energy for a variety of uses, not least to increase the mobility and hunting capacity of human communities. The archaeological site at Dereivka, 250 km south of Kiev, reveals that by about 6000 years ago people of the Sredni Stog culture were practising agriculture which included domesticated plants and animals. Of the latter, cattle, sheep, goats and pigs were commonplace but there is also evidence that these people kept domesticated horses. Horse bone remains increase in abundance when compared with bone assemblages at earlier sites in the region and horse meat probably featured prominently in the diet. There is also evidence to suggest that horses and dogs were used for some sort of ritual. Anthony *et al.* (1991) suggest that this highlights the horse's domesticated status and the presence of two perforated antler fragments that could be the cheekpieces of a bit indicate that the animal may have been ridden. This is confirmed by the detailed analyses of the teeth of the stallion that comprised the ritual burial. The anterior premolar teeth are bevelled in accordance with the use of a bit. Dating of the site at about 6000 years old also means that it preceded the invention of the wheel by some 500 years. The implications of horse-back riding for the social and economic organisation of early societies such as that at Dereivka were probably substantial. Increased mobility, at the very least, provides a means for increased contact with neighbouring groups and thus facilitates trade and the spread of ideas. As Anthony *et al.* point out, it may also have provoked increased warfare.

The domestication of animals also occurred in the Americas, as illustrated in Figure 2.7. The largest and most important domesticated animals are the llama and the alpaca. These are camelids and have two wild relatives which are the guanaco and vicuna. The guanaco is considered to be the wild ancestor of both domesticated species (Novoa and Wheeler, 1984) which provide wool, meat, hides and dung. The llama is also a beast of burden. It was domesticated about 6000 years ago in the Andes (Bahn, 1994), though hunting of the species was prevalent by at least 8000 years ago (Fiedel, 1987). The guinea-pig (*Cavia apera*) was also domesticated in the Andean region but, at 3000 years ago, it was much later than the llama and alpaca.

2.5 CONCLUSION

Interpreting the early history of human manipulation of plant and animal resources relies heavily on the evidence from archaeological sites. There is thus an inbuilt bias in the reconstructions because of the distribution of archaeological sites, which partly reflect the presence of conditions that facilitate preservation, and the intensity of research that has been undertaken to recover the evidence. The chronicle of how, when, where and why hunting–gathering was replaced by agriculture will assuredly change as new evidence comes to light and as new techniques for the dating and analysis of archaeological remains are developed.

What is already apparent is that hunter–gatherer societies at the end of the last ice age were highly organised and sophisticated. Many may have been sedentary. They had an intimate knowledge of local plant and animal populations and used several strategies to hunt animals via herding, the use of a dog and/or the use of fire. This organisation, or infrastructure, provided the setting in which agriculture developed. The motives for this are obscure and may always remain so. There are two schools of thought: materialism, which invokes social factors as the primary motivating forces, e.g. population growth; and environmentalism, which relies on a form of climatic determinism to create habitat and biotic changes. There is also the possibility that the real stimuli may have involved a combination of the two so it cannot be assumed that the two approaches are mutually incompatible.

The evidence to date suggests that there were a number of centres wherein the domestication of plants was relatively intense. Mesoamerica, the northern Andes and the Near East are examples. The Near East, in particular, produced some of the most important plant and animal domesticates. These were the ancestors of crop and livestock species that dominate modern agriculture, e.g. wheat, sheep, cattle. The Americas provided maize and the Far East provided rice.

The events in this 4000 year period during the early part of the present interglacial have given rise, first, to one of the most important agents of environmental change and, secondly, to the wherewithal to feed the Earth's current human population of approximately 5×10^9.

2.6 FURTHER READING

Bettinger, R.L. (1991) *Hunter–gatherers: Archaeological and Evolutionary Theory*. Plenum Press, London.

De Laet, S.J. (ed.) (1994) *History of Humanity. Volume 1. Prehistory and the Beginnings of Civilization*. UNESCO, Paris and Routledge, London.

Gebauer, A.B. and Price, T.D. (eds) (1992) *Transitions to Agriculture in Prehistory*. Prehistory Press, Madison.

Moore, A.M.T., Hillman, G.C. and Legge, A.J. (1992) *Abu Hureyra and the Advent of Agriculture*. Yale University Press, New Haven, Connecticut.

Smith, B.D., with contributions by C.W. Cowan and M.P. Hoffman (1992) *Rivers of Change: Essays on Early Agriculture in Eastern North America*. Smithsonian Institution Press, Washington and London.

Vavilov, N.I. (1992) *Origin and Geography of Cultivated Plants*. Cambridge University Press, Cambridge. (This is a collection of Vavilov's papers which have been translated by D. Löve.)

CHAPTER 3

Agriculture in Prehistoric and Historic Times

From the centres of origin (Figures 2.5 and 2.6) agricultural systems diffused into the surrounding areas and were adopted, often with modifications, by human groups that recognised their value. Why agriculture was adopted so widely and so comparatively rapidly is no more easy to ascertain than are the motives that stimulated its initiation (Sections 2.1 and 2.2). Materialism and environmentalism (defined in Section 2.1), possibly combined with inventiveness, probably both played a part, not only in the initiation of agriculture but also in its dissemination. There is the additional possibility that the inception and/or subsequent adoption of agriculture created a sequence of changes in society with the result that the two could only develop within a framework of mutual support. For example, if agriculture were to be initiated or adopted as a response to increased population, the resulting food security would have ensured its continuance and intensification. A similar relationship would have developed if the initiating stimulus was the need to provide a surplus; the advantages of that surplus, via trade, would have ensured the continuance of agriculture and provided an opportunity for population increase. This latter would have reinforced the need for agriculture. Should environmental reasons have underpinned the onset of agriculture, the food security produced would also have ensured agriculture's survival. Agriculture and population would thus have enjoyed mutual reinforcement. Fundamentally, the situation is little different to this today.

On a global basis, there is a dearth of evidence for the character of early agricultural systems and how they spread. The latter may have been by either stimulus diffusion, involving the transference and adoption of ideas, or as a result of migration whereby people moved into new areas taking with them already established practices and technologies. The dispersion of agriculture in the past probably occurred via both processes. Most evidence for early agricultural systems is available for the Near East and Europe; there is a rapidly growing body of evidence for the Americas but there are major gaps in and/or inaccessible data for much of Asia and Africa.

The ancient worlds of the Near East, Egypt and the Greek and Roman empires all flourished as a result of food security. It provided a stable basis on which military excellence, cultural and scholastic advancement, technological development, unique architectural styles and organised religion could prosper. These civilisations, along with equally advanced cultures in Asia, notably the Far East, were so successful that the vestiges of their activities remain to this day. In some instances these monuments constitute such important tourist attractions, and are consequently important wealth generators, that the modern countries wherein such monuments occur continue indirectly to benefit from the supporting ancient agricultural systems. Although some current literature on sustainable development and sustainable agriculture leans toward the view that all ancient food-production systems were environmentally benign because there were so few 'artificial' inputs, there is much evidence to the contrary. Soil erosion, in particular, was often a significant problem and ancient irrigation systems were subject to salinisation. Between AD 500, when the Roman empire was collapsing, and AD 1500 agricultural systems changed markedly (though for large parts of the world there is little information available). What is equally important, however, is the wave of exploration that began, taking Europeans into many other parts of the world. Both the lands they discovered and colonised and Europe itself were profoundly affected by the interaction. On the one hand, European agricultural practices were often imposed on the colonies. On the other hand, both Europe and its colonies exported and imported crop varieties, adding to each others' lists of cultivated species and changing the character of agricultural systems at the local level as introduced species vied with or usurped the roles of indigenous crops. Domesticated animals generally moved unidirectionally: out of Europe and into the colonies. The introduction of sheep and cattle, for example, has had profound repercussions for many agricultural systems in colonial lands.

In Europe, many agricultural innovations were occurring by 1600. After the deleterious impact of the Little Ice Age (Grove, 1988) *c.* 1300–1600, agricultural efficiency began to increase and paved the way for the so-called Agricultural Revolution of 1750–1850. Developments included new rotations, the introduction of new crops and land reclamation. Agriculture was beginning to become significantly more than a mere subsistent activity and in consequence human populations increased. In Europe, the improved food supply provided an essential foundation for the onset of the Industrial Revolution of the mid-eighteenth century. Agriculture provided a series of raw materials for industry and the production of a food surplus was one of several factors that allowed people to move out of the countryside into the towns where they constituted a source of labour for the emerging industries.

Industrialisation also led to the development of an embryonic chemical industry. This had major repercussions for agriculture, notably the production of artificial nitrate fertilisers. Despite such innovations, the outbreak of World War I in 1914 highlighted the precarious nature of the food supply in many developed countries. In the UK, for example, there was a heavy reliance on imported foodstuffs. Protectionist policies were subsequently adopted by governments in order to redress this situation. At the same time science and technology, especially through plant and animal breeding programmes, were becoming increasingly involved in the initiation of agricultural improvements. Agriculture was beginning to industrialise and to rely on fossil-fuel energy.

By the end of World War II, agrochemical industries expanded, capitalising on the stock of chemicals that had been built up by war-time research into biological and chemical weapons. Today agriculture in the developed world is closely controlled by political policies, and it is heavily dependent on fossil fuels through the use of mechanisation, artificial fertilisers and crop-protection chemicals (Figures 1.2, 1.6 and 1.7). In the developing world there is still a reliance on human labour though artificial fertilisers and crop-protection chemicals are becoming increasingly important in those countries that can afford them. Biotechnology and genetic engineering are likely to become the major instigators of agricultural change in the ensuing decades.

3.1 AGRICULTURE IN THE ANCIENT WORLD (TO AD 500)

From the centres in which plant and animal domestication occurred, agriculture spread into other parts of the world. Minc and Vandermeer (1990) go as far as to state that 'within 8,000 years following its emergence, at least half of the world's population had adopted agriculture as its primary mode of subsistence. Within that same time period, the world population increased by a factor of 30, from an estimated 10 million to 300 million.' While such figures are highly conjectural, it is true to say that for a large proportion of the world's population agriculture became the prime occupation and mode of food procurement. In the early centuries of agriculture, hunting also continued to be an important means of supplementing diets.

In view of the diversity of crops that were domesticated in spatially disparate centres, it is virtually impossible to generalise about the nature and spread of these early agricultural systems. It is equally difficult to report on how agricultural production was organised, what systems of food distribution were in operation and what strategies, if any, were employed to combat pests, notably insects. Archaeological sites throughout the world do, however, provide evidence of the crops grown, of land division and tenure arrangements, of crop storage facilities and processing technology. A great deal more information is available for certain civilisations of prehistory, notably those of Egypt, Greece and Rome, than for others such as those of Russia, Central Asia, India and China.

3.1.1 The Near East

As Figures 2.5 and 2.6 illustrate, one of the centres of the origin of agriculture was the Near East. This centre influenced not only agricultural development but also cultural advancement in three continents: Europe, Africa and Asia. In this centre, as discussed in Sections 2.1–2.4, barley, emmer wheat and einkorn wheat were the dominant cereal crops. Zohary (1989), for example, describes these as 'founder crops of Neolithic agriculture'. Not long after, a variety of legumes and flax joined the domesticated assemblage. From the beginnings of domestication approximately 10 000 years ago, these species dominated agricultural activities for many millennia. In the early stages of agricultural development the populations were aceramic, i.e. they had not yet begun to make pottery. Nevertheless there is evidence for a sedentary existence and it is from such beginnings that villages like Jericho and Beidha developed; these and many other villages in the region soon enlarged to become towns. Sickle blades, found in many archaeological sites, had wooden hafts and many are characterised by a sheen or gloss which archaeologists believe, on the basis of field experiments, indicates the harvesting of cereals. Harlan (1992b) states that there was also grinding and pounding equipment, including mortars actually ground into solid rock as well as portable equipment. No doubt these implements were also used to collect and process wild as well as cultivated cereals.

Harlan (1992b) also reports that there is evidence for the inclusion of domesticated animals in what he describes as agricultural rituals involving sacrifice. Obsidian tools and native copper working also date to this early period in the agricultural history of the Near East but there appears to be little evidence for the nature of land subdivision, if it existed, or the way labour and food dissemination were organised. The sedentary and concentrated nature of society, or at least a large part of it, does itself suggest some sort of economic order and probably also involved some sort of division of labour. According to Roberts (1992), Jericho, with its reliable spring, provided a natural focus around which people concentrated. Even as early as *c.* 8000 years ago it had vast water tanks that may have been used for irrigation. Within another 1000 years, pottery production was underway. Possibly the onset of permanent agriculture was an essential prerequisite, in its facilitation of a division of labour, for innovations like pottery and later bronze production.

Society, notably that in Mesopotamia, had by about 5000 years ago become sufficiently complex to be described by many historians (cf. Roberts, 1992) as representative of the emergence of civilisation, difficult though it is to define. Mesopotamia, an area centred on modern-day Iraq, includes the valleys of the Tigris and Euphrates rivers which were the most important food-producing regions. Here the Sumerian and Akkadian civilisations flourished, with their specialised craftspeople and religious centres, and which were responsible for the invention of writing. These civilisations date from approximately 5300 to 4000 years ago and overlap with that of ancient Egypt. Ur, Uruk and Kis were the great cities of these regions; they housed major temples in which a range of gods were worshipped. Clearly, religion was organised and, like the craft industries of the towns and villages, it was made possible by a highly productive agricultural base that exploited the fertile soils of the Tigris and Euphrates valleys via a well-developed irrigation system.

Further north in Subir, a state dependent on rainfed cereal agriculture also thrived in the Habur Plains of what is now north-eastern Syria. However, as Weiss *et al.* (1994) have demonstrated, early civilisations such as these were susceptible to collapse in the wake of climatic change that disrupted agriculture. In Subir, cereal production on moisture-rich alluvial soils on sediments deposited by tributaries of the Euphrates facilitated rapid urban development in three major centres between 2600 and 2400 BC. This produced a state-level society. Sumer also developed rapidly through contact with Subir. It became united with Akkad under the control of Sargon (of Akkad) who also expanded his empire into Subir during a period of imperialisation. Around 2200 BC the Akkadian-dominated habitations of Tell Leilan and Tell Brak in Subir were abandoned quite suddenly. Excavations elsewhere in the region confirm that urban abandonment was widespread. This Weiss *et al.* ascribe to abrupt climatic change and they suggest that regional desertification set in. A combination of increased aridity and the movement south of large numbers of Subir's population into Akkad and Sumer caused the irrigated agricultural systems to be overwhelmed. Despite a well-developed agrarian and urban infrastructure it seems likely that society collapsed as it was unable to adapt to rapid climatic change. Are there parallels here with the likely global warming of the early twenty-first century (see Section 10.5)?

3.1.2 Egypt

Agriculture also diffused southward from the Near East into Egypt, where another great civilisation flourished for several millennia, beginning about 5000 years ago with the protodynastic period that lasted until 2665 BC when the Old Kingdom was established (quoted in Roberts, 1992). The Egyptian civilisation's heyday was over by about 1000 BC. Hughes (1992) suggests that the ancient Egyptians practised a form of sustainable agriculture. This Hughes believes was the most important reason why no other ancient civilisation endured as long as that of ancient Egypt. The relationship between the longevity of the civilisation and its custodianship of the environment was examined two decades ago by Butzer (1976). Hughes (1992) takes this a stage further when he opines that 'The ecological attitudes and practices of the Egyptians were rooted in a world view that affirmed the sacred values of all nature, and of land in particular'. Both the importance of agriculture and appreciation of the land are, to an extent, manifested in the integration, rather than separation as occurred in the city states of Mesopotamia, of town and country. There are no walled cities, for example, in the Nile Valley. The river itself provided the wherewithal for sustainable agriculture since its annual flooding regimes provided nutrient-rich alluvium on a regular, predictable basis. Beyond the narrow plain that was replenished annually there was desert. The Egyptians were probably more acutely aware of where crops could grow and where they could not than any of the other early civilisations. That the valley land was at a premium is also implied by the fact that most of the tombs of the pharaohs, etc., in the valleys of the Kings and Queens, are located in the desert beyond the fertile strip. Not even the revered theocrats deserved such vital land!

Many of the gods of ancient Egypt reflect nature itself. Ra, the god of the sun, Nut the goddess of the sky and Hapi the Nile god amongst others, reflect the significance of the diurnal and seasonal changes that ruled the food-production system, and the respect that they were accorded. The production of barley and wheat, introduced from the centre of origin (Figures 2.5 and 2.6 and Table 2.5) to the north-east, is even immortalised in hymns to Hapi that were discovered in the *Pyramid Texts* (referred to in Hughes, 1992). In Sudan, at Esh Shaheinab, dated to *c.* 5400 years ago there is evidence of wheat and barley cultivation, and in Egypt, at Fayum and Badarian, dated to between 7500 and 5000 years ago, emmer wheat, barley and club wheat were being grown. The ancient Egyptians also devised various irrigation schemes to ensure an adequate moisture supply all year round and so take advantage of the annual flood. The engineering and technology associated with irrigation schemes may, some researchers believe, have been a paramount and revered occupation under the direct organisation of the Pharaoh. There is also the possibility that the administrative units of the Nile Valley, i.e. the local unit of organisation of land and people, were actually local irrigation units (Butzer, 1976). The canals and channels that constituted the irrigation systems, rather like stone walls, hedges or ditches, may well have provided the units of land tenure or ownership or at least marked the territory worked by a specific agricultural worker and/or family.

Despite a high degree of organisation, the production of food was not always constant. Periods of famine, for example, usually consequent on the failure or inadequacy of the Nile floods, are recorded in frescoes. In addition, there were vast granaries that were probably used to store cereals for lean years and as depots from which cereals could be exported in good years. There is also evidence that some environmental degradation occurred so the system was obviously not entirely sustainable. Some salinisation occurred, a problem often associated with irrigation (Chapter 9), not in the Nile valley where the river itself flushed out salts, but in some oases. Habitat demise, notably that of wetlands and woodlands, also occurred but, as Hughes (1992) points out, Egypt was still a highly productive region when the New Kingdom ended *c.* 1075 BC and as other civilisations of the Near East and the Mediterranean were achieving pre-eminence.

The early agriculture of the Nile Valley was based on crops originally domesticated in the Near East. Agriculture developed about 6000 years ago as the rate of post-glacial eustatic rise in sea-level decreased and caused an increase in deposition of fertile silt in the delta area (Stanley and Warne, 1993). Continuity of favourable conditions between Egypt and the Near East thus developed and instigated a switch from hunting and gathering to crop and livestock production. Interestingly, however, an independent locus of domestication was present in the desert of the southernmost part of Egypt about 110 km west of Abu Simbel. There is evidence for the cultivation of sorghum by about 8000 years ago (Wendorf *et al.*, 1992). As this species was not part of the Near East crop complex it seems likely that an African crop complex evolved independently, though the agricultural community subsumed sheep and goats from the Near East. Holmes (1993) suggests that people from this area, which became increasingly arid in the mid Holocene, migrated to the Nile Valley with their animals at about the same time as wheat and barley were being introduced through the Nile delta.

3.1.3 Europe

The possibility of a merger between two separate food complexes that is raised by the example of Egypt given above also highlights a fundamental question relating to the spread of agriculture from the Near East into Europe. According to Jones (1991), there are essentially two schools of thought: the diffusionist view and the migrationist view. The former espouses the idea that new techniques and tools spread from the centres of origin by a somewhat passive learning process; such innovations are transmitted through the diffusion of ideas as trade and commerce bring varied groups together. The migrationist process is a much more active, even aggressive, means of disseminating new ideas: it requires mass migrations as the stimuli to social and economic change. If the movements of people are involved then so is the movement of their genes, the patterns of which in modern people have provided some clues as to how innovations like agriculture spread. Where migrations of people are involved the process is known as a demic diffusion; it may also have a bearing on the origin of the Indo-European languages (Renfrew, 1987).

In the context of the spread of agriculture into Europe, Sokal *et al.* (1991) have compared observed genetic patterns with predicted patterns derived from the demic expansion hypothesis. There is a partial but significant correlation between genetic distances, calculated from human blood group data, etc., in over 3000 places in modern Europe, and the spread of agriculture based on radiocarbon-dated archaeological evidence. This relationship is depicted in Figure 3.1. It indicates that demic diffusion did indeed play a significant role in the spread of agriculture (see also Cavalli-Sforza *et al.*, 1993) though it was unlikely to have been the only mechanism. Demic diffusion does not necessarily imply waves of migrations of people into the periphery of Europe. The process was probably more peaceable and gradual than sudden and invasionary. Its root cause may have been population pressure prompted

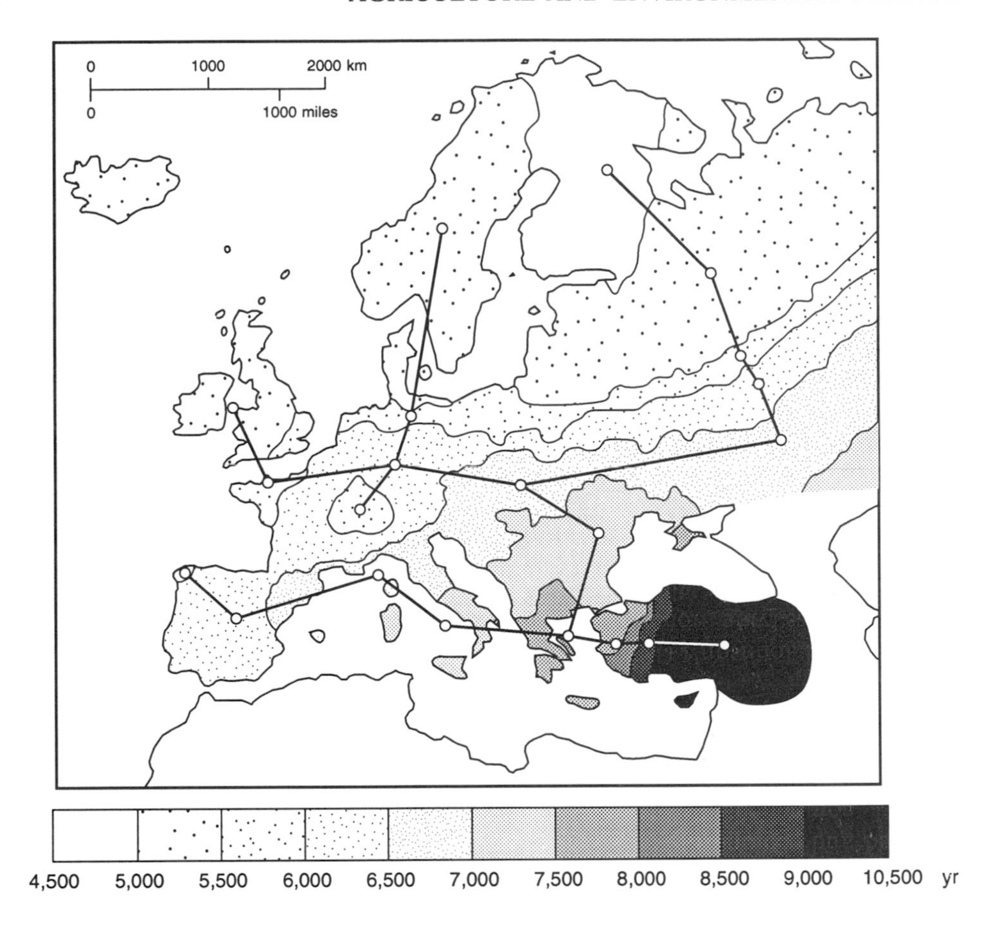

Figure 3.1 The spread of agriculture from the Near East into Europe based on radiocarbon-dated archaeological sites. Lines indicate the relationships between regions (adapted from Sokal *et al.*, 1991)

by food scarcity. When food production just met or fell short of local needs, necessity may have stimulated migration. Such migrators spread into non-agricultural areas where they interbred with local populations and so left genetic markers. Jones (1991) states that as a result 'The hunter–gatherers of Mesolithic Europe suffered a process of gentrification—or even yuppification—from the east . . .' According to Ammerman and Cavalli-Sforza (1984), the expansion of agriculture into Europe occurred at an average rate of 1 km per year, a rate consistent with a demic spread. Further work by Cavalli-Sforza *et al.* (1994) also confirms this. Their research also shows a similar pattern for the spread of genes and agriculture in Asia from the Near East.

The archaeological and palaeoecological record provides evidence, albeit limited, for the character of early agricultural systems in Europe. Chapman and Müller (1990), for example, have reviewed some of the evidence for the emergence of agriculture in Greece, Dalmatia and Italy. As Figure 3.2 illustrates, the earliest sites in Greece (as well as Sicily and Corsica) date to the centuries just after the inception of agriculture in the Near East. By 8000 years ago farming had spread to Dalmatia,

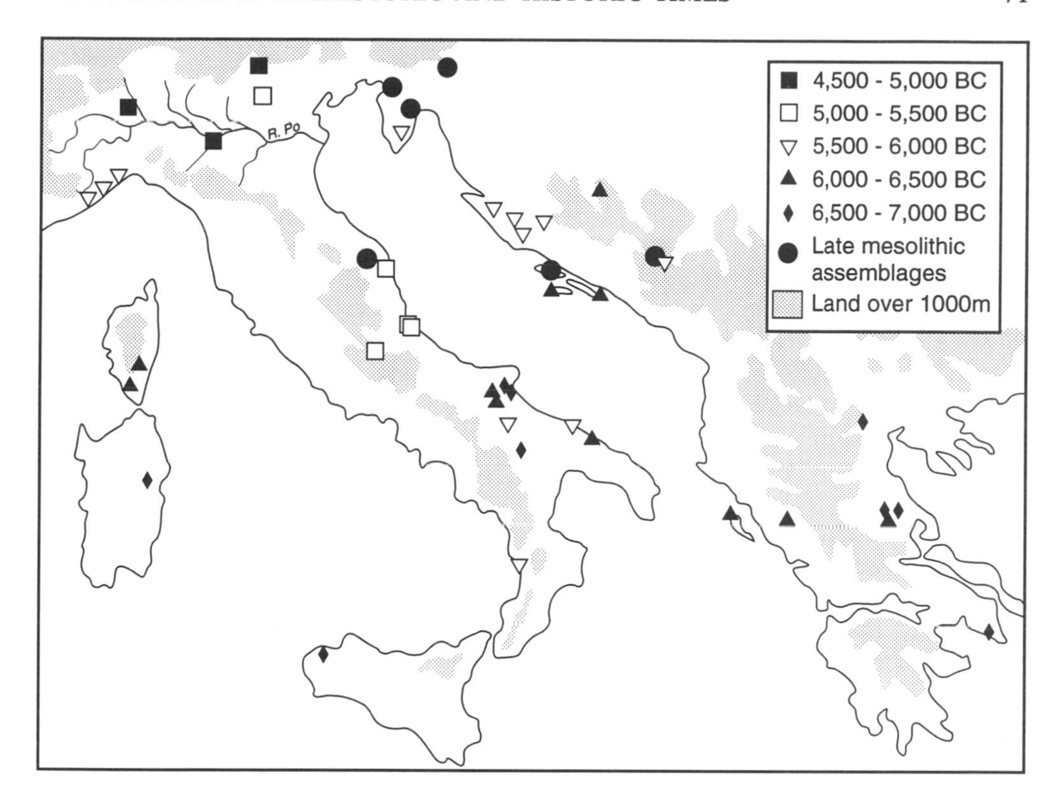

Figure 3.2 Early agricultural sites in Greece, the former Yugoslavia and Italy (adapted from Chapman and Müller, 1990)

particularly along the coast. All the components, i.e. domesticated plants, animals and pottery, are found together, which Chapman and Müller suggest indicates the adoption of a package of components. This contrasts with the implication from the Egyptian evidence (Section 3.1.2) for an agricultural complex derived from more than one source. Biagi *et al.* (1993) have examined the evidence for early agriculture and its impact in the Po Valley of northern Italy. Here the earliest settlements, which originated about 6000 years ago and whose inhabitants were cultivating einkorn and barley, are located close to the edge of the river banks. Those in the 5500 to 4800 year old age range occur instead on top of the river terraces where fertile loams were particularly suitable for cereal growing. Biagi *et al.* suggest that the later sites are representative of human communities with a more sedentary lifestyle that those of the river banks.

Agriculture reached the European periphery about 5000 years ago. MacNeish (1992) suggests that in Britain and parts of Scandinavia it took another 1500 years before true agriculture developed. In England, the onset of permanent cultivation occurred at about the same time as the construction of the megalithic monument of Stonehenge began. Whatever the association is between the two, it would appear that the food base provided by agriculture freed a proportion of the labour force to engage in monument construction, just as it did on a much larger scale in ancient Egypt (Section 3.1.2). Similar monuments, also known as megaliths, were constructed elsewhere in Europe; in Ireland there is Newgrange in the Boyne Valley group (O'Kelly, 1982) and there are numerous such monuments in the Iberian Peninsula (Chapman, 1981).

3.1.3.1 Ancient Greece and Rome

The rise to pre-eminence of a civilisation that gave to posterity not only temples and amphitheatres but philosophy, mathematics, literature and drama, was made possible by a well-organised agricultural system. The rise of the ancient Greek civilisation began about 3500 years ago, when the minoan palaces of Crete were constructed, and it ended around 150 BC by which time much of Greece had been annexed by Rome. Athens itself was sacked in AD 400. During its 2000 years of existence as a dominant power, ancient Greece witnessed many cultural and environmental changes, including much warfare.

According to Sallares (1991), there is no documentary or literary evidence to indicate the amount of land that was actually cultivated by the ancient Greeks. In relation to Attica, the region that comprises the hinterland of Athens, estimates of the area of cultivation range from 20 per cent to 50 per cent. Whatever the reality, it is likely that ancient cultivation exceeded the present-day extent since there are some areas of old, abandoned terracing. The most important crops were cereals, olives and vines. Many estates specialised in olive production; both olives and olive oil were mainstays of trade. Indeed, Herodotus states that Attica was the place of origin for the olive. This species also has the advantage that it will grow on land too poor to support a cereal crop though it grows better and produces a larger crop on more fertile soil where it probably competed with wheat, the favoured cereal of ancient Greece. According to Isager and Skydsgaard's (1992) interpretation of ancient texts, autumn sowing was standard with sowing by hand after the third ploughing of a fallow field. Occasionally, in times of difficulty, spring sowing may also have occurred. Homer's work attests to the use of draught animals for threshing, though stubble burning and/or composting of the stubble was practised.

Vine growing, cutting and pruning as well as tilling of the soil for weed control are all described in ancient texts. Theophrastus, for example, describes the cutting of vine roots to force them downward into the soil as a means of combatting drought, the application of manure every four years and the necessity of airing the soil. Other fruit crops mentioned in ancient texts (quoted in Isager and Skydsgaard, 1992) include pears, pomegranates, apples, almonds and figs, all of which were cultivated in orchards. There is also reference to the collection of wild chestnut, hazel and walnut. Millet and Italian millet, broad bean, peas, lentils, chick peas and vetch were also grown. Such an assemblage was fairly typical of the Mediterranean region in general and reflects the crop complex originally domesticated in the Near East (see Section 2.2 and Table 2.5).

Not a great deal is known about the role of animal husbandry in ancient Greece. Sallares (1991) suggests that meat was only widely consumed in classical Athens during religious festivals when it was paid for from the public coffers. Animals were certainly reared for draught purposes and some transport; they required the collection of fodder for winter feeding. Sheep, cattle, dogs, pigs and poultry were mentioned in ancient texts and horses were probably kept more as a status symbol than for practical purposes. Sheep in particular were kept for their secondary product, wool, rather than the primary meat product but, overall, high human populations in Attica meant that crop production was much more important than animal production.

Likewise, there is much written and archaeological evidence for agriculture in the Roman Empire, a political power that dominated the known world between 55 BC and AD 500. The Romans initiated changes in systems of land tenure which involved the spread of villas, estates and farms (Potter, 1987). Slaves, from the conquered lands of Europe, were used to provide labour. As in Attica of the ancient Greeks, the major crops in the Mediterranean zone of the Roman Empire were wheat, vines and olives. Even today there is evidence of olive storage pits alongside the roads that connected farms and estates. Rees (1987) states that the Empire's lands could be subdivided into two major categories: the Mediterranean zone with hot dry summers and the Northern Provinces characterised by a cool, wet climate and relatively short growing season. In the Mediterranean zone 'dry farming' was practised with arable crop, notably wheat, production on alluvial plains and the coastal zone; vine and olive production were relegated to less productive soils often on slopes while higher areas were used for summer pasture as part of a transhumance system. Arable and mixed farming were practised in the lowland zone of the Northern Provinces whilst the upland zone provided pasture.

The cropping system of the Mediterranean zone was similar to that of the ancient Greeks (see above), relying on a crop–fallow system but benefitting from the Roman's superior organisation, engineering feats and communications systems. Rees (1987) states that one crop was obtained every two years in order to take advantage of two winters' worth of rainfall. Weeding was frequent to conserve moisture by eliminating transpiration from crop competitors, stubble burning was practised to return minerals to the soil in the form of readily soluble ash, and the creation of a soil of fine tilth at the surface reduced evaporation from the lower levels of the soil and so conserved water. Drainage networks were also constructed in the Mediterranean zone and the Romans began many wetland drainage schemes elsewhere in Europe. In the English Fens, for example, the Romans were the first to initiate drainage schemes (Rackham, 1986). Another major drainage scheme was initiated in the Po Valley to increase the land suitable for agriculture.

Wheat, spelt, millet, legumes, fruit and vegetables were the most important crops in the Mediterranean zone where there is also evidence for an increased range of secondary animal products (Wacher, 1987). This means that livestock rearing was more important than it was in ancient Greece and that meat products were more prominent in Roman diets, at least in those of the elite. Animals like dogs and barbary apes were also kept as pets. In the Northern Provinces, the Roman Empire superimposed itself on the indigenous Iron Age communities, the characteristics of which have been described by Hedeager (1992). Village society was reorganised; small villages disappeared and were replaced with larger estates to improve the efficiency of food production. This asset was precisely why the Romans initially annexed the Northern Provinces, including Britain. They also contributed to increased efficiency by introducing improved implements, e.g. ploughs, scythes and sickles, as well as larger constructions like water mills.

The Roman Empire, because of its well-organised communications and trade, was also responsible for the first substantial wave of crop exchanges. For example, the

ʿnut (*Castanea sativa*) and the walnut (*Juglans regula*) were introduced into
ʾ vines (*Vitis vinifera*) may even have been cultivated. There is also
..ological evidence for the importation of exotic crops into London. Examples include peach, olive, fig, cucumber and a number of herbs including coriander, though these crops were never actually grown in Britain.

3.1.4 The Americas

As discussed in Section 2.2 (see also Figures 2.4–2.6), there was at least one centre in which agriculture originated in the Americas. This was Mesoamerica, focusing on Mexico, the developments in which have been discussed by Blanton *et al.* (1993). MacNeish (1992) opines that there was a further centre in the northern Andes, though Harlan (1992b) believes that this area was simply part of a much larger non-centre (see Section 2.2 for a definition) that occupied the Andean zone of South America (Figure 2.5). Some of the crops that were domesticated in these two regions are given in Table 2.5. To deal with all the developments in agriculture, post-domestication, in the entire region of Meso and South America requires at least one text in its own right. Here emphasis is placed on Peru where a wealth of archaeological sites provide evidence on which to outline the agricultural systems of the region and their changes up until AD 500. The later period will be considered in Section 3.2.

The standard chronological framework for the pre-Columbian history of Peru is given in Table 3.1. The dominance of the Andean mountain chain requires an analysis of each major zone which can be differentiated by altitudinal and topographical criteria. The three major zones are shown in Figure 3.3 and consist of the coastal region, the puna region which is a tableland, and the valleys of the highlands. Each is characterised by a distinct climate and each contained different biotic resources that have been manipulated through social and economic pressures (see Figure 1.1). The highland valleys were a zone of plant domestication (Section 2.2).

According to Pineda (1988), there was an important change in the society and economy in this central Andean region about 5500 years to 2800 years ago which, as Table 3.1 shows, includes the transition from an aceramic to a pottery-producing cultural system. The coastal dwellers increased in number, cultivating a crop complex dominated by squash, gourds, beans and cotton. Cotton was produced in order to make textiles. Agriculture also spread inland along river valleys where irrigation was practised as it was in the coastal zone. These food production systems supported what must have been a power structure and an organised religion that between them led to the distinctive architecture of the pyramid-temples. Further details on such developments are given in Wing and Wheeler (1988). In addition, a report on research findings at the coastal site of El Paraíso (Quilter *et al.*, 1991) shows that, about 3800 years ago, fish was the major protein source and was combined with domesticated crops, e.g. beans, peppers and squashes, and some wild plants such as specific sedges. Cotton was also a major crop and was used to produce fishing tackle as well as textiles. This latter is, Quilter *et al.* argue, a major reason why the society flourished and built its stone and adobe monuments. Only later in the early ceramic period was the technology that was used to produce cotton intensively employed to produce food. Such findings do, however, indicate that early societies were still dependent on agriculture though in this case the most important product was cotton and not food

Table 3.1 The chronological framework for Peruvian prehistory (based on Keatinge, 1988)

Period	Approximate date	Developments
ARRIVAL OF THE EUROPEANS		
Late Horizon	AD 1476–1534	Inca civilisation
Late Intermediate Period	AD 1000–1476	
Middle Horizon	AD 600–1000	
Early Intermediate Period	200 BC–AD 600	
Early Horizon	900–200 BC	
Initial Period	1800–900 BC	
DEVELOPMENT OF POTTERY		
Preceramic Period VI	2500–1800 BC	Textile production
Preceramic Period V	4200–2500 BC	
Preceramic Period IV	6000–4200 BC	Village agriculture
Preceramic Period III	8000–6000 BC	
Preceramic Period II	9500–8000 BC	First domestications
Preceramic Period I	?–9500 BC	

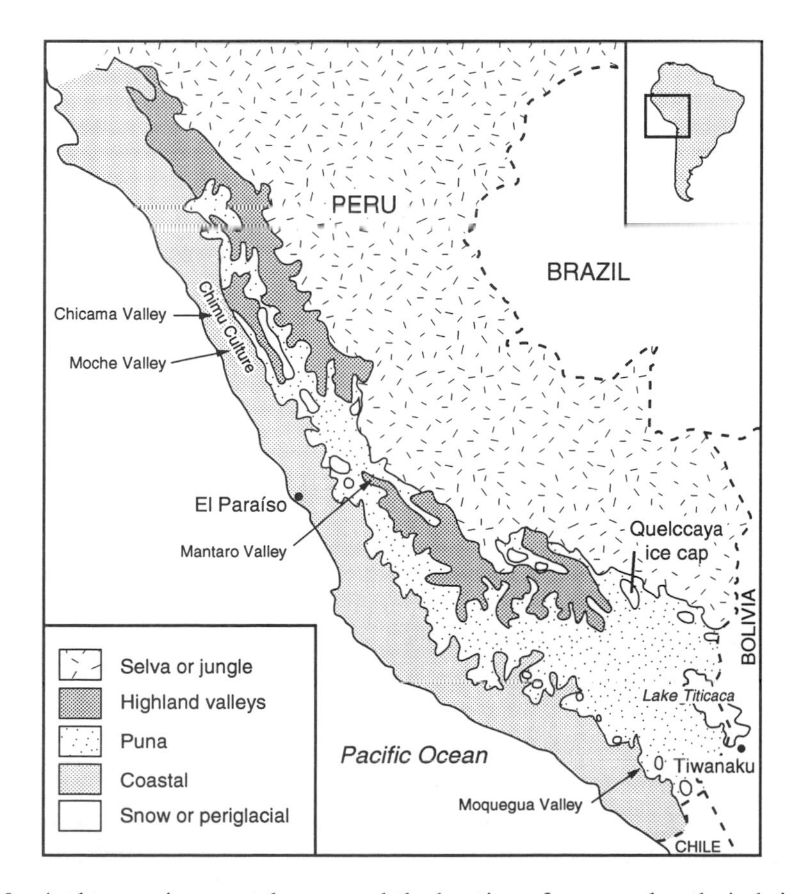

Figure 3.3 Andean environmental zones and the location of some archaeological sites in Peru

crops. This was only possible because of abundant food from marine resources. It seems that the shift to a ceramic culture (Table 3.1) reliant on agricultural food production came about because of the redeployment of a technology already well established.

A study in the Upper Mantaro Valley of the central Andes of Peru, by Hastorf (1993), has also revealed the nature of some of the changes that occurred in intermontane Peru in pre-Columbian times. Archaeological excavations provide evidence for five pre-Columbian agricultural technologies: stream irrigation, raised fields in tributary valleys, terracing on hill slopes, irrigation canals in the uplands and ridged fields on the hillsides of the lower puna. All indicate that local cultures were actively managing their environment, often to combat climatic and topographical conditions that were not conducive to high-productivity agriculture. However, Hastorf's research shows that in the period beginning *c.* 200 BC the Early Intermediate Period, and ending *c.* AD 900 at the end of the Middle Horizon (Table 3.1), only some of these technologies were in use; the rest came later as additional land was brought under cultivation. The people of this era operated socially at the level of the village and grew most of their crops in the fertile valley and immediate hillsides. Their diet was dominated by plant matter rather than meat. Quinoa, a goosefoot with seeds rather like rice and leaves like spinach, was the major crop with maize in second place. Hastorf suggests that together these accounted for about 65 per cent of the crop production, the remainder comprising potatoes, legumes and a variety of Andean tubers. Only after AD 1000 did production extend into the upland areas.

3.1.5 China

Roberts (1992) reports that at the same time as agriculture was beginning in the Near East (Section 2.2) people in southern China were also beginning to clear the forests for growing crops. Information from this area is, however, sparse and the so-called cradle of Chinese civilisation is considered to have been further north, centring on the Yellow River valley. The village of Banpo, near Xi'an in Shaanxi Province (the nearest city to the famous terracotta warriors), has yielded some of the earliest Neolithic artifacts in China. It is 6000 years old, by which time agriculture was well underway. Grain storage structures and palaeobotanical remains of millet and vegetables are in evidence.

According to Kaichen (1991), the early phase of Chinese agricultural history occurred prior to the third century BC, by which time ploughing, manuring and crop rotations were well established. Even earlier, by the time the Shang Dynasty was in power *c.* 1300 BC, some domesticated plant and animal species from the Near East complex, e.g. wheat, barley and goat, had already reached China (Harlan, 1992a). Millet and silk production (the latter from silkworms fed on mulberry) were already economic mainstays. Silk in particular encouraged trade and contact with the Greek and Persian empires and so resulted in a number of crop introductions. Harlan states 'alfalfa and grape were introduced in 126 BC; cucumber, pea, spinach, broadbean, chive, coriander, fig, saltflower, sesame, and pomegranate arrived from Iran at various times from the second to the seventh centuries AD.' Few crops, however, migrated in the reverse direction except, possibly, the millets.

Indigenous developments, coupled with the introductions, led to what Kaichen

(1991) describes as the developing stage of Chinese agriculture. It lasted until *c.* AD 1300. During this period, the characteristic regional trends in Chinese agriculture intensified. The three regions consisted of dry crop cultivation in the lower reaches of the Yellow River in northern China, rice paddy cultivation in southern China and the extensive grassland of the far north and west where pastoral systems were in operation. Some irrigation was practised in the north and soil-water conservation measures were employed. In the south, rice production was enhanced by the introduction of the curved-beamed plough and by improved water management. Not until after AD 1000 did population pressures increase to such an extent that land reclamation measures were instituted.

3.2 AGRICULTURE IN THE PERIOD AD 500 TO AD 1500

When compared with information on domestication, the inception of early agricultural systems and their spread in the ancient world, there is a paucity of information (at least accessible information) for the period AD 500 to AD 1000. This was the so-called Dark Ages of Western Europe. The rise to power of the Normans and their invasion of England in 1066 led to the compilation of the Domesday Book in 1086. This is a source of much information on the organisation of agriculture at the beginning of the Norman period. For the remainder of the Middle Ages, agriculture in the UK is reasonably well documented as is discussed below. As far as the Americas are concerned, the quincentenary of Columbus's discovery of America in 1992 prompted the publication of much literature on pre-European landscapes and land use. For other parts of the world, notably Africa, Central Asia and the Far East there is much less information than there is for the UK. Consequently, only the briefest comments are possible.

3.2.1 The United Kingdom

When the Romans left Britain in AD 410, the traditional view has been that there was social and economic disruption and general stagnation. This is one of the reasons why the period has earned itself the name 'Dark Ages'. The same is supposedly true of north-west Europe in general. However, the term 'Dark Ages' is more appropriate in terms of scarcity of information rather than in terms of cultural advancement. The traditional view is, thus, gradually becoming unacceptable, particularly in the light of recent archaeological evidence from both the UK and Europe. Instead of social disintegration there appears to have been considerable continuity from the earlier Romano-British period as well as innovation due to the advent of the Saxons from lands adjacent to the North Sea. Rackham (1986) states 'The Anglo-Saxons in 600 years probably increased the area of farmland, managed the woodland more intensively, and made many minor alterations. But this did not radically reorganise the woodland landscape.' The palaeoecological record provides evidence for changes in the amount of woodland, pasture and arable land but nowhere is this extensive. Woodland regeneration, for example, appears to have occurred in areas that were marginal for arable cultivation. In view of the exodus of the Romans, this is not surprising as many fewer mouths were left to feed. For the same reason, the intensity of land use declined.

According to Higham (1992), the lowlands of England were dominated by mixed agricultural systems in which sheep were the most important livestock. Many of the 'luxury' foodstuffs enjoyed by the Romans were no longer produced and barley was grown more extensively than hitherto in place of wheat. This Higham attributes to the production of grain-based alcohol. Rye was also extensively grown. Rather than causing fundamental changes in agricultural production and its organisation, the departure of the Romans and the advent of the Saxons had a greater impact on settlement patterns. The urban centres of the Romans declined in importance so that the economy became rurally based, focusing on small, scattered settlements and even individual farmsteads. Hall (1988), for example, has shown that areas of light, fertile soil, especially along the slopes of the major river valleys in the lowlands, were preferred for settlement; a practice begun much earlier in prehistoric times. Clay-rich soils, which were comparatively difficult to work, were avoided. A similar settlement pattern and economy was present in the English south-west (Todd, 1987) and the northern counties (Higham, 1987): possibly sheep herding, a major component of the mixed farming systems, was not conducive to nucleation. Before the Normans invaded, however, this dispersed settlement pattern had been replaced by villages and strip fields. This is the typical character of the rural landscape even today and one in which much effort is invested in preserving.

By 1066, when the Normans invaded, agriculture was intensifying. There were two major units of production: the demesnes, which were manors with lands that were not let out to tenants, and small producers. Sheep were important to both, whilst pig-keeping and livestock were more important to the small producers than to the demesnes, for whom arable agriculture was the mainstay (Harvey, 1988). Pigs especially were reared using the system of pannage whereby they were grazed in woodland pasture (coppiced or pollarded woodlands) for which a charge by the woodland-owning manor, by way of cash or pigs, was made. The Normans increased their revenues by raising rents from small producers and renting out the demesnes; pasture, meadow and woodland rights were often also rented out. This was a period of population increase as shown in Table 3.2. It was also a period characterised by a variety of farming systems as has been demonstrated by Campbell and Power (1989) and Power and Campbell (1992). The latters' results are given in Table 3.3 which shows that there were eight basic farming types; the most intensive occurred in the east and south-east adjacent to the major markets. Wheat, barley, oats and rye were grown in rotation with legumes. The latter were grown to enhance the nitrogen content of the soil and as a source of fodder for animals. The livestock kept included sheep, cattle and pigs with oxen and horses used for draught. The proportions of each that characterised demesnes varied according to the production potential of demesne soils and proximity to markets.

The importance of sheep farming was increased by the advent of Cistercian monks in 1128 and the emerging wool industry. Indeed, the siting of monasteries in rather remote upland regions, with few potential agricultural uses, led to the spread of upland pasture. This, plus the importance of wool as an export crop to Belgium and the Netherlands, is reflected in the prominence that sheep command in nearly all of the demesne-farming systems of Table 3.3. These data also reflect regional specialisation.

As agriculture expanded, the extent of woodland declined despite efforts by the

Table 3.2 Population change in the period 1066–1430 in England and Wales. Data from Hallam (1988) and E. Miller (1991)

Year	Possible population 1×10^6
1430	2.00–2.50
1377	2.50–3.00
1348	Black Death
1317	6.30–6.74
1292	6.52–7.20
1262	6.20–6.38
1230	4.96–5.12
1149	3.42–3.44
1086	2.20–2.20

crown via the designation of royal forests. Woodlands, nevertheless, continued to be a major source of revenue. There is also evidence to suggest that whilst the staple crops remained the cereals, there was an overall increase in the range of fruits and herbs being grown, often in gardens (Greig, 1988). Parsley (*Petroselinium crispum*) and mint (*Mentha* spp.), for example, were widely grown. Wild animals were still important components of the economy (Grant, 1988) and deer farming was an important addition to many agricultural demesnes and to the royal forests. Birrell (1992) states that these parks helped to conserve woodland and were not simply status symbols. Considerable management skill was necessary and the end product, venison, was high-quality meat. That there was pressure on the land to produce more food is reflected, to a certain extent, in the reclamation of wetlands such as the Fenlands of East Anglia. As a result this area became economically important and one of the well-to-do parts of England (Rackham, 1986).

The advent of the Black Death, which came to Britain in 1348, had a substantial impact on population numbers. As Table 3.2 shows, the population of England and Wales was more than halved. This inevitably led to change in agricultural production as is reflected in Figure 3.4. First, the amount of arable land, especially that producing cereals, declined by about 30 per cent in England. Campbell (1991) states that it was as much as 40 per cent in the south-west. Much land was converted to pasture, fallow frequencies increased and some land was transferred to the individual producer through leasing. Many of these trends had begun prior to the Black Death, partly due to the depression in grain prices in the 1330s. Figure 3.4 also shows that there were major changes in the ratio of livestock to crops. After 1350 there was a small increase in England, rising to a 16 per cent increase by *c.* 1400, which reflected a demand for pastoral products, particularly wool. Although there was a general economic recovery underway by *c.* 1500, it was not until the late sixteenth century that cereal yields began to recover their levels of the 1300s as demands for grain increased.

3.2.2 The Americas

Prior to the advent of Columbus in 1492, Denevan (1992) opines that the total population of the Americas in 1490 was between 43 and 65 million of which 24 million

Table 3.3 Demesne-farming systems in England, 1250–1349 (based on Power and Campbell, 1992)

System type	Dominant crops	Dominant livestock	Location
A Intensive mixed farming	Wheat in preference to rye or barley, grown in rotation with legumes to supply nitrogen and animal fodder.	High proportion of livestock due to fodder production. Horses more important than oxen. Cattle used for dairying. Sheep and swine also kept.	East Norfolk, East Kent, area near Peterborough. All had access to markets via rivers and the coast.
B Light-land intensive	Rye and barley dominate.	Horses outnumbered oxen. Dairy cattle were prominent. Sheep also kept.	Norfolk and Suffolk especially on light soils.
C Mixed-farming with cattle (less intensive than A or B)	Three-course system of cropping: winter corn, spring corn (mainly wheat) and fallow; some rye, barley, oats. Legumes less important than in A or B.	Mostly working animals: oxen provided most power. Cattle were kept for dairying.	Warwickshire, South Somerset, Home Counties.
D Arable husbandry with swine (a minority type)	Legumes particularly important and were used as fodder for animals.	Horses more important than oxen as working animals. Swine very important.	East Midlands, East Anglia, south-east UK around areas with woodland.
E Sheep-corn husbandry	Pasture very important. Assorted patterns of cropping: oats, barley, legumes.	Oxen used for draught. Sheep dominate livestock.	Not locally or regionally dominant. Access to grassland was essential.
F Extensive mixed-farming (see A and C)	Wheat, oats, some barley and legumes.	High stocking densities but cattle/sheep were most important.	Vale of York to Somerset.
G Extensive arable husbandry	Arable dominated: wheat, oats, little barley and legumes grown extensively.	Very few working animals. Oxen used for draught.	Geographically most widespread, especially in north, west and south-west England.
H Oats and cattle (smallest number of demesnes)	Oats dominated; some rye and barley but few legumes and no wheat.	Cattle and oxen dominated livestock; few horses, sheep and swine.	Near major urban centres; areas where other crops were unsuitable.

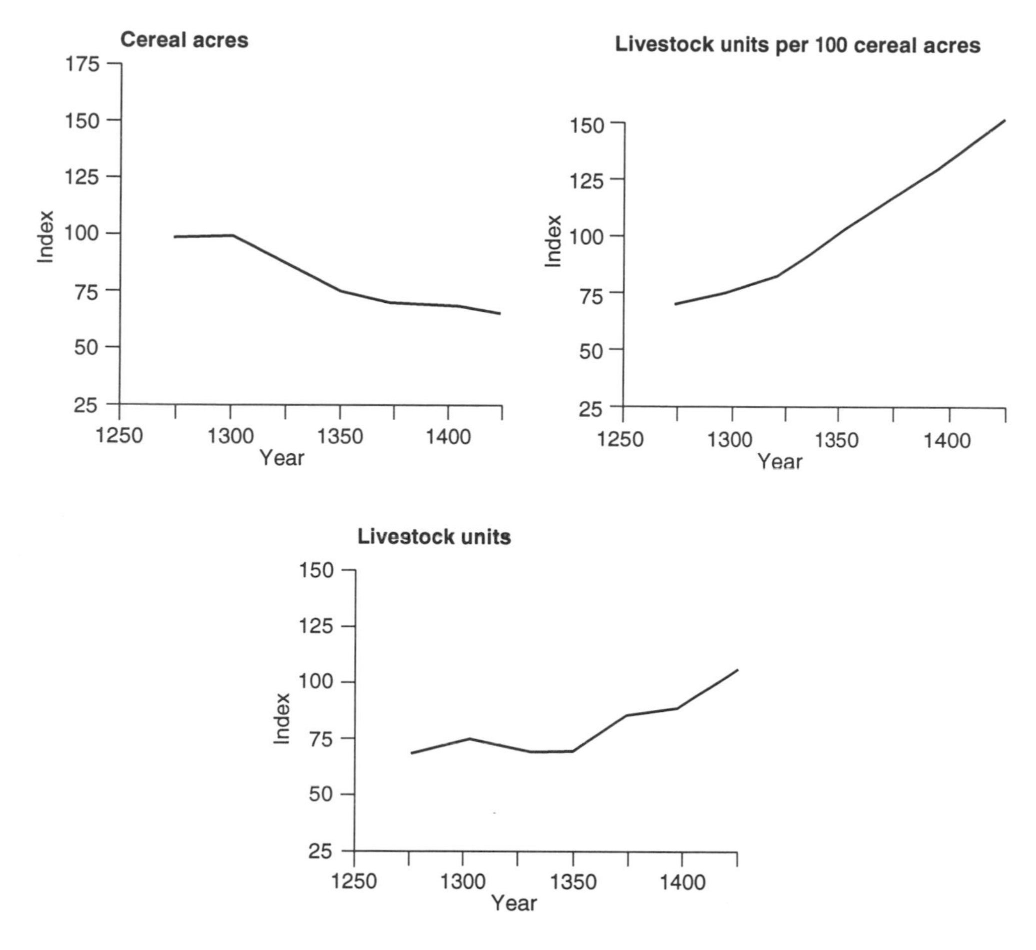

Figure 3.4 Demesne production trends, 1250–1449, in England (adapted from Campbell, 1991)

were in South America, particularly in the Andean region. A further 20 million occupied central America, four million lived in North America with three million occupying the Caribbean islands. All were supported by some form of agriculture. The popular myth that all pre-Columbian land/people relationships were environmentally benign has been scotched. Deforestation and soil erosion occurred just as they did in prehistoric Europe (Mannion, 1991). O'Hara *et al.* (1993), for example, provide evidence for three periods of accelerated soil erosion around Lake Pátzcuaro, Mexico. The first occurred between 3500 and 3200 years ago, the second between 2600 and 1400 years ago and the third around 800 years ago. Delcourt and Delcourt (1991) have also summarised some of the data for the impact of pre-Columbian agriculture in North America.

According to Doolittle (1992), native North American agriculture was concentrated in two areas: the south-west, from where it extended into Mexico; and the east, including most of the Mississippi basin covering the central plains and extending east to the Atlantic coast. In the south-west, agriculture was practised on the flood plains the desert streams and on upland slopes. Soils of the former were located in ;kets and were fertile whilst those of the latter were thin and base-poor. Water availability was a major problem generally and to combat this, canal irrigation, terraces and checkdams were constructed. Indeed, Minnis (1992) indicates that although the earliest domesticated species appear 3000 to 4000 years ago, sedentary agriculture did not occur until about AD 200 by which time irrigation was also being employed. Where the growing season was more than 200 days, two crops per year were produced, the second being reliant on irrigation. Such systems required management and Doolittle cites evidence for the abandonment of degraded systems and even, possibly, the demise of cultures that could not sustain their irrigation systems. The use of terraces, etc., was apparently not as widespread in pre-Columbian agriculture as were irrigation systems. Terraces were constructed to provide flat, cultivable ground whilst checkdams, constructed mainly of rocks, were used to protect downslope fields from flooding and sediment deposition. These features characterised the landscape only after AD 1000.

In the eastern region with its more varied climates, soils and biotic resources than the south-west, agricultural practices were much more varied. Delcourt and Delcourt (1991) suggest that early agriculture was concentrated mainly along the corridors created by the major rivers and Smith (1989, 1992) has summarised the major stages in agricultural development. These are given in Table 2.4. In particular, field agriculture expanded between 1700 and 800 years ago, after which maize was adopted as the major crop. Smith (1992) notes that prior to the introduction of maize, species such as goosefoot (*Chenopodium berlandieri*) and knotweed (*Polygonum erectum*) dominated crop assemblages (see Section 2.2) produced by the Hopewellian culture of the eastern woodlands though it is not clear how the crops were cultivated. Between AD 200 and 800 villagisation occurred as populations increased but they remained concentrated in river valleys. Doolittle (1992) debates the significance of slash-and-burn cultivation on the eve of European conquest, though it may have been important earlier as a means of woodland clearance. Smith's (1992) description of a maize-dominated Mississippian society based on field agriculture suggests that agriculture was a permanent characteristic of the landscape. Indeed, accounts of early European settlers indicate that annual cultivation with only short fallows was the norm. The intensity and relative permanency of agricultural systems in eastern North America is also suggested by the presence of relict raised fields (Doolittle, 1992) that are widespread in the region. The investment of effort necessary to construct such systems is too high for them to reflect shifting or extensive agriculture. Doolittle (1992) also draws attention to the likely significance of gardens as a source of variety of foods as well as seed beds.

Food production in pre-Columbian times in Mesoamerica and South America was just as sophisticated, and in some instances more sophisticated, than that of North America. In South America, the Andean region was the focus of agricultural production, continuing the processes begun 8000 years before with the first domestications (Section 2.2) and the cotton-using cultures of 5000 years ago (Section 3.1.4). In the Middle Horizon period of Peruvian prehistory (see Table 3.1) the state of Tiwanaku achieved pre-eminence on the edge of Lake Titicaca just inside the border of what is now Bolivia (see Figure 3.3). This state, which began its rise to power around AD 300, lasted for about 700 years until its maize-producing agricultural systems collapsed in the wake of climatic change (Ortloff and Kolata, 1993). The rate and magnitude of climatic change has been deduced from an ice core extracted from the Quelccaya ice-cap (Thompson and Moseley-Thompson, 1989) in the Cordillera Oriental mountain range in southern Peru (Figure 3.3). In its heyday, the Tiwanaku state was well placed to exploit a range of ecological zones in the plateau, the altiplani, in which Lake Titicaca is located and as far away as the Moquegua Valley of south coastal Peru (Figure 3.3). Control of the latter, Ortloff and Kolata suggest, was necessary to control lower-altitude arable land, locally known as yungas zones. They also state that 'the economic key to the functioning of the state was an integrated agricultural core area in the Lake Titicaca basin that operated by means of regional manipulation of land, labour and most especially, water resources.'

Kolata (1991) has described in detail many of the ways in which water resources were efficiently exploited. In the vicinity of Tiwanaku City itself, raised fields relied on ground, river and springwater. It is likely that a food surplus was produced and distributed throughout the empire. The success of agriculture, reliant as it was on the careful management of water resources, was due to accumulated knowledge involving responses to severe flooding–drought cycles that were typical of the altiplano region and the development of effective hydraulic engineering strategies. Despite this food security based on maize, quinoa, legumes or potatoes, the Tiwanaku state began to disintegrate around AD 800–1100.

Palaeoclimatic data from the Quelccaya ice-cap and palaeoecological data from lake sediments all point to the onset of chronic drought (Ortloff and Kolata, 1993). This would have had major repercussions in an environment that was already periodically subject to drought stress. The impact of this climatic change was regional and it affected above-ground and below-ground water resources to such an extent that hydraulic engineering was inadequate to mitigate the problem. Ortloff and Kolata also believe that the climatic change (possibly related to the Little Ice Age discussed by Grove, 1988) had similar consequences for cultures elsewhere in the Andes. The Chimu state, which developed along the north of Peru, particularly in the Moche and Chicama valleys (see Figure 3.3), is one such example. Like the people of Tiwanaku, Chimus were water-resource engineers, building a large canal to transport water from the Chicama River to the field systems of the Moche Valley that were experiencing severe water shortages. Other field evidence indicates that tectonic uplift may also have disrupted water supplies. Unlike Tiwanaku, however, Chimu expanded its territory north and instituted improved water resource management so that it survived. By about 1400, the central Andes had become a centre of considerable agricultural biodiversity. According to Zimmerer (1993), a dual production system developed under Inca rule with the greatest biodiversity being concentrated in the subsistent mode; the surplus-orientated mode, in contrast, relied on only a few crops. The most important crops of the latter mode were maize and potatoes but only a few of the many cultivars available were grown. Both major and minor crops were grown in the subsistent mode, including a wide range of potato and maize cultivars. Examples of minor crops include amaranth, ulluco and quinoa.

In Mesoamerica there is some evidence (Murdy, 1990) that late Mayan civilisation collapsed because of its inability to cope with soil erosion brought about by rapid population growth in the AD 400–600 period. However, pre-Columbian landscapes were dominated by the Aztec empire, the centre of which was in the basin of Mexico. Social, economic and agricultural infrastructures were well established long before the advent of the Spanish. Rainfed cultivation, known as *temporal*, was the most common type of agriculture throughout Mesoamerica with adaptations according to relief. In the coastal plain adjacent to the Gulf of Mexico and the piedmont region, Whitmore and Turner (1992) state that the landscape comprised a patchwork or mosaic of different cultivation types with scattered forest and scrub vegetation. Apparently, well-drained lands were cultivated during the wet season whilst lands subject to flooding were worked during the dry season. Terracing on slopes was commonplace while drainage ditches and canals allowed wetlands to be cultivated. Orchard–garden cultivation was also important. This provided vegetables, herbs and fruit though the region's agriculture remained dominated by maize cultivation. Other crops grown included beans, squash, maguey and tubers. Even the comparatively steep slopes of the Sierra Madre Oriental, the mountains separating the Gulf Coast from the Mesa Central with its volcanic basins and peaks, were cultivated using terrace systems.

The agriculture of the Mesa Central was intensive in order to produce sufficient food for the large populations of its many city states. Agriculture, Whitmore and Turner (1992) report, was not only the basis of trade and commerce but was also the source of tribute that the Aztec hierarchy demanded from its people. The higher peaks were left forest covered, from which fuel and wood products were obtained. The lower slopes and basins were each adapted to maximise productivity. Terraces were constructed on the slopes for rainfed agriculture; some irrigation was practised, especially at slope bottoms. Poor drainage in the basins themselves gave rise to wetland cultivation. Even today in Mexico City there are canals constructed by the Aztecs. They represent the remnants of a once more widespread network that linked the numerous islands in a large lake on which the city was built. On the slopes, the major crops were maize, beans and squash. In the wetlands, floating gardens consisting of an island, or field, anchored by trees at its edge, were common and highly productive. Whitmore and Turner indicate that the system was complex and reliant on hydraulic engineering to control water levels. Even the mud dredged from the canals was used as fertiliser. A great variety of crops were produced including such specialities as avocado, agave, Mexican hawthorn, various fruits and the cactus *nopal de grana* which acts as a host for the cochineal-producing insect. The Spanish invaders found an intensive, highly organised and efficient food-production system.

3.2.3 China

The first unification of China occurred around 220 BC when the construction of the Great Wall was begun. Kaichen (1991) states that agricultural production was still concentrated in the middle and lower reaches of the Yellow River. Soon after, drainage of some of China's southern wetlands began but only at a local level. As Table 3.4 shows, China's population was nearly 60 million by the year AD 2. By this time the iron plough had been adopted, which allowed more land to be cultivated (Section 3.1.5). By AD 1000, land reclamation on a large scale began in southern

Table 3.4 Changes in population and the amount of cultivated land in China (data from Kaichen, 1991)

Year	Population (10^6)	Approx. amount of cultivated land (ha 10^6)	Amount of cultivated land per capita ha
1982	1015.14	99.099	0.10
1949	541.67	98.370	0.18
1841	413.45	45.480	0.11
1766	208.09	52.020	0.25
1708	103.78	38.400	0.37
1393	60.54	56.910	0.94
755	52.91	95.770	1.81
2	59.59	38.622	0.65

China where rice-cropping techniques were improving and animal power was used to improve drainage. Increasing the amount of cultivated land was necessary because of high population growth. As a result marshland in river valleys was reclaimed; hillsides were terraced which, together with earth dykes, were also constructed to combat soil erosion and to conserve water.

By AD 1400 further changes were taking place to improve productivity. According to Kaichen (1991), population pressure made innovation essential. An integrated management system was developed in densely populated areas such as Hangzhou and Guangzhou. This involved grain production (often millet), mulberry trees on which to raise silkworms, and fish farming. China's conservation-orientated and recycling attitudes that are in evidence today thus have a long history. In this system, earth excavated to produce a fish pond was organised into piles around the pond in order to provide a base for the planting of mulberry trees. The silkworm waste was then used to feed the fish and sludge from the pond was used as fertiliser. Inlaid cropping, whereby two crops with different overlapping growing periods are grown together, was practised. This maximised both the trapping of solar energy and the use of the land. In north China, the practice of planting three crops in two years began. Chao (1986) states that the crops were millet, wheat and soy bean. In south China, two crops per year were obtained. Rice was planted in the summer with wheat, legumes or vegetables being planted in the winter. Double rice cropping was also practised in south China.

According to Chao (1986), the introduction of a new variety of rice during the Northern Song dynasty (AD 960–1126) represented a major agricultural improvement. The Emperor, Chen-tsung, brought a substantial quantity of champa rice from the coastal region of Fukien (adjacent to the Straits of Taiwan). This was a much higher yielding rice than indigenous Chinese types; it was also more drought resistant and earlier maturing. Its introduction represented something of an agricultural revolution since it was highly productive and the earlier maturation meant that champa rice could facilitate the production of two crops per year in areas that hitherto could only produce one crop. It also allowed the northward extension of crop growing. Thus rice became a staple crop, surpassing millet and wheat as the most important cereal. Further changes in Chinese agriculture took place in the sixteenth century as new crops were introduced from the Americas.

3.3 AGRICULTURAL CHANGES BETWEEN 1500 AND THE FIRST WORLD WAR

The period 1500–1914 was one of tremendous change in world agriculture. In Europe, and the UK in particular, there was a so-called Agricultural Revolution though there is little agreement about when it began. The increased production resulting from these innovations provided the necessary food resource that supported population increase. This in turn provided the essential labour for the Industrial Revolution. Agriculture itself also provided raw materials for embryonic food-processing industries. By the middle of the eighteenth century the fuel-powered urban-industrial ecosystems (cf. Odum, 1975; see Section 1.4 and Figure 1.7) were coming into existence. This development represented the onset of a new era in people/environment relationships as people became increasingly divorced from the land and food production. It was also a period of major changes in fuel energy relationships as fossil-fuel use began on a large scale. Eventually, such changes were to lead to the industrialisation of agriculture.

The era of European exploration, which began in the fifteenth century and culminated in the emigrations of the years between 1850 and 1920, also had major implications for agriculture. Never before, nor since, have so many crops and livestock been transferred from their centres of origin to new lands where agricultural systems were transformed in consequence. In Europe, the period 1500–1914 was one of deforestation as agriculture spread. The alteration of the natural vegetation cover was as severe as that which is presently occurring in the tropical forests of developing countries. In the newly discovered lands, systems of land tenure and ownership, along with produce distribution networks, were altered to suit the colonialists. Much agricultural produce, as well as other resources such as precious metals, was often annexed for the homeland.

3.3.1 The United Kingdom

According to Thirsk (1987), many of the innovations associated with the Agricultural Revolution of 1750–1850 had been underway since 1500. Certainly, Clark (1991) has shown that grain yields were beginning to rise by 1550, increasing by about 50 per cent over the subsequent two centuries. The steepest rise, however, occurred between 1750 and 1850 when yields increased from *c.* 85 million bushels to *c.* 200 million bushels. This was also a time of rapidly increasing population. Overall, between 1550 and 1884 grain yields per acre in England increased by approximately 150 per cent.

Apart from the debate about when it occurred, there is also some dissension about what exactly contributed to the Agricultural Revolution. Allen (1991), for example, advocates that the upturn in productivity was due to what he describes as a yeoman's agricultural revolution. This involved the efforts of small farmers who worked unenclosed fields, and was not, as is often concluded, the result of later enclosure and incorporation of small farms into large capital farms. The latter probably made the working of the land more efficient but this change, dubbed by Allen as the landlords' revolution, was only a contributory rather than dominating factor in the increase in agricultural production. Whatever the reasons, there is a considerable body of evidence to indicate that both grain yields and labour productivity increased by about

Table 3.5 Wheat and barley yields 1600–1800 in selected areas (data from Allen, 1991; Clark, 1991; Glennie, 1991)

Approximate date	Norfolk/Suffolk	Lincs	Herts	Oxon	Hants	Midlands
Wheat yield (bushels per acre)						
1600	11.4	11.8	9.7	15.0	11.1	N/A
1700	17.6	15.6	14.6	21.6	10.5	N/A
1800	22.7	24.3	20.7	21.9	21.5	20.7 (Av)
Barley yield (bushels per acre)						
1600	N/A	N/A	15.5	14.5	15.4	N/A
1700	N/A	N/A	28.0	18.3	16.9	N/A
1800	N/A	N/A	36.0	30.0	30.0	N/A

100 per cent between 1500 and 1900 (Allen, 1991). In north-eastern Norfolk, for example, which was a high producing region, wheat yields of 20 bushels per acre were achieved in contrast to 10 bushels per acre in low producing areas. According to Allen, the latter is equivalent to the yields achieved by traditional farming practices in parts of the developing world today. By 1800, however, high yields of 20–22 bushels per acre were being constantly achieved. The yields of other grains such as barley also increased. This trend is illustrated in Table 3.5.

Such yields were also common in other parts of Europe but only in England was productivity per labourer so high. It was between 40 and 50 per cent higher than elsewhere in north-west Europe which itself was a higher grain producer than eastern or southern Europe. Yelling (1990) suggests that several factors were involved in this transformation of English agriculture. These were enclosure (as defined above), landholding or land ownership, markets and land use and agricultural production. The period 1500–1900 witnessed substantial changes in all of these factors and it is thus difficult to isolate the role of any one factor as the dominant instigator of agricultural change. In relation to enclosure, this was predominant in the seventeenth century and is described by Allen (1991) as 'the amalgamation of peasant holdings into large capitalist farms'. This, in theory, increased efficiency so that productivity per unit of labour increased. There is clearly much evidence for the latter (see above) though there is disagreement as to whether employment on farms was increased or decreased as a result. The changes in land ownership and control can, Yelling (1990) believes, be divided into two main eras. The early period, *c.* 1500–1650, was when the lands of the Church and Crown were dispersed to the benefit of the 'yeoman' class, a bucolic upper middle class or gentry. By the 1700s the creation of large estates through enclosure had occurred, with landlords exercising a high degree of control over their lands and tenants. Capitalism with its inherent profit-making objectives may have been a major force in increasing productivity though this could only have been achieved within the constraints imposed by the environment and the degree of scientific knowledge.

According to Wrigley and Schofield (1988), the population of England increased from just under 3 million in 1541 to just over 6 million in 1751, with the most rapid increase occurring between 1601 and 1651. This undoubtedly provided an impetus for agricultural improvement but equally it could have been a response to increased productivity. Related to this is the developing infrastructure that linked markets and producers and which rendered the food-producing system more commercial. This may have spurred technological innovation (Yelland, 1990) and the expanding towns provided concentrated markets around which agricultural enterprises developed to provide the required products.

There were also changes in agricultural practices. Amongst the most notable of these was the inception of the Norfolk four-course rotation and other rotations. This involved wheat, turnips, barley and clover, which were grown in succession on an annual basis. In the fifteenth century, turnips were introduced into England from Holland where they were grown as a garden vegetable. When adopted as a field crop they provided animal fodder and could be produced on relatively poor soils. Most importantly, the combination of clover (a legume and thus a habitat for nitrogen-fixing bacteria) and turnips allowed the fallow period to be reduced whilst actually increasing the output of animal fodder. As increased numbers of livestock could be kept, the supply of manure, a major source of nutrients to replenish soils, increased. Inevitably, this and other similar rotations, contributed to the increased productivity of the period in question but, as Overton (1991) indicates, the relationship is far from straightforward. The importance of all of these factors in agricultural innovation simply reinforces the comments made in Section 1.1 and Figure 1.1 that agriculture is a manifestation of a wide range of cultural and environmental factors. Elsewhere in Europe agricultural productivity was also increasing but it was not quite as substantial as that of England. Hoffman (1991), for example, reports that periods of warfare and excessive taxation caused periodic declines in productivity in the Paris basin. He states, 'In the end, it took them [French farmers] three centuries to accomplish what the English did in two.' Political and fiscal stability must, therefore, also be added to the factors influencing agriculture. However, throughout Europe the improved food supply facilitated the continued expansion of population and underpinned the growing industrialisation that, in turn, prompted further agricultural innovations.

Between 1800 and 1914 cultivation increased in an attempt to keep pace with population increase. It encroached into marginal areas and by 1900 there was little land left that could be cultivated. By this time, however, the colonies and former colonies (see below) were expanding their agriculture at a considerable rate with Europe being the prime market on the one hand and the major supplier of migrants on the other hand. In England, and Europe generally, improvements to the land were being undertaken to increase productivity. The story was not always one of success. The famine in Ireland in 1845–1847, caused by the late blight fungus (*Phytophthera infestans*) attacking potato crops, is a case in point (Section 2.3.5). By this time, however, the character of European farming systems had changed significantly. In particular, the range of crops being cultivated had increased substantially due mainly to introduced species from the Americas. Many species were also introduced into Africa and Asia. The case of the potato has been discussed in Section 2.3.5 and elsewhere in Chapter 2.

McNeill (1991) has examined the reasons why and the mechanisms whereby crop plants from the New World became accepted, and often the mainstays of European agriculture. He suggests that the most important reason why species like maize and potatoes were accepted was because they produced more calories (energy) per unit of land area than indigenous crops. Thus they were more efficient converters of solar energy than crops like wheat and barley. McNeill maintains that potatoes were widely adopted across the North European plain because they provided a better source of calories than rye, the only grain that reliably ripened in the short, wet summers. The resulting increased productivity accommodated population growth. In addition, a potato crop could be grown on the fallow left between rye crops. Nor was storage essential as it was for grain because they could be harvested for consumption straight from the field. Potatoes rapidly became the staple food for the poor, providing sustenance for the majority of the labour force that was essential for the progress of industrialisation. McNeill (1991) states,

> If one looks for a measure of the importance of American food crops in Old World history, this is it. The surge of population and spread of industrialization in northern Europe, with resulting shifts of power since 1750, simply could not have followed their actual course without the nourishment provided by expanding fields of potatoes. No other single American crop played such a decisive role on the World stage; but maize in southern Europe and in Africa, together with sweet potatoes in China, also transformed the lives of millions of people.

The mechanisation of agriculture, which began in the 1800s with the growth of the iron industry, also contributed to increased productivity. New forms of enriching the nutrient content of soils were employed; from the 1840s chemical fertilisers came into use through the import of guano and nitrate from South America. Liming and marling became more widespread and even pulverised slag from iron works and ashes from coal were used to increase soil fertility. Following the Napoleonic Wars (1793–1815), these practices, plus the adoption of the Norfolk four-course and other rotations as well as land-drainage schemes, began the industrialisation of agriculture. Hitherto, farms had been relatively self-sufficient, relying on manure and nutrient recycling, but by the 1840s there were many inputs from outside the farm (or other farms), e.g. maize-seed, cotton-seed and cakes for animal feed. This became known as 'high farming' (Grigg, 1989).

The decline in agricultural prices from the 1870s to the 1930s, however, ensured that further changes in agriculture occurred and that 'high farming' collapsed. In particular, wheat and wool prices declined by up to 50 per cent; all other prices of agricultural products declined by between 20 and 40 per cent (Grigg, 1989). This had notable repercussions. It led to a major decline in fodder crops, peas and beans but an increase in livestock, due partially to a maintained demand for secondary products like milk, butter and cheese, a change to hay for fodder and the import of cheap grain as a fodder crop. The last of these also had implications for British agriculture. Imported grain was cheaper than home-produced grain and often of preferred varieties for brewing and bread-making. In consequence the cereal acreage in England and Wales fell by 42 per cent between 1875 and 1938, with a slight upturn during World War I (Grigg, 1989). Moreover, there were major changes in the pattern of meat consumption in Britain as a result of imported meat. According to Walton (1990), in the period 1851–1860 only 4 per cent of meat consumed was imported. By 1885–1889 this had increased to 27 per cent and a huge 42 per cent by the outbreak of World War I. All the advantages and innovations of the earlier 200 years had not given British agriculture a secure basis. As discussed in Section 3.4.1, significant changes occurred again with the advent of World War II and after 1974 when Britain joined the European Community.

3.3.2 The Americas

The period 1500–1914 witnessed tremendous changes in the agricultural systems of the Americas. In essence, the indigenous systems (see Section 3.2.2) were modified or displaced by agricultural systems that were introduced from the colonising nations of Europe. Thus began the export of biomass from the New to the Old World and although the USA and Canada have subsequently developed into major food-producing nations, the colonial legacy in Central and South America is still evident. Many of these nations rely on Europe or North America to provide markets for much of their agricultural produce. (The same is true of many African nations.) Today, the difference is that the produce in the developed world is shipped to earn foreign currency which is often necessary to repay debts to banks; in the developed world in the past, produce (and other commodities, e.g. wood and minerals) was obtained via slave labour and exported to pay dues to the 'mother country'. As Assadourian (1992) comments,

> *Población* denoted the first establishment of a European presence in the New World, essentially as a consequence of the conquest itself. This movement of people was to intensify notably during the second half of the sixteenth century, with a significant change in its nature: instead of *conquistadors*, the migrants became settlers whose function was to accelerate the implantation of a European colony.

Amongst the most important factors influencing Meso and Central American agriculture following colonisation by the Spanish (and Portuguese) was the system of tribute demanded by the overlords and the annexation of land by colonists. Both required an increase in productivity. Much of this was achieved through slave or feudal labour though diseases such as smallpox decimated native populations. Slaves were imported from Africa to the West Indian islands, Brazil and eventually into North America. In Meso and South America (chiefly Peru) an *encomienda* system was established. This consisted of a group of villages controlled by a Spaniard and for whom natives provided labour whilst he afforded them protection. Produce from the colonial lands, especially sugar introduced into the West Indies from Spain via the Canary Islands and later introduced into Brazil, was a major export. The introduction of this crop, which was brought to the New World on Columbus's second voyage in 1493, transformed the economies of the Caribbean Islands and Brazil. According to Mintz (1991), the first crop from the New World was sent to Spain as early as 1516. Even today, it remains a mainstay of Latin American exports.

Many other crops were introduced into Meso and Central America, as Crosby (1991) has discussed. These included wheat, other grains such as barley and millet, a variety of fruit crops including peach, pear, grapes and citrus fruits and many different sorts of vegetables. These displaced native crops to a certain extent and provided the colonial masters with familiar produce. Crosby also points out that with these crops came weeds. Clover in particular was successful in its new abode and quickly became a dominant species in pastures. Whilst crops made a considerable impact, imported livestock had an even greater effect on the New World's agricultural systems. Pigs, cattle and horses made their mark quickly; chickens, sheep and goats became important economically in the longer term. Pigs with their omnivorous proclivity adapted rapidly to the new conditions that ranged from the woodlands of New England to the grasslands of Argentina. Cattle also adapted well to meadow and savanna lands, and horses thrived once they reached the temperate grasslands and pampas of the Americas. Thus began cattle ranching, an activity that still dominates parts of North and South America. Not only were indigenous agricultural systems influenced by these introduced crops and livestock but so too were the native fauna and flora of the natural ecosystems. This is a story repeated in colonies elsewhere, notably Australia (e.g. Heathcote, 1988) and New Zealand.

In parts of Mesoamerica such as the Yucatán Peninsula a practice known as *milpa* was commonplace. This involved a form of shifting cultivation whereby plots of forest were burnt and planted with crops. In the following year about 60 per cent was cleared again and planted while a new area was cleared for the first time. Thereafter the land was left fallow. According to Ewell and Merrill-Sands (1987), *milpa* represents a system of fallow rotation. The major crops grown were maize, beans, cotton and squash with several minor crops including sweet potatoes, cassava, pineapple, sugar cane and tobacco. For example, maize plants will provide support for beans, which creep up their stems, and squash plants attach themselves to the lower leaves. The beans, via nitrogen-fixing bacteria, replace the nitrogen extracted from the soil by maize. In some places in the Yucatán there is evidence for irrigation. In colonial times the natives paid their tribute in maize and cotton whilst the colonists lived in towns and did not engage in food production. By *c.* 1900, *haciendas*, or landed estates, had developed throughout Mexico mainly to organise produce, notably hides, for export. The peasant farming activity, however, remained very important to provide food for the cities and the *haciendas* that were engaged mainly in livestock rearing. The two, somewhat complementary activities Ewell and Merrill-Sands describe as being symbiotic. By 1821, when Mexico gained its independence from Spain, landowners old and new sought to expand their holdings and this brought direct conflict with the *milpa* system. The disadvantaged Maya reacted in a rebellion known as the Caste War in 1847. Agricultural activities fell into decline in the Yucatán until the 1880s when plantation agriculture based on sisal production was initiated. This activity was a major provider of raw material for the binding twine that was used in the mechanical reaper invented in 1875. This machine itself represented a major advance in the mechanisation of agriculture in North America. As well as sisal, sugar cane, maize and cattle were produced in a *hacienda*-based system. Profits were made as a result of debt peonage, a form of enforced cheap labour. This continued until the Mexican Revolution in 1911 by which time many Latin American countries had well-developed agricultural systems that underpinned political stability and a degree of prosperity through export earnings.

In North America, the influence of the colonial powers on agricultural systems was also substantial. As in Mesoamerica and South America, indigenous systems were disrupted or displaced. Apart from changes to crop complexes due to introduced crops, animal husbandry, particularly the establishment of ranching, had profound effects on agriculture and the environment. This has been discussed by Bennett and Hoffmann (1991), who point out that ranching was introduced to the New World from Spain where it developed successfully in the medieval period. When the Spanish colonists entered the New World they found three extensive natural grassland areas for which their established Castilian cattle-rearing was ideally suited. These were the pampas of Argentina, the llanos of Venezuela and the Great Plains of North America. According to Bennett and Hoffmann, cattle were first brought from Spain to the Caribbean Islands from where they were imported to Mexico in 1521, and in 1549 cattle were imported directly from Spain to Brazil where Jesuits, also colonists from Spain, became cattle breeders. Two centuries later the Franciscans established themselves in the same role in North America. From Mexico, cattle were exported to Venezuela, Peru and North America. Between 1659 and 1682 the first ranches were established in what is now Texas (Bennett and Hoffmann, 1991) and by the 1830s cattle were installed in Oregon. Many herds were associated with missions.

Until the early 1900s the ranges were open and long cattle drives were common. Thereafter ranches were fenced and the grazing territories of cattle were curtailed. The southern states of the USA felt the impact of the Mexican Revolution (see above), however, as political instability ensued. Ranchers moved their stock to the USA where there was a ready market that was growing rapidly against the backdrop of World War I. Cattle ranching by this time was big business; the improvement of stock by breeding programmes, facilitated by fencing and range restriction, meant that the quality of meat was improving. The beginnings of the modern ranching industry were in place.

In addition to cattle ranching, the advent of European colonists post-1500 appreciably altered crop production, though a complete appraisal of this topic is beyond the scope of this text. According to Hart (1991), the early focus of agriculture was in the fertile limestone plains of south-eastern Pennsylvannia from whence European and/or modified European practices spread beyond the Appalachian frontier. In this area, soils and climate conspired to create a favourable environment for European colonists to mimic the food production systems of their homeland. Wheat was the first commercial crop and was produced in the early 1700s. Within 40 years it had become an export crop for European markets where it competed favourably in terms of price and quality (see Section 3.3.1). Maize, oats, rye, vegetables and orchard fruits were also widely produced for local consumption. After the American Revolution of 1775–1776 some farmers in the region began to change from wheat to beef cattle production as a response to the rapidity of urban growth on the coast. Farmers in areas like the Lancaster plain bought cattle driven from the west (see above), fattened them cheaply on maize and sold them to the markets of cities like Philadelphia, Baltimore and Washington. Farmers in the Eastern Seaboard region also benefitted from the Agricultural Revolution in Europe (see Section 3.1.1), particularly the establishment of new rotations that included clover and turnips. Crop and livestock productivity increased as a result.

From here, the crop assemblages, etc., were introduced into south-western Ohio (Hart, 1991) and thence into the central agricultural region, notably the corn belt. This occupied, as it does now, the Mid-West states from Ohio to Nebraska. The agricultural base comprised a rotation of maize, a small grain such as winter wheat, oats and a leguminous crop of clover and later alfalfa. This was still the case by 1914. The use of maize was particularly advantageous for pioneer farmers because it could be grown successfully between the remains of tree stumps on newly cleared land. The rotation was also good for efficient labour utilisation and soil-nutrient renewal. The demand for meat in the East coast cities, coupled with the relative ease of transporting animals on the hoof, led to the practice of fattening pigs and cattle. Such activities persisted until the 1950s (see Section 3.4.2). At various times in the period 1500–1914, and reflecting the spread of pioneers and railways from the East to the West, the differences in soils and climate that occur throughout North America and the introduction of crops from Europe, and Meso and South America, the crops that are traditionally associated with US and Canadian agriculture became established. The most important of these are wheat, cotton, tobacco, citrus fruits, peanuts and sugar cane.

3.3.3 China

The period 1500–1914 was one of tremendous change in the agricultural systems of China. These, like their counterparts in Europe and the New World, were influenced by introduced crop species, population growth and new techniques. The discovery of the New World and its crops led, by the second half of the sixteenth century, to the introduction of maize, peanut and sweet potato. The last of these especially was widely adopted. All three crops are high yielding and their cultivation helped to accommodate the continually rising Chinese population that forced the intensification of agriculture. Kaichen (1991) states that there were improvements in manuring techniques and rotations in the continued struggle to persuade the land to increase its productivity. The only real advantage that the Chinese had in this battle was the high investment of labour they could afford to make. Then, as now, China was faced with a land area of which only 20 per cent could be realistically cultivated.

McNeill (1991) asserts that the impact of American crop plants was as great as that in Europe. Maize, peanuts and sweet potatoes were all being cultivated by 1600. He states that 'peanuts were praised by an agricultural writer as early as 1538, and in 1594 the governor of Fukien, on the southern coast, touted sweet potatoes as an answer to a widespread failure of other crops in the province that year'. Apparently, China benefitted from the fact that the Spanish occupied the Philippines in 1565 and thereafter trade was established with Mexico. The new crops, adopted prior to this date, as the quotation above suggests, were successful in China because they were less labour intensive than the indigenous rice cultivation. These crops also suited the extant rotations and/or multicropping procedures that Chinese farmers employed.

Each of the three crops were grown in different locations and spread gradually. Maize was popular in the south-west where it was suited to lands newly cleared of forests just as it was in similarly cleared areas in North America (Section 3.3.2). Sweet potatoes were compatible with the thinner soils on the hillsides in rice-growing areas and did not require paddy-field construction on difficult terrain. McNeill (1991) reports that the increased food supply fuelled population growth but people remained rurally based even by 1900. This was in contrast to Europe and North America where urbanisation was intensifying in parallel with industrialisation. The discontent that epitomised the struggle between food production and population increase generated political revolt such as the Taiping Rebellion of 1850–1864. McNeill suggests that this more or less permanent food shortage caused the increased spread of these

American crops. China's agriculture remained traditional, largely subsistent and labour intensive until after World War I.

3.3.4 Russia

Although no mention has so far been made of agriculture in Russia in this text, it is appropriate now to raise the issue in view of the changes that occurred as the Soviet Union was formed in the aftermath of World War I (Section 3.4.5) and its subsequent rise as a world power. The brief comments that follow provide some information on the state of Russian agriculture prior to the Russian Revolution.

Leonard (1989) states that there was an escalation of grain prices in Russia in the 1700s. This, in turn, prompted an expansion of cereal cultivation, especially that of wheat, in southern Russia where the black earth soils were particularly fertile. As happened elsewhere, the improved food supply was accompanied by a population increase; between 1719 and 1859 the population tripled. Both demesnes (see Section 3.2.1) and land on large estates allotted to serfs (allotment land) contributed to this increase in productivity. Leonard opines that much of the increase in productivity was the result of the efforts of households rather than large landowners, i.e. that it was improvement literally at the grass-roots level. In the central and northern provinces of Russia soils were not as fertile as in the south. Rye was the major crop but income from it was too low for labourers to pay their *obrok* (payment to the landlord) and support their families. As a result people developed cottage industries, producing wooden utensils and artifacts and/or worked in towns during the winter months.

In the early 1800s increasing amounts of produce were being sent to market, and southern Russia commenced the export of grain which was transported via the Black Sea. Riasanovsky (1993) reports that the reliance on serf labour at this time was an obstacle to agricultural improvement. Landlords preferred payment rather than work for their tribute or rent and, as stated above, some of this was derived from enterprises other than agriculture. One result of this was the hire of free labour which became even more commonplace after the emancipation of serfs in 1861. Agricultural wages increased and some innovations occurred, including the introduction of fertilisers and some mechanisation. In addition to intensification, crop diversity increased. Rye and wheat continued to be staple crops and to occupy a substantial area but potato, sugar beet and vine cultivation were introduced. Potatoes in particular rapidly became a staple crop (see Section 3.2.2) and the introduction of a new improved breed of sheep led to the production of fine wool. In the 40 years between 1812 and 1853, the number of sheep rose from 150 000 to 9 million (Riasanovsky, 1993).

Nevertheless a crisis arose in the period 1861–1914 due to a substantial increase in population from 73 million in 1861 to more than 125 million in 1897 and 170 million in 1917 (quoted in Riasanovsky, 1993). Pressure on the land increased as a result, giving rise to a situation not unlike that in many developing countries today where high rates of population increase are giving rise to a landless poor. In Russia, land prices continued to rise steeply and although emancipated serfs (peasants) actually purchased land, the size of their allotments, along with those of tenant peasants, was dwindling markedly. These small land holdings in tandem with poor production methods, meant that families could not grow sufficient food to support themselves. In consequence, a series of agrarian revolts occurred toward the close of the nineteenth century, which culminated in the 1905 revolution. Russia had failed to capitalise on agrarian improvements and the onset of industrialisation as had other European nations. According to Roberts (1992), there were signs of improvements in the subsequent decade as co-operative movements were encouraged in order to make food production and dissemination more efficient than hitherto. The government encouraged migration to distant lands such as Siberia, so relieving pressure on the Russian heartland. The amount of land under cultivation also increased and state and Tsarian lands were sold off. The land reforms of Prime Minister Stolypin (1906, 1910 and 1911), designed to create a land-owning peasant class, and the consolidation of strips into workable lots contributed to these improvements though they also created antagonism between the 'haves' and 'have nots'. This was a crucial period in the agrarian history of Russia, and indeed in its political history, but it would appear that agriculture did not provide Russia with the advantages enjoyed elsewhere in Europe. Possibly, these other nations benefitted more by activity in their colonies and allied lands, such as the USA and Canada, then they did from agrarian policies at home.

3.4 AGRICULTURAL CHANGES FROM 1914 TO THE PRESENT

For the world as a whole, this 80-year period has been remarkable in terms of changes in both agricultural practices and policies. It has also been a time of change in rates of population growth and a period during which there has been considerable polarisation between the developed and the developing world. A synthesis of the events and transformations that have occurred during this time would fill several lengthy tomes. Certainly, the presentation in this section represents only a fraction of the information available.

Eclecticism dictates that emphasis needs to be placed on certain attributes and developments that have shaped modern agriculture. It could be argued that one of the most important changes of this period relates to the increasing political control of agriculture which is manifested in protectionist policies in the developed world. The Common Agricultural Policy (CAP) of the European Economic Community is a case in point. Such policies and their changes over time not only shape agricultural systems and their relationships to the markets but also contribute to the shaping of rural landscapes. In the developing countries, agriculture is not only a means of feeding rapidly increasing populations but also a means of generating income, through cash crops, for development. The markets for these crops are mainly in Europe, North America and Japan. Consequently, these nations have become net importers of developing world resources that include food energy, nutrients and the environmental costs of food production such as soil erosion and desertification. Many developing nations continue to produce cash crops for export even when they can barely meet the basic food requirements of their own people.

As well as the political aspect of agricultural systems, science and technology have made a major contribution to the improvement of productivity. The chemical stock derived from World War II biological and chemical weapons research provided a base for the fledgling agrochemicals industry in their search for crop-protection chemicals, notably pesticides. DDT and other organochlorine insecticides had a beneficial impact on productivity as they proved to be efficient reducers of insect populations.

Ecologically, however, they turned out to be detrimental and attention then turned to natural products that could act as crop-protection chemicals, including herbicides and fungicides as well as insecticides. The production and increasing use of artificial fertilisers also had, and continue to have, a major impact on agriculture. So too did the increasing sophistication of mechanisation, which also increased the tendency towards monoculture. All of these factors represent inputs from outside the agricultural systems and the increasing input of fossil-fuel energy. In the developed world, agriculture had become oil-based, though it was not a success everywhere. In the former USSR, for example, agricultural productivity since 1914 has barely kept pace with the needs of the people. Overall, the period around 1950 could be considered a watershed in global agriculture. As a result of the increasing reliance on fossil fuels and the growing inputs from science and technology, agricultural productivity increased without recourse to additional land. The other significant input of science to agriculture was plant and animal breeding which began in earnest in the nineteenth century with the work of Gregor Mendel and others on heredity. On a global basis, improved crop varieties have contributed substantially to increased agricultural productivity to such an extent that since the early 1970s it has been referred to as the 'Green Revolution'. The process of developing improved plants and animals is also being transformed by recent advances in biotechnology and particularly by genetic engineering. In consequence, the practice of agriculture has changed almost beyond recognition from the systems that were initiated 10 000 years ago.

3.4.1 Europe and the United Kingdom

The outbreak of World War I in 1914 found agriculture in the UK in a sorry state. The agricultural depression that began in 1870 was still in evidence. According to Offer (1989), Britain was importing about 80 per cent of its wheat and flour in 1913, mainly because the quality was better and the price lower than home-produced wheat. In many European countries weighty tariffs were imposed to safeguard home producers; in others, foreign imports were welcomed but home production continued. The onset of war, however, created major problems for the importation of goods. As a result, the prices of all agricultural produce escalated. By 1917 the British Government was intervening to guarantee cereal prices. For a while agriculture came out of the doldrums but it was a short-lived improvement. When the guarantee was rescinded in 1921 prices once again fell dramatically (Cooper, 1986). The impact was so great that the repeal of the Agriculture Act became known as the 'Great Betrayal'. Grain producers were hardest hit during the ensuing years of the depression, livestock producers less so whilst the price of market garden produce remained steady.

Grigg (1989) states that 'Protection for farming came in by the back door. In 1924 farmers were exempted from local rates and in 1925 British sugar production was granted a subsidy. But in 1932 more overt protection was offered.' This was the Wheat Act which provided wheat farmers with a deficiency payment. Ostensibly, it guaranteed a set price with the government making up any shortfall between the notional price and the price obtained on the open market by the farmer. The money to provide the deficiency payment was raised through a tax on flour. Another measure taken to protect British agriculture was the imposition of a system of import quotas on foreign wheat exporters. Two years later, a subsidy on fat cattle was provided and in 1937 a subsidy on the use of lime was established. Various Agricultural Marketing Acts were also passed to improve the marketing of milk, potatoes and hops. Grigg (1989) indicates that the cost of these measures was comparatively low, amounting to about 5 per cent of the value of the gross output of 1937–1938.

The outbreak of World War II in 1939 was a major turning point in the history of British and European agriculture. Food was scarce, re-emphasising the role of imports, and was rationed. In consequence, the government became the single purchaser of agricultural produce. Prices and subsidies rose and the 'Dig for Victory' campaign was inaugurated. Rationing, the enforced isolation from foreign food sources, as well as growing inequities between urban and rural populations, set the scene for the Agricultural Act of 1947. Politicians were unwilling to accept the UK's vulnerability to food scarcity in times of war. This Agricultural Act, thus, marked the beginning of intense protectionism that was later heightened by the UK joining the European Economic Community in 1973. This protectionism not only controlled the practice and distribution of agriculture but also fashioned the rural landscape of the UK and Europe (reviewed in Mannion, 1991). The availability of grants for land improvement, for example, has resulted in the destruction of hedgerows, woodlands, wetlands, moorlands, heathlands and downlands as well as the regrettable polarisation of agriculture and conservation.

In March 1957, the EC (now EU) was born when the representatives of six European nations signed the Treaty of Rome. These countries were France, Germany (the former FRG), Belgium, the Netherlands, Italy and Luxembourg. One of the aims of the EU was 'The adoption of a common policy in the sphere of agriculture' [Article 2 of the EU Treaty]. This aim was formalised as the Common Agricultural Policy (CAP) which was in operation by the mid 1960s. One responsibility of the CAP was to ensure that prices of agricultural produce in the EU were above world prices. It operates by the annual setting of target and intervention prices. If the price achieved falls below the target price to the intervention price, government agents are required to purchase surplus stock. The farmer is thus guaranteed the intervention price, surpluses are stored for later release and foreign produce, from places like the USA and Australia, is prevented by levies from flooding the market and lowering prices. The surpluses are the infamous wine lakes and butter mountains, etc., that cost the tax-payer as much to store as they do to produce, which may be exported with the aid of a further subsidy or distributed as food aid to developing countries. Overproduction became commonplace due to guaranteed prices and most efforts since 1977 to control this have been thwarted. Such efforts have included the imposition of milk quotas and a refusal to guarantee intervention prices for cereal produce over a given quantity (Avery, 1985).

This regulation of EU food commodities has tended to favour arable farmers. The combination of grants for land improvement and guaranteed prices for commodities like wheat and oilseed rape has ensured the loss of natural and semi-natural habitats and seriously questioned the role of farmers as 'custodians of the countryside'. The Least Favoured Area Policy initiated in 1975 has offered support to upland farmers in the least productive areas where sheep farming is prevalent. Since the 1980s it has been recognised that the CAP needs substantial reformation; this has been the cause of much acrimony amongst the member states as each attempts to safeguard the interests of its own farmers. In the late 1980s, 100 million hectares were classified as agricultural land. Details of farm size, cereal, milk, wheat, etc., production are given in Sasson (1990). Some of these details will be considered in Chapters 5–7.

Another turning point in the history of European agriculture was the attempt to introduce agricultural extensification in 1988. This scheme, implemented via an amendment to the 1985 EU Structures Regulations, came into effect on 1st July 1988 (Potter, 1988). It required each member state to construct and implement plans to bring about the extensification of crop and livestock production. In the UK, for example, this has been the responsibility of the Ministry of Agriculture, Fisheries and Food (MAFF) whose plans were outlined in MAFF (1987). Under the scheme, which dealt first with arable production, farmers are eligible for payments if they reduce the hectareage for cereal production by at least 20 per cent for at least five consecutive years (though there have been several changes in the rules since the scheme's inception so that the actual percentage has varied). This so-called 'set-aside' should not, however, be used to produce other crops that are already being produced in quantities that are 'surplus' to requirements in the EU. The land can be put to a variety of non-agricultural uses such as forestry or recreation or simply left fallow. There are similar policies operating in the USA (see Section 3.4.2). More recently, further reforms have been proposed (Ansell and Tranter, 1992). In May 1992, for example, EU ministers agreed to reduce or remove support for arable crops and to introduce an additional set-aside scheme requiring a duration of 20 years. Despite these measures, however, there has been little reduction in productivity as yield increases have compensated for the loss of productivity from set-aside land. Land use on set-aside land has focused mainly on equestrian activities; the expansion of forestry has not been very great. In terms of cost, Ansell and Tranter report that set-aside appears to have been less cost-effective than price support in 1988/89. What the longer term implications are for reducing the cost of CAP and for the provision of wildlife habitats remains to be seen. None of the interested parties, notably farmers, tax-payers and conservationists, appear to have been satisfied by set-aside policies.

Overall, the post-1950 changes in UK and European agriculture created a substantial food surplus. In so doing, fossil fuels were used to increase the production of food energy and agriculture was 'industrialised'. The control by government, via the protection of prices, also created visible landscape changes. Through the removal of hedges, to gain land and to enlarge fields, mechanisation was more effective than hitherto; the increasing use of artificial fertilisers not only contributed to the improvement of productivity but masked problems like soil erosion and created others, notably cultural eutrophication. These issues will be discussed in Chapter 8. In contrast with the crop rotations and complexes that had characterised agricultural systems since their inception, agriculture in the post World War II era tended towards monoculture. This was encouraged by mechanisation; machines were designed to deal with a specific crop. It was also encouraged by plant breeding and the availability of seeds of improved species which were cheaper to buy in large rather than small quantities. Crop-protection chemicals, especially those which are target-specific rather than broad-spectrum, also favoured monoculture. In addition, extensification policies have started a trend towards the production of so-called 'speciality crops' that command good prices because they do not add to EU surpluses. Examples include several oil-seed crops, prompted to a certain extent by current attempts to develop biomass fuels. The British and European agriculture of the 1990s is of a very different character to that of the 1890s.

3.4.2 The USA

In the 1830s Cyrus McCormick invented a reaper that was quickly adopted in the USA, notably in the Mid-West where wheat and maize were grown on extensive farms. Much of this produce was exported to Europe (see Section 3.4.1) where, by the outbreak of World War I, agriculture was depressed. By 1929 combine harvesters were being produced on a large scale in the USA where they were rapidly adopted by farmers, as they were in other extensive grain-producing countries such as Australia and Argentina. Productivity increased as a result; this is exemplified by labour requirements to produce wheat from one hectare of land. In the USA, 38 person-hours were necessary in the 1910–1914 period, declining to 10 by the 1955–1958 period when person-hours in England to undertake the same task were 33 (quoted in Grigg, 1992). Not only has labour productivity increased but so too has land productivity. This is illustrated by Table 3.6. Between 1958 and 1980 the total productivity of 17 principle crops rose from 252 million tonnes to 610 million tonnes, representing an increase of 142 per cent (Sasson, 1990). Only 3 per cent more land was cultivated in 1980 than in 1938. As Sasson points out, if yields per hectare had remained constant in that 40-year period, it would have been necessary to cultivate a further 177 million hectares of equally suitable land. The interesting implication of this fact is that the impact of science and technology to increase yields prevented the ploughing up of 177 million hectares and, by inference, all the environmental problems that such cultivation creates, including soil erosion and habitat destruction (Chapters 8 and 9). The application of science and technology, however, has created new problems such as cultural eutrophication and the adverse impacts of crop-protection chemicals (Chapter 8). According to Baltensperger (1993), this shift to high-technology agriculture was the underpinning cause of farm expansion in the Great Plains after 1959.

In common with the situation in Europe, government controls on agriculture shaped the character of food production in North America. Some of the legislation associated with agriculture has been reviewed by Hart (1991). In 1933 the Agricultural Adjustment Act was passed; its aim was to increase prices by reducing production. Its impact was, however, negated somewhat by the outbreak of World War II when US (and Canadian) farmers were encouraged to produce as much food as possible in order to support the war effort. After the war, the Agricultural Trade Development and Assistance Act continued to provide a vehicle for overproduction because it authorised the export of surplus foodstuffs to countries in need at subsidised prices. It was, thus, a form of food aid. Other efforts to use surplus production include the food-stamp programme, which is still in operation and which provides stamps that can be exchanged for food to those on public assistance, and the school-lunch programme. In addition, the Soil Bank programme which operated between 1956 and 1975 sought to initiate soil conservation measures by removing land from cultivation. Hart (1991) states that at its peak this programme removed 58 million acres from cultivation; it was, in effect, a set-aside programme.

Table 3.6 Growth of agricultural production in the USA between 1938 and 1980 (based on data in Sasson, 1990)

Crop	Average yield ($t\ ha^{-1}$)	Total production (thousands of tonnes)
1938 to 1940		
Maize	1.80	64 104
Wheat	0.96	22 453
17 cultivated crops*		252 033
1958 to 1960		
Maize	3.36	99 891
Wheat	1.67	35 883
17 cultivated crops*		391 388
1978 to 1980		
Maize	6.32	185 208
Wheat	2.22	57 016
17 cultivated crops*		610 293

* Maize, wheat, rice, barley, sorghum, oats, rye, soy beans, groundnuts, beans, potatoes, sugar-beet, flax, cotton, tobacco, hay and fodder-maize.

World economic conditions, however, were conducive in the late 1960s and 1970s to increased productivity. Exports to the USSR and China were encouraged whilst Japan's imports of wheat increased and devaluation of the dollar encouraged some developing countries to import basic foodstuffs. By the 1980s many of these markets had declined due to improved home producers and/or unfavourable dollar exchange rates. In particular, European markets had dwindled as EC policies (see Section 3.4.1) protected European farmers and the world recession of the early 1980s meant that many countries were forced to cut imports. As in Europe, food surpluses began to accumulate in the USA. Unlike Europe, food prices on domestic markets fell along with the price of farmland. The resulting vulnerability of farmers is, Hart (1991) suggests, an outcome of increasing specialisation, a situation also characteristic of European agriculture (Section 3.4.1). Just as an ecosystem becomes increasingly vulnerable as species diversity declines with specialisation, agricultural systems become less economically robust and susceptible to collapse when they veer towards monoculture.

In 1981 the Farm Bill was endorsed. This was essentially an interventionist measure whereby specified support was given to farmers, and products were supported in terms of price guarantees. The Payment in Kind programme (PIK) of 1983 sought to provide farmers with crop produce from surplus stocks equivalent to that which they could have grown. Sasson (1990) reports that there were other measures to curb production. These include the Acreage Reduction Program and the Paid Diversion Program. By August 1983 these measures had enabled 37 million hectares of arable land to be taken out of cultivation; this represented about 21 per cent of the total 180 million hectares of arable land. As a result the production of wheat, corn, sorghum, rice and cotton dropped. The expense to the government was huge and the PIK programme was abolished in 1984. Nevertheless, federal aid to agriculture reached \$18 billion in 1985, rising to \$26 billion in 1986. This reflects, as it does in European agriculture, the support of the tax-payer and the failure of market forces to regulate agriculture. It also indirectly reflects the vulnerability of agriculture to external influences like war and famine. In some respects it is surprising that such fragility is what sustains a world population of 5000 billion. With these, and many of the factors discussed in Sections 3.4.1, 3.4.3 and 3.4.4 in perspective, it is not surprising that discussions on the General Agreement on Trade and Tariffs in 1992–1993 have been so rancorous and concurrence so elusive. This underlines the comments made in Section 1.1 and Figure 1.1 and reinforces the concept of agriculture as a go-between that epitomises the people/environment relationship.

3.4.3 China

At the end of World War I China was experiencing acute food shortages. Agricultural production had not kept pace with population increase. Pearl Buck's novel, *The Good Earth*, published in 1931, highlights the lot of the Chinese peasant farmer during the early part of the twentieth century. It also illustrates what profits could be madc by enterprising individuals. After the demise of the imperialist regime, the People's Republic of China was founded in 1949. Thereafter land ownership was forbidden and farmers simply leased their land from the government. Only in 1993 was private ownership being encouraged and so far it extends only to houses or apartments. Since 1949 the Chinese Government has reformed agriculture several times. Figures for the Gross Output Value of Agriculture (GOVA) given in Table 3.7 illustrate the increase in productivity that has occurred since 1951. The category of sideline activities included township enterprises (until 1986). These, in turn, included industrial production at village level or below.

The control of the state is reflected in its setting of targets and was begun in 1979 with the reform known as *baogan daohu*, meaning the full responsibility of rural households. Contracts between the rural communes and the state guarantee fixed prices for crops and animals, quotas of which are set. Quantities produced over and above the target could be kept or sold by the producer. By 1982 this system was widespread in China. The system was successful and the standards of living of rural people rose markedly (Sasson, 1990) except in the north-west where mountain soils are poor and agriculture remains at a subsistence level. The *baogan daohu* was extended to cover products other than crops. This increased the possibilities for employment, and remuneration for goods encouraged enterprise. This has resulted in a slight fall in cropping or grain production (Table 3.7). Rice, maize and wheat are the staple grain crops but barley, millet and sorghum are also grown. Most of this output is consumed by the Chinese people and it has been no mean feat for the Chinese Government to ensure, through its agricultural policies, that its 1115 million people are adequately fed. Industrial crops, which include oil and fibre crops as well as tobacco and tea, occupied 20.29×10^6 ha in 1986 (Ren, 1991). This compares with 144.2×10^6 ha on which grain crops are grown. As a result of the success of the contract programme, albeit at the cost to the government of higher contract prices, a considerable amount of food is now available on the free market. According to Dao *et al.* (1991), the net income of peasants rose by 78 per cent during the 1978–1985 period, though it has slowed since. Today, each household is an independent management unit no longer answerable to the commune.

Table 3.7 The Gross Output Value of Agriculture (GOVA) of China, and its components, 1952–1986 (from Ren, 1991)

Year	GOVA (10^9 yuan)	Constituent components of GOVA: Cropping (%)	Forestry (%)	Animal husbandry (%)	Sideline activities (%)	Fishery (%)
1952	41.70	83.1	0.7	11.5	4.4	0.3
1957	53.67	80.6	1.7	12.9	4.3	0.5
1965	58.96	75.8	2.0	14.0	6.5	1.7
1975	128.50	72.5	2.9	14.0	9.1	1.5
1979	158.43	66.9	2.8	14.0	15.1	1.2
1984	337.70	58.0	4.1	14.2	22.0	1.7
1986	401.30	62.2	5.0	21.8	6.9	4.1

In 1987 the Bumper Harvest Programme was launched. Its aim was to increase productivity through an increased reliance on science and technology (Sasson, 1990). This was in line with the seventh five-year plan of 1986–1990 which sought to double the aggregate value of agricultural and industrial output by 1990 and to quadruple it by the year 2000. Improvements in education, rates of literacy and technical training were all components of this plan as was the policy of opening up the economy to the outside world. Today, China is rapidly developing but improvements in both agriculture and industry have been, and are continuing to be, achieved at great environmental cost as will be discussed in Chapters 8 and 9. There is now growing concern that biotechnology and genetic engineering are being widely applied but that there may be inadequate testing of transgenic organisms before their general use.

3.4.4 Russia and the former USSR

As in China, the post World War I period was a crucial one in the history of Russia. The overthrow of the Tsar Nicholas II brought the establishment of a communist regime and the eventual federation of the USSR which comprised some 15 republics. This finally broke up in 1991. Unlike its success in communist China, agriculture has not been so prosperous in communist Russia or the USSR. In fact, it seems likely that food shortages actually contributed to the unrest that characterised the disintegration of the federation.

According to Pallot (1991), collectivisation has been the mainstay of Soviet agriculture, especially since the 1930s when peasants were forced to give up their lands to collective farms known as *kolkhozi* on which they subsequently worked. In the USSR in 1989, state farms and collective farms together occupied 1293×10^6 hectares, of which 725×10^6 hectares comprised agricultural land (data in Pallot, 1991). Private enterprise had been more or less eliminated although an additional 50×10^6 hectares were brought into cultivation between 1940 and 1987 (Bater, 1989). Agricultural productivity did not keep pace with population growth though it had a substantial deterious effect on the environment. Witness, for example, the demise of the Aral Sea. In addition, agriculture became a major consumer of resources. According to Pallot

(1991), by the mid-1970s about 25 per cent of all new investment was in agriculture. Cereal production, a mainstay of Soviet agriculture, was also irregular due to drought and/or labour shortages. In 1981 and 1984, for example, yields were particularly low and throughout the 1980s cereals were actually imported into the USSR mainly from the USA. According to Sasson (1990), the USSR paid US$$7\times10^9$ for 28×10^6 tonnes of wheat and 27×10^6 tonnes of other cereals in 1984–1985.

Although the Soviet five-year plans always sought to improve agriculture, the goals were never really achieved despite several reorganisations of the administration of the agro-industrial system. Since 1989 the government has encouraged a return to individual farming through contracts between individuals, or groups of individuals, and the parent collective farm. This, however, has not improved productivity markedly. Gorbachev's later years in power did indeed open up possibilities for privatisation and enterprise even though there was some political opposition. The establishment of the agro-industrial bank, for example, provided the means whereby individuals could secure loans to purchase machinery. Similarly, the possibility of selling produce surplus to state quotas provided an opportunity to earn cash that could be used for agricultural improvements.

The disintegration of the Soviet Union in 1991 and the subsequent turmoil that has been created in many member states make it difficult to assess the impact of these reforms on Russian or former member states' agriculture. For Russia itself the condition of agriculture has been evaluated by Pockney (1993). He states that the aim of the current government of Russia is to increase productivity so that the country becomes once again at least self-sufficient in food production if not an exporter. Political unrest coupled with government attempts to turn the Russian economy into a capitalist one have not helped the cause of agriculture. According to Pockney, agricultural output declined by 5 per cent in 1991 and 12 per cent in 1992. Nevertheless, wheat and maize yields increased per unit area whilst there were major declines in the production of sunflower seeds and sugar beet. In addition, poor infrastructure has meant that insufficient fruit, vegetables and potatoes were reaching the major centres of population. Food shortages and the development of a black market have consequently ensued. In 1992 the average price rise of 70 basic foods was 2016 per cent as a result of a series of price explosions. Livestock farming was also badly affected in 1990–1992 mainly due to the soaring cost of grains, etc., for fodder. Despite state subsidies it became an unprofitable activity. Not only have meat supplies been cut drastically but so too have supplies of eggs, milk and cheese. The availability of processed foods also, inevitably, declined. Since the supplies of machinery, fertilisers and food-processing equipment have also dwindled it is difficult to envisage any major improvements in Russian agricultural production in the short term. Food security helps provide social and political stability whilst the latter provides an appropriate setting for efficient agriculture.

3.5 CONCLUSION

The foregoing sections clearly illustrate the substantive changes that have occurred in agriculture in many parts of the world. There is a distinct relationship between population growth and agricultural productivity, though other factors, such as the production of crops for trade, are also important. The development of agriculture and its efficiency have provided a firm basis on which civilisations have flourished and allowed other enterprises to prosper. The latter range from the first production of pottery in prehistoric times and the later use of copper, bronze and iron to the Industrial Revolution of the mid-eighteenth century. Today, agricultural productivity is allowing the silicon revolution to occur as well as the development of information technology, two of the biggest growth sectors in the world economy.

Throughout prehistory and most of the historic period the production of crops and livestock has largely been undertaken by manipulating soils and the environment, e.g. by introducing irrigation, and by adopting crop complexes and rotations that are inherently conservational, protecting both soils and crops from erosion and pests and diseases respectively. Post World War II agriculture is a very different enterprise, especially in the developed world where it is based mainly on monoculture and relies on a battery of inputs that derive from science and technology. These include artificial fertilisers, crop-protection chemicals and plant and animal breeding. Biotechnology is beginning to provide another powerful means of agricultural change. It contrasts with traditional approaches to agriculture, which rely chiefly on modifying the environment to suit crops, by facilitating the alteration of crops and livestock.

Post World War II agriculture also represents a departure from earlier agriculture in so far as it relies heavily on fossil fuels. Food energy is produced by using fossil-fuel energy to enhance the acquisition of solar energy. This has enabled some countries to become self-sufficient in food production and others to create wealth through the export of food. These same nations are those that have political and military pre-eminence. This position has been achieved through political manipulation of agriculture via protectionist and interventionist policies, though the more extreme forms of state control have not always been successful as is the case in Russia. Just as it did in the past, agriculture continues to generate, directly or indirectly, wealth and power.

3.6 FURTHER READING

Beckett, J.V. (1990) *The Agricultural Revolution*. Blackwell, Oxford.

Dodgshon, R.A. and Butlin, R.A. (eds) (1990) *An Historical Geography of England and Wales*, 2nd edn. Academic Press, London.

Grantham, G. and Leonard, C.S. (eds) (1989) *Agrarian Organization in the Century of Industrialization: Europe, Russia and North America, Part A and B*. In *Research in Economic History*, Supplement 5. JAI Press Inc., Greenwich.

Grigg, D.A. (1992) *The Transformation of Agriculture in the West*. Blackwell, Oxford.

Lewit, T. (1991) *Agricultural Production in the Roman Economy AD 200–400*. BAR International Series No. 568.

Roberts, J.M. (1992) *History of the World*. Helicon Publishing, London.

CHAPTER 4

Transitory Agricultural Systems

The majority of the world's food production takes place within permanent agricultural systems that are repeatedly manipulated to yield crops and/or animal products. There are, however, many parts of the Earth's surface which cannot, through environmental constraints such as inadequate rainfall or poor soils, be so intensively used or which are productive only under very different forms of management. If such areas are used for agriculture, as distinct from providing hunting grounds, the systems developed tend to be transitory. They appear and endure for short periods only in any given area which the agriculturalists leave to move onto another area. Such agricultural systems are thus spatially transitory but not temporally transitory.

The forms that transitory agricultural systems take are many and varied. This is because they tend to be the dominant production systems in what are generally considered to be marginal, or inhospitable, environments for permanent agriculture. Globally, the constraints on organic productivity (see Section 1.2) are water availability, annual temperature regimes and nutrient availability. These same constraints confine agricultural potential with the result that animal husbandry and/or crop production can only be successful under a limited range of conditions that must be complied with if the environment and the agricultural systems are to maintain their integrity. The implication of this is that such systems are fragile. This may be so for some transitory systems, as will be discussed below, but is not necessarily the case for all. In addition, for many transitory systems, which appear to have changed little since their inception many millennia ago, unique *genres de vie* and/or cultures have developed around them.

Transitory agricultural systems can focus on either animal husbandry or crop production. In some instances, they may concern both but one activity will dominate. Transitory pastoralist systems predominate in high latitudes, notably in the boreal and tundra zone of the northern hemisphere, in low latitude arid and semi-arid regions as well as in some mountainous regions. In all cases, external energy inputs are low; they are usually confined to human labour. These systems rely on the ability of specific animals to convert low-grade plant material into protein (Section 1.4). Moreover, with careful management the natural ecosystems that provide the setting for transitory pastoral systems remain relatively unchanged. Such agroecosystems, thus, represent manipulated natural ecosystems, the cohesion of which remains unimpaired (Section 1.3). The difference between the ecosystem and the agroecosystem is the replacement of part (or all) of the large vertebrate biota by a domesticated or semi-domesticated species such as reindeer, camels, cattle, sheep or goats.

A very different scenario exists in the case of transitory crop production systems which are known locally by many different names such as *milpa* (Section 3.3.2) in Mexico. Such systems are, in general, often referred to as swidden cultivation or slash-and-burn cultivation. They predominate in tropical forest and savanna regions where the limiting factors to organic productivity are nutrient availability and/or seasonality of water supply. The latter applies particularly to semi-arid regions. As with the pastoral systems, the external energy inputs are low, consisting mainly of human and possibly some animal labour. In consequence, energy ratios (Section 1.4) are high even when compared with those for permanent agricultural systems in the developed world (Chapters 5, 6 and 7). The problem of poor nutrient availability is overcome by the use of fire to ash the standing-crop biomass of the natural vegetation in which there is an accumulation of stored nutrients. The resulting ash is mineral-rich and is rapidly washed into the soil when rain occurs. Firing thus accelerates biogeochemical cycling as the ecosystem is transposed into an agroecosystem. Various combinations of crops are then planted for periods of between two and five years before the plot is abandoned. The fallow period ensues during which the natural vegetation recolonises and nutrients are once again stored within the biomass. The cultivators move on and repeat the procedure in a new area. The key to efficient and sustainable food production is the length of time that fallow is allowed to persist. Both environmental degradation and declining crop productivity may accompany shifting cultivation wherein the length of fallow is decreased.

Shifting cultivation systems vary substantially on a spatial basis. Whilst they generally occur in tropical regions, and involve crop complexes based on a range of species, rather than just one or two cultivated species, there are local adaptations that reflect the practitioners' understanding of local conditions. In many cases the practice of shifting cultivation is not haphazard; it is organised on a tribal or community basis and underpins specific cultural identities. Like many forms of transitory pastoralism, a great deal of shifting cultivation has been in operation for millennia. In contrast, however, new trends in shifting cultivation began to proliferate in the 1950s. These have accelerated as population increases in many developing countries have continued and produced a landless class of people with few options other than to go out into pristine forest regions to grow sufficient food to feed their families. Such shifting cultivators take advantage of the crude roadways, constructed by logging and mining companies, to invade forests that would otherwise be impenetrable. In some cases the colonisation of pristine land is actively encouraged by government resettlement policies and there are some forest areas that are particularly vulnerable because of their proximity to areas suffering environmental stress. The resulting environmental refugees may be forced to invade the forests and take up a life of shifting cultivation. In these cases the adoption of shifting cultivation is motivated out of necessity, which is the result of economic and political factors, rather than out of cultural tradition. All forms of transitory agricultural systems are, like all other agricultural systems, a product of the physical environment, including climate, and societal factors.

4.1 NOMADIC PASTORALISM

In general, pastoralists use land that cannot readily be cultivated to produce crops or from which it makes economic sense to generate animal products. Nomadic pastoralists fall into the former category and represent an extreme form of pastoralism that necessitates the migration of animal herds on a regular basis from one area to another. In some cases, movement may be over considerable distances of about 1000 km, in others it may be as little as 50 km. Indeed, there are some groups who may be cultivators for a period and then become nomadic pastoralists for several years. Even during this time they may plant crops. This also means that the distinction between permanent pastoralism/cultivation and nomadic pastoralism can be difficult to make. In addition, nomadic pastoralist economies are usually centred on the herding of more than one animal. Indeed, Galaty and Johnson (1990) state that

> Pastoralism usually involves 'complexes' of animals, combinations of species managed in a compromise between the particular needs of each. Pastoralists tend to diversify across species and ecological conditions, the result being greater resilience, insurance, and range of products than is the case if the number of species is diminished. But usually a herd complex revolves around a dominant species, which can be used to characterise pastoral systems and their geographical distribution.

The implication of this mixed species composition is that the pastoral system is well-managed and ecologically balanced through the varying requirements of the different livestock components. Thus, while such systems may appear fragile they are really stable systems within an environment which is fragile. Moreover, the integrity of the environment is only maintained through careful management which also generates wealth and status. Over the centuries these traditions have given rise to distinctive cultures.

Galaty and Johnson's (1990) comments on using the dominant animal species as a means of classifying nomadic pastoral systems are, however, apt at the local or medium scale. At the global scale another distinction can be made which reflects climatic controls. Nomadic pastoralists occur in arid and semi-arid environments that prevail in high latitudes, low latitudes and in some montane regions. Each of these major units can be subdivided according to the dominant animal species. For example, reindeer dominate high-latitude nomadic pastoral systems while camels predominate in arid lands. Examples of montane nomadic pastoralism include yak herding in Tibet, and llama and alpaca herding in the Andes. In each of these there are specific, and unique, energy flow characteristics. Nevertheless there are some generalisations that can be made about the energy characteristics of such systems. For example, most of the energy produced in the systems is used within the system; what little export of primary or secondary animal products occurs is usually part of a barter or exchange system to obtain other forms of food or essential goods. In general, there is little or no fossil-fuel energy in these nomadic pastoral systems which are, therefore, entirely solar powered (Section 1.3 and Figure 1.7). Similarly, nutrients are recycled within the systems rather than exported as is the case with most permanent agricultural systems. An important exception to this, however, are the reindeer-herding activities in northern Scandinavia where fossil-fuel energy is imported for use in the snow scooters, etc., that are used as an alternative to the traditional sleds. This production system also exports reindeer meat on a commercial basis. This does not alter the fact that nomadic pastoralist systems are a response to seasonal and spatial variations in the distribution of forage which, in turn, reflects climatic parameters, particularly rainfall. As Tivy (1990) has pointed out, these factors also determine whether nomadism is occasional, permanent and/or partial.

4.1.1 Nomadic pastoralism in high latitudes

Nomadic pastoralism in high latitudes comprises reindeer (*Rangifer tarandus*) herding. It is also a way of life that has declined considerably in the last 30 years as a sedentary way of life has been adopted. Reindeer herding is a major food production system in the land areas of the northern hemisphere that lie at latitudes above the Arctic Circle. In terms of natural vegetation communities (Figure 1.4) these areas are dominated by tundra where mosses, lichens, herbaceous species and dwarf shrubs predominate. To the south lies the taiga comprising birch, pine and/or larch (needleleaf) forests which may also be grazed. Throughout these zonal vegetation communities in Europe, Asia and Canada there are nomadic pastoralists whose economy is reliant on reindeer herds. Beach (1990) points out that the reindeer represents a fundamental constituent of many quite distinct cultures in the Arctic region. Indeed, he states that 'Certainly without the broad and bountiful distribution of the reindeer in the north it is doubtful that humans could have inhabited the circumpolar region so early and so successfully'.

It is generally believed that reindeer herding began with hunting strategies as human groups colonised the Arctic zone in the mid-Holocene. Beach (1990) lists the many and varied uses to which reindeer and their constituents can be put. The reindeer itself, apart from being the object of the hunt, has been tamed to act as a decoy to inveigle wild reindeer into the midst of the hunters; it has been used as a beast of burden for carrying accoutrements and for sled-pulling, and occasionally it has been ridden. Reindeer products include meat, which features strongly in the herders' diet, milk which may also be made into cheese, and hides that can supply clothing and shelter. Antler, as is often attested to in archaeological sites, may be used for a variety of practical purposes, e.g. spear points, spoons, knife sheaths, awls and even hand looms, as well as for esoteric purposes, e.g. immature reindeer antler is thought by some Asian peoples to enhance sexual prowess.

As Beach (1988, 1990) discusses, the physical characteristics of reindeer, which are known as caribou in North America, suit them well for a life in the harsh Arctic environment. The life-cycle of the animals, which mature in their third year and in which females reproduce up to the age of ten years, is taken into account along with the herds' social organisation, when herding strategies are established. Local variations in topography and vegetation are important determinants of the patterns of reindeer herding. Forest, tundra and coastal environments are each characterised by a specific pattern though the zones are not mutually exclusive and forest/tundra and tundra/coast combinations also occur. Occasionally, all three may be combined. The animals spend part of the year in each of the environments so the secondary productivity that they represent is the output of different types of primary productivity. Nevertheless, one of the most important sources of reindeer forage is lichen, the availability of which determines the carrying capacity of any given area or combination of areas. As a result, herders set the pattern of migration on the basis of the lichen quality and abundance. If this is overexploited, as grasslands may be in other parts of the world (Section 4.1.2), the carrying capacity for future years will be impaired. Upland to lowland, forest to tundra, tundra to coast movements, all of which follow a winter to summer pattern, are determined largely on the basis of year-to-year variations in plant productivity. Social and economic factors also feature in herding strategies and include access to slaughterhouses and markets, the sharing of pastures with neighbouring herds and the minimisation of losses due to predators such as wolves. In many parts of the world the business of reindeer herding is becoming increasingly market orientated rather than subsistence directed.

In relation to the dynamics of reindeer herding, a key factor is the ratio between reindeer and grazing land. This must be regulated within and between herding communities, otherwise carrying capacity will be exceeded and degradation of pastures will ensue. Internal adjustments and co-operation between communities is thus essential, though, as Bäck (1993) has discussed, social and ecological problems often arise. If herds become too large the animals will roam far and wide in search of forage. This makes rounding up difficult and labour intensive; it also means that dispersed animals are subject to easy predation so herd losses increase. Thus, beyond a certain size determined by carrying capacity, the gains of additional animals are counterbalanced by losses. Economic and social factors also contribute to the determination of herd sizes. If the latter increase, the environmental limits to productivity ensure that poor or lean years will dominate the herding system. Only large herders will be able to survive economically and small herders will be forced out of business. This impoverishes the herding community. There are also conflicts of interest with activities such as forestry, mining, hydroelectric power and tourism.

The reindeer-herding communities of the 1990s are very different to those of the 1940s, particularly in Europe and Asia. In Alaska, Canada and Greenland, however, reindeer herding has been introduced only in the last 40 years and is more or less superimposed on the eskimo hunting tradition. In Europe, reindeer herders occur in northern Scandinavia and northern Russia. The reindeer herders of Scandinavia are known as the Saami and they occupy an area, mainly north of the Arctic Circle, that is often referred to as Lapland. According to Fullerton and Knowles (1991), there are currently about 50 000 Saami, more than half of whom are in Norway, with a further 30 per cent in Sweden and the remainder in Finland and Russia, as shown in Figure 4.1. The majority of these people are now sedentary, or more appositely, the herds still migrate but families no longer follow them, only the herders themselves. Some of this sedentarism has been prompted by legislation restricting movement between Norway and Finland.

Fullerton and Knowles (1991) describe mountain, forest and coastal Saami communities. The first group are the main reindeer herders; even so only about 6000 people throughout Scandinavia are true nomads. Moreover, Beach (1990) refers to the 'herding industry' which is a reflection of its growing business rather than subsistence status. This is also manifest in the high technology that is now employed by herders. It includes snowmobiles, light planes, helicopters, as well as radio and satellite communications. As a result there is now a fossil-fuel subsidy to this agroecosystem though, obviously, this is nothing like as great as it is in grain production in Europe and North America (Chapter 5).

Amongst the many factors that influence reindeer herding (see Jones, 1988; Fullerton and Knowles, 1991), there are two that are worthy of further mention if only to illustrate how essential to good environmental and social management is an

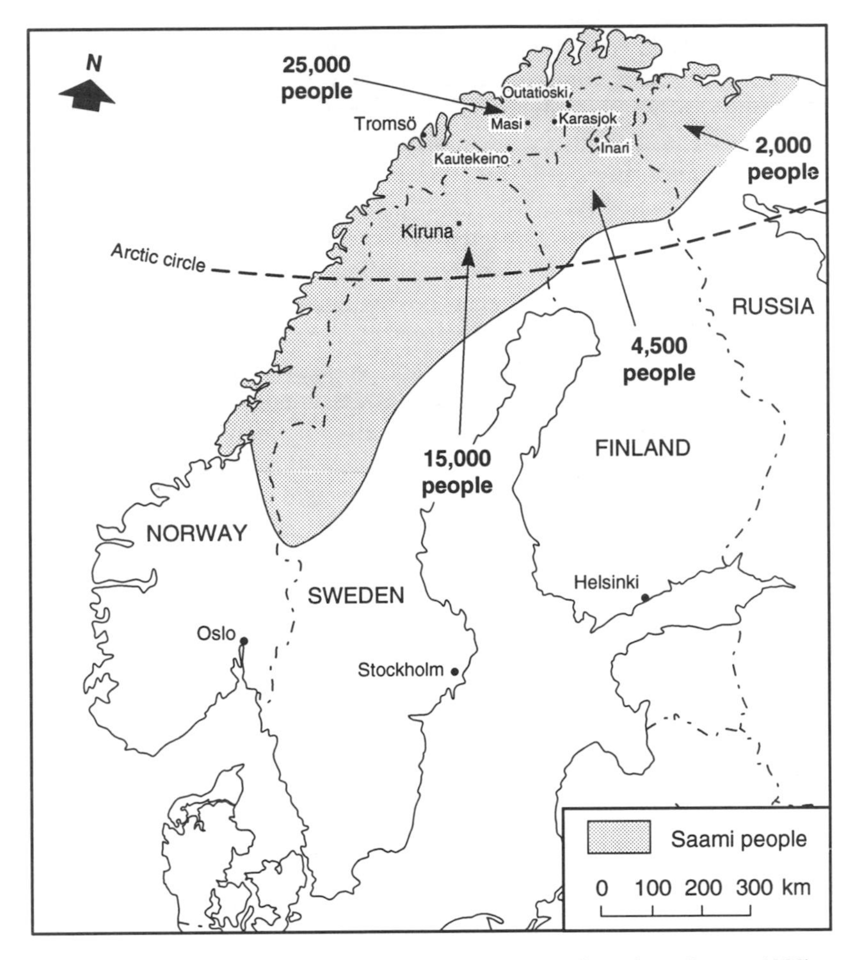

Figure 4.1 The distribution of the Saami people (based on Jones, 1988)

holistic approach. First, there is the issue of the Chernobyl reactor accident of 1986 which caused the emission of radionuclides into the atmosphere. Prevailing winds at the time carried clouds of emitted material west and north. In particular, the fall-out of caesium-137 caused food chains and webs to become contaminated. (Caesium behaves in a manner similar to potassium.) As a result, a ban on the marketing and consumption of reindeer meat, amongst other products of the contaminated zone, was imposed. Currently Saami communities are maintained with the help of government compensation schemes. Secondly, some conflicts arise between the Saami way of life and a number of modern development factors, e.g. hydroelectric dam construction, road construction, forestry, tourism and the protection of certain wild animals which predate the reindeer.

4.1.2 Nomadic pastoralism in low latitudes

Nomadic pastoralism in low latitudes reflects, as it does in high latitudes, environmental constraints on primary productivity. According to Gilles and Gefu (1990), nomadic pastoralism in low latitudes is confined to Asia and Africa. (This delimitation excludes transhumance which is practised in montane parts of Europe and the Americas.) In Africa, for example, there is evidence (A.B. Smith, 1992) for its existence in the Sahara, e.g. Libya, Algeria and Niger, from 7000 years BP. As in the case of reindeer herding, animal herds are moved from place to place to take advantage of seasonally produced forage. Mobility, and hence flexibility, is a key to the success of such systems provided they are well managed. Traditionally, management takes account of seasonal, or even interannual, rainfall and the forage that this produces. Both primary and secondary products are important and generate food as well as a means of bartering. However, many of these traditional agricultural systems are increasingly being jeopardised either by environmental factors, such as persistent drought and desertification, or by societal factors, notably land encroachment by crop farmers in good years and by the land hungry.

There are many different types of nomadic pastoralism in low-latitude regions. According to Gilles and Gefu (1990), the nature of the herding activity and the mobility that it requires influence the organisation of the nomadic society. Despite these influences, which might be expected to produce rather homogenous groups, there are, in reality, many variations. The location of the examples discussed here are given in Figure 4.2. What is common to the majority of nomadic pastoralists is, however, the way that land ownership and/or control is organised in relation to herd ownership. The former is usually collective. The advantages of this include access to wide areas, which becomes increasingly important the more unpredictable the water and pasture resources become, and the avoidance of overgrazing. Animal or herd ownership, however, is usually individual or family based. Much of the ecological balance that nomadic pastoralists usually achieve is the result of these counterbalances. There is also a great deal of pooling of labour as well as resources. Some of these factors, along with the differences that occur within nomadic pastoral systems, are illustrated by the examples given below.

Schusky (1989) has reviewed the characteristics of nomadic pastoralist activities in the African Sahel which borders on the southern edge of the Saharan desert. He points out that the variability of inter- and intra-seasonal rainfall means that only flexible, and hence mobile, pastoralists can survive. The herds are often mixed, comprising goats and camels as well as cattle and sheep. The first two are browsers of leaves and thorny shrubs whilst the other two graze grass. Thus different resources within Sahelian pastures are used. This is a principle that has parallels with multicropping and crop rotations in shifting cultivation (Section 4.2) and in traditional European agricultural systems of the seventeenth and eighteenth centuries (Section 3.3.1). The proportions of these animals will also vary temporally. In years of good pasture more grazers may be kept and the browsers traded; in poor years the reverse may occur. Schusky also points out that the pastoralists have a relationship with cultivators in the southern part of the Sahel. Often this is a mutually beneficial relationship that contributes to the overall stability of the pastoralists' and cultivators' systems. In years of poor rainfall, and hence poor pasture, herders will drive their animals onto the cultivators' lands. There they graze crop stubble whilst manuring the fields and the cultivators have the opportunity to supplement their diet with milk and meat. Sometimes, however, the pastoralists will only exploit the cultivators during lean years.

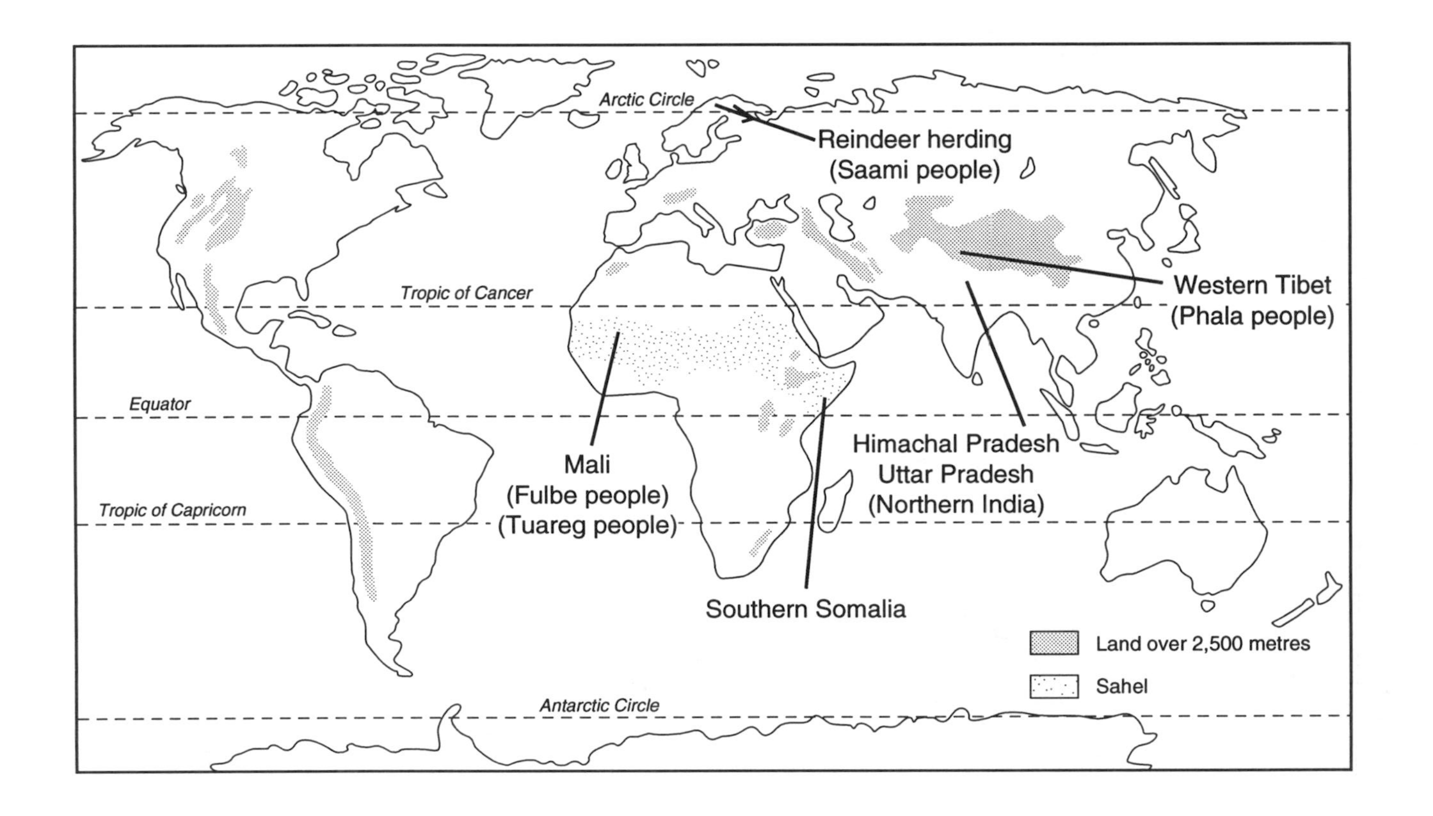

Figure 4.2 Map showing the location of case-study sites for nomadic pastoralism

Amongst the Sahel's nomadic pastoralists are the Fulbe people of the Doukoloma area of Mali (Figure 4.2) whose activities have been documented by Grayzel (1990). The Fulbe utilise a government-protected woodland located along the banks of the River Bani; some are cultivators that use the forest lands and others are pastoralist/ cultivators who herd sheep, goats and cattle. The cattle are exported whilst sheep and goats are consumed by the Fulbe themselves who practise a form of transhumance. Although there are permanent settlements, the animals and most of the people migrate between grazing areas of the forest, moving up to 200–300 kilometers. Grayzel describes the movement of the Fulbe people as 'a great outdoor ballet, the place and the pace of each performer is consciously and unconsciously coordinated with the overall movements of the entire group'. From October to December, a dry and relatively cool period, the cattle are herded into the forest to graze the understorey and ground vegetation. Simultaneously, the Bamana people engage in millet cultivation that is harvested in November. The animals are then allowed to graze the millet stalks. Moves to the south of the Bani River occur in December, by which time the millet stalks have been depleted and grass pasture is drying out. Crossing the river is really the start of the migration pattern into the grasslands. Here grazing is allowed until new groups approach which prompt movement and thus prevent overgrazing. Herds also move in response to water availability which reflects variations in the height of the water-table and/or the locations of wells. As the hot, dry season begins in April, movement south is halted and herds move only locally, if at all, and feed on bushes and trees as grass becomes depleted. During June and July, the beginning of the wet season, herds begin to move homeward in order to avoid mud and mosquitoes whilst animals retained in the Doukoloma area are allowed to enter the forest once again to graze on the rain-prompted grass. The first animals to return home are used to plough plots for the Fulbe themselves or are rented out to Bamana or others. In the ensuing two months the herds remain in enclosures in the permanent villages until the cycle begins again in October.

This system thus utilises interrelated ecological conditions that are determined largely by climate, in tandem with complex social interactions. These agricultural systems underpin highly organised societies that continue in their own distinct cultures because of this mutual support. The capacity of cattle to generate wealth is also reflected in inheritance, marriage, and indeed all family customs. As a result, herd sizes vary considerably but the factors discussed in the opening paragraphs of Section 4.1.2 continue to prevent over-exploitation of the pastures. Today, the major threats to this way of life are factors outside the control of the Fulbe, notably legal access to pastures by urban herd-owners who may be well-off business people rather than traditional grazers, and development schemes, e.g. ranching.

Other nomadic pastoralist groups are characterised by similarly complex ecological and social patterns. In common with the Fulbe, the continued operation of many such groups is threatened by external factors. For example, Bernus (1990) describes the practices of the Tuareg, in the Sahel, who comprise two distinct groups, each of which is based on a pastoral economy. The Iwellemmeden Kel Denneg occupy the northernmost part of the Sahel in Mali, Niger and Algeria (Figure 4.2). The Kel Geres occupy land to the south, which is less arid than that of the Kel Denneg, and where rain-fed agriculture is important. Both have long histories based on animal husbandry that has been managed by various tribes; the animals, especially camels, represent wealth and their management, etc., is intimately related to social and religious customs. As in the case of the Fulbe, neither the Kel Denneg nor the Kel Geres operate in isolation. The former, for example, need to obtain millet and sorghum from their subservient kinspeople, who cultivate these cereals on the edge of the rainfed agricultural zone, or from Hausa people further south. The chief movement of the Kel Denneg is to the north in summer to use grass pastures on the edge of the Sahara. Amongst these people are artisans who produce metal, wood, leather goods and jewellery. These can be exchanged for clothing and household textiles. The Kel Geres, in contrast, combine animal husbandry and agriculture. Camels are also important. The animals are grazed on the Hausa's southern, fallow lands to which they contribute manure, and then herded north again to graze summer pastures. Cultivation of cereals also occurs, and there are caravan links facilitating their supply to northern oases. The Kel Geres also obtain goods from southerly markets, in the Hausa areas, in exchange for crop and animal products and some of these are traded again in the oases for salt and dates.

However, the curtailment of movement across national boundaries is a major factor that is altering the lifestyle of these people. This began with colonial powers but has been continued by independent state governments. The drilling of deep wells and pumping stations has also resulted in distorted patterns of grazing when compared with traditional patterns. The granting of rights of access to pastures to anyone has also encouraged interlopers into the traditional Tuareg lands, including Fulbe. The equilibrium between pastures, people and water has, in consequence, been destroyed and antagonism created between the Tuareg and Fulbe. Simultaneously, Hausa people moved into traditional pasture lands, initiating cultivation in areas that were really marginal for rainfed cultivation. All of these developments were made worse by the drought of 1968–1973 and the later drought of 1984. Institution responses attempted to limit damage but the Tuareg (and others) adopted different movements of animals to cope. Trends towards sedentary lifestyles, along with lack of consideration for traditional customs, clearly do not contribute to cultural or ecological stability. Some of these issues have been discussed in detail by van Keulen and Breman (1990).

Herren (1992) has undertaken a similar analysis of camel pastoral systems in southern Somalia (Figure 4.2). Here the Garre and Gaaljacel peoples occupy interriverine areas where they raise camels, along with cattle and some crops, in a migratory agricultural system. A major source of income is the selling of camel milk to markets in Mogadishu. Over the last two decades, this has altered the pattern of migration to an extent, since it now takes account of where milk collection points are located. Moreover, wealth generation may have increased but camel management has become intensified and the animal stocks are not so easy to maintain. This is because animals are no longer allowed to roam so far and as a result pasture quality has diminished. Camel health has also deteriorated with increased mortality and reduced fertility. The wealth generated, mainly by owners of large herds, was used to increase the people's consumption of sugar and sorghum. At the same time, poor families with only small herds rarely consumed their own nutritious camel milk because of the market pressures to supply Mogadishu.

In contrast to some of these African examples of nomadic pastoralism, where recent government-led innovations (and droughts) have caused some disruption, the situation is quite different in the Western Tibetan plateau (Figure 4.2). According to Goldstein and Beall (1991), not even the imposition of direct Chinese control since 1959, or the initiation of often inappropriate development projects, have conspired to undermine the nomadic pastoralism of the Phala people. Goldstein and Beall's research on Tibet's Western plateau, or *changtang*, was based in the area known as Phala, some 500 km from Lhasa, where sheep, goats, horses and yak are kept. No cultivation is practised; at *c.* 6000 m OD this is impossible due to aridity and low average annual temperatures. Indeed, Goldstein and Beall suggest that it is these environmental characteristics that provide the nomadic pastoralism with its stability. The landscape comprises valley–peak topography providing adequate year-round pasture. The Phala move only about 40–50 km for animal grazing, though much longer distances are covered for the purposes of trade or salt collection. As in most other nomadic systems, the herds are owned by individuals or families and the grazing lands are commonly controlled, though in traditional Phala society in the days before Chinese rule, there was a lord who was in permanent residence in the area.

With the advent of Chinese domination in 1959, which brought with it the expulsion of the Dalai Lama, some changes came about within Phala society but the overall structure and activities of these people changed little. Instead of control of grazing, etc., by the lord, local officials were made responsible. Although the wealthy nomad class was more heavily taxed than hitherto, they remained independent and in charge of their animals; they retained the ability to pay other nomads of less wealthy means to be their servants and shepherds and they were allowed to continue selling their products. The 'Cultural Revolution' that began in 1966 instigated widespread changes in the Chinese and Tibetan society. In Phala society wealthy nomads, after a quelled uprising, had their animals and property confiscated, becoming, along with all other nomads, members of a commune. As a result, all decisions about the production system, as well as its ownership, were transferred from the household to the commune. Otherwise the way of life continued as before. The products of the system, e.g. live animals, milk, meat, butter, hair and wool, which were produced surplus to each commune's requirements were transported to villages a month's travel away. Here they were bartered for grain with other communes and government offices.

In the early 1980s, the new government of Deng Xiaoping began to open up China's economy. The role and influence of communes in village life came to an end and the production system revolved around the household unit once again through reforms known as *gendzang* (meaning 'complete responsibility'). In Phala the dissolution of the commune meant that each nomad family received its equal share of the herds which comprised yak, goats and predominately sheep. These people were also exempt from tax until 1990 and restrictions of religious practices, which had been imposed as part of the 'Cultural Revolution', were lifted. The pastures were allocated, as they still are, to small groups of households or *dzugs* that enjoy exclusive usufruct (the use and profit but not the property) rights. The major difference between pre-1959 practices and those of the post-1981 period is that there is no system for accommodating changes in herd sizes between *dzugs*; the modern system is thus less flexible than the old one. This has caused shortages of pasture for some *dzugs* whilst others have an excess of unused pasture. Such problems are now generally addressed at the local level but there are no overall guidelines. However, the migration pattern and field production system has not changed a great deal, largely because the harsh environment is not easily manipulated. There are two primary moves and several secondary moves. In the autumn, each *dzug* moves to a pasture area that has been left fallow since the start of the growing season in May. This is a primary move and secondary movements include those within the pasture area to capitalise on the variability in pasture quality. By the end of the year the pasture is usually exhausted so the Phala return, with the sheep and goat components of the herd, to the home-base. Yak, however, are moved higher up into the mountains to graze on *Kobresia* spp., which are types of sedge. The herders form a camp for about four months while the yak are moved about the sedge meadows. In addition, special sites are established for the birthing of goats and sheep near to the home-camp but with tents, etc., to allow herders to remain on site. At the home-base, permanent tents predominate along with storage facilities. These are overseen by elderly Phala all year round if they can be spared from herd minding.

The surplus animal products are still used to purchase essential grains and other commodities. Wool is the most important product and now cashmere is growing in significance. Most of the products are compulsorily sold to the government. However, there is some trade with farmers along the edge of the *changtang*; with traders who visit the Phala communes in the summer to exchange products and labour for animals and animal products; with other nomads with whom livestock, e.g. horses may be exchanged; and, recently, with the town of Shigatse which can be reached within two to three days by lorry. Goldstein and Beall (1991) state that despite an 8 per cent decrease in herd size, the Phala are better off economically now than in the early 1980s mainly because of increasing prices for wool and animal products in general. They conclude,

> In contrast to the bleak future facing nomads in most other parts of the world, nomadic pastoralism on the Tibetan *changtang* is flourishing. Because the *changtang's* severe environmental conditions preclude agriculture and because the nomads' livestock products earn the TAR (Tibet Autonomous Region) a substantial proportion of its foreign currency, no attempt has (or is) being made to end or diminish this ancient way of life.

Success such as this is not, however, characteristic of other nomadic pastoralist systems. A case in point is that of the Gujars of Himachal Pradesh and Uttar Pradesh in northern India. Their system, described by Gooch (1992), comprises buffalo herds from which dairy products are sold in nearby towns. The Gujars are pastoralists who practise transhumance, using winter pastures in the Himalayan foothills and summer grazing in the upper ranges. Their winter camps are located in the forests of western Uttar Pradesh. The pasture and forest resources are both diminishing, mainly through forest management policies and environmental degradation. Despite their traditional role as providers of milk the Gujars are gradually being marginalised. This is occurring because apple orchards are being planted in some areas of their traditional pasture in the upper ranges of the Himalayan foothills. These orchards are also encroaching on forests whilst reafforestation programmes are contributing to the loss of pasture. In addition, increasing debts often prompt Gujar families to borrow money against the next year's milk yield. Their plight is not a happy one but so far no government initiatives have been presented to allow the Gujars to continue in milk production even when they are willing to do so from a permanent base.

In terms of the energetics of nomadic pastoral systems the few studies that have been undertaken indicate that they are very efficient. Energy inputs comprise solar energy, human labour and some animal labour. Giampietro *et al.* (1992) have compared the energetics of four different systems of livestock production, including beef production, in a nomadic pastoral system representative of the Sahel. Only a small percentage of gross primary productivity is transferred into secondary productivity. The values are higher for the production of 1 kg of edible protein in a closed New York state feedlot, for example, but this also requires 513 MJ of fossil-fuel energy. Whilst the nomadic system is energy efficient, with no fossil-fuel subsidies, it cannot support high human population densities. Such systems provide a means of efficiently extracting protein from a poor primary source that would not produce an edible food crop otherwise. Moreover, it is apparent from the examples given above that nomadic pastoral systems do not operate in isolation but are intimately connected with adjacent arable farming communities.

4.2 SHIFTING CULTIVATION

Two of the most important similarities that exist between nomadic pastoralism and shifting cultivation, which is also known as swidden, are the spasmodic and extensive use of land and the lack of a fossil-fuel input. Both agricultural systems represent ways of efficiently manipulating environments that would not be particularly productive under arable farming. Shifting cultivation like nomadic pastoralism, varies enormously from one part of the world to another as the examples given below attest.

As Smil (1991) points out, shifting agriculture probably played a major role in the early development of agricultural systems (Chapter 2). Indeed, the palaeoecological records from many parts of the UK, for example, reflect small-scale clearances for cultivation and subsequent woodland regeneration (reviewed in Mannion, 1991). Shifting agriculture was part of the continuum that embraced hunting and foraging on the one hand and permanent cultivation on the other hand. It may have been an important means whereby domesticated plants (and animals) were introduced into areas beyond the centres of domestication. Today, shifting cultivation is still a very important way of life for millions of people in Africa, Latin America and Asia. Indeed Robison and McKean (1992) suggest that worldwide 250 million people rely on this form of agriculture. The characteristics of the systems and the crops cultivated vary spatially but there are many similarities as has been discussed by Weischet and Caviedes (1993). In most of the areas where shifting cultivation is practised the major limitation on primary productivity is nutrient availability. This is overcome in natural vegetation by the rapid breakdown and recycling of organic matter so that the impoverished soil nutrient store is partially bypassed. Where shifting cultivation is undertaken natural vegetation is cleared, collected and allowed to dry. Burning then ensues which has the effect of breaking up the soil surface ready for cultivation, of clearing the slashed undergrowth, and of reducing pest populations. Most importantly, firing ashes the biomass and although nitrogen, as nitrate, is rapidly lost other mineral nutrients are made available for plant growth. Thus mineral cycling is speeded up and the shifting cultivators take advantage of these temporary conditions to grow crops. Some of the major trees, particularly those that yield a useful oil or fruit crop, are often left standing; otherwise a range of crops is cultivated. Smil (1991) indicates that between two and five crops may comprise staples but that as many as 50 crops may be grown in total. Some of these will be for fibre and medical purposes as well as for food. This diverse array of crops, to a certain extent, mimics the natural vegetation that it replaced. Smil also indicates that the number of people supported by shifting agriculture is much higher than it is for pastoralism. The latter is commonly 1–2 people/km^2 whilst shifting cultivation carries 15–55 people/km^2. This means that the amount of food energy produced is high in comparison with nomadic pastoralism systems and is more reliable.

4.2.1 Shifting cultivation in Africa

Shifting cultivation in Africa occurs in two of the continents' major biomes: the tropical forests and the savannas. According to Schusky (1989), between 25 per cent and 33 per cent of African farmers are swiddeners. Moreover, he believes that there are two types of swiddeners: those that practise shifting cultivation in order to supplement other, more permanent cropping activities but which are inadequate to support them and their families, and those that have been forced out of their 'traditional' homes with little option but to eke out a living by encroaching into forests, etc. These latter people often do not know how to manage the land as they lack the accumulated knowledge of many generations of swidden agriculturalists, nor do they have access to sufficient land to ensure adequate fallow periods. In Africa, as in Latin America and Asia, there are growing numbers of these marginalised people who are creating ecological degradation because they have little alternative. This widespread problem prompted Myers in 1982 to state that the people 'tend to advance upon the natural forest in waves: they operate as "pioneer fronts" pushing even deeper into forest tracts, leaving behind them a mosaic of degraded croplands and brush growth where there is no prospect of a natural forest re-establishing itself, even in impoverished secondary form'. More than a decade later in 1993, Myers refers to the role of the 'shifted' cultivator in accelerating deforestation, an epithet which distinguishes between the displaced and the traditional swiddener. Clearly, this problem is growing and is now a major cause of deforestation (e.g. Witte, 1993), a theme which will be discussed more fully in Chapter 9.

In relation to the practice of traditional shifting cultivation, it is undertaken on the basis of an intimate knowledge and understanding of the environment. It is one of the many traditional farming systems of Sub-Saharan Africa which have recently been described by IITA (1992). Locally, shifting cultivation is referred to as bush-fallow cultivation which is an umbrella term for a range of farming practices. The point is made that true shifting cultivation, involving the movement of family and/or village communities, is now unusual in Africa generally (apart from the invaders of the forest margins; see above). Bush-fallow cultivation, as the name suggests, requires a period of no cultivation that is long enough for the natural vegetation to recolonise and replenish the fertility of the soil in the form of nutrients and organic matter. Whilst it relies on fallow for its maintenance, it is an intensive agricultural system when compared with true shifting cultivation. Moreover, bush fallow can be distinguished from grass fallow. The former tends to be associated with a tree crop and is characteristic of humid regions like West Africa; the latter predominates in the drier and more seasonal savanna regions. Both bush and grass fallow rely on the increased nutrients that enter the soil when the rains come after slashing and then burning of the vegetation in the previous dry season. To prepare the ground, unsophisticated tools, e.g. digging sticks and hand hoes, are used. The enhancement of the soil nutrient pool means that fertility is highest in the first year of cultivation, resulting in the highest crop yields. According to IIAT (1992), the staple crops are usually grown in the first year, and in Nigeria's forest and moist savanna zones (see Figure 4.3), where the IIAT is located, the dominant crop is yams which are planted on mounds or along ridges. This variation in microtopography expedites drainage and provides adequate soil near the tubers to encourage their growth. Attaching the yams to stakes is also important to increase the leaf area exposed to the sun (see reference to leaf area index (LAI) in Section 1.4) and so foster photosynthesis. As the ground is not completely covered, weeds invade and need to be eradicated. The harvest takes place seven to nine months after planting and must be undertaken with prudence to avoid damage to the tubers. This high labour requirement means that yam cultivation is only appropriate when the soil is at its most fertile in order to achieve the best crop possible.

Subsequent crop yields decline markedly as the nutrient pool becomes depleted. The crops grown after the first season depend on individual farmers' preferences and their evaluation of soil capability. For example, cassava is often grown toward the end of the cultivation period because it tolerates low fertility and is not labour intensive. Where grass fallow is important, under drier climates, sorghum, millet and legumes such as groundnuts or soya beans are grown. Throughout the cultivation period intercropping is implemented. This makes more ecological sense than monoculture and has its parallels with mixed animal herds in nomadic pastoralist systems (Section 4.1). Each species has different nutrient requirements so mixed cropping makes the best use of the soil resources available; species numbers decline as the cropping events proceed. Intercropping in rows tends only to be implemented if animals or machinery are to be used for harvesting. Consequently crops are thoroughly intermixed except occasionally for some cash crops.

Figure 4.4 is a representation of a typical African farming system in the tropics. The homestead at the centre has garden cultivation which is permanent and given over to the production of a wide range of crops for diverse purposes. Soil fertility is maintained by the continual addition of organic matter from crop and household refuse. Here, women family members undertake most of the work. In the next zone, permanent cultivation of crops is still possible for similar reasons though the diversity is reduced. This zone merges with one where semipermanent cultivation occurs. Crops of high value e.g. yams, maize and vegetables, are grown in each of these three zones. The outer two zones are occupied by intensive and finally extensive shifting cultivation. Slash-and-burn cultivation is practised with increasingly long fallow periods at increasing distances from the homestead. In the outermost zone(s) cassava is commonly grown because it requires less attention than other staples. Intercropping is common throughout and in the outer zone, which the male members of the household will need to visit only infrequently, certain wild species may be shielded from burning because they yield a valuable product, e.g. fruit or oil. As Okigbo

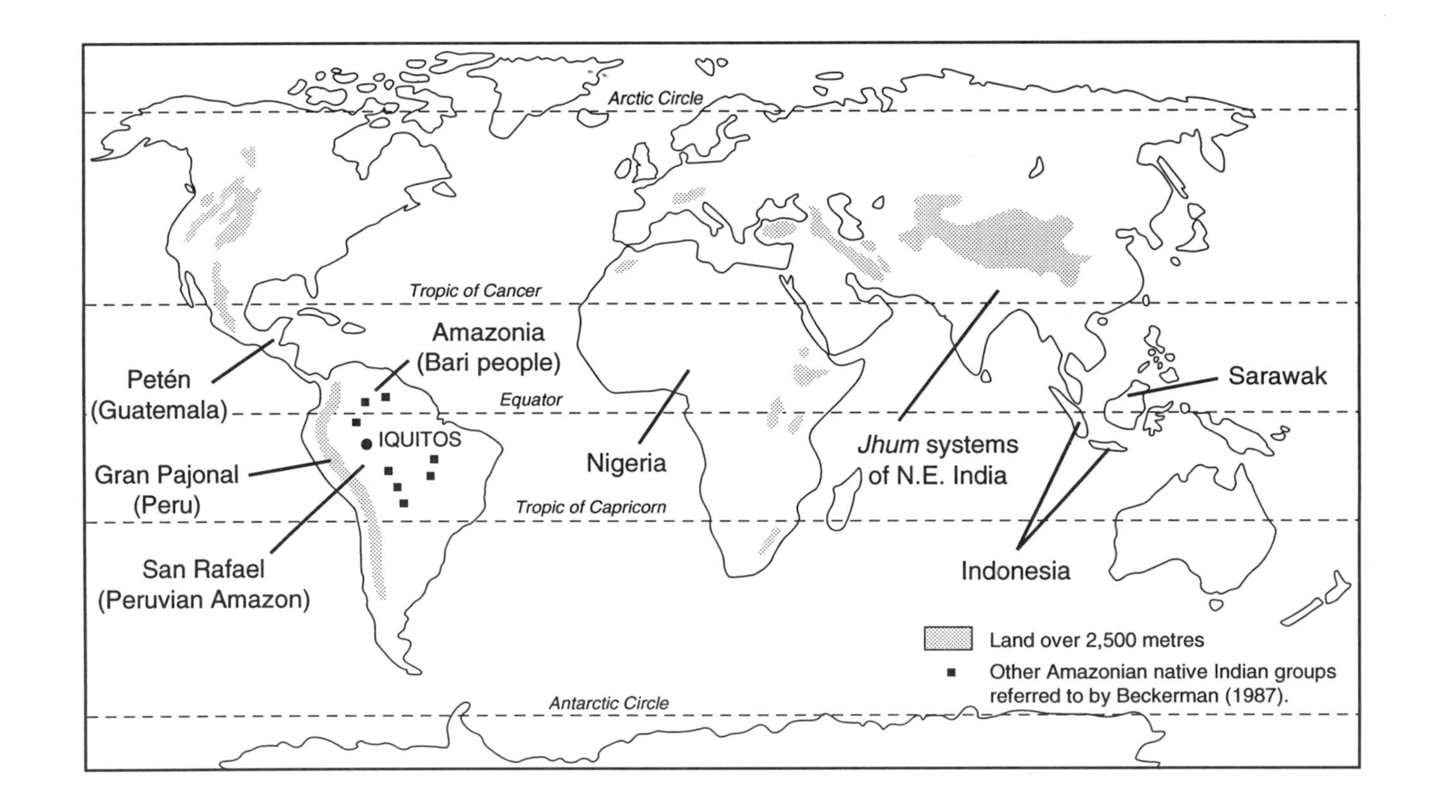

Figure 4.3 Map showing the location of case-study sites for shifting cultivation

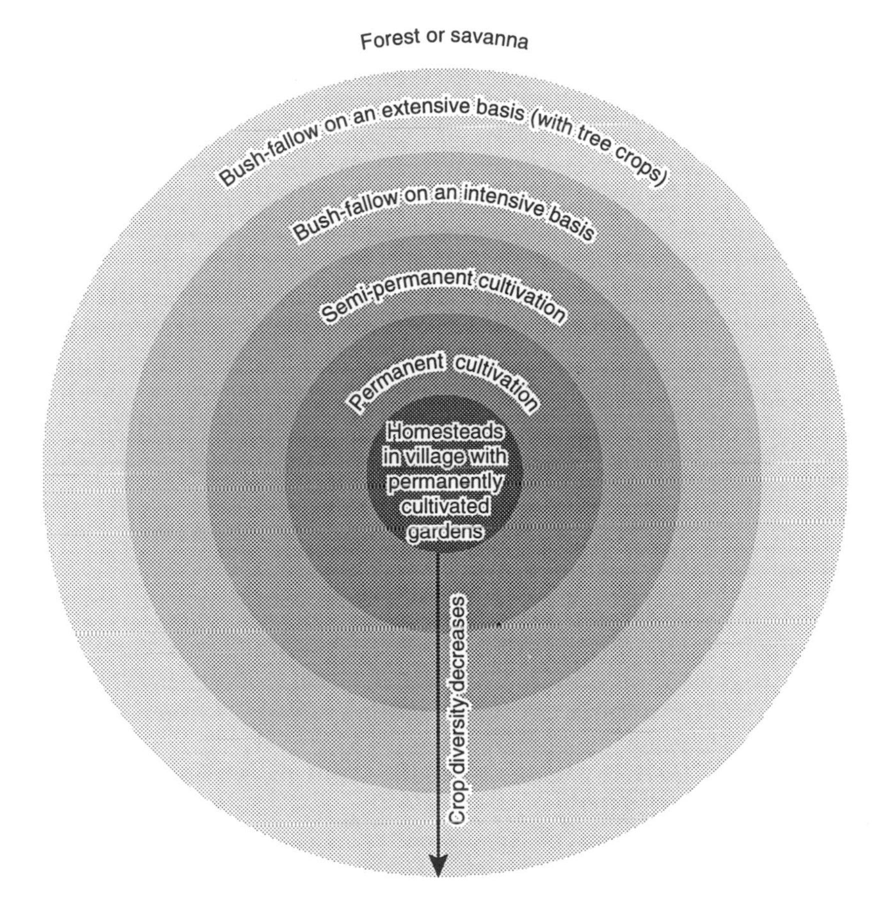

Figure 4.4 A generalised representation of a tropical African farming system that includes shifting cultivation (adapted from International Institute of Tropical Agriculture, 1992)

(1991) has discussed, there are many variations on this model but similar divisions of labour, land management practices and agroecological conditions prevail.

In principle, this system is ecologically sound and sustainable. Where crop productivity has declined it is usually a response to shortening of the fallow period, which is itself a product of social issues, particularly population growth, and a growing lack of land for the expansion of existing systems. Apart from a shortening of the fallow period, deforestation has increased, including the removal of potentially valuable fruit and oil trees. IITA's (1992) research shows that where cultivation in the outer zones (Figure 4.4) exceeds two cropping seasons there is a substantial fall in crop productivity. Apart from declining soil nutrients, there is a change in competition from weeds; species that are difficult to control invade. This is particularly the case in forested environments where hardy grasses replace woody species. Continued burning ensures that woody species do not recolonise and are, eventually, eradicated. Consequently, the ecosystem's biogeochemical cycles are permanently altered; the storage component in the biomass is considerably reduced and even the grasses themselves may be slow to regrow. The problems that ensue in relation to soil properties have been discussed by Lal (1987). The declining period of fallow causes increased susceptibility of soils to erosion, to such an extent that Tropical Africa has one of the most severe soil erosion problems on arable land in the world. The reduced incidence of fallow also encourages pests and diseases to flourish and these too contribute to reduced productivity. As the IITA (1992) points out, such trends are set to accelerate in the future which is all the more reason why research into increasing the viability of bush-fallow systems is essential. Moreover, these trends reflect the dynamism of agricultural systems as they are required to adapt and, ideally, accommodate changing social and economic conditions, just as they have done over the last 10 000 years. In this case, however, it is difficult to establish how these agricultural systems can persist indefinitely though innovations such as improved mulching, different crop combinations, alley cropping or agroforestry (see Section 7.5), which involve the cultivation of staples between tree crops, may bring short-term gains, as indeed may crop improvement through biotechnology and genetic engineering (see Chapter 10).

4.2.2 Shifting cultivation in Latin America

In Latin America, shifting cultivation is particularly prevalent in the Amazonian rain forest where it is the traditional form of crop production. It is also very variable. Villachica *et al.* (1990), for example, indicate that where local opportunities such as the presence of a large town or city occur, shifting cultivation becomes more market-orientated than subsistent. In addition, many areas of tropical rain forest in Latin America are being encroached by the landless poor or the 'shifted cultivators' as they are referred to by Myers (1993; see Section 4.2.1). For example, many of the people who were resettled, as part of the Government's official *Programa Integrazao Nacional*, along the roads built to open up the Amazon between 1964 and 1977 were ultimately forced into shifting cultivation (Browder, 1988). This is only one of many deliberate resettlement programmes in Latin America which is encouraging deforestation on lands never hitherto used for shifting cultivation and by people with little or no farming experience. An equally important example is that of Guatemala (Figure 4.3), particularly in the northern zone called the Petén. Here the rural population is increasing by a massive 5.5 per cent annually. The soils, deriving from the limestone bedrock, are thin and deforestation is occurring apace as settlers colonise. In large part this is a response to the rapidly diminishing ratio of agricultural land per capita. Southgate and Basterrechea (1992) state that the ratio declined from 1.11 ha per capita in 1964 to just 0.70 ha per capita in 1979. Government policies to open up the area have also contributed to Petén's problems. Rather than providing for the landless poor as it was intended, the resettlement programme has allowed large-scale ranching to expand. Such enterprise was only possible by the comparatively rich (Colchester, 1993).

Beckerman (1987) has discussed traditional shifting cultivation in Amazonia (Figure 4.3) which includes slash-and-burn as well as slash-and-mulch systems. As the name suggests, the latter does not involve burning of the forest; the cut biomass is allowed to rot instead. Population densities supported by such agricultural systems are typically under 1 per km^2. This is less than for shifting cultivation systems elsewhere (see Section 4.2). The clearance methods are much the same as those

Table 4.1 The nutritional value of two important crops in Amazonian swidden agriculture (adapted from Beckerman, 1987)

Crop	Water	Carbohydrates	Protein	Ash	Fibre	Fat	Calories
Cassava (manioc)* (raw peeled roots)	62%	35% (mostly starch)	0.8%	1%	1%	0.2%	150/100 gm
Bananas	75%	22% (mostly sugar)	1.2%	1%	1%	0.8%	90/100 gm

* Contains prussic acid (cyanide) in various proportions depending on the variety; this must be removed during food preparation. Manoic gives the highest yield of calories per unit land area per unit time of any cultivar.

described for Africa in Section 4.2.1. First there is cutting of undergrowth, and then the felling of most of the trees. After being allowed to dry, the slashed material is burnt to clear the land for planting, to restrict weed growth, to depress pest populations, and to provide nutrients by ashing the biomass. For seven native Indian groups discussed by Beckerman, the average field size varied between approximately 0.6 and 2 ha, with a mean around 0.4 ha. The major crop planted was cassava (manioc), possibly with bananas or plantains and frequently with maize as a secondary crop. The characteristics of these crops are given in Table 4.1. These are the staple crops though many other crops, each occupying only a small area, will also be cultivated to provide for a wide range of each family's needs. How the crops are distributed in any one field will vary between monoculture, zonation and intercropping. Harvesting proceeds as the need for the food arises; only enough is collected for a few days at any one time and the harvester is usually the woman of the household. Replanting occurs at the same time. This procedure continues for between three and four years during which time productivity declines. Replanting ceases though some harvest continues to be obtained from residual crop plants and long-lived trees that were never cleared. Forest species invade and eventually the forest regenerates possibly to be cleared again in 20–30 years. This type of agriculture provides mainly carbohydrate; protein is derived from fishing and hunting.

As a result of the rapidly declining productivity in swidden fields, cropping occurs for only a short time. This means that the swidden landscape is dynamic as fields are abandoned and new ones cut. Beckerman (1987) states that a traditional Bari Indian group, for example, comprising about 50 people will generally have about 8 ha of land under cultivation at any one time. This will be out of a total territory of about 150 km^2, representing a ratio of productive to total land of about 1:1700. For the Bari, the selection of plots for cultivation will depend chiefly on soil characteristics. They prefer the *birida* which is a brown alluvial soil with a high sand content. They also prefer forest that has never before been cleared. In terms of productivity, there is much temporal and spatial variation depending on the crop and the age of the fields. On average, the Bari obtain approximately 18 t ha^{-1} yr^{-1} (Beckerman, 1987) which is about average for Amazonian swiddeners as a whole.

A detailed study of shifting cultivation, centred on the village of San Carlos de Rio Negro in the Amazon territory of Venezuela, has been reported by Jordan (1989).

Here, experimental and control plots were established to monitor the changes in nutrient cycles and productivity that occurred during slash-and-burn agriculture. In relation to net primary productivity, there was little variation over the four year monitoring period at the control site. However, under cultivation which involved a crop complex dominated by cassava (manioc, *Manihot esculenta*) with pineapple, plantain and cashew, net primary productivity decreased from 5633 kg ha^{-1} in the first year to 4140 kg ha^{-1} in the third year. For the edible parts of the crop, the manioc roots, net primary productivity declined from 1465 to 700 kg ha^{-1} over the same period. Estimates of energy ratios were also calculated based on inputs of field and processing labour (including the removal of cyanogenic compounds; see Table 4.1) and outputs comprising the caloric content of the processed manioc roots. A value of 13.9:1 was obtained which is much higher than Er values for mechanised agriculture, e.g. Tables 5.4 and 5.5. This value falls within the range 11:1 to 15:1 reported by Smil (1991) for small grains but is lower than the 20:1 to 40:1 he reports for root crops. However, Jordan (1989) makes the point that the San Carlos energy ratio takes no account of the energy contained within the forest that it was necessary to burn in order to release the required nutrients. Including values for this in the San Carlos energy ratio reduces it to 0.005:1. Jordan states, 'Thus when the services of nature are included, shifting cultivation becomes very energy inefficient. Those services are gathering nutrients from a dispersed and inaccessible state in soil, and concentrating the nutrients in a form which can be readily utilised by crops'.

One variation on shifting agriculture that has been discussed in relation to Amazonia is the agroforestry that is practised near the Peruvian village of Tamshiyacu (Figure 4.3), near Iquitos (Padoch *et al.*, 1985). This differs from traditional shifting cultivation in several ways. First, not all the slashed vegetation is burnt; instead much of the woody biomass is converted to charcoal and sold in Iquitos. Secondly, the crop complex is geared more to the commercial than subsistence level. For example, the annuals, rice, cassava, pineapple, peaches and tomatoes are grown along with plantain, cashew and uvilla which are semiperennials. In the second year, many of these crops are replanted along with crops like peach palm, umari and Brazil nut. Many of these crops are for the Iquitos market, as are the perennial tree crops that may still be harvested some 30 years after initial planting. The latter still require the fields to be tended. However, most of the staple crops cease to be harvested after about three years and land is left fallow for about six years before the cycle begins again. In this system both the charcoal and food produced are exported to Iquitos. More recently, Pinedo-Vasquez *et al.* (1992) have compared, using various economic formulae, the revenues possible in the area around the village of San Rafael in the Peruvian Amazon (Figure 4.3) from a variety of land-use types. The net revenue (after expenditure for labour and other costs is taken into account) is greatest for swidden-type agriculture. At US $893, the revenue for agricultural produce is higher than for the US $617 income from the selling of latex and fruits collected within the forest or the US $481 which is the value of the merchantable timber extracted from a swidden plot and which can be obtained from the plot in any case. The combined value of US $1374 makes swiddening even more profitable than keeping the forest intact and just harvesting a portion of its renewable resources. As Pinedo-Vasquez *et al.* point out, 'rural populations in the region can be expected to continue converting forested land to swidden agriculture unless alternative land uses become more attractive [than shifting cultivation] economically.' This case study also illustrates that shifting cultivators react to both the natural and social factors with which they are faced. The shortcomings of nutrient cycles are overcome by slash-and-burn techniques whilst the proximity of a market results in changes to the range of crops grown. This capacity to adapt has also been discussed by Jones and O'Neill (1993). They have modelled the shifting cultivators' response to increased population pressure and increased crop prices, i.e. changing market values. Both result in a reduction of the fallow period and an increase the amount of land under cultivation, as do higher interest rates on agricultural loans. Thiele (1993) has also examined the dynamics of shifting cultivation systems in frontier regions in Bolivia and shown them to be responsive to natural and cultural stimuli.

As in many African communities that practise shifting cultivation, there are pressures in some Latin American swidden systems that are creating environmental degradation and declining productivity. This is illustrated by the problems being experienced by the Campa Indians of the Gran Pajonal area of central Peru (Scott, 1987). Here primary and secondary forest is intermixed with savanna and grassland. The Campa Indians have been forced to move into the more inaccessible parts of Gran Pajonal because of pressure from settlers from the coastal and Andean cities of Peru where conditions for the poor are appalling. As a consequence, the new colonists have altered the land-use system, employing a fallow of only seven years. This contrasts with the Campa garden (*chacra*) plots on which cassava is cultivated with plantains, bananas, sweet potatoes and beans. Garden plots of *c.* 1 ha are felled, cultivated for about three years and then abandoned. A fallow period of at least 15 years is required for the system to be sustainable. Clearly, the recent adoption by the newcomers of a seven-year fallow is likely to create problems for the maintenance of soil fertility.

Not only do these examples illustrate the dynamic nature of shifting agriculture and its substantial role as an agent of environmental change but they also reflect the delicate balance that is essential between periods of cultivation and fallow. These examples also illustrate the impact of market forces on shifting cultivation and, most importantly, the impact of the displaced urbanites and landless poor.

4.2.3 Shifting cultivation in Asia

The rain forest regions of Asia are no more sacrosanct from the 'shifted' cultivators than are those of Latin America and Africa. In some instances government resettlement policies have been the root cause of forest degradation. One of the largest relocation schemes is Indonesia's Transmigration Programme (known as *Repelita*) whereby the landless are resettled from the heavily populated islands of Java, Bali and Madura into the outer islands, e.g. Sumatra and Kalimantan, of the Indonesian archipelago. Begun in 1978, the programme consists of a series of five-year plans. By 1988, 1.2 million people had been relocated and the final goal is reported to be 15 million people by AD 2000 (Repetto and Gillis, 1988). For many resettled families, plots of land have proved to be too small or soil fertility has been lost rapidly through leaching and soil erosion. In consequence, many have taken to shifting cultivation in the tropical forests where they are adding to the forest demise officially sanctioned as part of the government's resettlement programme. Despite such large-scale efforts to reduce the pressure on Java's landscape, there is abundant evidence (e.g. Smiet, 1992) to demonstrate that its few remaining forests are under threat from landless people.

On a more salutary note, Cramb (1993) has reported on the importance of shifting cultivation to the sustainability of farming systems in Sarawak, east Malaysia. Here some 20 per cent of the total land area, or 2.5 million ha, is used for shifting cultivation. The practice involves each family clearing between 1 and 3 ha of forest by axe, firing it to ash the biomass and then planting hill rice which is intercropped with maize and cassava. Unlike most of the examples of shifting cultivation described in Sections 4.2.1 and 4.2.2, the plot is only utilised for one year before it reverts to fallow for a period of between 5 and 15 years, during which time secondary forest develops. According to Cramb (1988), only about 5 per cent of the land cleared annually for shifting cultivation is in primary forest so the problem of the landless causing forest degradation is not as acute in Sarawak as it is elsewhere. Cramb (1993) also assesses whether shifting cultivation, which continually reuses secondary forest, is also sustainable. His conclusions do not support the government view (Sarawak, 1978 quoted in Cramb, 1993) that population pressure is reducing productivity through a shortening of the fallow period. For example, there is no statistical evidence to show that either the total area planted with the major staple, hill rice, or its yield per hectare has changed a great deal since 1960.

However, to examine the question of sustainability further, Cramb (1993) undertook a detailed analysis of two villages and their agriculture. He also makes the point that it is only possible to judge the sustainability of the agricultural system as a whole and not just that of its component parts. In Sarawak perennial cropping, to produce rubber, cocoa and pepper, is just as important as the shifting cultivation. Although the villages Batu Lintang and Nanga Tapih are located in the Saribas District of south-west Sarawak, which 40 years ago was considered to be environmentally degraded, shifting agriculture is still practised. Data for the demographic characteristics, land-use types and crop output are given in Table 4.2. Of the crop types shown, hill rice is produced by shifting cultivation with a low labour requirement per hectare and low output. Pepper is the most intensive permanently cultivated crop with correspondingly high labour requirements and output values. Rubber is intermediate.

During the time period represented by this study, population densities and commodity prices altered. The villages reacted to these external stimuli in different ways. First, Batu Lintang experienced out-migration; while the population of Sarawak increased overall by 32 per cent, Batu Lintang actually reduced its population. Conversely, the population of Nanga Tapih increased by 20 per cent. This difference Cramb attributes to the higher level of educational achievement in Batu Lintang that provided opportunities for people to seek non-agricultural work elsewhere. Changing commodity prices also occurred during this period; pepper and rubber prices, the latter to a lesser degree, experienced an escalation followed by a slump and then another escalation. Table 4.2 represents the adjustments made in these two communities to these external stimuli. True shifting cultivation, i.e. hill rice production, increased in 1984–1985 (see total output values in Table 4.2) as cash crop prices declined and then fell again in 1988–1989 as cash crop prices recovered. Cramb (1993) maintains that the changes in hill rice production were used to cushion

Table 4.2 Demographic, land use, crop output and income statistics for the villages of Batu Lintang and Nanga Tapih Sarawak (adapted from Cramb, 1993)

	Batu Lintang			Nanga Tapih		
	1979–1980	1984–1985	1988–1989	1979–1980	1984–1985	1988–1989
Demographic statistics						
No. of households	30	31	33	19	20	22
Total population	185	183	182	110	111	132
Population density (persons km^{-2})	14.0	13.9	13.8	18.0	18.2	21.6
Hill rice						
% Households planting	93	84	76	100	90	68
Area planted (ha)	36.30	33.00	13.00	27.00	34.00	14.90
Total output (t)	22.21	38.04	20.88	8.89	13.38	10.69
Output/ha (t)	0.61	1.15	1.61	0.33	0.39	0.72
Pepper						
% Households planting	97	81	88	100	90	100
Area planted (ha)	6.03	4.97	7.09	6.15	5.52	9.45
Total production (t)	8.24	7.19	10.43	10.56	6.26	16.20
Output/ha (t)	2.16	1.76	2.73	2.55	1.37	3.14
Rubber						
% Households tapping	87	32	79	79	75	23
Total output (t)	12.33	5.32	17.55	1.96	2.18	1.40
Output/household (t)	0.47	0.53	0.68	0.13	0.15	0.28
Income (Malaysian dollars)	Total (%)	Total (%)	Total (%)	Total (%)	Total (%)	Total (%)
Rice	298 (27)	1049 (41)	639 (20)	599 (20)	705 (31)	520 (11)
Pepper	70 (29)	817 (32)	1348 (41)	1859 (61)	1001 (44)	3932 (19)
Rubber	604 (23)	206 (8)	761 (23)	149 (5)	120 (5)	86 (2)
Other	561 (21)	500 (20)	522 (16)	455 (15)	430 (19)	430 (9)
Total	2633 (100)	2567 (100)	3269 (100)	3062 (100)	2255 (100)	4968 (100)

declining incomes as cash crop prices fell. The two systems, perennial cropping and shifting cultivation, thus enjoy a reciprocal relationship to keep the overall system stable but dynamic. In addition, the overall decline in area planted with hill rice does not, Cramb opines, necessarily imply instability. Rather the increases in productivity per hectare reflect improved management.

Pepper productivity mirrored price changes but was sustained overall indicating that the interrelationships between the crops were stable. The ability of such a commodity to recover is, Cramb suggests, an indication of resilience within the system. Rubber output also mirrored price rises and falls. At Batu Lintang, for example, there was a major decline in 1984–1985 as the price fell and an upturn in 1988–1989 in response to a revival in the market. Like pepper production, rubber remained an important output of the village. For Nanga Tapih, rubber production, never a major activity, was always fairly low and stable. The data for farm income also reflect most of these trends as would be expected. Both villages enjoyed a relatively stable economic environment though that of Batu Lintang, because it relied less on pepper than Nanga Tapih and had higher rice production, was more stable. Thus shifting cultivation could be considered to be the major anchor of the agricultural systems in this area. It must not, however, be forgotten that Batu Lintang's population density of 13.9 per km^2 was an advantage over that of Nanga Tapih at *c.* 20 per km^2. The integrity of the agricultural system, when considered as an amalgam of shifting and permanent agriculture, thus remained intact over a decade within which there were major accommodations to circumstances. Unfortunately, Cramb does not provide any data on environmental variables, such as soil erosion, to indicate if the agricultural system was being kept stable at the expense of the soil resource, though over a decade and with few artificial inputs it is unlikely that the depletion of such a resource could go unnoticed. In an earlier paper, however, Cramb (1989) suggested that there could be a substantial move of labour from shifting cultivation into cash cropping if the price of rice fell markedly. This would reduce the amount of shifting cultivation practised though it is unlikely to be eliminated in the foreseeable future. A similarly stable and sustainable swidden system incorporating rubber production with upland rice has been reported by Dove (1993a) from West Kalimantan, Borneo.

The role of shifting cultivation as one of several land-use systems within a landscape is also the subject of a detailed study in north-east India (Figure 4.3) by Ramakrishnan (1990, 1992). Here shifting agriculture is known as *jhum* and it is practised at both low and high elevations on slopes of between 20 and 40 per cent. The other forms of agriculture with which it is associated are valley agriculture, which involves the monoculture of rice, and terrace agriculture. The latter has been introduced as an alternative to *jhum* and involves a cropping pattern similar to that of *jhum* on terraces constructed in the hillside. This introduction was considered to be essential because of pressure on the food-production systems due to rapid population increase. Each *jhum* cycle comprises slash-and-burn cultivation for a year and then fallow for 15 years. The cycle, in the past, has usually been longer at between 20 and 30 years. The size of a plot is between 1.0 and 2.5 ha and each family consists, on average, of two adults with two to four children all of whom work on the plot. This region of north-east India enjoys a monsoon climate which, in turn, influences the *jhum* cycle. The dry winter months, for example, are the time when slashing of the undergrowth occurs; clear felling ensues usually being undertaken by the males of several families which indicates the communal nature of the enterprise. Burning takes place at the beginning of April, before the start of the monsoon, to ash the biomass *in situ*. Burning may even be repeated to leave the demarcated field on the hillside ready for sowing.

Seed sowing takes place after the first rains which provide soil moisture. The crops vary; cereals dominate long *jhum* cycles whilst perennial and tuber crops dominate short *jhum* cycles. As many as 35 species in total may be cultivated. To begin with, seeds of pulses, vegetables, cucurbits and cereals are mixed with dry soil and then spread after the burn. This ensures a good mixture and, obviously, means that crops are not planted in rows. Other crops, notably maize and rice, are sown using a dibble stick, amongst the established crops. A range of species, e.g. ginger, tapioca and banana, which provide food supplementary to that of the staple crops, is planted throughout the growing season. In addition, the leaves of *Ricinus communis* are collected for feeding silkworm caterpillars. The range of species grown thus provides a varied diet for the people whilst attempting to exploit the nutrients in the soil efficiently. As they mature, the crops are harvested. The practice continues for only a year though some perennial crops, e.g. banana, pineapple and orange, will provide a harvest for several years after the fields have been abandoned. Throughout the cultivation period weeding is essential.

The *jhum* cycle at higher elevations is similar though the natural vegetation of pine trees is not removed; only the lower branches are slashed. Because of the higher altitude, the crops grown differ from those on the lower elevation *jhum*. Tuber crops like potato are planted on the ridges of soil that are placed along the slope. Maize is also planted along the ridges. With a *jhum* cycle of 15 years, cropping is undertaken for only one year. Pig and vegetable manure are used to maintain fertility. With a *jhum* cycle of only five years, cropping occurs for two to three years but the range of crops cultivated is restricted to potato, maize and cabbage. Moreover, inorganic nitrogen–phosphorus–potassium fertilisers are required to maintain productivity.

In contrast, the valley agriculture consists of the monoculture of rice, though it is subsidiary to *jhum* because it can only be undertaken where topography permits between hill slopes. Permanent cultivation is possible because the soil is replenished by nutrients washed out from slopes at higher altitudes. One or two crops are produced annually from plots of between 0.5 and 0.75 ha per family. Terrace agriculture, on the other hand, has not proved a viable alternative to *jhum*. Although the cropping pattern is similar, a substantial input of artificial fertiliser is required. When the scheme was initiated, government agencies supplied the fertiliser but now the farmers themselves are responsible for paying. As a result, the system has not been successful, especially when soil loss and the consequent declining area for cultivation are taken into account. Ramakrishnan (1990, 1992) has also compiled a wide variety of data, which encompass economic as well as environmental data, to compare the efficacy of these various systems. Some of these data are given in Table 4.3.

In relation to the *jhum* cycles, sediment loss is greatest during the five-year cycle at both low elevations and high elevations. The relatively high sediment loss values indicate that erosion is a problem generally. Water loss also increases when the fallow shortens. The figures for energy output show that the 10-year cycle is the most

Table 4.3 Data on the *jhum* system of agriculture in north-east India (adapted from Ramakrishnan, 1992)

JHUM	Output/input energy ($MJ\ ha^{-1}\ yr^{-1}$)	Nutrients released: Ash ($kg\ ha^{-1}\ yr^{-1}$)	Nutrients released: Blow off* ($kg\ ha^{-1}\ yr^{-1}$)	Runoff from low elevations (cm): High	Runoff from low elevations (cm): Low	Sediment loss from high elevations ($kg\ ha^{-1}\ yr^{-1}$)
30-year cycle	34.1	3217	1511	29	23	–
10-year cycle	47.9	2677	1590	34	23	49.7
5-year cycle	46.7	643	264	37	30	54.9
Terrace	6.7					33.9
Valley (2 crops)	17.8					
Fallow—10 year cycle				19	0.8	3.5
Fallow—5 year cycle				27	1.1	1.9

* Lost from the system through soil erosion.

efficient. The energy efficiency for the terrace agroecosystem is low because it requires inputs of fertilisers. This innovation has not provided a suitable or popular alternative to *jhum*. The environmental problems caused by *jhum* are mainly due to the enforced shortening of the fallow period. Ramakrishnan (1990) suggests that it should be altered rather than abandoned or replaced with terracing. He advises that the minimum length of the cycle should be 10 years but where this is impossible due to population pressure an alternative would be to grow plantation/horticultural crops with forest on part of the land with limited *jhum*. Most importantly, it is necessary to avoid the importation of fertilisers, etc., and encourage the use of locally produced manure and/or biofertiliser. There is some potential for improving the efficiency of existing *jhum* systems by changing the crop mixture and bringing together the expertise of several different villages to improve crop organisation and management. With a five-year cycle, however, the system is clearly not sustainable unless it can be bolstered through the incorporation of agroforestry. This provides both products and services; the former include fruit whilst the latter include improved soil and water retention capacity. Problems of a similar kind have been experienced by shifting cultivators in other parts of the world. The Lua' farmers in northern Thailand, for example, have been forced to reduce the length of fallow and to encroach into primary forest (Kunstadter, 1987). In addition, there is evidence from lowland south-west Papua New Guinea (Eden, 1993) to show that intensification of swidden agriculture, for whatever reason, prevents forest regeneration and maintains the fallow as savanna.

4.3 CONCLUSION

Both nomadic pastoralism and shifting cultivation are widespread agricultural practices. Apart from reindeer herding, which takes place in the Arctic zone, most nomadic pastoralism and shifting cultivation take place in the tropics. Both are almost entirely dependent on solar energy, their only subsidy being that of human and, possibly, animal labour. This has led some workers (e.g. Smil, 1991) to refer to people who practise these ways of agriculture as solar farmers. The lack of a fossil-fuel input makes these systems energy efficient. Moreover, nomadic pastoralism allows high-quality protein to be produced from poor-grade vegetation.

Globally, there are many variations in nomadic pastoralism and shifting cultivation, both of which have developed over the millennia since the inception of agriculture. The spatial variations reflect environmental conditions and cultural traditions. This reinforces the comments made in Section 1.1 in relation to the many and varied components that comprise and influence agricultural systems. Moreover, these traditional systems are no less sophisticated than are high-technology agricultural systems (Chapters 5 to 7). For millennia, as far as it is possible to judge, they have constituted sustainable agricultural systems. Currently, however, many are experiencing major problems which are fundamentally social and/or economic in origin. Overgrazing and shortening of the fallow period have resulted, often causing environmental degradation. The most important underpinning cause of this is increasing population pressure within swidden communities. This is manifest in the example of shifting agriculture in Sarawak (Section 4.2.3) where one of the factors in the maintenance of sustainability in the system focusing on the village of Batu Lintang was out-migration of people. Such problems are exacerbated by the increase in landless poor, often originating in the ghettoes of developing world cities. These 'shifted cultivators' are creating a different problem to that which is threatening the sustainability of established shifting cultivation. On both counts though, tropical forests are under threat.

In the case of nomadic pastoralism there are other hazards. These include land encroachment by cultivators, government policies to restrict free movement across national boundaries and, ironically, some forms of development plans. There is a global trend towards sedentarisation. Another general characteristic of nomadic pastoral communities is that they do not operate in isolation; they are dependent on cultivators at the periphery of their grazing lands to provide grain and vegetable crops. Shifting cultivators, in contrast, operate as relatively self-contained village units unless they are near settlements that provide a market. Both, however, are forms of extensive agriculture whose continued existence is jeopardised for many reasons but particularly by population increase.

4.4 FURTHER READING

Colchester, M. and Lohmann, L. (eds) (1993) *The Struggle for Land and the Fate of the Forests*. The World Rainforest Movement, Penang, The Ecologist, Sturminster Newton and Zed Books, London.

Galaty, J.G. and Johnson, D.L. (eds) (1990) *The World of Pastoralism*. The Guilford Press, New York and Belhaven Press, London.

Gliessman, S.R. (ed.) (1990) *Agroecology: Researching the Ecological Basis for Sustainable Agriculture*. Springer-Verlag, New York.

Paine, R. (1994) *Herds of the Tundra: A Portrait of Saami Reindeer Pastoralism*. Smithsonian Institution Press, Washington, DC.

Schusky, E.L. (1989) *Culture and Agriculture*. Bergin and Garvey Publishers, New York.

Stone, J.C. (ed.) (1991) *Pastoral Economies in Africa and Long Term Responses to Drought*. Aberdeen University African Studies Group, Aberdeen.

CHAPTER 5

Settled Agriculture: Arable Systems

Agricultural systems that require ploughing, tillage or some form of soil preparation prior to crop planting are known as arable agricultural systems. This is the traditional definition which, to a certain extent, is inadequate because so-called 'no-tillage' crop production systems have been developed since the 1970s when direct drilling seed-emplacement techniques were first established. This chapter is concerned with both traditional arable and no-tillage agricultural systems (though the latter is also referred to in Section 10.4), which are collectively referred to as arable agriculture.

Of all the world's agricultural systems, the arable systems are the most diverse. Globally, a huge range of cultivars is grown to provide food for direct human consumption, feed for animals, fibre and speciality goods such as biomass fuels, beverages, etc. According to the World Resources Institute (1992), the world's croplands cover an area of 1477×10^6 ha; this is about 11 per cent of the world's total land area (see Table 1.1). In addition, arable agriculture is one of the components of agroforestry systems, the other major component comprising trees which permanently occupy part of the agricultural system (see Section 7.5). Whilst arable agricultural systems may be diverse in relation to the range of crops grown, they are also diverse in terms of the way in which crops are grown. Some arable systems are monocultural wherein the focus is on one crop only; this is particularly the case with high-technology or industrialised arable systems that predominate in the developed world. On the other hand, arable systems may concentrate on the production of a range of crops, as is the case in many subsistence systems in the developing world. There are also variations in relation to the nature and amount of energy that enters or is put into the system. There is a continuum ranging from very low energy to very high energy input systems. At the low energy input level the energy source, other than solar energy, is human and/or animal labour. Conversely, high energy input systems are those that are fossil-fuel energy intensive.

The spatial dimension adds yet another variable that determines the nature of arable agricultural systems. They occur in all but the most climatically hostile regions, from the boreal zone to tropical rain forest areas. Even the hot deserts are not exempt as irrigation agriculture is becoming increasingly widespread. Here the limiting factor of water availability is overcome by technology though, as discussed in Chapter 9, this is rarely without environmental consequence. Only the tundra and polar zones remain unaffected by arable agriculture. In the rest of the world the nature of the arable agriculture depends on the factors listed above, i.e. monoculture versus polyculture, subsistence versus commercial or cash-cropping agriculture, and energy-input characteristics, which operate within the constraints imposed by the physical environment (see Sections 1.2 and 1.3).

As a consequence of these many variables, it is virtually impossible to devise a satisfactory classification scheme for arable agriculture, or indeed for agriculture in general. That given in Figure 1.5 represents a major oversimplification but is included in this text chiefly because it illustrates the relationships that exist between the Earth's climatic zones (Figure 1.3), vegetation types (Figure 1.4) and agricultural types. A perhaps more appropriate classification scheme is given in Figure 5.1, which is based on the components and inputs of agricultural systems, notably crop types and numbers, production types and energy inputs. It is relevant to arable, pastoral and mixed farming systems. The examples given below relate to many of the combinations that are possible in arable agriculture.

5.1 THE PRODUCTION OF FOOD FOR DIRECT HUMAN CONSUMPTION

As discussed in Section 2.3, the cereals are the world's most important crops. Data for their productivity are given in Table 5.1 (see also Table 2.6) which shows that, in order of total production, wheat is the most important cereal, followed by rice and then maize. The other principal crops grown for human consumption include sugar cane, the tubers/roots of potato, cassava, sweet potato and yam, the pulses, various oil crops of which only a proportion is for human consumption, vegetables, fruits, berries and nuts. Most of these are cultivated in all of the possible combinations given in Figure 5.1. A complete appraisal is, clearly, impossible in a text such as this so the case studies that follow will serve to illustrate the many different types of agricultural systems in existence.

Why cereal crops have become, and continue to be, so important has been discussed by Wibberley (1989). The reasons are given in Table 5.2. More than half of global wheat production takes place in the developed world (Table 5.1), notably in the USA, Canada, the EU and the former Soviet Union (Figure 5.2). Here, wheat production is undertaken on a high-technology or industrialised basis. Most of the remainder is produced in China and India where there is still a considerable reliance on human and/or animal labour but with limited inputs of crop-protection chemicals and fertilisers. The threefold increase in wheat production that has occurred worldwide in the last 50 years is, according to Pickett (1993), due to an expansion of the cropping area and increased yields (Table 5.3). The latter are due to improved varieties, produced via plant breeding programmes, in combination with improved methods of cultivation. In the long term, further improvements are more likely to derive from biotechnology and genetic engineering (Sections 10.1 and 10.2) than from expansion of the cultivated area.

In much of the developed world wheat is a primary crop, some of which is exported to other developed countries, such as Japan, and to developing countries, such as China. One of the world's largest producers of wheat is the USA where it is the major crop in many central states of the Great Plains. Here dryland farming is practised on what was once extensive grassland. According to Riebsame (1990), approximately 90 per cent of the Great Plains is now agricultural land, i.e. comprises farms and ranches. Thus of 1.31×10^6 km^2, which is the total area of the Great Plains, 1.18×10^6 km^2 is agricultural and of this approximately 75 per cent, or 0.89×10^6 km^2, is cultivated, mostly for wheat. A general account of cropping systems on the Great Plains is given by Smika (1992) while a detailed account of dryland wheat farming in

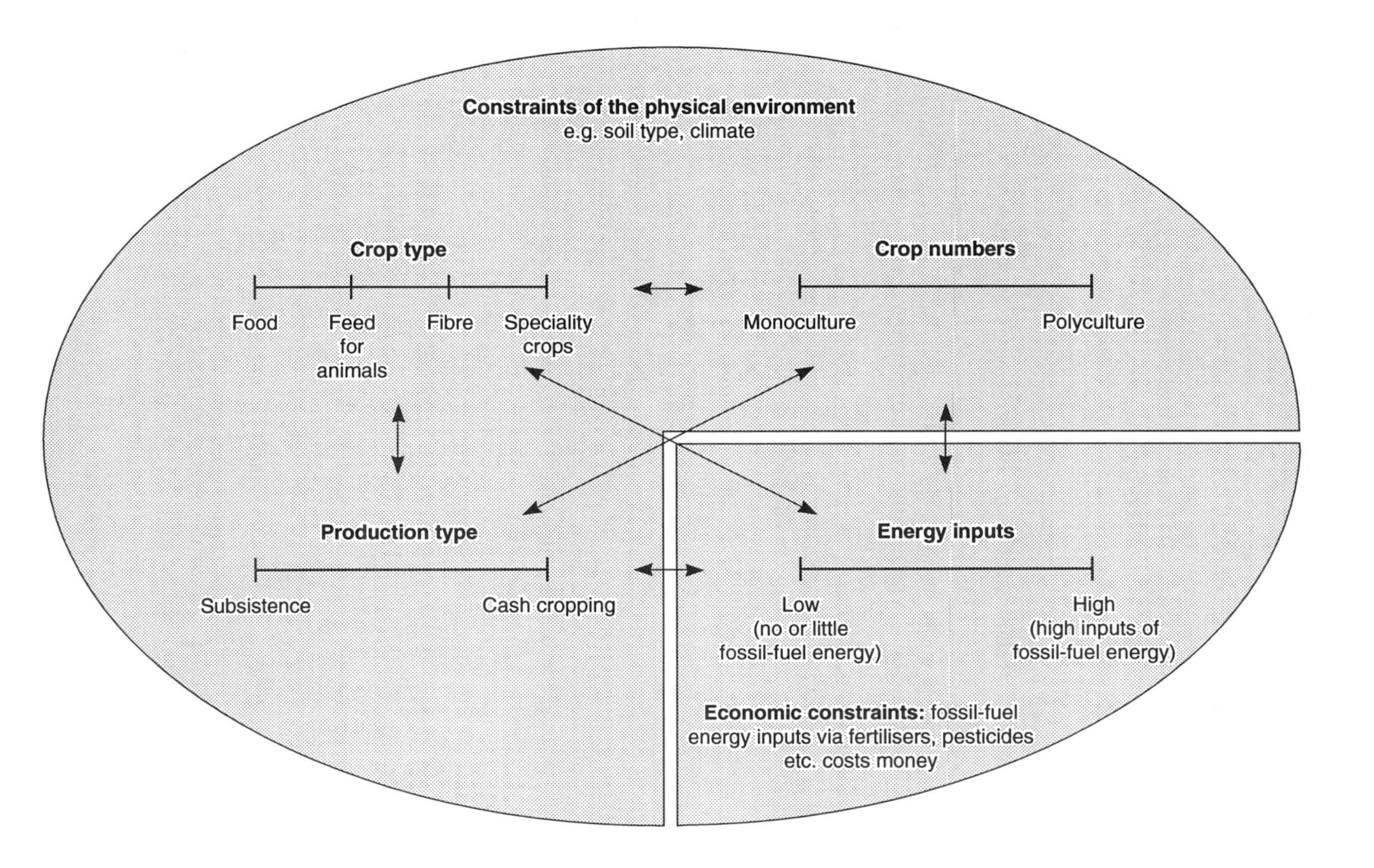

Figure 5.1 A potential scheme for classifying agricultural systems

Table 5.1 Production data for the world's major crops (abstracted from Food and Agriculture Organisation, 1993)

	1990	1991	1992
Wheat production ($\times 10^6$ Mt)			
Africa	13.682	17.547	13.308
North and Central America	111.539	89.944	100.415
South America	16.954	15.168	14.589
Asia	203.229	207.365	214.742
Europe	131.368	133.226	115.498
Oceania	15.255	10.869	15.172
Former Soviet Union	101.891	72.000	89.925*
Total World	592.918	546.199	563.649
Rice production ($\times 10^6$ Mt)			
Africa	12.422	13.545	14.011
North and Central America	9.182	9.167	10.109
South America	13.315	15.331	16.343
Asia	480.772	474.768	479.588
Europe	2.404	2.305	2.257
Oceania	0.879	0.774	1.162
Former Soviet Union	2.166	1.986	2.006*
Total World	521.140	517.875	525.475
Maize production ($\times 10^6$ Mt)			
Africa	33.790	32.646	23.963
North and Central America	226.474	214.452	263.479
South America	32.191	36.806	46.106
Asia	132.717	133.926	132.301
Europe	44.697	63.030	52.824
Oceania	0.385	0.382	0.375
Former Soviet Union	9.886	9.759	7.362*
Total World	479.140	491.001	526.410
Potato production ($\times 10^6$ Mt)			
Africa	7.314	7.836	7.531
North and Central America	22.779	23.465	23.885
South America	10.395	11.055	10.645
Asia	65.502	65.238	68.848
Europe	96.477	84.031	83.119
Oceania	1.488	1.443	1.440
Former Soviet Union	63.632	64.861	73.024*
Total World	267.586	257.929	268.492
Pulse production ($\times 10^6$ Mt)			
Africa	6.869	7.353	6.913
North and Central America	4.398	4.952	3.803
South America	3.010	3.608	3.707
Asia	25.691	26.544	26.134
Europe	7.802	7.156	7.031
Oceania	1.418	1.891	1.731
Former Soviet Union	9.006	5.210*	8.135*
Total World	58.194	56.714	57.455

* Estimate.

Table 5.2 The advantages of cereal cultivation (adapted from Wibberley, 1989)

A	USES
(i)	Cereals have a wide variety of uses, e.g. food for direct consumption by humans, livestock feed, industrial raw materials (examples include materials for brewing and for industrial and domestic use).
(ii)	There is a high demand because cereals provide a source of carbohydrates and fibre as well as some protein, fat, minerals and vitamins. In addition, prices have been stable for some time and there is plenty of scope for trade.
(iii)	Cereals can be easily stored.
(iv)	Cereals can be easily transferred, either from place to place as a component of regional/international trade or into other goods or cash.
B	DISTRIBUTION
(i)	Cereals are adaptable to a wide range of growing conditions. Adaptability has been enhanced by plant breeding.
(ii)	Cereals are an important and sometimes essential component of arable agricultural systems.
(iii)	The production of a surplus cereal crop is politically advantageous.
C	PRODUCTION
(i)	Total crop failure is unlikely so cereals have an inherent reliability. In addition, the labour requirement per tonne is relatively low.
(ii)	Cereals lend themselves to mechanisation because the plant parts harvested are above ground and widespread cereal cultivation, especially in the developed world, has justified the development costs of machinery.
(iii)	Compared with other crops, the necessary investment per hectare is low.
(iv)	Compared with other crops, cereals have good energy ratios, i.e. the output of energy in relation to the input of energy. Cereals also give a good yield in relation to seed planted.

Sedgwick County, Colorado, is given by Späth (1987). Here, as part of a wheat–fallow system, winter wheat is planted in early September and harvested using combine harvesters (usually with commercial rather than farm-based crews) in early July. The grain is transported to storage facilities which are individually, co-operatively or centrally owned. Thereafter the land is left fallow for some 14 months, during which time mechanical weed control takes place. Such practices are geared to the conservation of soil moisture and to minimising soil loss by wind and water erosion. The 14-month fallow period allows sufficient moisture to accumulate so that the risk of failure of the subsequent crop is reduced.

Späth states that several tillage systems are used, including maximum tillage, stubble mulching, minimum tillage and ecotillage. Ecotillage involves the application of herbicides to control weeds instead of mechanical control. Stubble mulching, whereby stubble from the previous crop is left in the field and which may be ploughed in, is used on *c.* 50 per cent of farms. However, ecotillage is particularly effective at conserving water and nitrogen availability, and because fewer mechanical weeding operations are required it is less fossil-fuel energy intensive than other forms of tillage. It also ensures that soil nutrients are not removed by weed competitors. Other measures to effect environmental conservation include contour terraces where sloping ground is cultivated, to control water supply locally, though this is not common. Wind erosion is minimised by maintaining large clods that provide a rough surface to

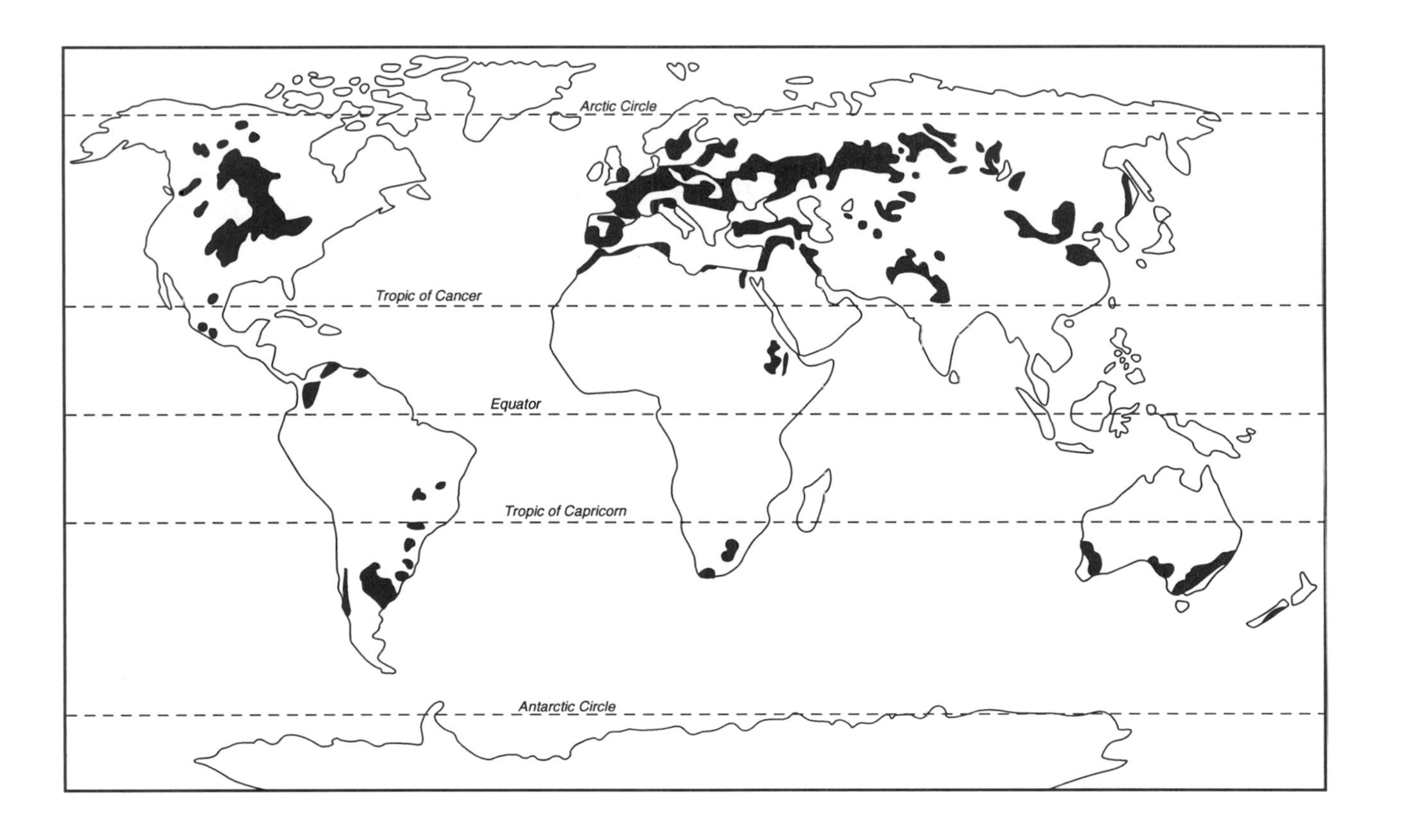

Figure 5.2 Wheat-producing regions of the world

Table 5.3 Changes in the production of some staple crops in relation to changes in the area under cultivation (abstracted from Food and Agriculture Organisation, 1992)

	Production × 10^6 t			Area × 10^6 ha		
	1990	1991	1992	1990	1991	1992
Wheat						
Developed countries	358.029	305.028	319.713	128.786	119.709	120.416
Developing countries	234.889	241.092	243.936	102.935	102.606	99.591
Rice						
Developed countries	25.623	24.183	26.772	4.385	4.279	4.518
Developing countries	495.516	493.693	498.703	142.303	142.691	142.650
Maize						
Developed countries	272.273	278.809	308.989	44.693	45.740	48.088
Developing countries	206.867	212.192	217.421	82.729	83.987	84.178
Potato						
Developed countries	187.723	177.296	184.841	11.322	11.138	11.408
Developing countries	79.863	86.634	83.651	6.406	6.646	6.623
Pulses						
Developed countries	20.799	17.246	19.258	11.407	11.196	10.747
Developing countries	37.395	39.468	38.197	56.224	59.330	56.717

trap entrained soil particles, by providing shelter belts, by reducing the field length in the dominant wind direction and by retaining a vegetation cover on the soil, e.g. by stubble mulching. Comparisons between the water-use efficiency, a key issue in Great Plains agriculture, and yield of continuously grown wheat and fallow–wheat show that both are much higher for fallow–wheat systems. As Späth states, 'Summer fallowing represents the single most important practice in wheat production under semiarid conditions'. Sixty-year averages for yield are 497.3 kg ha^{-1} for continuous wheat cropping and a huge 1424.6 kg ha^{-1} for the fallow–wheat system. Späth has also provided data on energy budgets for Sedgwick County farms. These are summarised in Table 5.4 which shows that the energy efficiency is closely related to fuel inputs. If these can be reduced by reducing the number of mechanical operations then the energy ratio can be improved, i.e. more energy is produced in the agricultural system for every unit of energy input. In addition, the fewer mechanical operations that are necessary, the more straw/stubble is left in the field. This provides protection from wind and water erosion as well as a means of recycling nutrients. These considerations are not fully reflected in the energy ratios of Table 5.4 but are nevertheless important issues in establishing sustainable agricultural systems.

Several additional studies have been undertaken on the energy flows associated with high-technology wheat production. Koizumi *et al.* (1990), for example, have examined the efficiency of solar energy utilisation in Japanese agroecosystems that involve double cropping. In this study, conducted in upland fields in Ibaraki Prefecture in central Japan, the three double-cropping systems consisted of upland-rice with barley, peanuts with wheat, and dentcorn (a type of maize) with Italian rye grass. For the peanut–wheat crop the annual net primary production was between 24.4 and 25.3 MJ m^{-2}, which represents 0.5 per cent of the global solar radiation received. The data can also be expressed in terms of dry matter production. For

Table 5.4 Energy budget data for wheat-producing farms and various tillage practices in Sedgwick County, Colorado, USA (adapted from Späth, 1987)

	Field-energy budget (kcal)			
	Fertilised wheat		Unfertilised wheat	
	Five tillage operations		Six tillage operations	
Total fuel energy	28 096 471.5	(19.96%)	55 098 759.5	(70.89%)
Total labour energy	5 440.5	(0.04%)	105 932.8	(0.14%)
Total embodied energy	216 151.1	(0.15%)	61 340.7	(0.8%)
Total seed energy	21 262 500.1	(15.10%)	19 440 000.0	(24.92%)
Total fertiliser energy	89 317 728.3	(63.44%)	0.0	(0.00%)
Total herbicide, etc.	0.0	(0.00%)	0.0	(0.00%)
Total transport energy	1 727 689.9	(1.31%)	3 317 606.5	(4.25%)
Yield energy	$1\ 567\ 350 \times 10^3$		$1\ 458\ 000 \times 10^3$	
Total input energy	$140\ 786 \times 10^3$		$78\ 024 \times 10^3$	
Total output energy	$1\ 567\ 350 \times 10^3$		$1\ 458\ 000 \times 10^3$	
Output/input ratio (Er)	11.13		18.69	

Energy ratios for various tillage practices

	No. of mechanical operations	Er
Clean tillage	5–8	5–15
Semiclean tillage	4–6	10–15
Stubble mulching	4–5	15–20
Minimum tillage	2–4	20–30
Eco-tillage	1–3	20–35

example, for each 1.0 MJ of global solar radiation absorbed, 1.3–1.5 g of dry matter for cereal crops, including wheat, are produced. (Note that there is a difference between global solar radiation received and global solar radiation that is absorbed; see Section 1.3 and Figure 1.8.) This, however, is with the help of 1500 g m^{-2} of nitrogen fertiliser. The total annual net production is in the order of 1500 g m^{-2} (15 000 kg ha^{-1}), which is comparable to that of evergreen broad-leaved forests in the warm temperate zone. However, data for the efficiency of solar energy utilisation show that the forage crops, i.e. dentcorn with Italian rye grass, are twice as efficient as wheat, with values of 0.5–0.6 per cent for the latter and 1.0–1.2 per cent for the former. The values for the forage crops are higher than any values so far reported for natural ecosystems. Unsurprisingly, however, the fact that about 90 per cent of the energy is removed from the agroecosystem means that the remaining 10 per cent, in the litter, stubble and roots, is inadequate to maintain fertility.

The role of fossil-fuel energy in wheat production has also been considered by Tsatsarelis (1993) who has investigated winter wheat production in Greece. The production systems are technology and fossil-fuel dependent as they include seed-bed preparation, drilling, fungicide, herbicide and fertiliser applications as well as harvesting by combine. Table 5.5 gives energy budget data for this system and variations that include straw baling and irrigation. The energy ratios reflect increased energy transfer efficiencies where straw baling and irrigation are involved though the output from this type of system is not all food energy. This increased efficiency

Table 5.5 Energy budget data for soft winter wheat production in Greece (based on Tsatsarelis, 1993)

INPUTS		
Fuel	Common operations	5 026–9 508
Fertilisers		8 105–12 449
Crop-protection chemicals		193–305
Machinery		954–1 238
Seed		1 815–2 570
Total energy		16 120–26 070
Fuel	for straw	2 176
Machinery		794
Total energy		2 970
Fuel	for irrigation	1 300–2 600
Machinery		150–300
Total energy		1 450–2 900

SYSTEM	INPUTS (MJ per ha)	OUTPUTS (MJ per ha)	Er
Common operations	16 120–26 070	37 800–75 600	2.35–2.90
Common operations + irrigation	17 570–28 970	45 360–90 720	2.58–3.13
Common operations + straw	19 090–29 040	111 510–173 880	5.84–5.99
Common operations + irrigation + straw	20 540–31 940	119 070–189 000	5.79–5.92

reflects the high values associated with fertiliser input which accounts for nearly 50 per cent of the total energy input. Mechanised activities, e.g. seed-bed preparation and drilling, are also energy intensive. Any efforts to make this system more energy efficient would need to focus on these two aspects of the agricultural system.

Data such as these highlight the importance of fossil-fuel inputs into agricultural systems in the developed world. Recently, however, Bonny (1993) has examined the energy consumption characteristics of wheat production in France for the period 1958–1990 and discovered some interesting underlying trends. Her data show that the energy intensity of agriculture increased between 1959 and 1977 at an approximate rate of 3.8 per cent per year. Thus, over this 18-year period, an increasing amount of energy was required to produce each unit of agricultural output. After 1977, the situation changed with a decrease in energy intensity of *c.* 1.6 per cent per year. This can only mean that agriculture was becoming more energy efficient, a condition that had occurred a few years earlier in other sectors of the economy. This change in energy-consumption pattern in agriculture Bonny attributes to increased efficiency prompted by the oil-price issues of the early 1970s. Energy conservation measures include improved methods for greenhouse heating, improved fuel consumption by tractors, improved machinery such as crop sprayers and fertiliser spreaders which allowed more land to be treated per fuel unit than hitherto, and improved methods of forage drying. Clearly, there is scope for further enhancement of energy efficiencies. In view of the data given in Table 5.4, which shows that fertiliser use constitutes a major component of the energy input to wheat production, a substantial saving will ensue if nitrogen-fixation by crop plants themselves is made possible by genetic engineering (Section 10.2). Improvements in crop health, the efficacy of crop-protection chemicals, mechanical efficiency and irrigation will also contribute to improving energy efficiency.

Arable agriculture is not, however, always characterised by high, or even any, inputs of fossil-fuel energy. Sorghum, for example, is the major cereal crop produced in the developing world and is often grown as a component of intercropping systems. As Figure 5.3 illustrates, most sorghum production occurs in Africa and Asia, though a proportion, albeit small, is also produced in the developed world, notably in the USA, often as one component of cereal-producing systems. Of the many species of sorghum that exist, the grain crops are *Sorghum bicolor* (Langer and Hill, 1991). The seed of grain sorghum is also known as dura but because it contains no gluten it is not suitable for making bread. Instead, it is made into a porridge or batter by grinding the grain into a flour and mixing it with fat or water. In addition, sorghum may be fermented to produce alcohol and it may be fed to pigs and poultry. As a forage crop, varieties of *Sorghum saccharatum* are cultivated to provide food for cattle, particularly in North and South America, Europe and Australia.

One of the major advantages of sorghum as a crop is its resistance to drought, which makes it appropriate for growing in areas unsuitable for other cereals. It has, in consequence, become the most important cereal in semi-arid regions, including those with an erratic rainfall. Doggett (1988) suggests that the cultivated sorghums originated from the wild species known as *Sorghum bicolor* subspecies *arundinaceum* in the Ethiopia–Sudan region. As a C4 plant (see Section 1.2), sorghum is well suited to hot and dry environments though it is only moderately tolerant of salinity. The various processes involved in sorghum cultivation have been detailed by Paul (1990) who describes the majority of the sorghum production systems in the tropics as subsistent rather than commercial. In particular, he describes the role of sorghum in intercropping systems which involve the cultivation of two or more crop types in the same field at the same time. The aim is to enhance total crop productivity by using crop combinations that complement each other in terms of their nutrient requirements (see also Section 4.2.2) and which maximise the often-limited resources of the available soils. This reflects what Loomis and Connor (1992) describe as niche differentiation and what Ong and Black (1994) refer to as complementarity in resource utilisation.

The two most important considerations which determine intercropping combinations are the length of the growing seasons and the adaptation of crops to specific environments (Vandermeer, 1989). With average annual rainfalls of less than 600 mm, early-maturing and drought-tolerant crops such as sorghum and millet are the most suitable. Where there is rainfall in excess of 600 mm per year, cereals and legumes may be included. In Latin America, for example, intercropping systems are dominated by maize and beans along with sorghum. According to Paul (1990), *c.* 6×10^6 tonnes of sorghum are produced in Central America for both animal feed and direct consumption. It is also produced using high- and low-input agricultural systems (Figure 5.1). In Mexico, Costa Rica, Panama, the Dominican Republic and the Pacific coastal plains, sorghum is grown under monoculture for animal feed using imported hybrid seed and mechanisation. Its production is, thus, more akin to the commercial production of wheat described above. In the interior of Central America, however, in the cultivated basins, valleys and hillslopes of Guatemala, El Salvador,

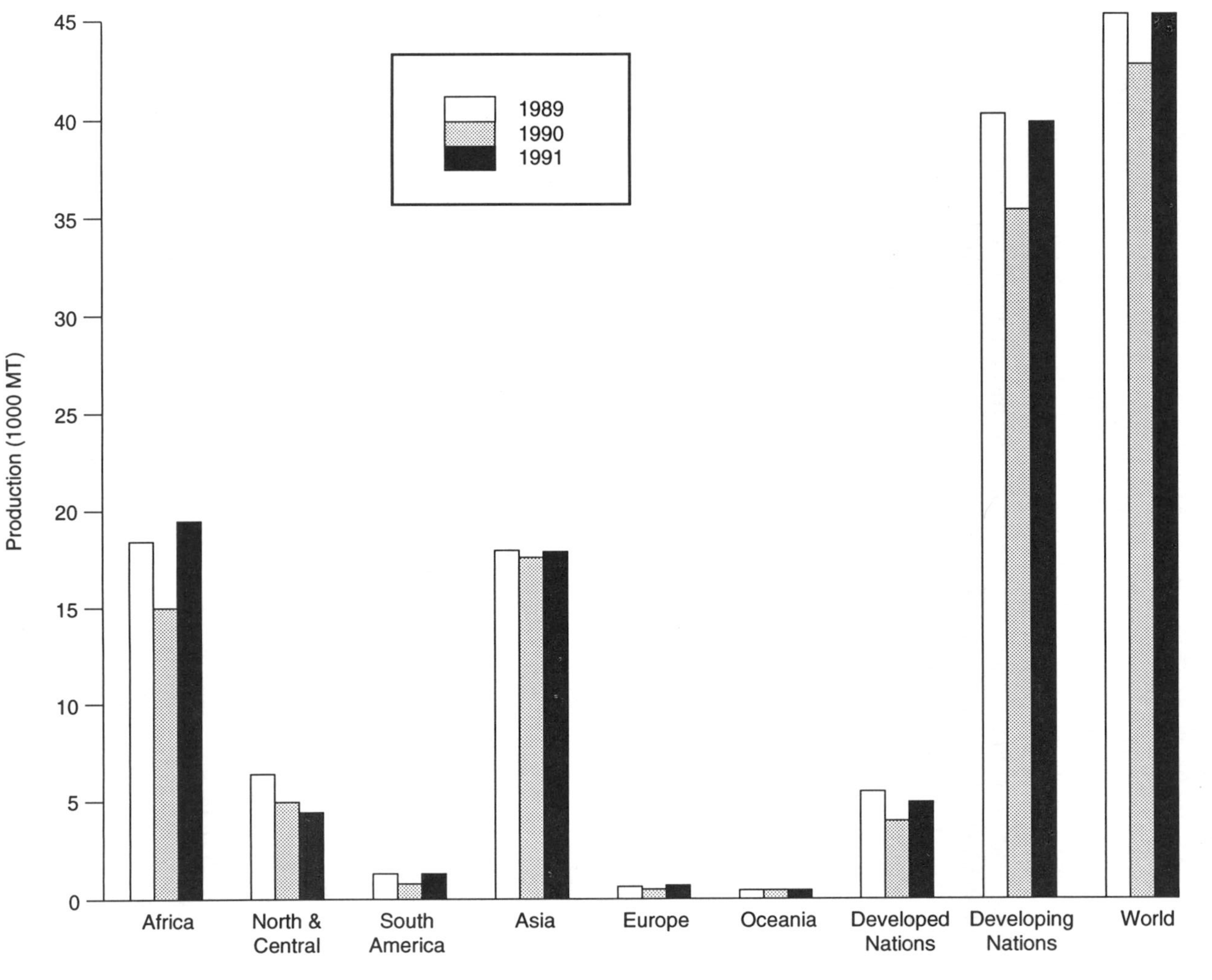

Figure 5.3 The amounts and distribution of sorghum production, 1989–1991 (based on Food and Agriculture Organisation, 1992)

Honduras and Nicaragua, small farmers with rented landholdings of less than 7 ha (and sometimes less than 3 ha) grow *c.* 30 tonnes of sorghum annually in subsistent multiple-cropping systems. Local stocks of the crop, which are prone to attack by insect pests and disease, are cultivated. This means that productivity is low but, overall, the system is balanced and has many similarities to that of the shifting cultivation system of south-west Sarawak discussed in Section 4.2. In the latter, hill rice production acted as a buffer for farm incomes that relied on fluctuating pepper prices. In Central America, Paul reports that sorghum production is the buffer whereas maize productivity is variable. Generally, maize is the most important of the three crops used in the intercropping systems. Sorghum may be second or even third in importance after beans, e.g. lima bean (*Phaseolus* spp.) or cowpea (*Vigna* spp.). When the maize harvest is good, sorghum is used for animal feed; when the reverse occurs, sorghum is used for human consumption. It thus acts as a form of protection against a maize harvest failure in much the same way that hill rice production in south-west Sarawak hedged against widely fluctuating pepper prices.

The nature of the intercropping systems themselves has been detailed by Hawkins (1984) and Paul (1990). Maize and sorghum are planted, either together in the same row or in alternate rows, in May, when the wet season begins. The crops compete and sorghum will depress maize productivity when the two are planted in the same row rather than alternately. In mid-August, maize attains physiological maturity and bends below the ear. As a result the sorghum is less shaded and grows rapidly, encouraged by the second wet season. It flowers in November and is harvested, along with the maize, in December/January. Thus, differences in crop physiology, light and water requirements are exploited. Paul, however, suggests that planting the two crop types in the same row is not as efficient as inter-row cropping because of interspecific competition. In addition, there are some advantages to planting the sorghum crop much later than the maize, in June or even August. By this time the maize is established and does not suffer so much from competition. Increases of 46 per cent in productivity can be obtained though sorghum productivity declines because of its reduced growing season. Declines in productivity may be as much as 41 per cent because sorghum seedlings have to compete with maturing maize plants with established root systems. The addition of nitrogen fertiliser, experimentally, gave increased yields but adds cost and energy to the production system. In another variation lima beans are included in the agricultural system. Planted with maize in May, the beans are harvested in August by which time the maize is doubled over and there is additional nitrate in the soil (due to the bacteria associated with the leguminous bean crop). In September, newly planted sorghum, especially if it comprises improved varieties that are not sensitive to photoperiods (day length), can do well under these conditions. Both maize yields and sorghum yields increase.

The intercropping systems described above obviously relate to the production of maize as much as they do to sorghum. Intercropping systems of a related type in Mexico have been described by Gleissman (1988, 1990). Here, near Tabasco, intercropping utilised a maize–beans–squash complex. As in the sorghum intercropping system, the beans constitute a means of enhancing nitrate in the soil system which, in turn, benefits the other crops. Squash helps to control weed growth through the presence of its broad horizontal leaves which shade the soil. Overall, maize yields are increased by as much as 50 per cent when compared with its monoculture.

Although the productivity for beans and squash was lower than if they were cultivated singly, the overall advantages make it an attractive low-input agricultural system.

The use of leguminous crops to encourage nitrogen-fixing bacteria, and so enhance nitrate availability in the soil, is also known as 'green manuring'. Historically, it was part of traditional rotation systems in Britain and Europe in the Middle Ages (Section 3.3.1). Today, it is a practice that, first, underpins many subsistent arable agroecosystems such as those discussed above and, secondly, is a major component of arable agricultural systems that are 'industrialising', i.e. those agricultural systems that are becoming increasingly commodity-oriented and commercial but which still rely, to an extent, on traditional methods. These are what Turner and Brush (1987) describe as mixed-technic systems since they employ traditional technologies, including high labour inputs, and modern technologies that involve fossil-fuel inputs. Many of the agricultural systems in Asia fall into this category. Examples include rice production in the Punjab of north-east India (e.g. Leaf, 1987), Bangladesh (e.g. Ali, 1987) and many other regions (e.g. Hargrove, 1990).

One such example is that of agriculture in Upper Egypt. As discussed in Section 3.1.2, agriculture has a long history in Egypt where the Nile and its silt have traditionally been the mainstays of crop productivity. The construction of the Aswan Dam in 1964, however, altered the course of Egypt's agricultural (and other) development because it facilitated tight control of irrigation in the Lower Nile valley and prevented silt deposition. The former prompted the pumping of groundwater in Upper Egypt in order to ensure water availability. This, in turn, prompted double cropping and meant that applications of artificial fertiliser became essential to compensate for the nutrient deficiency resulting from the curtailment of silt deposition. Hopkins (1987) has described the agricultural systems that occur around the village of Musha, 400 km south of Cairo, as being typical of Egypt as a whole. In particular, he makes the point that there are three strips parallel to the Nile, each characterised by well-defined features. The strip nearest the river consists of a raised bank where village agriculture produces palms, bananas and some vegetables. Fishing provides a source of protein and a product to trade. The outermost strip occurs along the margin of the desert and contains villages, constructed as resting places for caravans, as well as quarries, cemeteries and monasteries. In between, there are basin villages of which Musha is one. These villages are built on mounds that are above the old level of the flood water. The soils are dark and heavy but productive. Such villages attract labour from the other two strips. Around Musha itself, strip fields predominate, most of which are held under freehold tenure.

The chief summer crops are cotton, maize and sorghum. Wheat, beans, lentils, chickpeas and bersim (a type of clover) are the main winter crops. The rotation cycle, with a change from summer to winter crops in November and vice versa in April, operates over two years. For example, in the first summer cotton may be grown; it will be followed by wheat in the first winter and then maize, sorghum or fallow in the second summer. Finally the legumes will be grown. Limited amounts of fruit and vegetables will also be cultivated and all farmers keep animals, notably water buffaloes and cows which provide dairy produce. In relation to technology, animal and human labour predominate. Traditional implements, e.g. the short-handled hoe, sickle, digging stick and shallow plough, are widely used though tractors have been important since the late 1960s. Tractors may be owned, by farmers or families with sufficient holdings to make ownership worthwhile, or rented. Other important inputs are artificial fertiliser, some pesticides and animal manure. The first two of these, like the tractors, represent energy inputs which are supplemented by the pumping machines required to extract and disseminate irrigation water. Irrigation systems are government controlled and there is much government intervention in the marketing of produce, especially cotton. The system thus consists of combinations of contrasts: freehold tenure of land with government control of irrigation and, to some extent, marketing in tandem with traditional land preparation allied with mechanisation and artificially produced crop-protection chemicals and fertilisers.

Rice is also an important crop of Egyptian agricultural systems. According to Herdt (1989), it occupies about 20 per cent of the area planted to summer crops and the total production of approximately 2.45×10^6 t. Average yields are *c.* 5800 kg ha^{-1}, which are comparable with yields in north-east China (see below). However, the environmental problems that accompany irrigation (Chapter 9), along with the ever-increasing demand for improved productivity, mean that Egypt's agricultural systems must adopt conservation practices. The management of artificial fertilisers, for example, has been discussed by Hamissa and Mahrous (1989) in relation to the timing of applications, the rate of irrigation and the rice varieties being cultivated. Other potential improvements and experimental work in progress, notably at the International Rice Research Institute in the Philippines, have been examined by Hargrove (1990).

Another example of a mixed-technology agricultural system is that of rice production in north-east China. One crop per year is produced via irrigation of 1.2×10^6 ha; the average yield is 5100 kg ha^{-1} (Wen and Pimentel, 1992). Most of the energy required for planting and harvesting comes from human labour but there is a subsidy which derives from irrigation and artificial fertiliser. Wen and Pimentel suggest that the efficiencies of both of these factors could be increased with a resulting expansion of rice production and improved water resource use. The latter could be effected by employing irrigation techniques which do not provide for a water cover throughout the growing season; instead water is applied periodically five or six times per year. Yields of between 5300 kg ha^{-1} and 7500 kg ha^{-1} can be so achieved whilst simultaneously conserving water which could be used to extend the area under cultivation by as much as 40 per cent. Wen and Pimentel also suggest that the cultivation of the common duckweed (*Azolla filicaloides*) could contribute to productivity and reduce the need for artificial fertilisers. This is a floating pteridophyte (fern type) with an associated blue alga in a symbiotic relationship. It grows well in paddy fields, along with rice, where it serves to control weeds which it engulfs. Rice yield is increased because the algae fix nitrogen and so enrich the soils while the *Azolla* itself provides a feed for pigs, chickens and fish. It can also be ploughed into soils as a form of green manure. Measures such as these are particularly important for improving crop productivity, and at the same time protecting the environment, in a country with such a large and growing population (see also the discussions by Tao, 1991a, b; Zhao, 1994).

Many crops other than those described above are produced from arable agricultural systems. These include roots, tubers, fruits, other cereals such as rye and millet, and vegetables. Most of these are also produced as components of mixed farming systems (Chapter 7). The examples given above, however, illustrate the diversity of crops that are produced via arable agriculture worldwide.

5.2 THE PRODUCTION OF FEED FOR ANIMALS

Many arable agricultural systems are given over to the production of feed for livestock. Such systems are most prevalent in the developed world where meat consumption is highest, though there is some production in the developing world. The range of crops grown for this purpose includes cereals, legumes, roots and tubers. As in the case of arable crops for human consumption, the systems of production vary considerably; some are fossil-fuel energy intensive whilst others rely heavily on human labour. The production of feed for animals can be subdivided into two types: fodder crops and forage crops. Fodder crops are usually harvested and then fed to animals while forage crops are directly grazed by domestic livestock. In some cases, the crops are transferred from their point of origin to be consumed by livestock elsewhere. This may even involve international import and export. Alternatively, the crops may be consumed *in situ*. This means that there is overlap with the mixed farming systems considered in Chapter 7.

Globally, the most important fodder crop is maize. Worldwide nearly 526×10^6 tonnes were produced in 1992 (Table 5.1). Of this, approximately 66 per cent is fed to livestock; of the maize production in the USA (*c.* 263×10^6 t in 1992) a massive 87.8 per cent is fed to livestock (Brenner, 1991). The figures for several other developed nations are similar: 80.2 per cent for Japan, 82.5 per cent for Canada and 82.8 per cent in France, though in France this percentage is declining (Brenner, 1991) due to competition from imported crops that attract only a low import duty. As Table 5.6 shows, the situation is quite different in developing countries where a much lower, but nevertheless, increasing proportion of total maize productivity is used as animal feed.

In the USA, which dominates world maize production (Table 5.1), the crop is produced in high-technology agricultural systems, particularly those that predominate in the so-called 'corn belt' states of Iowa, Illinois and Indiana in the Mid-West. A typical maize-producing agricultural system has been described by Loomis and Connor (1992). Crop production centres on maize and soy bean with small amounts of leguminous forages, oat, wheat and some pasture. The output consists of high-quality animal feed which is used for the production of pigs, beef cattle and dairy products. The farms that replaced the original tall-grass prairie have experienced, and generally benefitted from, such technological innovations as seed-quality improvement, mechanisation, artificial fertiliser applications and crop-protection chemicals. Consequently, fossil-fuel inputs are high.

Crop rotations are employed, e.g. maize with soy bean. The maize is sown in April or May after field preparation which involves the ploughing-in of earlier crop residues and the application of fertilisers impregnated with pre-emergence herbicides. Pest and weed control measures may be necessary during the 120-day growing period and the maize is harvested using combines. The grain is dried artificially in company-owned or farm-owned elevators. A proportion of the crop is sold; the remainder is fed to animals. Similar methods, but with reduced pest protection, are used to produce soy bean which is nearly all exported from the region to be used as animal feed elsewhere.

Table 5.6 Data on changes in maize utilisation (data abstracted from Brenner, 1991)

	1983–1987 % used as			% growth rate for food			% growth rate for feed		
	Food	Feed	Other	1961–1970	1970–1980	1980–1986	1961–1970	1970–1980	1980–1986
Developing Nations	40	50	10	3.5	3.7	1.8	7.7	7.7	2.1
Developed Nations	6	78	16	2.0	2.7	1.6	3.1	2.7	0.3
Former USSR and Eastern Europe	4	72	25	–0.1	–0.1	–0.4	–1.6	8.6	3.1
World Total	20	66	14	3.3	3.6	1.8	3.8	3.7	0.7

The animals raised in the corn-belt farms receive animal health treatment which represents an additional energy input, though Loomis and Connor (1992) do not provide any overall energy budgets. For maize production itself, however, an energy ratio of 6.2 is quoted, based on an analysis of a farm in Indiana in the mid-1970s. Energy use may have become more efficient in the ensuing two decades, as it has for wheat production in many high-technology systems (Section 5.1). Nevertheless, the most significant energy input in both USA maize production and European wheat production (Section 5.1) is that related to fertiliser use.

Soy bean is another major crop grown mainly, but not exclusively, for animal feed. Domesticated in north China (L.T. Evans, 1993), soy bean has a huge range of uses. It is consumed directly, and produces flour as well as oil which can be used for cooking and many other products including non-food items, e.g. soap. It has a comparatively high protein content, 36 per cent of the whole seed (Langer and Hill, 1991), and it is widely used as animal feed. According to Borget (1992), soy bean was primarily a food of the Far East, where it was used as a protein curd from ground seed and in various fermented forms, until the 1920s. China was the main producer until 1930 but its introduction into the USA at the beginning of the century led to its rapid adoption. Today, the USA is a major producer, as reflected in Table 5.7, though its prime use is as an animal feed. Overall, its versatility has led to the rapid rise in its production. According to Langer and Hill (1991), its spritely adoption in the USA was during World War II when there was a shift from butter to margarine consumption. Today, soy bean provides *c.* 15 per cent of all the edible oil consumed. After oil extraction, the remaining soy bean cake is widely used for feeding livestock. Like maize, soy bean is grown under a wide range of conditions. In the USA and Europe there are substantial subsidies of fossil fuel whereas in developing countries, the major producers being those of Latin America (see Table 5.7), production is mainly in agricultural systems characterised by low or no fossil-fuel inputs or those with mixed technology (cf. Turner and Brush, 1987; see Section 5.1). In the USA and Europe, the agricultural systems wherein soy bean are cultivated are similar to those described above for maize and for wheat (Section 5.1). In the tropics, in developing countries, Borget (1992) states that soy bean is cultivated in alternation with other crops, notably cereals, or as one component of mixed-cropping systems (Chapter 7).

Other crops grown to provide food for animals include wheat, barley, peas, field beans, as well as various roots and tubers, e.g. beets, mangels, turnips and potato, and some brassicas, e.g. kale and rape. Many of these are produced as components of mixed farming systems (Chapter 7) and all are grown for direct human consumption as well as for animal feed. The use of cereals grown specifically to produce silage for animal feed, in Europe, has been discussed by Stark and Wilkinson (1992). Crops of wheat, barley and legumes that are converted into silage in Europe are grown in high-technology agricultural systems with high inputs of fossil fuels. The role of other fodder crops has been discussed by Pollott (1992) in relation to agriculture in the UK. In many other parts of the world, livestock are often fed on the remains of crops that cannot be used for direct human consumption. One such example is sugar cane. In Cuba, 11 different animal feeds produced from sugar cane amounted to nearly 5×10^6 t in 1988 (Perez, 1990). These feeds comprise various types of molasses, trash and yeasts and are produced to support poultry, pigs, dairy cattle and beef cattle. The sugar mills, which produce the raw material for feed production, now produce their

Table 5.7 Production data for soy bean (abstracted from Food and Agriculture Organisation, 1993)

	Area cultivated $\times 10^6$ ha				Production $\times 10^6$ t			
	1979–1981	1990	1991	1992	1979–1981	1990	1991	1992
Africa	0.352	0.370	0.486	0.419	0.326	0.608	0.677	0.492
North & Central America	28.145	23.661	24.438	24.596	56.195	54.304	56.299	61.881
USA	27.561	22.869	23.476	23.626	54.961	52.416	54.066	59.780
South America	10.928	17.699	15.356	15.273	18.010	33.049	28.278	32.471
Asia	9.719	12.643	12.209	12.637	10.339	16.761	15.306	16.178
China	7.506	7.564	7.045	7.240*	8.266	11.008	9.718	9.707*
Europe	0.495	1.034	0.710	0.835	0.623	2.457	1.950	1.999
Oceania	0.050	0.049	0.040	0.030	0.085	0.077	0.062	0.051
Former USSR	0.852	0.827	0.808	0.800*	0.494	0.844*	0.920*	0.940*
Developed countries	29.402	25.470	25.859	26.150	55.613	57.436	58.780	64.422
Developing countries	21.138	30.881	28.187	28.440	29.033	50.704	44.712	49.589
World	50.540	56.351	54.046	54.591	86.072	108.141	103.492	114.011

* Estimate.

own livestock products, e.g. pigs, broilers, sheep, poultry, for local consumption. Cattle-rearing is a separate enterprise. This makes good use of material which was often hitherto discarded or burnt, though it is still necessary to import cereals to mix with the sugar-cane products in order to provide a balanced feed. The production of sugar cane itself is 68 per cent mechanised (Perez, 1990) and it is a crop produced in monocultural agricultural systems. In Cuba, sugar-cane production is a mixed-technology agricultural system. It is also a crop from which biomass fuels are being produced.

Of the world's root and tuber crops that are cultivated to produce animal feed, the potato is the most important. Originally domesticated in the South American Andes (see Section 2.3.5) the potato is now widely grown (Table 5.1). Of the 261×10^6 t grown annually, approximately 6.5 per cent is used for animal feed, mostly in Western Europe where it currently accounts for *c.* 20 per cent of the total potato crop. According to Horton and Anderson (1992), this represents a decline from about 50 per cent before World War II. However, approximately 33 per cent of the current potato crop in Eastern Europe is used for animal feed. In both Eastern and Western Europe, though more so in the latter, potato production is highly mechanised. All of the stages in production, which includes seed-bed preparation, fertiliser application, planting, pest control, harvesting, handling, and grading, involve an energy subsidy. These stages have been described in detail by Beukema and van der Zaag (1990) and Witney and McRae (1992). According to the former, one hectare of potatoes (50 tonnes) can be produced and stored in 20–40 person-hours. This contrasts with the 200 person-hours required three decades ago when one hectare produced only 30 tonnes. The high level of mechanisation responsible for this change does, however, mean that the farms producing potatoes are usually highly specialised and concentrated in specific areas. The versatility of the potato and its high nutritional content have resulted in much effort being expended to find suitable varieties for developing countries. According to FAO (1991), much of this work is based at the International Potato Centre in Lima, Peru, and includes the biotechnological improvement of existing varieties.

5.3 THE PRODUCTION OF FIBRE AND OIL CROPS

In comparison with food crops, there are relatively few fibre crops. Cotton is by far the most important of the world's fibre crops which include flax, hemp, jute, sisal and a number of locally produced crops, such as New Zealand flax (*Phormium tenax*) and kapok (*Ceiba pentandra*). These contribute but little to the world's total fibre production, as shown in Table 5.8.

Cotton production has been important since prehistoric times (see Section 2.3.6) though it was not the first natural product to be exploited for its fibre value. The use of wool, flax and silk all predate the initial use of cotton which dominates the range of plant fibres produced commercially today. According to Watson (1990), cotton accounted for *c.* 85 per cent of world fibre consumption pre-1950. Thereafter, the development of synthetic fibres, which are oil-based products, reduced cotton's share of the fibre market to just under 50 per cent in the mid-1980s. Currently, there is a rapidly growing market for organically grown cotton, i.e. cotton produced without chemicals, and that grown as coloured varieties. According to Pleydel-Bouverie

Table 5.8 Production data for fibre (abstracted from Food and Agriculture Organisation, 1993)

	Area cultivated 1000 ha		Production 1000 t	
	1991	1992	1991	1992
Flax: World	1 028	336	699	673
Former USSR†	740 *	72 *	270 *	250 *
Developed Countries	913	226	437	417
Developing Countries	116	110	262	257
Hemp: World	266	260	199	196
India†	68	63	38	35 *
Developed Countries	111	108	70	67
Developing Countries	155	152	129	129
Jute: World	2 276	1 968	3 653	3 135
Bangladesh†	593	505	962	898
Developed Countries	23	21	46	49
Developing Countries	2 252	1 946	3 607	3 085
Sisal: World	540	504	420	383
Brazil†	300	271	234	210
Developed Countries	6	6	8	8
Developing Countries	534	498	412	376
Cotton lint: World	N/A	N/A	20 668	18 430
China†	N/A	N/A	5 663	4 528
USA	N/A	N/A	3 835	3 527
Former USSR	N/A	N/A	2 410	2 046
Developed Countries	N/A	N/A	7 013	6 413
Developing Countries	N/A	N/A	13 655	12 018
Other fibres: World	N/A	N/A	401	406
India†	N/A	N/A	155 *	156 *
Brazil	N/A	N/A	71 *	70 *
Developed Countries	N/A	N/A	50	50
Developing Countries	N/A	N/A	352	357

* Estimate.
† Major producer.

(1994), several fashion designers and textile companies are now combining high fashion with environmentalism!

According to Langer and Hill (1991), cotton requires an average temperature of at least 21–22°C during the growing season, when a minimum of 500 mm of rainfall (or irrigation) is also required. It is, thus, rarely grown above latitudes of 40°, as shown in Figure 5.4. The cotton-producing countries use a variety of agricultural systems, ranging from high- to low-technology systems, to produce cotton. These have been described by Munro (1987) and Hearn and Fitt (1992). In low-technology agricultural systems, hand or ox ploughing is the first stage in land preparation, though in some areas ploughing may be preceded by ridging in order to provide microtopography that is best suited to water retention, distribution and drainage. Burning of weeds may also precede ploughing, especially if there is a heavy infestation. Implements used in such low-technology agricultural systems include long-handled chisels, hoes, and wooden and light-metal ox ploughs. Such techniques are typical of family-based operations in many parts of the developing world, e.g. the Yemen, Tanzania, Malawi

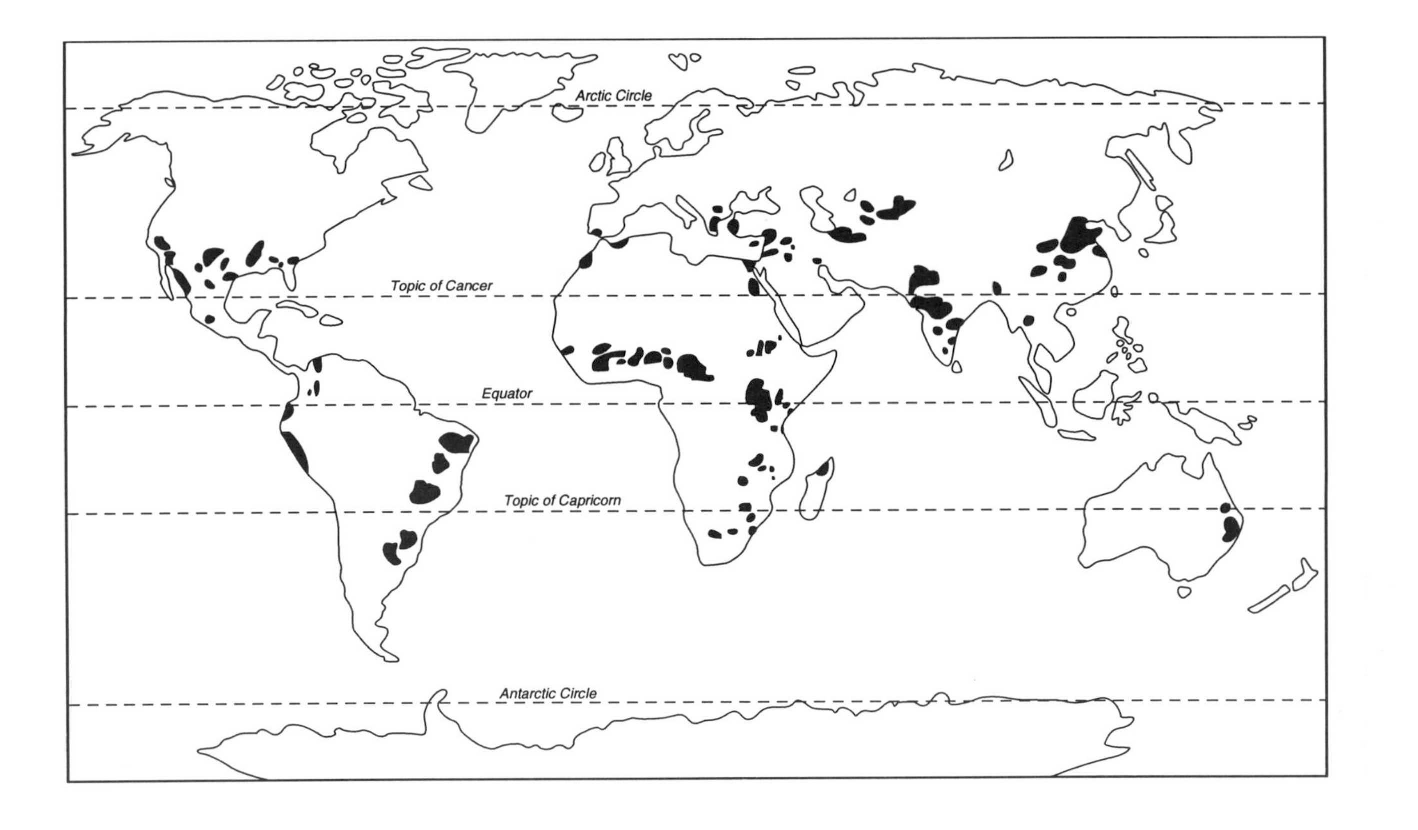

Figure 5.4 The world's cotton-growing areas

and Uganda. Hand sowing also characterises low-technology systems. Seed planting involves the creation of a small hole which may be made with a digging stick. As many as 40 seeds may be inserted (though four or five are recommended) and then covered over with soil that is pressed down by foot. The seeds are usually planted in rows, or along ridges if irrigation is used, and cotton may comprise one component of an intercropping system, as occurs in parts of Nigeria where it is grown with millet (Munro, 1987). Thereafter, weed control is particularly important because cotton is susceptible to competition. As it grows slowly after germination and does not achieve full ground cover until eight or more weeks later, cotton is easily shaded out or nutrient starved, especially of nitrogen. In low-technology systems, weeding is mainly by hand tools both before planting and as the crop grows. Even small-scale cotton growers are beginning to use herbicides to assist in weed control. In addition to weed control, thinning of the germinated seedlings is necessary to select the most vigorous seedlings and even out the density. Some gap-filling may also occur, though as Munro (1987) points out, the value of this is equivocal because it produces mixed-aged stands that, in principle, require differently timed treatments.

Unlike many of the crops discussed in Sections 5.1 and 5.2, cotton does not benefit very much by being grown in rotation. Munro (1987) reports that its yield may even be depressed by other crops, e.g. sorghum. Conversely, however, some crop yields, e.g. maize and tobacco, improve when they are grown in rotation with cotton. Such rotations are more common in irrigated agricultural systems, e.g. Egypt (see Section 5.5), than they are in rainfed agricultural systems though cotton is rarely if ever grown exclusively in low-technology agricultural systems where mixed-cropping or intercropping predominate. Cotton does, however, respond to nitrate and phosphate fertiliser applications which prompt growth and increase yields, though the relationship between fertiliser applications and yield is not as straightforward as it is for many food crops. Just as it is important to effect weed control, it is very important to exert insect and disease control since cotton is very susceptible to both. There are some low-technology options available including the destruction of crop residues by burning or burying. These methods help control pests and diseases specific to cotton, e.g. the boll weevil, leaf miner and the bacterial blight of cotton. Other methods include the timing of sowing to avoid peak pest populations, care with intercropped groups to avoid other crops that may be hosts to pests that are not just cotton specific, the control of plant density, water management and the timing of the harvest. These have been discussed by El-Zik *et al.* (1989) and Summy and King (1992) who also point to the use of cultivars that have been bred, by conventional plant-breeding techniques, to have some resistance to pests and diseases, though Watson (1990) suggests that such efforts have not been particularly successful in producing cultivars that offer much resistance to the major pests. Several cultivars, including seven multi-adversity resistance types, have been evaluated by Poswal (1993) who reports that some of these have increased resistance to bacterial blight as well as giving high yields. Indeed, the need for resistance, plus the high commercial value of cotton, has resulted in intensive efforts to manipulate cultivar characteristics by genetic engineering (Section 10.2). The agrochemical company Monsanto, for example, is about to market transgenic cotton seeds with inbuilt insect resistance. This marks a new departure in agriculture from cultural pest control as well as chemical and biological pest control. The latter two, because of cotton's commercial value, are currently big business, as discussed by DeVay *et al.* (1989) and Sterling *et al.* (1989).

Integrated pest management strategies, which involve all types of disease and pest control, are also widely used in cotton production, especially in high-technology enterprises. Using examples of cotton-producing enterprises in the Lower Rio Grande Valley of Texas, USA, Summy and King (1992) have examined the efficiency of cultural control in insect pest management strategies. They state that in the last 20 years increasing emphasis has been placed on cultural controls as important components of integrated pest management schemes. The resulting reduced inputs of chemical pesticides allows increased biological control of pests and reduces the fossil-fuel energy subsidy. Low-technology activities, in contrast, rely mostly on cultural or traditional methods of pest and disease control. In China, for example, cotton is grown using direct seeding methods. However, Ren (1991) reports that in the 1950s a new method of cultivating seeds was developed. This is still in use today and involves growing individual seedlings in a nutrient cube which is eventually transferred to the field. With three decades of experience and abundant labour, nearly 80 per cent of China's cotton is grown in this way, mainly in the Yangtze River Valley. The advantages include bringing the growing season forward by 15–20 days which allows leeway to manage pests, increased yields of between 10 and 30 per cent, and the ability to plant cotton in saline water from which the seedling is protected by the nutrient cube.

Harvesting in low-technology cotton-growing agricultural systems is still undertaken by hand. Munro (1987) states, 'Although picking is one of the most costly operations in growing cotton, it is often the least efficient. It needs a large number of people for a short time, and this often means recruiting temporary unskilled labour'. Generally, a sack-sized bag is tied around the picker's waist to allow both hands free for picking. The first pickings comprise grade A cotton; the second pickings are grade B. Packing ensues and the cotton is transported to market or to the ginnery. The latter houses machines, called gins, which separate the lint from the seed. The first such machine, the saw gin, was invented by the American Eli Whitney in 1793. It revolutionised, as well as commercialised, cotton production (see quote by Harlan in Section 2.3.6) particularly in the USA. In consequence, cotton became the major cash crop in the USA in the area known today as the 'cotton belt' (Schulman, 1991) which lies to the south of the 35th parallel (Figure 5.4). Whitney's invention, thus, changed the character of US agriculture in the southern states. According to Frisbie *et al.* (1989), localised cotton production and the cotton industries it sustained were transformed into large-scale cotton production as a cash crop. Markets in Europe, especially in the UK, also grew rapidly as the Industrial Revolution focused on the textile industry and mechanised both cotton and woollen cloth production. The mechanisation and industrialisation of agriculture occurred more or less in parallel so that, quite rapidly, all of the stages in cotton production that are described above became machine-based in the major cotton-producing regions of the world. Munro (1987) observes that nearly all the cotton produced in the USA is harvested by machine; planting, as well as the application of crop-protection chemicals and fertilisers, is also mechanised. As a result, labour costs are low but the agricultural systems are heavily subsidised by fossil-fuel inputs.

As stated above, ginning separates the seed from the lint. The latter is used in the textile industry while the seed is used for two purposes: it is the basis of the following year's crops and cottonseed oil may be extracted with the residue being used to produce cotton cake. According to Langer and Hill (1991), cotton seed contains 18–24 per cent oil and 16–20 per cent protein. The oil contains linoleic, palmitic and oleic acids which are used to manufacture margarine and cooking oils. The remaining seed can be compressed into cotton cake for consumption by livestock, though the 1–2 per cent of the phenol gossypol that the cotton seed contains must be removed, usually by heat treatment, because it is toxic to pigs and poultry. Hulls extracted from the seed may be milled to give 'hull bran' which can be used as a filler in animal feed (Munro, 1987). Cotton is, thus, a dual-purpose crop, though its major value is as a cash crop for the textile industry. Indeed, Hearn and Fitt (1992) indicate that the value of the USA's cotton crop is about US $\$6000 \times 10^6$.

Of the other fibre crops produced globally, jute is the most important. As Table 5.8 shows, approximately 2×10^6 ha produce *c.* 3.4×10^6 t annually. The world's largest producer is Bangladesh. Jute (*Corchoris olitorius*) fibre from the stem is used for making sacks and mats whilst the leaves and shoots are a source of food. As in the case of cotton production in China, jute production is labour intensive. According to Denton (1993), the main areas of jute cultivation are located at the latitudes just to the north and south of the Tropic of Cancer (23°30′N) and between longitudes 86° and 92°E. Here there is high relative humidity, an annual rainfall of 1500 mm or above with at least 250 mm occurring during each of the months of March, April and May, and a mean temperature range of 18–33°C. Overplanting usually takes place and then the crop needs thinning. This is practised to ensure that uniform stands occur as these are most suitable for fibre production. The leaves of the plants removed may be eaten locally as a vegetable. After harvesting, the fibre is extracted from the stem. Its versatility means that it can be marketed and exported as a cash crop. Several different types of thread can be produced, including one that resembles wool. In Bangladesh, cottage industries, often sponsored by relief agencies such as Oxfam, have developed based on jute fibre which is used by Bangladeshi women to make knitwear.

Another important group of crops grown for a variety of purposes is that of the oil crops. These are varied but as Table 5.9 illustrates, amongst the most important in terms of volume produced are groundnut, sunflower seed, rapeseed and sesame. Soy bean (Section 5.2) and cottonseed (see above this section) are both more important than any of these. As Hatje (1989) has discussed, the most important oilseed-producing areas are in the temperate zone, particularly in North America, Europe and China. Oilseed crops comprise two basic products: oil and oilcake. In many cases the oil can be put to a range of uses, some of which are non-food uses, e.g. in the manufacture of paints and varnishes, nylon, soap and lubricants. In the early 1990s oilseed crops are also providing a means of producing biomass fuels for powering motor vehicles. As stated above in relation to cottoncake, and in Section 5.2 in relation to soy bean meal, the residue left after oil removal may be used as an animal feed.

Groundnuts, alias peanuts (*Arachis hypogea*), are, in contrast to the majority of the oil seeds referred to in Table 5.9, a tropical crop though they are widely grown in temperate regions. Domesticated nearly 4000 years ago in the Ancon–Chillon region of Andean Peru (quoted in Evans, 1993), there are two cultivars: upright and

Table 5.9 Production data for oil-seed seeds (abstracted from Food and Agriculture Organisation, 1993)

	Area cultivated 1000 ha		Production 1000 t	
	1991	1992	1991	1992
Sunflower seed: World	16 725	17 641	22 666	21 645
Europe†	3 868	4 374	6 681	6 737
Former USSR†	4 488	4 802*	5 643	5 479*
Developed Countries	10 275	10 622	14 867	13 748
Developing Countries	6 449	7 018	7 800	7 897
Rapeseed: World	19 602	20 736	27 900	26 661
China†	5 800	5 950*	7 436	7 653
India†	5 722	7 065*	5 152	5 840*
Europe	3 457	3 273	9 584	8 051
Developed Countries	7 220	6 808	14 601	12 390
Developing Countries	12 382	13 929	13 299	14 271
Sesame seed: World	6 546	6 945	2 268	2 433
India†	2 670*	2 500*	750*	750*
China†	750	575	436	440
Developed Countries	1	1	1	1
Developing Countries	6 545	6 943	2 267	2 432
Linseed: World	3 606	3 138	2 604	2 104
India	1 148*	1 160*	339	250*
China†	170*	170*	510*	520*
Europe	227	247	275	276
Developed Countries	1 549	1 274	1 212	853
Developing Countries	2 058	1 864	1 392	1 252
Groundnut: World	20 333	20 609	23 975	23 506
India†	8 350*	8 600*	7 428*	8 200*
Developed Countries	961	960	2 460	2 180
Developing Countries	19 372	19 649	21 515	21 326

* Estimate.
† Major producer.

prostate, both of which are annual. According to Langer and Hill (1991), groundnut plants have little resistance to drought, require aerated soils with abundant available calcium, and require rainfall throughout the growing season. As a member of the Leguminosae, groundnut enjoys a symbiotic relationship with the nitrogen-fixing bacteria that inhabit its root nodules. This makes it a useful species to grow in crop rotations and it can be grown on soils of low fertility which are unsuitable for other crops (Ashley, 1993). Groundnut is produced in both high- and low-technology agricultural systems. For example, in the USA, one of the world's major producers, there is a high fossil-fuel subsidy whereas in India and China, the other two major producers, low-technology systems make use of the abundant and cheap labour.

The methods of groundnut cultivation in dryland Africa have been described by Ashley (1993). Land preparation involves producing a soil of medium tilth with ridging. The seed is most viable if it is removed from its shell only just before planting. This keeps damage by insects and fungi to a minimum. This can also be achieved by using soil disinfectants though they will kill any *Rhizobium* bacteria and thus reduce nitrogen fixation. Rosette virus disease can also be a problem which, Ashley states, develops best under conditions of low plant density. As Coffelt (1989) has discussed, much effort has been expended in breeding programmes to produce disease-resistant cultivars; rosette virus resistant cultivars have been produced in Senegal, Nigeria and Malawi and many cultivars with resistance to other diseases, e.g. leaf spots, have been produced in the USA. Seed planting is usually in rows along the ridges. Weed control is essential as, like cotton, groundnut is particularly susceptible to competition. Weed control must begin during land preparation to ensure a competition-free environment for the groundnut seedlings. Planting in rows facilitates hoeing by hand and slight lifting of the plant encourages the penetration of the pegs (roots) into the soil. Harvesting can be achieved by machine or hand. The whole plant is lifted and left in the sun to dry for a few days. The pods are then stripped off and can be shelled by hand or machine. The latter is most common even in Africa. In the USA all the stages in groundnut production are mechanised.

As a crop, groundnut is versatile. The vines and leaves provide a high-protein hay for livestock; the shells can be fed to livestock, burnt as fuel, as an alternative to diesel, and made into particle board (Coffelt, 1989). The nuts themselves can be used to fatten animals though human consumption is the major use in the form of flour, peanut milk and peanut butter, and directly as roasted nuts. The nuts can also be crushed to produce oil and the remaining meal is protein-rich. Both are suitable for human consumption. The oil is used for cooking, as a fuel and as a food constituent. According to Coffelt (1989), the groundnut contains 26 per cent protein, 2.7 per cent minerals and vitamins, 43 per cent oil and 24 per cent carbohydrate. This, coupled with its versatility, makes it an excellent species for subsistence or cash cropping. Coffelt, referring to groundnut crop energy characteristics, asserts that it is a net energy producer and accumulates 3.26 times the energy necessary to grow the crop. He does not state what type of agricultural system to which this figure refers but the value derives from Virginia, USA, so it presumably refers to a high-technology system. Apparently, the production of the equivalent amount of energy in a soy bean crop would require three or four times the land area.

Rapeseed and sunflower seed are also amongst the world's major oil crops. As Table 5.9 shows, sunflower seed is particularly important. It is a crop of the temperate zone, though sometimes it is also grown in cool-tropical regions under irrigation. According to Fick (1989), both the oil and the meal that can be obtained from the seed command important world markets. The oil, especially, is of high quality and usually obtains higher prices than soy bean oil or rapeseed oil. Most of it is used as table or cooking oil. Sunflower (*Helianthus annuus*) originated in the Americas though it is not clear where it was first domesticated. L.T. Evans (1993) believes that it was domesticated by Native Americans in the west. In terms of its agronomy, sunflower is grown in high- and low-technology agricultural systems. In Europe and the former USSR, the major producers, sunflower is produced using mechanisation, crop-protection chemicals and artificial fertilisers. According to Langer and Hill (1991), the seed is sown in wide rows to achieve a suitable plant density. This is important in the penultimate stage of production as the large inflorescences require space for drying prior to harvesting. Optimum spacing ensures that plants are not too slow to dry and that they do not produce small seeds which can occur if spacing is too close. Weed control is required and fertilisers improve productivity. The seeds mature in about four months but harvesting is not carried out until the heads are dried; this may be speeded up by using chemical desiccants such as paraquat. The crop is harvested by combine or it is cut and threshed. Fick (1989) has described the various cultivars that have been developed to improve oil productivity and to increase resistance to various diseases.

Rape (*Brassica napus*), another oilseed crop, has a long history of use, though precisely when and where it was domesticated remains unclear. It is, however, directly referred to in Indian Sanskrit manuscripts that are 4000 years old (Langer and Hill, 1991). The cultivars grown for oil (some cultivars are grown for forage) are planted in the autumn or spring. The former gives a longer period for growth and usually results in higher yields than spring-planted rape. Harvesting takes place after drying, or artificial desiccation, and usually requires combine harvesting. The seeds contain 38–44 per cent oil which is obtained via crushing. The value of the oil depends on its composition, especially on the proportions of the various fatty acids. Indeed, manipulation of the oil content has been a prime objective of plant breeding programmes, particularly the reduction of the erucic acid content which can have detrimental effects on the growth rate of animals. Usually, erucic acid is 20–45 per cent of the rapeseed but there are now cultivars, e.g. Canbra which is produced in Canada, which contain little or no erucic acid (Langer and Hill, 1991). Rapeseed oil has a variety of uses. The most important of these are for industrial purposes and for human consumption as a vegetable/cooking oil and in the production of margarine. In industry, rapeseed oil is used for lubrication and in the manufacture of soap. The meal remaining after oil extraction can be fed to livestock.

In the last few decades, the development and use of biomass fuels has increased. In particular, many of the oil crops referred to above are increasingly being used as a basis for biomass fuel production. As Quick (1989) discusses, this is not a new idea as vegetable oils have been viewed as potential fuels for diesel engines since the early 1900s. It is in this context, diesel engines, that biomass oils have most potential as vehicular fuels. In the UK, for example, there have recently been experiments with rapeseed oil as a fuel for diesel buses in Reading. The major advantage of such fuels is that they contain little sulphur and so do not contribute to 'acid rain'. They are, thus, considered to be cleaner than fossil fuels. In Europe, the adoption of a set-aside agricultural policy (through the EU's Common Agricultural Policy), is also an encouragement to oil-crop production because they provide an alternative to overproduced conventional crops.

5.4 THE PRODUCTION OF BEVERAGES

Many crops are grown for a variety of purposes other than for their calorie or energy content. The most important of these crops are those grown for the production of beverages: coffee, cocoa and tea. The major beverage crops are products of the tropical zone and as such they are often very important cash crops in developing countries.

Coffee, L.T. Evans (1993) records, was domesticated in north-east Africa, probably in Ethiopia where wild populations of the plant still occur in the vegetation communities of the high Ethiopian plateaux. According to Coste (1992), there are several cultivated species of coffee, the most important of which are *Coffea arabica* L. and

Coffea canephora Pierre. Robusta coffee (*C. robusta*) is considered to be a variety of the latter and there are several varieties of *C. arabica*, e.g. *C. arabica* L. var. bourbon (B. Rodr) Choussy, which is very productive. In terms of volume of production, *C. arabica* is the most important species. It is an evergreen shrub which grows to between 8 and 10 m in height, as is *C. canephora* which has a higher caffeine content than *C. arabica*. The ecological preference, particularly in relation to temperature and rainfall regimes and soil type, of the two species helps to explain their distribution as crops. *C. arabica* prefers environmental conditions that are similar to those of its centre of origin. It requires a combination of latitude and altitude that provides a dry season of four to five months and an annual rainfall of 1500–1800 mm. These requirements render it most suitable as a crop of high-altitude tropical regions, as in Kenya and Tanzania in Africa, where Coste (1992) states that it is cultivated on terraces between 1300 and 2100 m. It is also grown on the plateaux of Cameroon as well as in the states of São Paulo and Parañа in Brazil and in many of the Andean nations, e.g. Colombia, and Central American nations. In contrast, *C. canephora* produces best in an equatorial climate where the temperature is equable at 24–26°C all year and there is abundant rainfall for 10 or so months. The ecological requirements of both coffee species have been discussed in detail by Coste (1992) and Cambrony (1992). Cambrony states that *C. canephora*, of which the majority of plantations comprise the robusta type, grows best in hot, humid areas, e.g. Cameroon, Angola, Zaire, Indonesia and northern Brazil.

Coffee, according to Coste (1992), is usually grown on land previously occupied by forest. Thus, initial land preparation requires clearance of the trees, often by burning. Then the plantation must be designed to produce optimum yields. Factors such as slope, competition for light and water, ease of equipment manipulation, management costs and pest control must be taken into account. Most plantations have geometrical designs with the coffee bushes being arranged in squares or rectangles. On sloping land, planting is most appropriate along the contours. Coste also draws attention to new methods of intensive coffee production that have been developed during the last decade. In Costa Rica, for example, densities of 7000 plants per ha, as compared with the more traditional 4000–6000 plants, are possible with intense pruning. *C. arabica* responds well under these circumstances producing many branches and fruits; yields average 4000 kg/ha.

Coffee plants may be grown in nursery beds or in individual pots. These are planted in rows in holes created manually or with hole-boring machines in marked positions. Prior to planting, the roots are inspected and any damaged parts are removed. The roots may also be dipped into a slurry made from clay and water when they are removed from the nursery. This helps their establishment more so than any other treatment such as planting with a root ball. Transplanting from nursery to plantation is best carried out at the start of the wet season. The young plants may need some protection during their early weeks. This is usually effected by placing palm leaves or branches over the seedlings. Between the bushes, a leguminous crop may be grown to protect the soil from erosion though small-scale producers may intercrop with several different types of crop. For example, Njoroge and Kimemia (1993) refer to beans, maize, potatoes, sorghum and millet as well as fruit trees such as macadamia, bananas and mangoes being intercropped with coffee in Kenya. Soil mulching, i.e. covering the soil with straw, grass, dead weeds or compost, is also practised for the same reason, though not in areas susceptible to frost. Pruning is another important task; it is necessary to produce a robust and well-balanced plant and it encourages fruit protection. The various pruning strategies have been described by Coste (1992) and Cambrony (1992). Coffee also requires the addition of fertilisers. In particular, nitrogen and potassium need to be added as the coffee plant requires an abundance of both. This demand increases as the bush ages, especially in the third and fourth years after planting. The legumes grown in association with coffee (see above) provide much of the nitrogen in a process known as 'green manuring'. Farmyard manure is also used where it is available, as is composted waste from processed coffee; otherwise mineral fertilisers are used though this increases the cost of production. Despite encouraging a ground cover of leguminous plants between the coffee bushes, weeds do occur and weeding is necessary. This must be quite intensive in the early stages of growth. Coste (1992) also points out that the use of herbicides is increasing. The costs increase but are partly offset by a reduction in the labour force as hand-weeding is no longer necessary. Coffee-bean harvesting is also becoming increasingly mechanised. In addition, irrigation may be used to provide coffee plantations with a reliable water supply. This occurs in Yemen and parts of India, Kenya and Brazil. Both sprinkler and gravity methods are used.

Pest and disease control is also essential to ensure high productivity. According to Cambrony (1992), there are many pests and diseases to which coffee is susceptible. Occasionally some of these can reach epidemic proportions, e.g. *Hemileia vastatrix* which is the fungal disease rust and *Hypothenemus hampei* Ferr. which is the berry borer. Chemical pesticides are usually applied to control pests. Again, this increases costs and means that the input of fossil fuels is increased. Consequently, a great deal of energy is expended to produce a crop that has very little energy value! Harvesting takes place when the crop is ripe. Cambrony notes that a growing period of six to eight months is necessary for *C. arabica* whilst *C. canephora* requires ten to eleven months. Harvesting usually takes place at the end of the rainy season to avoid having to dry the beans. It is labour intensive as no mechanical means of selecting mature beans has yet been devised. At the processing factory, cleaning and sorting occurs. In relation to yield, this varies from plantation to plantation and country to country. Coste (1992) quotes average yields in Brazil of 350–400 kg ha^{-1}, though yields as high as 1200 kg ha^{-1} have been obtained in Cameroon and Kenya where improved varieties and high mineral fertiliser inputs are used. Coffee trees have long lives of 50 years or more, though diseased or impaired trees may need to be replaced as their yields will be low.

As Table 5.10 shows, tea is another important beverage crop. Most of it is produced in Asia where China is both a major producer and a major consumer. According to Weatherstone (1992), the tea plant, *Camellia sinensis* L. O. Kuntze, originated in the south-east part of the Tibetan plateau, with the Assam variety occurring in north-east India. It is likely that the species, along with its many varieties, is indigenous in the forests of south-east Asia. The origins of tea-drinking remain enigmatic though there are accounts of it dating to the fifth century AD in China from where Turkish traders carried it westward. By the tenth century a regular tea trade was established within and between China, Tibet and Mongolia but it was not until the seventeenth century that tea consumption spread into Europe, beginning with Russia whence it spread by caravan. Dutch traders also brought tea to Holland

Table 5.10 Production data for coffee and tea (abstracted from Food and Agriculture Organisation, 1993)

	Area cultivated 1000 ha		Production 1000 t	
	1991	1992	1991	1992
Coffee: World	11 217	10 927	6 111	5 919
Africa	3 375	3 351	1 170	1 216
Central America	1 733	1 662	1 142	1 048
South America	4 687	4 438	2 841	2 654
Brazil	2 767	2 510	1 525	1 298
Asia	1 364	1 431	899	953
Tea: World	2 503	2 531	2 607	2 473
Africa	203	191	336	300
South America	58	58	64	64
Asia	2 161	2 205	2 079	1 991
China	1 122*	1 172*	742*	703*
Former USSR	77*	73*	120*	109*

* Estimate.

from where it was imported into England in the 1650s. Some considerable time later, in the 1820s, a British army general was made aware of tea consumption by the natives of Assam and botanical inspection showed the plant to be a relative of the Chinese species. Tea cultivation began in Assam and other parts of India under the auspices of the British in the mid-1830s, using local and Chinese varieties. Thus began the tea plantations of Assam, Kumaon and the hills of south India and Sri Lanka. Weatherstone (1992) states that the tea industry in India today employs about 1.5×10^6 estate workers, reflecting its significance to India's economy. As in India, tea exports are a major earner of foreign currency for China. Findlay *et al.* (1993), for example, state that in 1989 tea exports earned China $US 421 million. India's foreign earnings (probably) amounted to *c.* $US 450 million. Data in van de Meeberg (1992) show that India exported 87 000 tonnes, Sri Lanka exported 68 000 tonnes and China exported 84 000 tonnes of black tea in 1990. The major importers are the UK, USA, other European countries and the former Eastern block.

According to Takeo (1992), the two main cultivated varieties of *C. sinensis*, *C. sinensis* var. *sinensis* and *C. sinensis* var. *assamica,* produce nearly all the world's green and black tea respectively. The two varieties are different in terms of their chemical content which is why they are used for different purposes. *C. sinensis* var. *assamica* has a high content of flavanols which would give a bitter taste to green tea. Consequently, it is used to produce black tea which requires drying and several other processes prior to exporting. Green tea, conversely, is brewed from the leaves without any processing other than steaming, to reduce the moisture content, rolling and drying.

Thc field operations to produce tea have been described by Willson (1992a). The major tea-producing regions occur in areas with diurnal temperatures in the range 14–32°C, preferably with long hours of sunshine. Approximately 150 mm of rainfall per month is necessary, with an annual total of *c.* 1800 mm; below 1500 mm irrigation is necessary. As in the case of coffee, tea plantations occupy land previously occupied by rain forest. Trees can be removed by ring barking though Willson states that burning should be kept to a minimum so as not to raise the pH of the soil; tea generally prefers acidic soils, often podzolic in nature. Land preparation may also involve the construction of terraces and drains to control soil erosion and Willson advocates the planting of a cover crop as soon as these measures have been completed. Such factors help to prevent erosion in the early stages of plantation establishment, they may increase the opportunities for mulching thereby adding nutrients and organic matter, and they may provide shade for the young tea plants. Seedlings, produced in a variety of ways, e.g. from seed or by vegetative propagation, are placed in pre-marked locations at which a hole has been created. The optimal densities are those that give a complete ground cover; wide spacing intensifies the weed problem and soil erosion. Similar problems occur in coffee plantations (see above). Planting is usually in rows, following the contours of sloping land. When the seedlings are inserted into the holes, fertiliser is usually added as the hole is infilled. Stakes are used to give the seedlings strength and shading may be necessary using a ground crop.

Pruning, fertilising and weed control, as well as pest and disease control, are all necessary operations. Today, herbicides are used to control weeds and yields have increased as a result because there is minimum disturbance to the root systems. Herbicide use does, however, increase costs and add a fossil-fuel subsidy. Both nitrate and phosphate fertiliser applications increase yields. The most important pests are those that destroy the leaves. One example is that of blister blight leaf disease caused by the fungus *Exobasidium vaxans* Massee. This and other pests are usually controlled using integrated pest-management strategies that are designed for specific plantations (this is a similar approach to that used for pest control in cotton; Section 5.3). They include cultural, biological and chemical methods and they are detailed in Arulpragasam (1992) and Muraleedharan (1992). In relation to harvesting, the best black tea is derived from the actively growing bud (Willson, 1992b). Fine plucking involves the collection of two leaves and a bud; coarse plucking involves the collection of more than two leaves and a bud. Quality control is the remit of the picker where manual plucking is practised. The interval between plucking is known as the plucking round, and may be as little as four or five days and as much as five weeks depending on the weather and general plantation conditions. Conversely, the harvesting of green tea is much less complex as the quality does not vary as much as it does for black tea. Many of the operations necessary for tea production are now mechanised, the financial advantages of which depend on the cost of local labour. Harvester machines, for example, are available for plucking. After harvesting, the tea is sent to tea factories for processing, packaging, etc. Eventually, the packets of tea arrive on shop and supermarket shelves many thousands of miles from their place of origin.

5.5 CONCLUSION

The various types of arable agricultural systems discussed above represent only a limited range of those which actually characterise some of the world's diverse array of agricultural systems. The production of fruits and vegetables, for example, is not included though some examples of these are given in Chapter 7, which deals with mixed agricultural systems. Despite this limitation, the arable systems herein presented represent some of the most important agricultural systems in the modern world. The cultivation of cereals, for example, not only sustains a large proportion of the world's population but also provides an important component of world trade as well as underpinning a large proportion of the processed-food industry. Cereal production, both directly and indirectly, provides food security in the developed world in particular. It also makes a major contribution to the financial and political stability and political pre-eminence of developed nations. In developing nations, cereals provide a staple food for indigenous populations. It is probably fair to say that more energy, including human and animal labour as well as fossil-fuel energy, is used in the cultivation of cereals than in any other crop or group of crops. This energy subsidy may be manifest in a variety of forms, e.g. pesticides, artificial fertilisers, mechanisation, etc. The basic aim is to improve the capture and storage of solar energy within the edible component of the crop. In the most energy-intensive so-called industrialised agricultural systems energy inputs exceed energy outputs though the latter is consumable food energy.

Other arable agricultural systems may be high- or low-energy systems that are monocultural or polycultural but instead of providing food for direct human consumption they provide fodder and/or forage for animals. In the developed world, maize is the most important fodder crop and is used to feed cattle and pigs. Crops other than cereals are also grown for animal consumption, e.g. soy beans, potato and some brassicas. Animals may also be fed on the waste products of arable agriculture, e.g. the various feeds produced from sugar cane. The animal products derived from these agricultural systems provide an important source of food energy and protein though the bulk of consumers is located in the developed world.

In contrast, the production of fibre crops provides no (useful) energy. World fibre production from agriculture is dominated by cotton which is an important cash crop in several developing countries as well as a major crop of the USA. It is energy intensive, particularly because it requires much protection from pests and diseases, but only produces relatively little food energy in the form of cotton cake which provides animal feed. The main value of the crop is as a fibre source. Some crops are also grown for their oil content of which a proportion is processed for human consumption and so has a food-energy value. In the last few decades, however, there has been an increasing trend to produce oil crops as a source of biomass energy to replace fossil fuels in vehicle propulsion. Beverage crops, notably coffee and tea, are also produced for human consumption but they contain little energy or nutritional value. Generally, coffee and tea are cash crops produced in plantations in developing countries mainly for export to developed countries.

Overall, the examples given above attest to the diversity of the world's arable agricultural systems. The production data, notably the areas under cultivation, given in Tables 5.3 and 5.7–5.10 also attest to the importance of these systems as a means of bringing about environmental change. As economic circumstances change so too will production. Consequently, these agricultural systems, along with all others, are dynamic and so contribute to the dynamism that characterises environmental change.

5.6 FURTHER READING

Anthony, K.R.M., Meadley, J. and Röbbelen, G. (eds) (1993) *New Crops for Temperate Regions*. Chapman and Hall, London.

Bushuk, W. and Rasper, V.F. (eds) (1994) *Wheat: Production, Properties and Role in Human Nutrition*. Chapman and Hall, London.

Loomis, R.S. and Connor, D.J. (1992) *Crop Ecology: Productivity and Management in Agricultural Systems*. Cambridge University Press, Cambridge.

Pearson, C.J. (ed.) (1992) *Field Crop Ecosystems*. Elsevier, Amsterdam.

Rowland, J.R.J. (ed.) (1993) *Dryland Farming in Africa*. Technical Centre for Agricultural and Rural Co-operation, Wageningen and The MacMillan Press, Basingstoke.

Willson, K.C. and Clifford, M.N. (eds) (1992) *Tea: Cultivation to Consumption*. Chapman and Hall, London.

CHAPTER 6

Settled Agriculture: Pastoral Systems

Pastoral agricultural systems occupy approximately 50 per cent of the world's land area. Their distribution is given in Figure 1.5 which shows their approximate correspondence with the world's semi-arid regions of the tropical zone and those parts of the cool temperate zone where temperature and/or water availability limit primary productivity. These are the world's rangelands whose natural vegetation is grassland or savanna (Figure 1.4). To these areas must be added the lands of the mainly dairy herds of highly populated regions such as Europe and the New England and Great Lakes regions of North America, as well as those upland areas of the temperate and tropical zones where pastoralism is the only possible agricultural activity.

Worldwide, there are currently 3160×10^6 grazing animals comprising cattle, sheep, goats, buffaloes and camels (pigs are excluded from this figure because they are rarely free-ranging), and a further 60.843×10^6 horses (Table 6.1). These animals provide a variety of commodities which include meat, milk and milk products, blood, hides, wool and labour. As food products they provide most of the world's dietary protein. A substantial proportion of these animals are reared under conditions of low fossil-fuel subsidy, especially in the developing world. Some of these agropastoral systems have already been described in Sections 4.1 and 4.2 which detail some of the world's nomadic pastoralist activities. Those which involve permanently settled groups are considered here, though occasionally the distinction between the two becomes blurred as, in times of crisis, settled agropastoralists may adopt a nomadic lifestyle temporarily. In other agropastoral systems, notably those in the developed world, there is usually a fossil-fuel energy subsidy in the form of transport, pasture improvement via fertiliser addition and the often intensive use of animal health products. There are thus parallels between these pastoral agricultural systems and the high- and low-energy-subsidised arable agricultural systems discussed in Chapter 5.

In addition to the range animals there are others that make major contributions to the world's source of protein. The most important of these are pigs and poultry. Both are usually farmed relatively intensively. In the case of pigs, they are usually penned, even in agricultural systems that have little fossil-fuel subsidy. They are fed either on specially grown grains such as maize (see Section 5.2) or on household and agricultural crop waste. Poultry, notably chickens, ducks, geese and turkeys, are similarly produced as components of high- and low-technology agricultural systems. Indeed, battery production, the most intensive mode of production, has been widely criticised on the grounds of causing unnecessary suffering. Worldwide, sheep farming is also very important. Meat and wool are the major products in both developed and

Table 6.1 The numbers of grazing animals worldwide (data abstracted from Food and Agriculture Organisation, 1993)

Animal	Numbers $\times 10^6$
Cattle	1284.19
Sheep	1138.36
Pigs	864.10
Goats	574.18
Buffaloes	147.52
Horses	60.84
Camels	17.02

developing countries. Where pastures are poor, goats also provide a source of wool, milk and meat.

A variety of other animals are husbanded but only on a minor basis if they are considered in the global context. Examples include the llama and alpaca which are produced for labour and wool in some of the Andean nations of South America. In New Zealand and Western Europe there has been an increase in the last decade in deer farming to provide venison, a low-fat substitute for beef. Similarly, there has been an upsurge in the husbandry of the Angora goat as a consequence of the increasing demand for mohair, a quality product required in the clothing industry as an adjunct to wool. There is even the growing possibility that animal farming will take a novel turn as sheep, cattle and pigs are genetically engineered to produce substances of medical value. Sheep have already been engineered to produce certain types of human proteins, e.g. α1-antitrypsin, in their milk; transgenic pigs can produce human haemoglobin and it may eventually be possible to use the organs of engineered pigs for transplanting into humans.

6.1 THE PRODUCTION OF MEAT (BEEF)

The production of meat globally is by either extensive grassland farming, or ranching, and intensive grassland farming. The former occurs on the world's rangelands, as illustrated in Figure 6.1, whilst the latter occurs mainly in the temperate zone, notably in areas where there are high population densities. Meat production on an extensive and intensive basis also occurs in high- and low-technology agricultural systems. In all cases, there is less energy available from the end product than there is from arable crops produced on the equivalent amount of land. This is because of the additional transfer of energy that takes place between the primary producers (the forage species in grazing lands) in the agroecosystem and the primary consumers, i.e. the animals. These relationships are considered in Section 1.4 which examines the energy transfer characteristics of animal production systems. Table 1.6 also details the variations that occur between farmed animals in relation to protein and energy yields.

The quantity and quality of the primary production, thus, largely determines the quantity and quality of the secondary production, the other major factor being the ability of the grazing animal to convert the primary production into secondary production. In the context of forage quantity and quality, the dominant control is

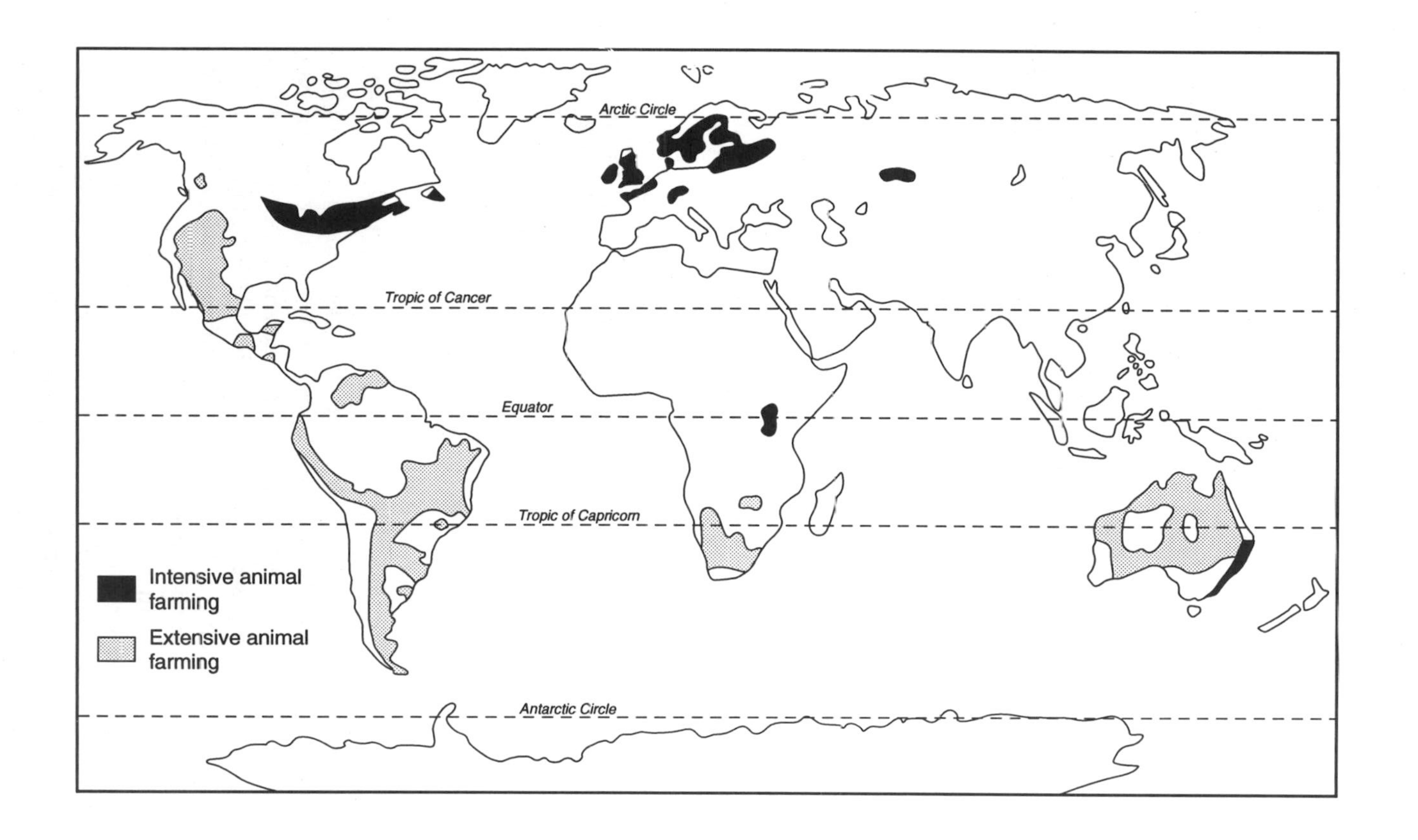

Figure 6.1 The permanent pastoral systems of the world

Table 6.2 The distribution of cattle worldwide (data abstracted from Food and Agriculture Organisation, 1993)

	Numbers of cattle $\times 10^6$	
	1991	1992
Africa	189.10	187.50
Ethiopia	30.00 *	30.00 *
South Africa	13.50	13.60
Sudan	21.03	21.60
Nigeria	15.14	15.70
North & Central America	163.57	162.82
USA	98.90	99.56
Canada	128.43	130.02
Mexico	31.46	30.16
South America	272.23	277.50
Brazil	150.00	153.00
Colombia	24.35	24.77
Argentina	50.08	50.02
Asia	391.98	397.75
India	193.33	192.65
China	78.66	82.76
Bangladesh	23.50	23.70
Pakistan	17.71	17.75
Turkey	11.38	11.97
Europe	119.95	114.02
France	20.97	20.93
Germany	19.49	17.13
UK	11.84	11.62
Oceania	32.35	32.65
Australia	23.66	23.60
New Zealand	8.10	8.45
Former USSR	115.76	111.94
Developed countries	397.91	389.53
Developing countries	886.96	894.66

* Estimate.

climate, as discussed in Section 1.2. In the case of the world's rangelands (Figure 6.1) the dominant climatic factor is the range and distribution of rainfall that occurs annually. Apart from camels, of which there are 17.019×10^6 animals worldwide and which are not primarily reared for their meat, the animals dominating the rangelands are cattle and sheep. Both are reared for meat though sheep are often reared for wool, as is discussed in Section 6.3. Both of these animals are husbanded in high-technology and low-technology agricultural systems.

The distribution of cattle is given in Table 6.2 which shows that the greatest number of cattle are kept in Brazil, followed by the USA. In Asia, the largest producer is India, and the second largest producer is China where cattle are kept as components of mixed farming systems (Chapter 7) as well as being the products of rangelands. Australia is the largest producer in Oceania. Of the world's cattle, approximately 20 per cent are used for beef production of which some are culled adults that were used for breeding, labour and/or milk. Cattle thus dominate the secondary productivity of some of the world's most extensive rangelands. The floristic characteristics (i.e. the species present) of these disparate areas are individual though in terms of their physiognomy (vegetation structure) they have many features in common. The same is true of Africa's rangelands which, although they carry fewer animals than those of North or South America, are nevertheless crucial to the food supplies, local wealth and national economies of many Sub-Saharan African nations. This has already been discussed to a certain extent in Section 4.2, in relation to nomadic pastoralism.

As Tivy (1990) points out, the rangelands in the semi-arid zone are characterised by seasonally distributed precipitation: often the growing season, when there is adequate soil moisture, is quite short and there may be long periods of drought. These characteristics all deter plant growth to which the number of animals grazed must be adjusted. Too many animals will deplete the plant cover so that the soil becomes susceptible to erosion and the possibility of desertification is enhanced. These issues are discussed in detail in Chapter 9. Conversely, however, grazing is essential to the maintenance of a diverse plant cover. West (1993), for example, has discussed this in the context of rangeland biodiversity. As ever, in environmental matters it becomes a question of balance. The grazing herds of domesticated species must, in a pastoral agroecosystem, mimic as far as possible the numbers, movement and grazing pressure of herbivore populations that characterise such areas under natural conditions. The degree of aridity and associated availability of soil moisture will influence greatly the so-called carrying capacity of any given rangeland. The carrying capacity defines the optimum number of grazing animals that can be supported without impairment of the cover or composition of the vegetation. Consequently, those areas with a very low rainfall, *c.* less than 1000 mm per year, and those areas where the rainfall may be 1500 mm but which is erratic, have much lower carrying capacities than rangelands with a low but reliable rainfall.

In general, cattle dominate the rangelands of less arid areas, especially where there is a proximal and/or export market. This is particularly the case in the USA, several South American countries and Australia. The latter two rely heavily on exporting their produce. All, to a greater or lesser extent and in contrast to related agroecosystems in Africa and parts of Asia, employ artificial methods of manipulation which include the control of pasture composition and animal health products. Jarrige and Auriol (1992) have analysed beef cattle production figures and have shown that there is a far from even distribution worldwide. Europe and North America, especially the EU and the USA, dominate production. The EU and the USA account for 16 per cent and 24 per cent respectively. The former USSR accounts for a further 16 per cent, with 17 per cent in Latin America, particularly Argentina and Brazil. In Brazil, cattle ranching has expanded considerably during the last century, with a major increase occurring since 1960 (Hemming, 1994). The grasslands are created at the expense of tropical rain forest and in comparison with other cattle-producing regions their productivity is low. This will be discussed again in Chapter 9. Jarrige and Auriol also point out that tropical and semi-arid regions produce less than 20 per cent of the world's beef despite the fact that they are responsible for more than 55 per cent of the total cattle population (Table 6.2). There is a corresponding diversity in consumption which relates to number of cattle per person in any given region and the beef production per head of stock. The amount of beef supply per inhabitant, for example, ranges from less that 1 kg per year in east Asia to 24 kg in western Europe, 71 kg in North America and 95 kg in Australia and New Zealand. These values are paralleled by those for calorie intake. Grigg's (1993a) analysis of FAO data shows that people in the developed world obtain as much as five times more calories from livestock products than do people in developing nations. Whilst this generalisation includes dairy products, it also holds true for beef consumption. For example, Grigg shows that bovine meat (cattle and buffalo) provides 15.2 per cent of the calories per capita per day in the developed world (161 calories out of a dietary total of 1057 derived from livestock) and 9.7 per cent of those in the developing world (21 out of a dietary total of 217 derived from livestock).

Of the largest producers of beef cattle, the USA, Argentina, Brazil, Australia and New Zealand use ranch-style methods on rangelands. In these areas beef cattle are raised from suckler herds. Cows are bred to produce calves which suckle all their milk, weaning when they are between 6 and 10 months old. According to Jarrige and Auriol (1992), the efficiency of this system's conversion of food energy into edible products is low: possibly as little as 2–3 per cent for gross energy. Such values, however, reflect the increased links in the food chain when compared with crop production and thus the relative inefficiency of energy transfer as well as the amount of energy which is converted into inedible components of the animal. Nevertheless, these animals still manage to produce protein from lands that carry a poor-quality forage (see Section 1.4) though they require a large area of land per head when compared with the lush pastures of northern Europe. As Jarrige and Auriol (1992) have discussed, only about 30 per cent of the cattle in Western Europe are suckler cows whereas they dominate the herds of North America, Latin America, Australia and New Zealand. Cattle were introduced to these lands from Europe by colonists, beginning with the Americas in the fifteenth century and Australia in the eighteenth century. (Cattle were originally domesticated in the Near East at about the same time as wheat and barley, and in India; see Section 2.4 and the review in Mannion, 1991).

According to Payne (1990), the first cattle to arrive in the Americas were those brought by Columbus on his second expedition in 1493. They were taken to the island of Hispaniola and 28 years later the first cattle were taken to the mainland, to Vera Cruz in Mexico. These animals originated in Spain and were eventually introduced throughout the Americas. Some 200 years later, British and other European cattle were introduced into North America from whence they gradually spread southwards. The introduction of zebu (*Bos indicus*), a humped species which originated in western Asia, led to its dominance in the south-east USA, along with crossbreeds. Today, zebu-types and their crossbreeds, as well as Iberian and north European cattle, dominate the beef cattle rangelands of the Americas. As Meissner and Morrison (1991) point out, however, the beef industry did not really flourish until the 1880s when refrigerated ships became available, allowing the development of meat exports to Europe. An additional growth period occurred just after World War II consequent on good beef prices and an increase in demand from the home market in North America (Jarrige and Auriol, 1992).

Of the beef cattle in the USA, about 60 per cent are components of mixed farming systems in herds of moderate size which comprise *c.* 50 animals. Jarrige and Auriol (1992) state that they graze cultivated pastures and consume the by-products of cereal production (see Section 5.2), mainly in the north-central area of the USA. The remaining 40 per cent of the USA's beef cattle occur as large herds (an average of 300 cattle) in the south-central and western states where they graze the rangelands. The majority of these animals spend their final days in a 'feedlot' where they receive high-energy diets to increase the meat content and quality. The organisation and operation of US feed lots have been examined by Perry (1992). The majority comprise farmer feeders, i.e. they are privately or group owned, and have a handling capacity of less than 1000 animals. These predominate in the Corn Belt (see Section 5.2) where crops are produced specifically for animal feed. Although they are much fewer in number, commercial feedlots handle about 50 per cent of the total cattle marketed. These may be privately or corporately owned and can cope with *c.* 8000 animals. They predominate in the south-east and south-western states. In these feedlots, pre-mixed feed is delivered to feeding troughs mechanically, usually in large sheds with slatted floors to facilitate the collection of faeces and urine.

Animals may be transported large distances from their grazing areas to feedlots, with auction houses in between. Generally, only cattle graded as US Good or US Choice are taken to feedlots where they are kept for between 90 and 149 days (Perry, 1992). Whilst in the feedlot, animal health products are commonly used. Antibiotics, for example, increase weight gain by improving the efficiency of feed conversion. The feeds themselves are varied. Most are cereals, e.g. maize and barley, but waste products from other crops are also widely used, e.g. potato wastes and molasses. These are the basic 'rations' used in feedlots. To these are added concentrates which are high-energy feedstuffs such as oil meals and grains. These are processed in a variety of ways to provide a concentrated source of energy and nutrients. Feeding programmes exist for each type of animal, its age and the desired end product. Ultimately the meat is marketed, usually to packer-buyers who organise buying to keep in step with the efficient operation of packing plants.

The entire operation is geared to maximise the efficiency of meat production which essentially requires the optimum manipulation of energy flows. The end product in the case of North America receives its energy from two sources: the forage of the rangelands and the rations and concentrates, which are crops and crop products, that are administered in the feedlot. The use of adequate shelter and animal health products is also focused on increasing energy transfer from the primary food to the secondary meat. The energy flows in the system are thus complicated, especially when the fossil-fuel inputs to the feedlot component are considered. These issues have been discussed by Giampietro *et al.* (1992) who have compared the fossil-fuel energy, land and labour requirements of beef production under different agricultural systems. Where imported feed is used, 1143 MJ of energy per kg of edible protein produced per year are required. This compares with 513 MJ for a closed feedlot wherein all the feed is produced within the farm, and 243 MJ for an organic feedlot in which there are no artificial fertiliser or crop-protection chemicals used. At face value, the latter system is the most energy efficient. However, it requires 287 m^2 of land per kilogram of edible beef protein produced. This contrasts with 35 m^2 for the feedlot with imported food and 136 m^2 for the closed feedlot. The nutritional and energetic relationships under various types of beef cattle production have also been examined by Owens and Geay (1992) and Baker *et al.* (1992).

As Figure 6.1 shows, rangelands occur extensively in Latin America. Their history, in terms of the development of cattle ranching, is similar to that of North America.

The production of beef cattle in the temperate zone, notably in Argentina and Uruguay, occurs in the extensive grasslands known as the pampas. Here, 90 per cent of the land is used for livestock production though there is some mixed farming (Camillo and Schiersmann, 1992). Table 6.2 shows that Argentina had approximately 50.02×10^6 head of cattle in 1992 whilst Uruguay had 9.5×10^6 head (FAO, 1993). The weight of the average animal is about 410 kg. This is lower than that of feedlot animals in the USA which average 520 kg for steers (Perry, 1992).

The pampas region is characterised by beef farming which is managed extensively on ranches or *estancias*; herds vary in size from 200 to 2500 head, and ranches vary from 400 ha or less to 5000 ha or more (Camillo and Schiersmann, 1992). The pastures, some of which are periodically flooded, are grazed continuously or nearly so. Their net primary productivity varies between 1000 and 2000 kg of dry matter per hectare. Thus, at the top end of this range one head of cattle requires 2 ha of land. Once they are weaned, calves are retained on the *estancia* for a period referred to as a stocker period. How long this is depends on the requirements of the farms engaged in the final fattening. According to Camillo and Schiersmann, 80 per cent of the animals are transported to farms in the central sub-humid pampa for fattening; the remaining 20 per cent are retained in the breeding region. In the former, pastures are managed to produce a series of annual or perennial grasslands that the animals can graze sequentially. Winter pastures comprise rye, oats or barley whilst summer pastures comprise sorghum or maize which are grown without artificial fertilisers. Pastures dominated by legumes, e.g. lucerne and various clovers (*Trifolium* spp.), and grasses, e.g. tall fescue (*Festuca arundinacea*) and orchard grass (*Dactylis glomerata*), are also used. Grazing regimes are rotational involving 10 animals per hectare, though this can rise to as many as 20. Pasture forage may be supplemented by pasture hay, maize silage or sorghum silage. When meat prices are high, especially in relation to grain prices, maize and oats may be fed directly to the animals. Animal health products are widely used to control the incidence of disease. Of the beef produced in Argentina, only 10 per cent is exported, compared with 45 per cent in 1920. This trend has resulted from increased local consumption and competition from Brazil which is now a major beef exporter to Europe.

Australia is also a major exporter of beef (Squires and Vera, 1992), the majority of which is produced in tropical arid and semi-arid regions. Most beef cattle production takes place on an extensive basis, often on marginal land that is unsuitable for crops. Grazing animals (introduced from Europe) have, however, exacted an environmental toll, having created significant environmental problems, some of which are discussed in Section 8.2 (see also Friedel *et al.*, 1990; Wilson, 1990). In contrast to the examples of the USA, Argentina and Uruguay given above, beef cattle production in Australia involves breeding and fattening ready for slaughter *in situ*. The predominant species is the droughtmaster, a crossbreed between the zebu and several European breeds, e.g. *Bos taurus* (Payne, 1990). In the savanna zone of northern Australia various strategies have been employed to improve beef productivity, which is constrained mainly by water deficits during the dry season. For example, McCown and Williams (1991) report that crossbreeding with the zebu produced an animal more tolerant of drought and poor forage than European breeds. The provision of non-protein nitrogen supplements in the form of herbage during the dry season also improves herd quality and substantially reduces animal mortality. Manipulation of pasture vegetation can also improve productivity. Either the replacement or supplementation of native grasses and legumes by more nutritious species can be undertaken. McCown and Williams report that most effort has been spent in identifying suitable legumes because of their ability (via symbiotic bacteria) to increase the nitrogen content of the soil. One such legume, the Caribbean stylo (*Stylosanthes hamata* cv. Verano), has proved particularly valuable. In New Zealand, McChesney *et al.* (1981) report that beef meat production requires an input of only 95 MJ kg^{-1} of protein as compared with 348 MJ kg^{-1} in Britain.

Beef cattle production also occurs in Africa, where there are two separate production systems. The first is a traditional system centred on communal land; the second is more akin to the commercial beef ranching described above and is often owned or managed by people of European ancestry (Squires and Vera, 1992). According to Tacher and Jahnke (1992), approximately 95 per cent of the beef produced in Africa derives from traditional systems, particularly those in inter-tropical Africa. Some of these are based on nomadism and have been discussed in Section 4.2. Others are permanent agropastoral systems and many occupy semi-arid environments that require adaptation to drought (e.g. Scoones, 1992). These variations, together with differences in forage quality, disease incidence and animal species, give rise to a wide variation in animal and meat quality and quantity. For example, the live weight of three-year-old male cattle may vary between 152 kg and 364 kg. Even at the higher end of this range it does not compare favourably with the average 520 kg in the USA (see above), though weights of ranched animals in systems akin to those of the USA in Kenya can reach *c.* 430 kg (Tacher and Jahnke, 1992). This wide variation in environmental and cultural conditions as well as agricultural practices makes it difficult to generalise about Africa's beef cattle production.

Apart from the nomadic agropastoral systems, cattle are kept in the tropical lowlands as components of mixed farming systems. They are integrated with crops to provide labour and manure and these will be considered in Chapter 7. Cattle are also a dominant feature of the highlands of East Africa, notably in Ethiopia, at altitudes above 1500 m. Transhumance is often practised, exploiting the plains, plateaux and pastures above 2300 m. Many of the cattle are kept for their milk rather than their meat. In all of these systems the mortality rate is high, possibly as much as 40 percent in the 0- to 1-year age group and 15 per cent in the 1- to 2-year age groups. Drought and disease are the main causes of high mortality. A further example of highland cattle herding with transhumance is that of the El Kala region of Algeria, which has been discussed by Homewood (1993). During the three months of the winter, cattle are moved to forest pastures in a transhumance system known as *achaba*. During the rest of the year cattle graze upland pastures. They are the prime source of wealth and income and during the winter the animals benefit from barley and other fodder crops cultivated by the herd-owners who have permanently located settlements.

Detailed examples of animal husbandry systems in Africa's Sahel zone have been provided by Le Houérou (1989). He defines four types of animal production systems: sedentary agropastoral, short-range transhumant agropastoral, medium- and longer-range transhumant, and nomadic and agro-sylvo-pastoral systems. The latter two are discussed elsewhere, in Sections 4.2 and 7.1 respectively. In relation to the agro-pastoral system, Le Houérou cites the examples of central Mali. Here, villages concentrated around the main rivers of the Senegal, Niger, etc., practise rainfed crop production. Beyond the cultivated fields are the rangelands on which the animals are kept during the wet season. They are grazed on crop residues and stubble during the dry season. By the age of five years, males may achieve a weight of 300 kg though as much as 20 per cent of this may be lost during the dry season. Short-range transhumant agropastoral systems are also associated to a certain extent with crop production. In parts of Niger, Mali and Burkino Faso, millet is grown as part of either a shifting or fallow agricultural system (Le Houérou, 1989). Cattle comprise 75 per cent of the associated livestock and will cover distances of between 50 and 200 km in order to exploit temporary ponds during the dry season.

The occurrence of drought in parts of tropical Africa has led to the development of many mitigating strategies in traditional agricultural systems. One such example is that of the Masai people of Tanzania and Kenya whose livelihood is largely dependent on cattle. These are now concentrated in a much smaller area than in the 1890s when both drought and a major outbreak of rinderpest caused major changes in their lifestyle (Maghimbi, 1991). Instead of being nomadic and/or transhumant, many Masai have been obliged to settle on a semipermanent basis along rivers. Such environments are not conducive to cattle rearing because of the relatively high incidence of disease and the inconvenience of flooding. The construction of small-scale dams, which is currently being undertaken, may help alleviate these problems. However, D. J. Campbell (1991) suggests that such initiatives, along with others, will not contribute to the preservation of Masai herding and will encourage movement to Kenya's cities, thus accentuating urban deprivation. Other problems are also interfering with good rangeland management in the transhumant systems of southern Kardofan in Sudan. Here, cattle, goats and sheep are kept but carrying capacity is being exceeded because of encroachment of cultivators into the rangelands and the difficulties of following traditional transhumant routes because of the civil war (El Wakeel and Sabah, 1993).

The raising of beef cattle is becoming an important aspect of Chinese agriculture. Tuan (1993), for example, has drawn attention to the substantial increase in beef production (as well as other meats and animal products) that has occurred since 1978 when major reforms for the rural sector, notably decentralisation, were initiated (see also discussion in Ash, 1993) in post-Mao China. From a value of 16.8 per cent of the gross value of China's agricultural sector in 1979, beef cattle products accounted for 26 per cent in 1988 (Tuan, 1993). Moreover, the annual average growth rate of livestock production was about 9.3 per cent. Although pork is the focus of China's livestock production, because of the relatively efficient conversion rate of primary into secondary productivity, beef production has grown by more than 12 per cent annually in the 1978–1988 period. Emphasis has been placed on not only increasing livestock numbers but on improving the quality of production (e.g. Tao, 1991b). The major pastoral regions of China are in the northern and western provinces. In Xin Jiang, for example, grassland pastures produce draft and breeding cattle with an average body weight of 400 kg per head (Jin *et al.*, 1990). This is similar to the body weight of animals in Argentina but is approximately 100 kg less than for beef cattle in the USA (see data given above). To improve overall productivity, Jin *et al.* suggest that the stocking rate should be reduced to improve the quality of the pastures and reduce animal mortality. Further improvements could be achieved by importing grain, or transporting animals to grain-producing areas, for fattening prior to slaughter.

Schemes in other parts of China for improvements in pasture quality, and hence improvements in animal productivity, have also been evaluated (see reviews in Brown and Longworth, 1992; Hu *et al.*, 1992). Michalk *et al.* (1993a, b), for example, have reported on trials with various pasture grasses and legumes in the rangelands of Hainan Island off the south coast of China. Here, range improvement is desirable because low rainfall and poor soils have prevented the expansion of crop agriculture and because cattle rearing is well established. The experiments of Michalk *et al.* indicate that there is potential for improvement, provided the legumes and grasses are carefully selected for their resistance to various diseases, drought and fire as well as for tolerance to specific soil types.

Worldwide, many studies have been undertaken on methods with which to improve rangeland productivity. Everhart (1991) has reviewed the various factors that need to be taken into account, Haferkamp *et al.* (1993) have examined the impact of mechanical treatments and climatic factors on rangelands in the Great Plains of the USA, and Hart *et al.* (1993) have discussed the impact of grazing patterns. Haferkamp *et al.* (1993) advocate the use of models by livestock managers who must adjust herd sizes to forage availability and quality. This, in turn, is linked to moisture availability which can be particularly variable in semi-arid environments such as Montana. They found that mechanical disturbance of the soil surface improved the productivity of forage, particularly that provided by annual grasses, though it is not clear why this is so. Hart *et al.* (1993) also report that the availability of water is an important factor in rangeland utilisation. Their comparisons of different grazing regimes, i.e. continuous and rotational grazing, on pastures of 24 ha with that of a 297 ha pasture showed that the latter is not fully exploited if there are areas further than 3 km from water. Animals on the 24 ha pastures gained more weight than those on the larger pasture because they roamed less. These results lead Hart *et al.* to conclude that balanced livestock distribution is as important for efficient rangeland management as are correct stocking numbers.

These studies reflect some of the many factors that contribute to the intricate relationships within rangelands. Coupled with climatic unpredictability, the behaviour of pasture species under different management techniques makes it difficult to obtain maximum secondary productivity without, in some way, impairing delicate ecological balances. When this happens environmental degradation ensues and causes a marked decline in biological productivity. This is one of the characteristics of desertification, a major consequence of rangeland overexploitation, which is discussed in Section 9.3.

6.2 THE PRODUCTION OF WOOL AND SHEEP MEAT

Many of the factors that relate to beef cattle production (Section 6.1) also relate to the production of wool and sheep meat. Like cattle, sheep are ruminants or grazing animals that are reared on the world's rangelands and grasslands on either an extensive or intensive basis. Except in the USA, sheep are the dominant livestock type in all the world's arid lands, where they are components of both nomadic and sedentary pastoral systems (Arnon, 1992). As Table 6.1 shows, there are approximately 1138.0×10^6 sheep in the world. Of these, the majority are in Asia, particularly in China, and in Australia and New Zealand (Table 6.3) where they are reared on an extensive basis for both their meat and wool. According to FAO (1993),

Table 6.3 The distribution of sheep worldwide (data abstracted from Food and Agriculture Organisation, 1993)

	Numbers of sheep $\times 10^6$	
	1991	1992
Africa	206.34	205.99
South Africa	32.58	32.11
Ethiopia	23.00	23.20
Sudan	20.70	22.60
Algeria	18.50	18.60
North & Central America	19.45	19.43
USA	11.20	10.80
Mexico	5.88	6.18
Canada	0.78	0.91
South America	104.56	100.72
Uruguay	25.99	25.70
Argentina	26.50	23.71
Brazil	20.30	19.50
Asia	342.99	344.05
China	112.82	111.14
Iran	44.68	45.00
Turkey	40.55	40.43
India	44.21 *	44.41 *
Europe	144.13	138.24
UK	30.15	28.93
Spain	24.63	24.63
France	10.64	10.58
Oceania	218.41	200.33
Australia	163.24	146.82
New Zealand	55.16	53.50
Former USSR	133.22	129.62
Developed countries	540.72	512.34
Developing countries	627.78	626.02

* Estimate.

approximately 4.632×10^6 tonnes of wool were produced in 1992 worldwide. Of this, nearly 1×10^6 tonnes came from Australia and New Zealand where sheep-rearing constitutes a very important sector of the agribusiness industry. Elsewhere, e.g. in North Africa and parts of western Asia, sheep meat is produced on a subsistence basis and is a mainstay of local diets.

Originally domesticated in the Near East (Section 2.4) from the mouflon, sheep spread into Asia and Europe in tandem with the spread of agriculture. They arrived in the New World with the advent of the Europeans, reaching the Americas in the fifteenth century and Australia and New Zealand in the eighteenth century. Today, sheep are components of high- and low-technology agricultural systems from the small to the large scale. Equally they may be produced for subsistence or as a cash commodity, i.e. meat and/or wool, and, as Table 6.3 shows, they are almost equally distributed between the developed and developing worlds. Gatenby (1991) also distinguishes between sheep-rearing in the semi-arid areas and humid regions of the tropics. In the former, flocks are large and will graze vast rangelands to take advantage of seasonal forage production. In the humid tropics, in contrast, flocks are usually small and are rarely moved large distances, as is the case in many parts of the temperate zone, e.g. the UK. Sheep are not always husbanded on their own. In Africa and western Asia they are often kept with cattle and goats which graze semi-arid rangelands, and in arid areas they may be kept with camels (Payne, 1990). This is more typical of low-technology traditional agricultural systems than of high-technology systems which usually concentrate on the rearing of one type of animal.

Pagot (1992) and Gatenby (1991) have described the various ways in which sheep are managed in tropical regions (Figure 6.2). The characteristics of the migratory pastoralists are similar to those described in Section 4.2. Gatenby states that flock sizes range between 50 and 200 sheep and the purpose of their husbandry is to provide meat. Smallholder systems generally occur where there is sufficient rainfall to produce crops and sheep are kept as an addition. Only about 10 animals are kept and they are managed in a variety of ways: free grazing, shepherded grazing, tethering and stall feeding. This is a mixed farming system that is widespread in the tropical world, particularly for meat production. According to Charray *et al.* (1992), sheep farming in the humid zone of Africa has made significant advances due to recognition of the fact that sheep can be more productive than cattle and also due to the establishment of a number of government-sponsored development projects. One of the major drawbacks, however, is the relatively high occurrence of diseases and parasites. The West African Dwarf sheep, for example, lives in areas prone to tsetse-fly infestation. Consequently it is always at risk from trypanosoma, a disease produced by these blood parasites.

Ranching has also become established in Kenya and Botswana (Gatenby, 1991) and is modelled on some of the extensive sheep production systems of lowland Europe. Large flocks of more than 200 animals are kept on a comparatively large area of land that is enclosed by a fence. The density of animals per hectare varies but it is low, possibly as low as one animal per hectare. The stocking rate depends on the carrying capacity of the land, which in turn depends on the quality, seasonality and distribution of forage. The grazing of the animals is managed. For example, animals may be excluded from parts of the ranch to allow the vegetation to recover. This is rotational grazing whilst creep grazing allows lambs access to pasture of a better quality than that for more mature sheep. In order to obviate the need for the migration of animals, as was the case before the establishment of ranching, the ranches must be of an adequate size to provide year-round grazing. Many of the ranches of East Africa produce fine-woolled breeds, some of the wool of which is exported.

As Table 6.3 shows, Spain has one of the largest numbers of sheep flocks in Europe. The sheep graze several different ecosystems and so require many different management strategies (Montserrat and Fillat, 1990). For example, in the Pyrenean region of Burgos–Navarra, the slopes of the Ebro Valley are clothed in oak woodlands, below which there are grasslands. Both ecosystems are grazed by sheep on an extensive basis. Elsewhere, transhumance is practised, e.g. in the Cameros of Spain's Central Mountains. More intensive sheep-producing systems characterise the Léon–Castilla region in north-west Spain, south of the Cantabrican mountains. Both wool and milk are produced by the Churra breed which is kept in flocks of between 100 and 140 animals on heathlands dissected by valleys. The sheep are also allowed to graze on the stubble of crops grown in the same area and on forage crops. Elsewhere

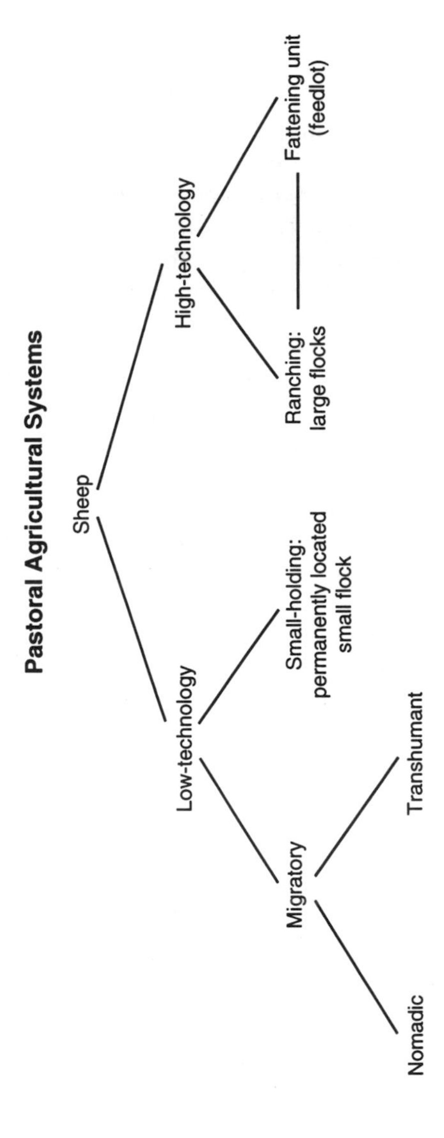

Figure 6.2 Sheep production systems

in the region, Entrefina Castellana sheep are reared for their meat and milk. They graze either wooded pastureland or moorland. In the moorland areas, lucerne and sainfoin are also cultivated. The sheep are allowed to graze the stubble remaining after the crop is harvested and their diet is supplemented in the winter by hay produced from these crops. These systems are intermediate between the extensive ranching and traditional grazing systems of the dry topics and the intensive production that occurs in other parts of Europe.

Sheep ranching is also a major component of Australia's agricultural sector. It dominates the arid and semi-arid zones which are grasslands, savannas or semi-deserts. Livestock had been introduced into these lands by the late 1880s, a century after the initial colonisation of Australia by Europeans (Heathcote, 1987; Friedel *et al.* 1990; Hobbs and Hopkins, 1990). The evolution of Australia's landscape in the absence of any hard-hooved animals, and indeed few large herbivores, may be one of the reasons why the livestock sector has created many environmental problems. According to Heathcote (1987), there were 80 million sheep in Australia by 1888, the heyday of livestock production which he describes as 'the golden age of ranching'. The expansion was the result of increasing migration into Australia, which created a home market for meat and wool, as well as a growing market in Europe. Some of the environmental problems ensuing from this growth are considered in Chapter 9.

Currently, Australia's 163.24×10^6 sheep occupy extensive tracts of land. According to Pagot (1992), the most extensive ranching in the world today is probably that of Australia's Northern Territory. Properties may cover several thousands of square kilometres and are owned mainly by multinational companies rather than individuals. Stocking rates are very low, e.g. one animal per 20–100 ha, depending on the quality of forage and its availability. In Western Australia stocking rates have increased from 5.5 sheep per pasture hectare in the early 1960s to 6.1 in 1990–1991 (Kelly and Marshall, 1993). During this period the productivity of sheep farms has increased by 2.7 per cent, resulting from a 10 per cent improvement in lambing successes and an increase from 4.6 kg to 5.9 kg in greasy fleece weight per adult sheep, usually of the merino type. However, as Kelly and Marshall point out, this improvement in productivity has been achieved by increasing the area of land under pasture from 3.1×10^6 hectares in 1960–1961 to more than 7×10^6 hectares in 1990, and a more than 100 per cent increase in sheep numbers from 17.2×10^6 to 38.4×10^6. Not only is an enhancement of animal quality and health necessary to achieve these increases but good pasture management is also required. Factors such as forage quality, quantity, carrying capacity and seasonality are essential ingredients of management strategies that pertain to individual ranches in order to allow for local conditions. Computer-based models are now widely used by ranch managers to assist in making management decisions, e.g. HERDECON devised by Stafford Smith and Foran (1991). Animals are sometimes herded to fattening areas (i.e. the equivalent of the US feedlot, see Section 6.1 and Figure 6.2).

In contrast to these extensive sheep-producing systems are the intensive systems that characterise much of Europe, including the UK (Croston and Pollot, 1994). According to Slade (1990), the sheep flock of the UK has increased by more than 42 per cent during the 1980s, making the UK one of the major producers of sheep meat and wool (MAFF, 1990) in Europe (see Table 6.3). This increase has been concomitant with a decline in imports from the Commonwealth countries of Australia and New Zealand, increased exports to France and support from the EU Common Agricultural Policy (CAP). This has included provision of subsidies for hill sheep and a variable premium system that guaranteed the prices achieved by store lambs. However, changes to CAP support in the late 1990s and beyond may once again alter the structure of the sheep industry in Europe in general. Surprisingly, however, this increase has occurred when per capita consumption of sheep meat has actually declined in the UK though not to such low levels as to exclude limited imports from other parts of Europe and New Zealand. Brown and Meadowcroft (1989) describe the productivity of UK sheep as poor though they are not explicit about why this should be the case. They state that 'sheep have justified their presence on the hills because of low land values and a lack of alternative enterprises. In the lowlands their poor performance has been tolerated because of their beneficial effects on soil fertility and because they can be used to graze the ley break in the rotation without a large commitment to capital or labour.' They also consider the improvements in productivity that have already been achieved and which indicate that further gains could be made. These comprise improvements in the weight of lambs sold per ewe (this is the product of the number of lambs reared per ewe and their average sale weight) and sold per hectare, enhanced disease control, winter feeding and housing, and year-round rather than seasonal marketing. Clearly, secondary productivity is not at its optimum in the case of the UK sheep industry. This contrasts with sheep production in New Zealand where there is a low energy input of 116 MJ kg^{-1} of protein as compared with 465 MJ kg^{-1} of protein in the UK (McChesney *et al.*, 1981).

Sheep flocks in the UK are either upland or lowland. In either environment, rainfall is rarely a limiting factor in pasture quality and quantity. Thus, many of the problems associated with sward seasonality that characterise arid lands are avoided (see above and Section 6.1). Nevertheless, the quality of the pasture is all important as a determinant of secondary productivity (Gordon and Illius, 1992). Various types of grassland management have been discussed by Brown and Meadowcroft (1989) and include both rotational and creep grazing (see above). Upland and lowland pastures are grass-dominated with a mixture of leguminous species such as white clover (*Trifolium repens*). The latter, in combination with perennial rye-grass (*Lolium perenne*), provides good-quality grazing (Brown and Meadowcroft, 1989). In addition, the application of nitrate fertilisers to pastures can result in increased sheep productivity through the generation of quality pasture.

The many factors that can be manipulated in the field to improve productivity have been discussed by Brown and Meadowcroft (1989) and Treacher (1990). Increasing the weight of lambs sold per ewe or per hectare (see above for definitions) requires the provision of quality forage. Other characteristics, such as the rate of conversion of primary into secondary productivity, relate to the metabolic efficiency of specific breeds and animal health. The field factors that can be manipulated are listed in Table 6.4. In addition, forage crops may be included in the feeding programme. Turnips, rape or potatoes are valuable in this context, particularly during the late summer when the growth rate of pasture grass is low. Lambs that are newly weaned in June and July also benefit from such crops, while forage rye will bring forward the grazing season by between two and four weeks in the spring prior to the onset of grass growth. Animals obviously benefit from forage crops during the winter. Another aspect of improving animal productivity is animal health. Factors such as

Table 6.4 The field factors that can be manipulated to improve sheep productivity in the UK (based on references quoted in the text)

A PASTURE COMPOSITION
Italian rye-grass (*Lolium multiflorum*) gives higher yields than other grasses but, as a biennial, it is most appropriate for short-term leys (pastures) of two years. For longer leys, perennial rye-grass (*Lolium perenne*) is most appropriate. White clover (*Trifolium repens*) is a useful ley legume though red clover (*T. pratense*) should be avoided because it may lead to a reduction in the fertility of grazing ewes. Legumes increase the available nitrate.

B HEIGHT OF SWARD
Numerous trials in the UK indicate that there is a direct relationship between the height of the sward and animal productivity, e.g. below 3 cm height, lamb growth is restricted due to difficulties in the animals securing sufficient herbage. There is also evidence to show that sward height gives the best results when it is varied throughout the year between *c.* 4 cm and 8 cm.

C USE OF ARTIFICIAL FERTILISER
This increases the primary productivity of leys because it supplements the nitrate that is available naturally (possibly also phosphorus and micronutrients). Applications of fertiliser should be carefully managed to avoid cultural eutrophication (see Chapter 8). Ley grass growth is particularly well stimulated if nitrate fertiliser is applied around two weeks before soil temperatures reach 5.5°C which activates grass growth.

D STOCKING DENSITY
This relates to A, B and C above and whether the pasture is upland or lowland. Optimum stocking density can only be achieved if forage availability and the animals' forage requirements are co-ordinated continually. Stocking density will also vary according to the grazing patterns employed, e.g. if rotational or creep grazing is used.

trace-element deficiencies, metabolic problems and internal parasites, the most common problems associated with ley-reared sheep (Brown and Meadowcroft, 1989), need to be addressed.

According to Webster and Povey (1990), increasing attention is being paid to the finishing of lambs as a result of consumer demands for high quality and the desire to maximise profits. This has led to changes in finishing practice and a decline in fodder-crop feeding in the field. For example, cereals are increasingly being fed to newly-weaned lambs because they provide a good source of energy and can be manipulated by feeding as individual grains, to avoid the production of soft fat. Slaughter weight can be achieved in 12 to 14 weeks. Growth rates can also be accelerated if sources of high-quality protein, e.g. fish meal, are included in the feed. In addition, indoor finishing systems are gaining favour and are based on concentrate foods, e.g. oil cake or on silage with supplements. No doubt genetic engineering will lead to the breeding of more productive sheep than those of the present. Murray and Oberbauer (1992) also point out that fat quality and quantity can be manipulated using pharmacological methods, though this is not necessarily desirable. Recent occurrences of bovine spongiform encephalopathy (BSE) in cattle and scrapie in sheep are the results of manipulating the feeding programmes of grazing animals. Concerns are rightly being expressed about the possibility of transmitting these diseases to humans through the food chain.

6.3 THE PRODUCTION OF MILK AND OTHER DAIRY PRODUCTS

According to Grigg (1993b), milk is the second most important livestock product, after meat, which dominates food intake when it is measured as calorie consumption. Moreover, its consumption is mainly related to income though the prevalence of a condition known as primary adult lactose malabsorption, which means that digestion is inhibited, in west and central Africa, and parts of China and south-east Asia, precludes its importance in these regions. However, although the consumption of milk (and dairy products) per capita is highest in the developed nations, it constitutes a very important source of calories per capita in many developing countries. Here, Grigg (1993b) points out that it provides 41.7 per cent of the calories that are derived from livestock products. This compares with only 26.4 per cent for developed countries. The significance of milk in diets worldwide reflects its production from a range of sources. The most obvious source is cattle which provide the majority of the world's milk and which are kept in a wide variety of livestock-producing systems. These range from the nomadic, as discussed in Sections 4.1 and 4.2, to the most intensive, as will be examined below. As well as this distinction, cattle rearing may be a sole enterprise, as in the case of several Sahelian groups (Section 4.2), or intense dairy farming, or it may be a component of mixed farming. The latter are considered in Chapter 7. As well as cattle, sheep, goats, buffaloes and camels are reared for their milk. Camels, for example, have already been referred to in Section 4.2. Chamberlaine (1990) also points out that in many tropical regions fresh liquid milk is not always the most important product. It may be converted into products with a longer life, e.g. fermented and concentrated milks and ghee, a clarified butter made from cow or buffalo milk. There is also a general consensus that dairying in tropical countries is in a period of growth. This is because of rapidly increasing markets as urban centres expand and the success of western-style dairying in regions like the Caribbean which do not harbour traditional pests.

According to van den Berg (1990) and Matthewman (1993), there are four systems of dairy farming that characterise the tropical zone. These are nomadism and transhumance, ranching, mixed farming and specialised dairy husbandry. The latter accounts for less than 5 per cent of animals, with mixed farming systems (Chapter 7) accounting for more than 50 per cent. The production of milk on ranches is really an adjunct to the rearing of cattle for beef (see Section 6.1); the cows suckle their calves but they will be milked once per day. In what van den Berg describes as dairy-developing countries, milk production is often concentrated in and around urban areas which constitute the main market. Cattle (possibly buffaloes and goats as well) are kept in sheds in large or small numbers. The milk produced is either sold from the shed or it is ferried around neighbouring streets and sold on the doorstep. Occasionally, the animals themselves may be ushered through the streets and milked at the door of each customer. Van den Berg argues that the only real advantage of this type of milk production system is that the distance, and hence time for transport, between animal and consumer is short and so there is less wastage. There are, however, many disadvantages. For example, feed has to be imported and animals are often slaughtered when their milk supply runs dry because it is more financially advantageous to do this than send them back to rural areas where they fetch only low prices. Other disadvantages concern hygiene. As an alternative to the keeping of cattle in cities, some governments have contributed to the establishment of cattle colonies in areas around the cities. These colonies comprise sheds, housing for the cattle owners and basic services, e.g. water and electricity as well as veterinary services and the possibility of collective feed purchases. These cattle colonies proved unsuccessful, e.g. in India and Pakistan, and the demise of urban dairying will probably be a consequence of urban growth and the lack of accommodation for animals.

Many developing countries have successfully introduced structured dairy industries, and others are just beginning to do so, usually with government aid and often with foreign aid. The situation in India, for example, has been reported by Aneja (1990). Today, India is the third largest producer of milk in the world, a situation which has developed through what Aneja describes as a 'white revolution'. This refers to 'operation flood', which began in 1970 and was the world's largest dairy development scheme aimed at providing India with a secure dairy industry. It has given rise to 6×10^6 milk producers who produce 50×10^6 t of milk per year. The basis of this industry is crop residues. Its success has been due to the drying of the milk surplus produced in the winter to compensate for the deficit in the summer and the organisation of milk producers into co-operatives. These co-operatives organise and provide collectively-purchased feed and veterinary services. Some 1.5×10^6 litres of milk are daily transported to dairies from more than 1000 village co-operatives in the Anand district alone. In addition, 400 000 kg of concentrates per day are produced in cattle-feed factories from whence they are distributed to the villages, whilst the same lorries return the milk to the dairies. The milk producers themselves may keep only a few animals. Indeed, Aneja states that 40 per cent of landless agricultural workers are milk producers, utilising crop residues provided by the farms for which they work as a component of their wage. The total number of villages organised into milk-producing co-operatives throughout India is now 61 000 and it is increasing at a rate of 4000 per year. One of the most important aspects of this development is the way in which the industry is geared to local resources rather than the superimposition of western-style dairying on unsuitable environments. Such an approach, whereby the local relationships between land, cattle, forage/fodder quality and quantity and labour are prime factors, is essential for the establishment of successful dairying industries in developing countries.

The importance of physical environmental factors, as well as social issues such as land tenure, education and technical support, pricing policies and availability of financial aid, have been discussed by Ruiz (1990) in relation to milk-production systems in Latin America. Together with the Caribbean, Latin America houses 16 per cent of the world's population of milk-producing cattle which produce 7.4 per cent of the world's milk production. There is, however, not only considerable variation between regions in this vast area but also much potential for the improvement of milk production. Despite this variation, a basic distinction can be made between production systems in the highlands and those in the lowlands. The characterisitics of these regions in relation to milk production are given in Table 6.5. Specialised dairying, for example, occurs throughout the valleys of the Andes and in the upland regions of Central America. In the lowland regions, beef-cattle production systems predominate (see Section 6.1), though dual-purpose, i.e. milk and beef, production systems are increasing in importance. There are also localised, successful dairying industries in tropical lowland areas, notably in the Caribbean region and around the

Table 6.5 The characteristics of milk-production systems in the highland and lowland zones of Latin America (based on Ruiz, 1990)

	Highlands	Lowlands
Altitude (m)	1000–3500	Below 1000
Rainfall (mm)	1000–1600	600–1000 (dry areas) 1000–5000 (wet areas)
Average temperature (°C)	15–21	27
Seasonality	Low	5–7 months drought in dry areas
Farm size (ha)	30–55	Variable
Herd size	90–180 animals	100–1000 animals
Pasture type	Temperate grasses and legumes	Low-quality grass
Stocking rate (animals per hectare)	*c.* 3–4	*c.* 1.5
Milk production (kg cow^{-1} yr^{-1})	2150	880
Calving rate (%)	80–85	Not available
% of animals milked	20	30–50

cities of Mexico City and Lima, because of the availability of agricultural waste products in the case of the former and the large markets created by the high populations of the latter. The production, in Cuba for example, of 11 different animal feeds from sugar-cane waste, some of which is used to support dairying, is a case in point (see Section 5.2).

Milk production is also an important component of Egypt's agriculture. According to Rowntree (1993), milk and milk products account for 35 per cent of the value of goods produced on the farm; 80 per cent of the milk produced is consumed or processed on the farms. Unlike the examples given above, the buffalo is the chief milk-producing animal in Egypt. Buffalo herds provide about 66 per cent of Egypt's milk whilst native cattle provide the remainder. Rowntree reports that milk is produced under two different systems. The main system comprises small farmers engaged in traditional local-scale mixed farming. Each farm will have only two or three animals but collectively these farms produce about 80 per cent of the total milk production. The remaining 20 per cent is produced by commercial dairy herds located on the edges of major cities or by large-scale state-owned farms. In semi-urban systems, the animals are usually slaughtered after one lactation for similar reasons to the urban cattle of India and Pakistan (see above). Milk productivity overall is low in comparison with that in developed nations but cheap labour means that it is economic to produce. According to Ward (1993), the average lactation yield is about 670 kg per cow (excluding milk taken by the calf) whilst buffalo have average lactation yields of 1300 kg per animal. The home processing of milk is undertaken by the women of the farms, who produce cheese, butter and ghee. Alternatively, milk is processed by the state-owned Egyptian Milk Company which provides milk products for sale in the urban centres.

Scope for improving yields, especially in cattle throughout the tropical world, has been discussed by Maule (1992). He has drawn attention to the potential of crossing indigenous breeds with other tropical and temperate breeds. Many of these indigenous breeds have not previously been considered for breeding because their success is not well documented. Nevertheless, there are several instances of substantial improvements in milk yield which have occurred as a result of interbreeding. For example, the introduction of Sahiwal cattle from India into Kenya and their breeding with the East African zebu. The crossbreed is a good milk provider, eventually averaging 1630 kg during a single lactation period. A similar project, known as the Criollo project at Santa Cruz, Bolivia, set out to establish a Criollo herd for breeding with Holstein and Brown Swiss cows. The programme has resulted in total milk yields of more than 1500 kg in a lactation season. These data not only show how interbreeding can increase productivity but they also highlight the necessity of preserving the genetic diversity of domesticated animals (see also Cunningham, 1993). The same is also true of crop plants and there is much importance being attached to preserving not only crop varieties but also their wild relatives (see Chapter 10).

The countries with the highest production and consumption of milk (and other dairy products) in the world are the developed nations. According to Grigg (1993b), milk provides on average 278 calories per capita per day, or 8.2 per cent of all calories consumed per capita per day. The corresponding values for developing nations are 58 calories per capita per day and 2.4 per cent. Of the developed nations, the highest consumption is in Australasia, North America and Western Europe, with values of 357, 356 and 322 calories per capita per day respectively from milk, cheese and yoghurt; in Eastern Europe and the former USSR the value is also high at 249 calories per capita per day (quoted in Grigg, 1993a). In many cases, milk is produced from cattle which graze pastures. Consequently, many of the conditions that apply to the rearing of beef cattle (Section 6.1) are also relevant to milk production. These include the quality and quantity of the sward, the use of concentrated feeds, grazing management and animal health control.

The milk-producing systems of the UK, which typify those of much of Western Europe, have been described by Thomas *et al.* (1991) and Slater (1991). According to Slater, milk accounts for *c.* 20 per cent of the total sales of food products from UK farms. Its significance, however, to the UK agricultural sector is not just as a product of a climate well suited to grass production but reflects systems of subsidy and guaranteed payments that are components of the Common Agricultural Policy (CAP). Moreover, until recently the milk was marketed via five Milk Marketing Boards. It is interesting to note that so significant are these economic and political factors to present-day milk production, that Slater's book does not begin to discuss the production of milk in the field until Chapter five which begins on page 82! In most instances, dairy farms also grow crops that provide valuable fodder or forage. In the west of the UK, grass and root crops are most suited to the wetter climate while cereal crop production is better suited to the drier eastern and south-eastern areas. Slater states that on small farms of 40 ha or less, especially those in high rainfall areas to the west of the country, recent trends have revolved around specialisation on grass and the elimination of other livestock enterprises. In the east of the UK, cereals, kale and sugar beet are used to supplement feed from the pasture. The cereal crop also provides litter for the animals' stalls, the scarcity of which in the wetter UK regions is a problem though sawdust and wood shavings may be substituted. Slater (1991) also points out that there have been some trials with keeping dairy cattle housed throughout the year, substituting mechanically delivered feed for grazing. This is similar to the feedlots in the USA that are used for fattening beef cattle (see Section 6.1). Apparently, success is only possible if at least 300 animals can be accommodated to justify the cost and operation of the necessary machinery.

The primary concern of all dairy farmers is to achieve high milk yields. This is only possible if the quality and quantity of the pasture is high. According to Baker *et al.* (1991), the wide range of UK grassland types can be classified, as either permanent grasslands (described in detail by Green, 1990; Rodwell, 1992) or long- and short-term leys (pastures). As in the case of sheep production (Section 6.2), a sward of perennial and Italian rye-grasses provides a high feed value for dairy cattle. Varieties of these are the favoured species for leys in both upland and lowland situations. Permanent grasslands, however, will contain a much wider variety of species than sown leys but their overall value as feed still relates to the content of perennial rye-grass. This can be encouraged by management, e.g. frequent grazing and fertiliser application. Baker *et al.* have also discussed the increasing use of white clover as a ley component (see Section 6.2) and as a means of reducing artificial nitrate fertiliser applications.

Since the mid-1970s the use of nitrate fertiliser has increased and has given rise to improved pasture and hence improved milk production per cow. The management of fertiliser application has been discussed by Baker *et al.* (1991) who point out that average rates of nitrate fertiliser applied to dairy-farm grasslands is approximately 200 kg N ha^{-1} (though Slater, 1991, states that it is 250–300 kg N ha^{-1}). In general, with applications of between 200 and 300 kg N ha^{-1} the yield of grass increases linearly by 15–30 kg dry matter ha^{-1} for every kilogram of fertiliser applied. Above 300 kg N ha^{-1} the response declines, though an increase still occurs until it levels off at between 500 and 850 kg N ha^{-1}. Baker *et al.* (1991) states that 'The point where the response drops to 7.5 kg dry matter of grass per ha for each kg of nitrogen which is applied is defined as the target yield and target nitrogen rate for dairying'. Apart from the fact that after a certain level of application it is apparent that few gains are to be made, there is also the problem of cultural eutrophication as drainage systems and groundwater become contaminated with nitrate-rich runoff from agricultural land. This problem has been referred to in Section 5.1 and will be discussed in detail in Chapter 8.

An additional and crucial factor in the efficient production of milk is the organisation of grazing. Much the same principles apply to dairying as to beef-cattle ranching (Section 6.1) and sheep production (Section 6.2). Controlled grazing is essential to maximise productivity. This requires an understanding of grass phenology (see also Section 1.4) which concerns the species' life-cycles. For example, grass grows rapidly in the spring with a further flush in the autumn, both depending on weather conditions such as increases in temperature in the spring and moisture availability. There are also substantial changes in the feed value of grass during the year; it may vary from lush and green, but purgative, to dry and straw-like. Slater (1991) believes that the key to managing these factors is to cut or graze the grass sward before it becomes too dry, to allow it sufficient time between grazings to recover and to ensure that nutrients, especially nitrogen, do not become deficient. Mayne *et al.* (1991) advocate several best practices which focus on a compromise between the performance of individual animals and milk output per hectare. The two are mutually exclusive because milk yields decline as the competition between animals increases. Grassland management can be achieved using various types of grazing strategies, e.g. rotational grazing which requires the ley to be rested for at least two weeks before regrazing, continuous grazing and strip grazing whereby a limited grazing area is set for a day and is controlled using electric fences. Although there appears to be little difference in productivity between these strategies, rotational grazing provides some advantages. These include improved control of sward height which is best kept between 6 and 8 cm, and the use of some paddocks for the production of silage.

Although grass is the cheapest way of producing optimum yields in dairy cattle, many of the factors described above may mean that some supplementary feeding is necessary to substitute for grass during periods of shortage. In addition, the attainment of higher yields than the ley allows will require feed supplementation. However, some concentrates may simply lead to increases in body weight rather than significant increases in milk yield. As with the grass management, the inclusion of concentrates must be undertaken with care to achieve the best economic results. Mayne *et al.* (1991) also report that the use of feed supplements other than concentrates, e.g. straw, hay and forage crops, can also have variable results. During the winter, the herd should be divided into feeding groups as differently aged animals have different requirements. Each group requires varying proportions of bulk feeds, concentrates and silage.

Similar techniques are employed in the management of New Zealand's dairy cattle herds (Thompson and Poppi, 1990) to ensure high daily and yearly milk production. Commercial dairy farms produce about 550 kg ha^{-1} yr^{-1} of milkfat. This requires 4000 kg yr^{-1} of dry matter derived from pasture. The quality of the latter needs to be especially good in the spring months to stimulate high milk production in preparation for calving. Thompson and Poppi (1990) also advocate that sward height should be controlled to ensure not only high milk production but also to encourage underground tillering which increases the underground biomass containing nutrients and starch ready for later regrowth. In North Island, Askin (1990) reports that pastures recently incorporating white clover (see above) have proved successful, not least because of the nitrogen fixation that ensues. However, experiments with mixtures of subtropical and temperate species indicate that management would be difficult. As in the UK, rye-grass has proved to be particularly successful as a pasture species. Indeed, Askin recommends that in New Zealand's North Island, pastures for dairying should be made up of 20 kg of rye-grass, as much as 5 kg of a short-rotation species of rye-grass and 3–5 kg of huia or pitau white clover (both are commercially produced cultivars of white clover—*Trifolium repens*).

Dairy farming is also a characteristic of parts of Siberia where Ruvinsky *et al.* (1992) reports there are *c.* 10×10^6 ha of hay meadows and 19.5×10^6 ha of pastures. Most of this is in the eastern region while the western region is dominated by arable land. Although much of this land is marginal, or not highly productive, the large extent of it, 58.5 million ha in total, means that Siberia contributes 16.6 per cent of the gross agricultural production of Russia (7.7 per cent of the gross agricultural production of the former USSR). Cattle (and other animals) are grazed extensively as the rangelands have a low carrying capacity. Per capita consumption of milk and dairy products is approximately 265.8 kg per year. This compares with 54.5 kg of meat and meat products, 75 kg of potatoes and 103.8 kg of bread and illustrates the significance of milk products in the local diet. Average milk production is, however, low in comparison with that of the temperate zone. Ruvinsky *et al.* quote values of 2000–2600 kg per cow per year for the period 1980–1989 which represents substantial increases on the production figures of the previous decade. This improvement in productivity has been achieved by increasing the amount of feed, notably cereal produced within Siberia, used as a supplement to pasture forage. Improvements in animal health care and breeding programmes have also contributed to increased production though there still remains scope for further improvement.

6.4 OTHER TYPES OF PASTORAL AGRICULTURAL SYSTEMS: GOATS, DEER AND CAMELIDS

As Table 6.6 illustrates, 95 per cent of the world's goats are in the developing world, particularly in Asia. They are reared mainly for milk, from which other products such as cheese are made, and sometimes for skins and fibre. According to Devendra (1990), goat meat is highly prized in many developing countries where demand exceeds supply. This has led to a significant increase of *c.* 25 per cent in goat populations over the last decade. The milk from goats is usually consumed where it is produced; for many rural communities it is the most important source of protein. Lactation yields vary considerably, depending on the species and the environment in which they are reared (Pagot, 1992). Amongst the highest milk producers is the Jamnapari goat of India with a lactation yield of between 200 and 562 kg. Other high milk producers are the Mamber goat of Israel and the Damascus goat of Cyprus (Devendra, 1990).

Goats are reared in a variety of agricultural systems. Within villages, for example, goats may be tethered in cultivated plots to graze or they may be directly fed crop residues, as well as organic wastes from kitchens. Animals may be integrated with cropping systems, e.g. plantation or orchard crops where there is a ground cover of vegetation. Extensive grazing may be operated where common land is available and/ or where the forage quality is poor and the land can be used for little else. Flock sizes tend to be large under these circumstances and may provide a family's only means of survival. The goats may also be herded with sheep but the density of animals will be low compared with other goat-rearing systems, varying between one and four animals per hectare. Such systems are akin to those described for extensive cattle (Section 6.1) and sheep (Section 6.2) production systems and they may be associated with nomadism (Section 4.2). Extensive goat grazing (along with sheep) occurs along the north-western coastal zone of Egypt in combination with the production of rainfed barley. Van Duivenbooden (1993) reports that the grazing activity is important for maintaining the vegetation cover. Grazing removes some of the shrubs' leaves, thus reducing transpiration and conserving water which can be used later, so extending the growing season.

Kuznar (1991) has discussed a different type of extensive goat-rearing system in the Central Andes of South America. Here, goat herders practise transhumance in order to capitalise on the spatial and temporal variations that occur in the availability of forage in an environment with a steep altitudinal gradient. On average, 30 goats are kept, along with *c.* 30 sheep and 5–10 cows. The goats are kept mainly for meat though the milk of females that have lost offspring is used to make butter and cheese. During the wet season (November to April) the lower high sierra zone (at elevations

Table 6.6 The numbers of goats worldwide in 1992 (data abstracted from FAO, 1993)

	No. of goats $\times 10^6$
Africa	167.945
Nigeria	24.0
Ethiopia	18.1
North and Central America	15.743
Mexico	11.008
South America	22.209
Brazil	12.0
Asia	345.411
India	117.0
China	95.032
Europe	14.992
Greece	5.832
Spain	3.0
Oceania	1.403
New Zealand	0.70
Former USSR	6.479
Developed countries	30.745
Developing countries	543.435
World	574.181

of 2500–3200 m) provides abundant grass for grazing animals for approximately three months. It is cropped by June when the animals are moved to the upper high sierra (at elevations of 3400–3800 m) to graze perennial shrubs and grasses. Some families have several residences located in both zones while others have a permanent residence with arable plots. Other possibilities for rearing goats involve intensive and semi-intensive systems. The latter involve limited stall feeding in combination with grazing on a daily basis. Intensive systems are of two types: the grazing of cultivated pastures with a high carrying capacity, and zero grazing based on stall feeding with crop residues and concentrates. Whatever system is used, the daily energy requirement is similar to that for sheep. A 30 kg liveweight animal, for example, requires an energy supply of 5.44 MJ kg^{-1}. Although goats are well known for their wide-ranging tastes in food, the highest milk production (and best meat production) results from a balanced diet containing the necessary protein and minerals as well as sugar and carbohydrate.

In contrast to the rearing of goats on an extensive basis (see above) or on a subsistent basis where a family may keep just a few animals to provide milk, etc., there is a growing trend in intensive goat rearing. In New Zealand, for example, there are farms with goat flocks of between 500 and 1000 animals (Daly, 1990). These are being reared mainly for fibre: cashmere, mohair and castigora, though there are increasing exports of goat meat to Asia. The animals are reared under an intensive grazing regime. Daly reports that there has also been an increase in milking goat populations though these are associated with smallholdings rather than large-scale agricultural enterprises.

Deer farming is a specialised type of pastoral agriculture that has enjoyed an increase in popularity since the mid 1970s. It is an enterprise characteristic of

Table 6.7 The numbers of deer farms and deer in the major deer-farming nations in 1985 (based on data in Reinken *et al.*, 1990)

	No. of farms	No. of animals
Austria	400	?
Sweden	141	4 000
Switzerland	76	2 000
Australia	400	15 000
New Zealand	3 000*	400 000*
UK	150	6 000

* Data quoted in Daly (1990).

developed nations, particularly Germany, the UK, New Zealand and Australia. Apart from the aesthetic value they add to deer parks, which are quite different enterprises to deer farms (Nahlik, 1992), deer are prized for their meat. Venison is rapidly increasing in popularity because it provides a novel, reduced-fat alternative to beef; it is also versatile and can be smoked. Both red deer (*Cervus elephus*) and fallow deer (*Dama dama*) are the species being farmed. The numbers of animals involved are considerably lower than for other types of pastoral agriculture. Reinken *et al.* (1990) distinguish between wild deer and farmed deer simply on the basis of sex ratios. In wild deer populations it is 1:1 while in farmed deer there is a sex ratio of 1 male to 20 or 30 females. These animals are not domesticated in the sense that sheep, cattle and pigs are domesticated.

Table 6.7 gives various data on the major deer-farming nations. New Zealand, in particular, houses the largest number of animals in a successful enterprise that has developed since the early 1970s. Daly (1990) reports that the first deer farms were established in 'bush and high country tussock grasslands' where the secondary productivity of domesticated animals was low. The success of these initial enterprises coupled with the popularity of venison led to expansion so that today deer farms are established on prime lowland grazing land on both North Island and South Island. On the latter, deer farming is even moving into the eastern coastal lowlands where there are irrigated pastures (Daly, 1990). As a consequence of this success, fuelled in part by an expanding export market for venison, deer-management strategies are well established. Grazing patterns, for example, have been detailed by Thompson and Poppi (1990) and relate to the deer's seasonal breeding calendar. As red deer hinds calve in late spring after the increase in pasture growth stimulated by rising temperatures, there is a need for pasture management to maintain an optimum sward height (this has been discussed in Sections 6.2 and 6.3 in relation to sheep and cattle grazing). The fawns are secreted for up to six hours per day; only the hinds graze and then return to suckle their young. Even in short pastures fawns are difficult to see, so mowing of the pasture must be undertaken with great care. During the winter months the fat content of stags declines by *c.* 25 per cent. They, along with hinds, require supplementary feed, e.g. silage, grain or processed deer nuts. The maintenance of good health is essential. Other practices that have yielded high returns for New Zealand deer farms include single-sire mating, rotational grazing and good deer-farm design.

Reinken *et al.* (1990) have described the deer management policies that prevail in Europe. They state that the nutritional requirements of a 'fallow deer unit', comprising a hind, plus calf, plus yearling, is *c.* 20 200 kJ ME day^{-1} from January to March, increasing to 20 600 kJ ME day^{-1} until slaughter in August/September, and falling to a low during the winter. As is the case in all pastoral systems, the essence of good management requires that animal food requirements should match, as far as possible, the forage provision. With deer, the mismatch means that there is a surplus of forage in spring, as there is in New Zealand (see above), a shortage in August and September and an even bigger shortage during the winter. To avoid, or at least minimise the use of expensive concentrates during the late summer, the surplus spring forage should be collected. In winter the use of supplementary feed may be essential. Other components of management strategies include weed control to improve forage quality, the application of artificial fertilisers to encourage forage growth, and adequate fencing to prevent animal escape and mortality on roads.

Another group of animals that provide meat and/or fibre but which are of minority importance when considered on a global basis, are the New World camelids. Also known as llamoids, these animals are the alpaca and llama, the domesticated species, and their wild relatives, the vicuña and guanaco. The llama and alpaca are particularly significant in the agricultural economies of Peru and Bolivia. According to Fernàndez-Baca (1990), the value of these animals is due to their sustainability for grazing high-altitude pastures above 4000 m. These areas could otherwise not be utilised to generate animal products. The largest of these animals is the llama of which there are two types: the *ccara* with a smooth hairy coat and the *chaku* with a woolly coat. The fibre produced from llamas is coarse and the animal is also used for meat and as a pack animal. Alpacas are smaller than llamas and are usually reared for their fine fibre fleeces and for meat. The two types are the *suri*, which has long straight hair, and the *huacaya* with a short curly fleece. The other two camelids are wild animals and are not widely reared in captivity, though the vicuña does produce fine fibre akin to wool.

Fernàndez-Baca (1990) reports that alpacas can utilise between 10 and 50 per cent more of the food energy in pasture than sheep, indicating that they have a more efficient digestive system than other ruminant animals. He states, 'The ability of llamoids to utilise crude fibre so efficiently may explain their ability to utilise the poor-quality, highly lignified pastures found at high altitudes.' Herds of llamas and alpacas vary between a few hundred, in small community farms, to several thousand in big livestock enterprises. Most animals belong to the former farm type where they are herded together regardless even of species. This means that mating occurs between llamas and alpacas. On larger enterprises herds of between 200 and 1100 animals are segregated according to hair colour, age and sex, with 30–40 per cent comprising females of breeding age. A further 30–35 per cent will be castrated males for fibre production. The stocking rate is usually low, 1.5–2.0 animals per hectare. Shearing may take place every year or every other year. Fibre production by alpacas is very variable, though Fernàndez-Baca states that 1.8 kg animal^{-1} yr^{-1} is average; the range is 0.9 kg to 4 kg animal^{-1} yr^{-1}. Llama fleece weight is about 1.8 kg yr^{-1}. In relation to meat production, alpaca meat is not unlike mutton. The primary productivity of the high-altitude pastures is the only source of food for these camelids and there is rarely any feed supplement. A greater understanding of camelid life-cycles, feeding and reproductive habits, etc., coupled with improved management, would probably lead to increased productivity. Some interest has been generated in the UK because of overproduction of traditional livestock products. Success with cashmere production from goats is encouraging livestock producers to diversify further and to consider fibre production from camelids (Russel, 1994).

6.5 CONCLUSION

The pastoral agricultural systems discussed above reflect the many ways in which secondary productivity and certain fibres are generated. In many of the examples presented above, particularly those from rangelands, it is evident that extensive cattle, sheep or goat grazing is the only way of producing food for human consumption from an area characterised by low primary productivity which is often of poor quality. Such environments are, however, particularly vulnerable to degradation if management practices are poor. All aspects of management must focus on matching the requirements of the animals with the quality and quantity of forage. Optimum productivity can only be achieved if grazing patterns reflect resource availability. Similarly the integrity of the resource will only be maintained if the carrying capacity is not exceeded.

In intensive pastoral agricultural systems, the same principles apply whether the objective is beef, sheep meat, venison or wool production. These systems are usually characterised by a higher fossil-fuel energy input than are extensive systems. Most of this input is in the form of artificial fertiliser which is applied to enhance pasture quantity and quality. A high fossil-fuel subsidy is also characteristic of those intensive and extensive animal producing systems that utilise intensively-grown feeds and/or concentrates *in situ* or after transporting the animals to feedlots for fattening.

For many people in the developing world the animals they own represent wealth. Large, good-quality herds reflect prosperity though in times of drought the size and herd quality can decline quite substantially. In addition, the animals provide the only source of protein, usually in milk or dairy goods produced from the milk. This is also the case where arable farmers in the developing world keep just a few animals for their milk.

Overall, the world's population relies heavily on comparatively few domesticated animal species for a large proportion of its protein and a significant amount of its fibre. Some of this is produced from nomadic pastoral systems (Chapter 4) but the majority is produced from settled pastoral systems, some of which may employ transhumance. Whatever the type of pastoral system, its output is directly related to the abundance and quality of the primary producers, which in turn are related to climatic conditions.

6.6 FURTHER READING

Arnon, I. (1992) *Agriculture in Dry Lands: Principles and Practice*. Elsevier, Amsterdam.

Blakely, J. and Bade, D.H. (1994) *The Science of Animal Husbandry*, 6th edn. Prentice Hall, Englewood Cliffs, New Jersey.

Bonney, M. (1993) *The World of Sheep and Goats*. Just Print, Carmarthen, UK.

Grant, W. (1991) *The Dairy Industry: An International Comparison*. Dartmouth, Aldershot.
Mack, S. (ed.) (1993) *Strategies for Sustainable Animal Agriculture in Developing Countries*. FAO, Rome.
Speedy, A.W. (ed.) (1992) *Progress in Sheep and Goat Research*. CAB International, Wallingford.

CHAPTER 7

Settled Agriculture: Mixed Farming Systems

Mixed farming involves the integration of crop and livestock production. Unlike many of the animal production systems discussed in Chapter 6, and some of the crop production systems discussed in Chapter 5 (e.g. the so-called 'corn on the hoof' production in the Mid-West of the USA, in which crops are grown specifically for animal feed), truly integrated mixed farming involves food crop, fodder crop and livestock production. Except for the more extreme climatic zones, mixed farming enterprises occur throughout the world and there are, consequently, many permutations of the major variables. This makes any classification on the basis of components impractical. The balance between food crops, fodder crops and livestock is particularly variable. In some instances only a few animals may be kept just to provide the farmer with essential animal products. Conversely there are mixed farming enterprises in which livestock, e.g. pigs or poultry, are the dominant products. Many mixed farming systems incorporate speciality crops, e.g. medicinal herbs or fibre crops.

As in the case of arable and pastoral systems (Chapters 5 and 6), mixed farming activities may enjoy a substantial fossil-fuel energy subsidy and thus qualify as high-technology agricultural systems. Such enterprises occur mainly in the developed world and are usually less diverse, in terms of the array of crops and livestock they produce, than are their counterparts in the developing world. In the latter, mixed farming systems may comprise 20 or more components, the objective being to produce marketable commodities, i.e. cash crops, as well as essentials. The majority of mixed farming systems in the developing world operate at or just above subsistence level although this is beginning to change as agriculture in general becomes commercialised. There may be a small fossil-fuel subsidy, though this is set to increase as the use of crop-protection chemicals and artificial fertilisers expands. The continuum of energy inputs that characterise arable agricultural systems (Chapter 5) also characterises mixed farming systems, the productivity and operation of which are likewise constrained by the nature of the physical environment. However, concerns about the use of chemicals in agricultural systems are causing a change in the attitudes of some farmers in the developed world in response to consumer demand. So-called organic farming practices are, consequently, emerging which involve little or no artificial inputs.

In so far as it is impossible to classify mixed farming systems on the basis of their components, this chapter is organised on the basis of energy flow characteristics, i.e. high-technology and low-technology systems. Within this broad subdivision, and in view of the fact that mixed farming systems are especially diverse in the developing world, further subdivisions are adopted here on a regional basis simply because they are expedient. It is also worth mentioning that the precursors of many of the high-technology agricultural systems that characterise much of the developed Old World, e.g. Europe, originated from the mixed farming economies of the historic, and even the prehistoric, periods (Sections 3.1.3, 3.2.1 and 3.3.1) with their crop rotations to control pests and use of livestock to provide manure. These traditional agricultural systems often incorporated forestry, in so far as woodland lots were a very important component of manorial and estate enterprises. This was because woodlands provided habitats for game and grazing land as well as fuel and building materials.

To a certain extent, and because of the versatile products that combining agriculture and forestry generates, a comparatively new type of mixed farming has developed in the last 30 years or so. This is called agroforestry. It is considered in a separate section of this chapter because of its widespread adoption as a sound agricultural practice. It is, for example, an important component of sustainable agricultural development for reasons that are examined in Section 7.5 of this chapter and again in Chapter 10. Agroforestry does not necessarily include livestock but because it may involve a combination of forestry and agriculture it is discussed here as a *bona fide* type of mixed farming.

7.1 MIXED FARMING IN HIGH-TECHNOLOGY AGRICULTURAL SYSTEMS AND ORGANIC FARMING

Ecologically, mixed farming is a much more balanced type of agriculture than is specialised arable or pastoral agriculture. At the very least this is because the variety of plants and animals within the agricultural system is a more direct counterpart of the natural ecosystem that agriculture replaced. In parallel with the increasing use of fossil fuels in specialised agricultural systems, mixed farming enterprises in the developed world have become increasingly mechanised and reliant on crop-protection chemicals and artificial fertilisers. Nevertheless, and in contrast to specialised agricultural systems, the interrelationships between the livestock and crop components of the mixed farms mean that there is at least some cycling of nutrients within the system. The animals provide manure in their stalls or in the field. In specialised agricultural systems there is only a continuous output of nutrients in the crop and livestock harvests; nutrient replacement is usually almost entirely via artificial fertilisers from outside the farm. A further advantage of mixed farming is that the labour requirements can be evened out on an annual basis and where the farm is not too large it can be operated by a family unit. The dominant crops grown in European and North American mixed-farming systems are the cereals and roots, much of which are produced specifically for animal feed. Typical livestock comprise cattle, sheep, pigs and poultry.

An example of a typical mixed farm in the UK is that of Home Farm in the village of Abberton, Worcestershire, for which the energy flow characteristics have been detailed by Taylor *et al.* (1993). Here, there are three major activities: dairying, beef cattle and cereal production, the characteristics of which are given in Table 7.1. Of the crops produced, both types of wheat, barley and beans are exported. The silage produced is used on the farm as a feed for the dairy and beef cattle. Hay, when made,

Table 7.1 The characteristics of Home Farm, a mixed farm near Abberton, Worcestershire (based on Taylor *et al.*, 1993)

Livestock	Dairy cattle			Beef cattle		
No. of animals	145			150		
No. of animals dry at one time	33			N/A		
Milk production per animal per day	116 litres			N/A		
Concentrated feed per animal	1.2 t yr^{-1}			1.5 t in year 1 0.6 t in years 2 and 3		
Farm-produced silage	Yes			Yes		
Arable crops	**Biscuit wheat**	**Milling wheat**	**Winter barley**	**Beans**	**Silage**	**Hay**
Hectarage	35	66	38	18	71	undefined
Yield (t yr^{-1})	187	334	196	43	1200	60

Other products
Straw = 500 t yr^{-1}

Energy characteristics
Inputs = 13 378 GJ
Cycled = 4207 GJ
Outputs = 21 191 GJ

is also used as a feed supplement. Of the 500 tonnes of straw produced per annum, 20 per cent is used for animal bedding whilst the remainder is sold for off-farm use. Taylor *et al.* have examined the inputs and outputs of the system in terms of energy flows; these are illustrated in Figure 7.1. Although some 7873 GJ of solar energy is captured yearly (the difference between the inputs and outputs in Table 7.1), the system relies heavily on fossil-fuel inputs, the largest of which is that associated with the use of artificial nitrate fertiliser (referred to in Section 5.1). However, there is a corollary of this which concerns the carbon dioxide balance. The output of this gas is 2.7 times the amount used by the photosynthesis of the crop plants. If nitrous oxide and methane production are also taken into account, Taylor *et al.* demonstrate that 3.8 times the amounts of heat-trapping gases are emitted than are absorbed. Moreover, nitrogen from artificial fertilisers contributes to cultural eutrophication (Chapter 8). These comments relate to arable and pastoral systems where artificial fertilisers are used, and are not confined to mixed agricultural systems. The results of Taylor *et al.* confirm comments in Chapter 10 in relation to global warming.

Mixed farming is also a characteristic of Danish agriculture. According to Finley and Price (1994), Danish farms produce between four and six different crops in conjunction with one type of livestock. The latter are either dual-purpose dairy cattle or pigs and provide the basis for a substantial export trade. Denmark ranks first as the world's largest exporter of bacon and fourth as an exporter of dairy products. The most important crop produced is barley which comprises both winter and spring varieties. This occupies approximately 60 per cent of farm land whilst a variety of

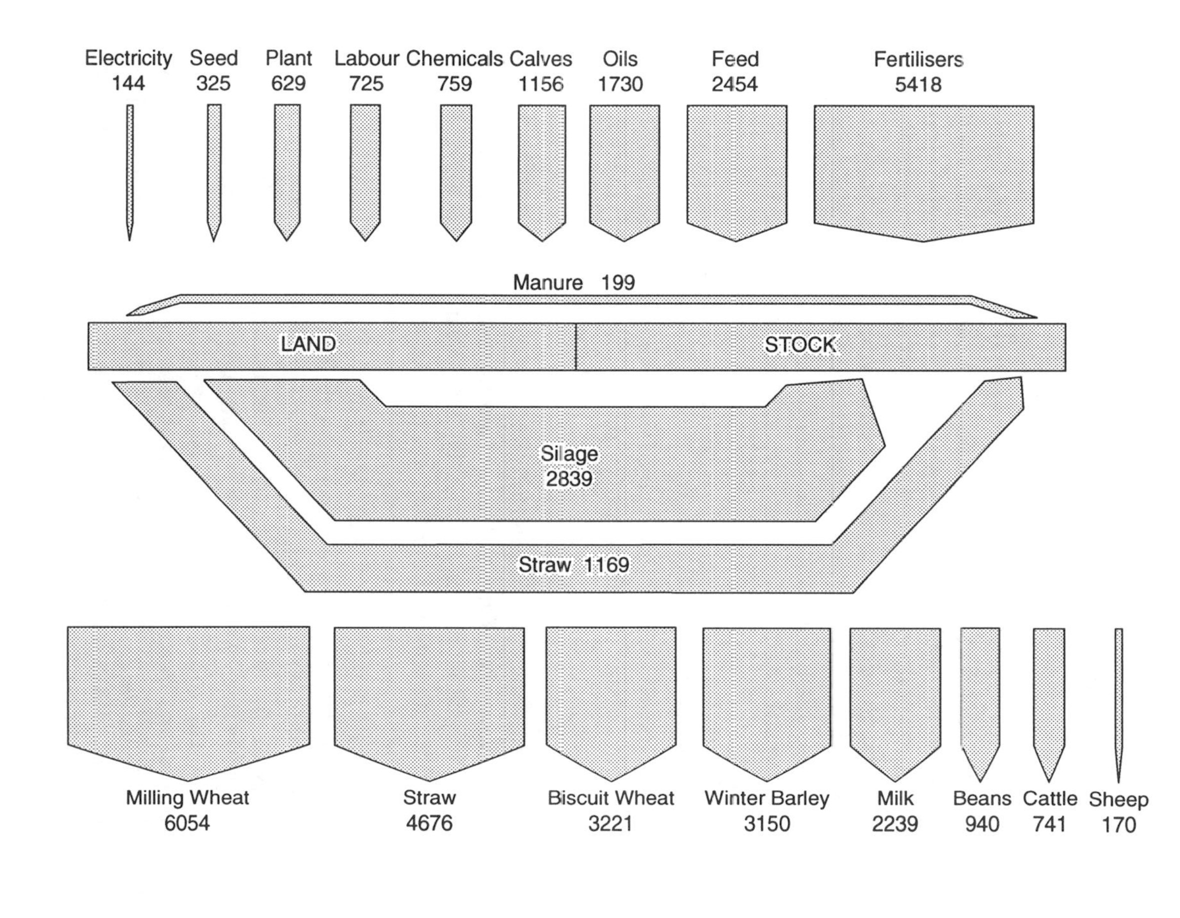

Figure 7.1 Annual terrestrial energy flows for Home Farm, Abberton, Worcestershire (based on Taylor *et al.*, 1993). Values are GJ of energy transferred

crops (including wheat, rye, oats, fodder beets, sugar beets, canola, potatoes, grasses, legumes, silage corn and various seed crops) are produced on the remaining 40 per cent. Most Danish agriculture is typical of high-technology agriculture, employing artificial fertilisers, farm mechanisation and irrigation during dry periods. Similar practices prevail in Sweden and the Netherlands.

Mixed farming is also important in Japan where only 14 per cent of the land area is suitable for cultivation and there is a population of 123 million (Bowring and Kornicki, 1993). However, the development of mixed farming is comparatively recent (post World War II) as, traditionally, Japan has been a rice and vegetable producing country. Indeed it is quite remarkable, in view of Japan's fragmented farm-ownership characteristics and the fact that the average farm size is less than one hectare, that the country produces sufficient rice, the staple crop, to satisfy national demands. However, as Witherick and Carr (1993) state, it is possible to classify Japan's agriculture into two broad groups: that of the central zone which encircles the major metropolitan markets and that of the outer zone where agriculture is less intensive than in the central zone but is still commercial and specialised. Both contain mixed farming enterprises, many of which are operated by part-time farmers who have additional employment in cities. This, in part at least, is due to a US-enforced policy in the immediate post World War II period that no individual could own more than 12 ha in Hokkaido (it was originally 1 ha) and no more than 5 ha elsewhere in Japan. Even today, fifty years after the end of World War II, an average farm-holding of 10 ha has not been achieved. Whilst this type of ownership may spread wealth, in so far as land is scarce in Japan and home-grown produce commands adequate prices via protectionist policies, it is not conducive to the optimisation of agricultural production.

In Japan's central zone, the location of which is given in Figure 7.2, dairying, in combination with vegetable, fruit and flower production, is dominant. The animals, pigs and poultry as well as cattle, are mainly factory farmed and are not free-range. In this respect, these agricultural systems are similar to those in Europe and North America though individual farms are small and rice growing is undertaken by part-time farmers on small paddy fields. The crops grown for fodder include maize, clover and Italian rye-grass which are supplemented with feed concentrates. These activities dominate the alluvial lowlands and alluvial terraces as well as reclaimed wetlands. On the lower slopes of the uplands fruit trees, tea bushes and mulberry bushes (for silkworms) are cultivated. Adjacent to the Pacific coast, with its mild winters, crops can be grown all year round. These comprise vegetables, including salad crops, soft fruit and some orchard fruits. Mixed farming is also important in the Tohoku region of northern Honshu (Figure 7.2). Here, rice is still the dominant crop but in the post World War II period there has been an expansion of intensive livestock farming, fruit production and mushroom cultivation, all of which may be practised by individual farmers. In Hokkaido, livestock production accompanies root vegetable (sugar beet and potatoes) production as well as wheat, bean and barley cultivation. Barley is cultivated mainly for beer. There is a substantial input of fossil-fuel energy to all of Japan's agricultural sectors, notably through mechanisation, crop-protection chemicals and artificial fertilisers. Further major increases in efficiency and productivity are unlikely to be achieved without some rationalisation of farm size.

For a variety of reasons, not least of which are the consequences of artificial

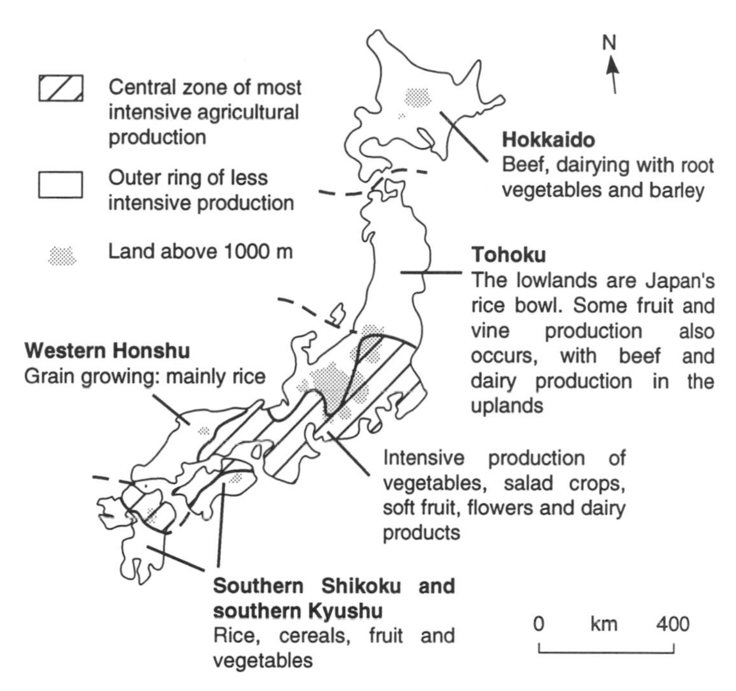

Figure 7.2 The major agricultural divisions of Japan (based on Witherick and Carr, 1993)

fertiliser and crop-protection chemical use, there has been a growing trend towards so-called organic farming in developed countries in the last few decades. The basic aim of this type of agriculture is enshrined in the most widely accepted definition of organic farming, that of the United States Department of Agriculture (USAD, 1980):

> Organic farming is a production system which avoids or largely excludes the use of synthetically compounded fertilizers, pesticides, growth regulators and livestock feed additives. To the maximum extent feasible, organic farming systems rely on crop rotations, crop residues, animal manures, legumes, green manures, off-farm organic wastes, mechanical cultivation, mineral-bearing rocks, and aspects of biological pest control to maintain soil productivity and tilth, to supply plant nutrients and to control insects, weeds and other pests.

This description is reminiscent of the traditional agricultural systems of pre-Industrial Revolution times examined in Sections 3.1.3, 3.2.1 and 3.3.1 though modern organic farming systems also use modern practices.

The term 'organic farming' is, according to Lampkin (1990), intended to reflect the interrelationship that exists between the farm biota, its production and the overall environment. It is thus an holistic approach to agriculture in the same tradition that the systems approach is to geography or that Gaia is to the Earth as a whole. The operation of negative feedbacks in organic farming is, in principle, optimised so that as little interference as possible with the wider environment can take place and create positive feedback. In addition, the term 'organic agriculture' is often used synonymously with biological, ecological and sustainable agriculture. The holistic nature of organic farming implies a relationship between crop and livestock products. Whilst

Table 7.2 The principles of organic agriculture according to the International Federation of Organic Agricultural Movements (1990)

A	To produce food of high nutritional quality in sufficient quantity.
B	To work in harmony with natural systems rather than attempting to dominate them.
C	To encourage and enhance biological cycles within the farming system by adopting practices to maintain populations of micro-organisms and soil flora and fauna and plants and animals.
D	To operate, as far as possible, a closed system in relation to organic matter and nutrients, i.e. to encourage recycling.
E	To provide livestock with conditions of life that allow them to perform all aspects of their innate behaviour.
F	To avoid pollution that may result from agriculture.
G	To maintain the genetic diversity of the agricultural system and its surroundings. This includes the protection of wildlife habitats on and off the farm.
H	To allow agricultural producers an adequate return and satisfaction from their work as well as a safe working environment.
I	To consider the wider social and ecological impact of the farming systems.

this need not always be the case the majority of organic farms are involved with crop and livestock production. Consequently, organic agriculture is included in this section on mixed farming.

According to a special report from the British Organic Farmers in conjunction with the Soil Association (1992), the organic agriculture movement began in the 1920s. Its fundamental tenets are the same now as they were then though the ways of upholding these tenets have changed as agricultural science has developed appropriate innovations, e.g. direct drilling. These fundamental principles are given in Table 7.2 and concern the sustainability of agriculture, an issue discussed in Section 10.4. In order to ensure that the term 'organic' is applied to appropriate produce and so ensure that the consumer receives the correct information, there are various standards that must be complied with before certification of goods is granted. These are detailed in Blake (1990), Byng (1992) and British Organic Farmers in conjunction with the Soil Association (1992). In the early 1990s in the UK, estimates and surveys of organic farms, in terms of number and extent, suggest that 25 000 ha are fully organic and that a further 24 000 ha are undergoing conversion (Redman, 1991). Cudjoe and Rees (1992) suggest a total figure of 55 097 ha. These figures represent a value of 0.4–0.5 per cent of all the UK's agricultural land. This may not represent a great deal of land but the extent of organic agriculture looks set to increase as the millennium approaches. Of this approximately 50 000 ha, about 18 per cent is concerned with intensive horticulture producing fruit and vegetables, possibly with some livestock. The majority of organic farms, in contrast, produce cereals, field-scale vegetables and livestock.

In an examination of the impact of organic agriculture on the landscape of Britain, the British Organic Farmers in conjunction with the Soil Association (1992) have reviewed the available literature on organic farming practices. In relation to mixed farms, the crop rotations depend on soil, climate and management. Typically such a rotation might involve three to four years of grass with a clover ley followed by one year of wheat and then a year of oats with an over-wintering green manure (a nitrogen-fixing crop). In the following year spring beans would be produced and a crop of barley (undersown with a grass–clover ley) would be employed to build up the fertility of the soil enabling arable crop production. Consequently, the proportion of grass with clover leys on any one farm is rarely less than 50:50 and may be 70:30. Many such farms may include leys that are maintained for more than four years or there may be permanent pasture which itself may be species-rich, thus making a contribution to wildlife conservation. The crops grown during the arable part of the rotation will also be varied so that each depletes the store of available nutrients in differing proportions. For example, it is customary that a mixture of autumn- and spring-sown cereals are grown; often these are undersown with grass, clover or some legume in order to enhance nitrogen fixation and thus the availability of nitrate in the soil. Not only do these practices provide for efficient nutrient use and replenishment, they also contribute to landscape diversity. Most importantly, they reduce the amount of land left bare at any one time and so reduce the risk of soil erosion. Moreover, the leys and permanent pasture can be grazed by sheep and/or cattle which, by 'free-range' grazing, contribute toward the replenishment of essential nutrients.

There are, however, disadvantages to such systems. These are economic rather than ecological. For example, varied expertise is necessary to deal with varied livestock and crop products. There are problems associated with mechanised harvesting; what makes ecological sense in terms of crop rotations and variations is not necessarily conducive to keeping costs low unless equipment is owned or hired collectively. Moreover, the relatively low demand for organically produced foodstuffs in the UK means that prices in the supermarkets are high in comparison with 'chemically produced' foods; sometimes the price difference is as much as 800 per cent. The operation of market forces may eventually tip the balance toward the organic farmer. There are, of course, prices and prices; in Europe, for example, the real 'cost' of producing a loaf of bread via conventional agriculture is masked by subsidies, guaranteed prices and, most importantly, by disregarding the environmental costs, e.g. soil erosion, eutrophication, etc. A loaf of bread produced via organic farming methods could cost less than the conventionally produced loaf in the longer term. Agricultural protectionism, as is reflected in the EU Common Agricultural Policy for example, is not necessarily a guarantee of food security in the future.

As Goering *et al.* (1993) have discussed, there are a number of movements other than, but similar to, organic farming that are occurring as alternatives to industrialised agriculture. These are summarised in Table 7.3. Of particular note, as one of the more widespread practices, is biodynamic farming which relates to a philosophy known as anthroposophy. This holistic philosophy involves all aspects of life such as religion, art, education, etc. (Koepf *et al.*, 1976; Sattler and Wistinghausen, 1992). Like organic farming, livestock numbers relate to their ability to maintain and improve soil fertility in order to support crop rotations. In addition, in an holistic context, the farm must be capable of producing all the feed required by the animals. The choice of livestock and crops must relate to local conditions, i.e. climate and soil type as well as the influence of external factors, i.e. biochronological rhythms which are dictated by the sun, moon, stars, etc. Biodynamic preparations are also applied to stimulate growth, decomposition, etc. According to Sattler and Wistinghausen (1992), most of these preparations are plant-based, e.g. a cold extract made from valerian (*Valeriana officinalis*) is added to compost to regulate the release of phosphorus. The exothermic reactions provide warmth, and spraying crops with the extract can reduce the impact of early frost damage. Similarly, animal health is controlled

Table 7.3 Types of alternative agriculture (based on Goering *et al.*, 1993)

A BIODYNAMICS
This approach to agriculture is based on the philosophy of Rudolf Steiner (1861–1925). It began in Germany in the 1920s and has since spread throughout Europe and North America. It differs from conventional agriculture in that it is based more on a qualitative-ecological approach rather than on an analytical-quantitative approach. It differs from organic farming (see text and Table 7.2) in so far as it embraces a way of life and biochronological influences. Natural processes, e.g. organic matter decomposition and nutrient transfers, are instigated and stimulated by the use of herbal preparations.

B PERMACULTURE
This consists of a network of individuals who share information and suggestions on *genres de vie* that are sustainable and well suited to a specific place. The aim is often to combine the best of both traditional and modern farming practices. It embraces the principles of organic farming in an integrated, self-sustaining way of life.

C ECO-VILLAGES
Such developments are notable in Scandinavia where groups of like-minded people are amalgamating to establish eco-villages. Their aim is to keep human impact on the environment to a minimum whilst accommodating the villages' basic needs locally, e.g. the food is produced locally using organic methods. Recycling of waste, local energy production and collective responsibility for all activities within a village framework characterise this type of agricultural system.

homeopathically. As in many farming enterprises, biodynamic systems may use ground mineral rocks as an alternative to artificial fertilisers to help maintain soil fertility. In relation to productivity and income generated, a study on Swiss biodynamic farms indicates that they compare favourably with conventional farms under comparable environmental and market conditions (Sattler and Wistinghausen, 1992). This is confirmed by Reganold *et al.* (1993) who have compared biodynamic and conventional farms in New Zealand in relation to soil quality and financial performance. Sixteen farms in total were compared, comprising eight pairs of biodynamic and conventional farm types. Overall the biodynamic farms were characterised by enhanced physical and biological quality, i.e. they contained more organic matter and more earthworms, had thicker top soil and better soil structure, with lower bulk density making soils more penetrable, than conventional farms. When adjusted to income per hectare, biodynamic farms also performed well in relation to conventional farms. The issues and practices of organic and related agricultural systems will be examined in relation to sustainable agriculture in Chapter 10 and are referred to in Section 7.2 in relation to agriculture in China.

7.2 MIXED FARMING IN LOW-TECHNOLOGY AGRICULTURAL SYSTEMS: ASIA

Mixed farming dominates small-scale agriculture in most of the developing world except where climate conditions are unsuitable. Even in these areas, as discussed in Sections 4.1 and 6.1, extensive and often nomadic animal-rearing is firmly linked with crop production in land adjacent to the ranges. Much of this mixed farming, in contrast to plantation agriculture which usually involves cash cropping, is practised at a subsistence level though commercialisation is occurring rapidly as development proceeds. According to Finley and Price (1994), approximately two-thirds of the world's farmers operate subsistence farms which support between 30 and 40 per cent of the world's population. A large proportion of these farmers are in Asia where in many nations, but especially in China as a component of post-Mao reforms, major changes are taking place in the agricultural sector.

The socio-economic context of food production and patterns of food production in China have been examined by Leeming (1994). China's struggle to feed a large population on a comparatively small amount of arable land continues; while the former continues to grow, albeit at a reduced rate, the latter is being consumed by the expansion of housing and industrial plant, road and rail networks and land degradation. The low prices commanded by grain, due to compulsory purchase by the state of 31 per cent of the total grain output which is used mainly in the cities (state subsidised/operated factory canteens, etc.), has led to a limited increase in animal products. These comprise mainly pigs, poultry and eggs with some fish and milk which are produced alongside arable crops. A study of the agricultural energetics of the village of Fengjiacun in Shandong Province, eastern China, provides an example of the components and linkages in a typical Chinese agricultural village (Odend'hal, 1993). This is illustrated in Figure 7.3 which shows that nearly 50 per cent of the crop energy produced is channelled into the animal population. This, in turn, provides traction though the majority of the energy, through animal-produced goods, is exported from the village. Odend'hal suggests that energy savings could be made by reducing the amount of crop refuse used as cooking fuel. This could be implemented by employing a chipper to pulverize the crop refuse for briquette production and by using more efficient stoves than those used at present. Whatever crop residue was unused as domestic fuel could then be used to augment the organic matter in the soils as well as to replace valuable nutrients otherwise lost in smoke. However, since no values are given for the inputs of 'culturally derived' energy, i.e. fossil fuel and fossil-fuel products, it is not possible to present a full analysis of the energy transfers or their efficiencies for Fengjiacun. Nevertheless, the data given in Figure 7.3 illustrate what a large proportion of the solar energy trapped in this agricultural system is used to produce animal labour and animal products.

China's efforts since the 1960s to feed its large population and eliminate starvation and famine have achieved remarkable results. Even so, the problem not only of feeding a growing population, even if it is growing at a much slower pace than in the pre-1960 period, but one that is also altering its food requirements in response to growing incomes and standards of living, is unlikely to diminish in importance as the millennium approaches. Already, China's farmers have adopted western agricultural practices. According to Cheng *et al.* (1992), the importance of organic fertilisers from plant and animal sources, which was the mainstay of Chinese agriculture, took second place to artificial fertilisers in the early 1980s. Today, China's use of inorganic NPK fertiliser is about the same as that of Western Europe, 2.5 times that of the USA and 4.4 times that of India. It is, however, still heavily supplemented by organic residues. This state of affairs is alarming in so far as it means that a large and growing input of fossil fuel is essential to sustain China's population. This is illustrated by Cheng *et al.*'s examination of industrial energy use in China between 1965 and 1988. During this time, energy used in agriculture alone increased by a factor of nine. Much of this

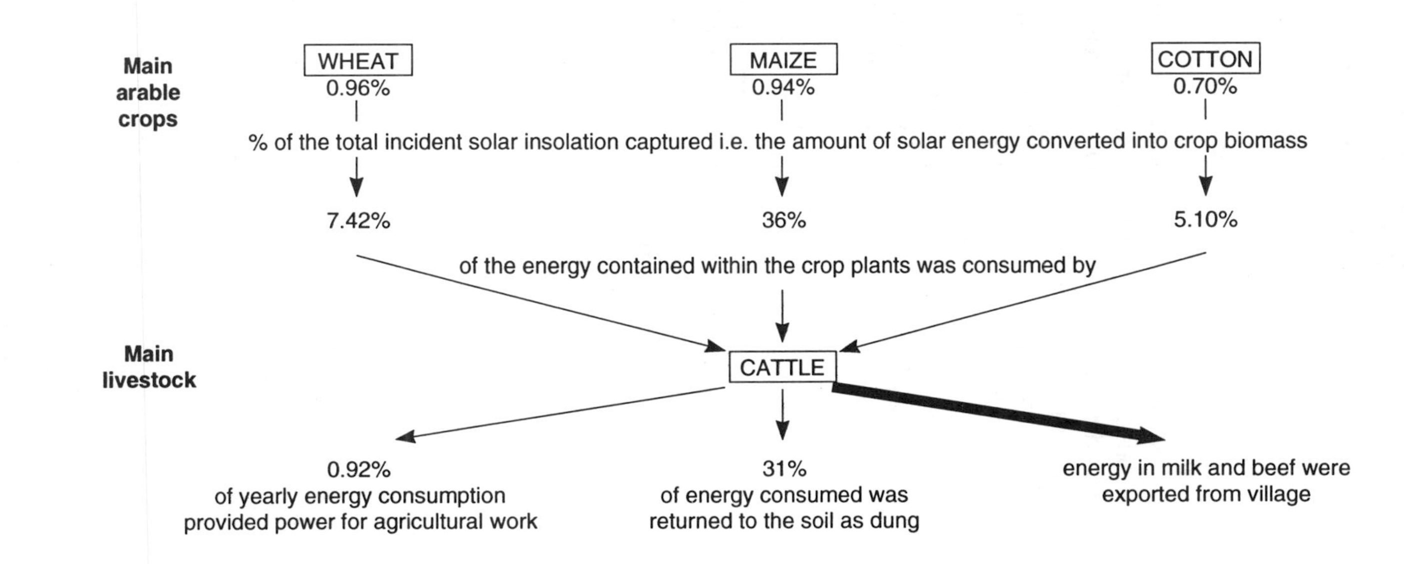

Figure 7.3 The linkages and chief energy transfer characteristics of Fengjiacun village, Shandong Province, China (based on Odend'hal, 1993). Inputs of fossil-fuel energy, e.g. fertilisers, etc., are not included. Other losses from the system include burning of the wheat straw in the field and the use of cotton roots and stalks for domestic fuel

increase was due to the manufacture of artificial fertiliser which, in 1988, accounted for 83 per cent of the energy used in China's agriculture. Moreover, apart from the serious consequences of using so much fossil fuel, there is the problem of cultural eutrophication caused by fertiliser runoff from agricultural land (see Chapter 8). In the light of these concerns, Cheng *et al.* argue that China's farmers must combine traditional and modern technologies in order to not only sustain and improve yields but also to develop sustainable agricultural systems. Despite the heavy reliance on artificial fertilisers, and the growing importance of mechanisation, China's farmers still widely employ traditional practices, such as those described in Table 7.4, many of which rely on mixed farming. Indeed, King in 1911 described these very same practices as the 'backbone' of Chinese agriculture. Almost a century later, such *modi operandi* are considered to be the linchpins of sustainable agriculture as it is known in the West (see Section 10.4). In China, it is referred to as ecological agriculture.

Cheng *et al.* (1992) state that there are five main components which underpin ecological agriculture. These are, first, the adoption of an holistic approach requiring the use of a range of resources in any given locality and, secondly, the multi-dimensional use of space, e.g. via intercropping and relay cropping (see Table 7.4). Thirdly, production systems should be fully integrated to facilitate internal recycling, e.g. combined crop, livestock, aquaculture (fish in ponds and paddy fields) and microbial production systems. Fourthly, elements of environmental management should be included to safeguard the wider environment, e.g. reducing the reliance on inorganic fertilisers. Finally, diversification is important, not only in relation to crop and livestock production but also in relation to affiliated activities such as food processing and other non-agricultural rural employment. Clearly, such aims are difficult to achieve in total but China's demonstration ecological villages illustrate some of the possibilities. They focus on mixed farming and incorporate biogas digesters which efficiently convert raw manure and other organic waste products into methane and slurry-sludge. The methane provides fuel for use in village homes, for cooking and lighting as well as the heating of water, greenhouses and animal housing. It is also used for the pumping of domestic and irrigation water, egg incubation and food processing industries. The slurry-sludge is used as a fertiliser, food for plankton which in turn provide food for fish, and as an additive in concentrates for pigs and cattle. The crop component of these villages varies with location and soil type. Traditionally, grain crops have dominated such enterprises; although they still do so there is more variety than there was in the 1970s as an increased range of fruits, vegetables and mushrooms is now produced. Leguminous forage crops are also included in the crop-production system to provide fodder and to enrich the soil with nitrogen. A variety of livestock are raised. In addition to cattle, pigs and poultry, fish are often an important source of protein. Rabbits, ducks and goats may also be reared. The livestock generate the manure for the biogas digesters, the slurry from which is used to fertilise the crops, or the manure itself may be applied raw. The livestock products give rise to income from their sale in urban areas. Most villages have fish ponds; even individual restaurants may have fish ponds. The fish are sustained by biogas slurry and the sludge from the fish ponds is used as fertiliser for the crops. The introduction of herbicides and tractors, etc. has also allowed a shift in the use of the labour force in such villages. Instead of tending the fields, for example, labour has transferred to village-based business operations such as food processing

Table 7.4 The traditional 'ecological' practices employed in Chinese agriculture (based on Cheng *et al.*, 1992)

A MANURING
Manure from animals and/or humans is stockpiled and applied at appropriate times to crops.

B COMPOSTING
Crop-processing wastes, household wastes, general organic wastes are composted, i.e. digested with soil or subsoil and then applied to crops as appropriate. This allows multiple cropping but is labour intensive. It also lengthens the growing season.

C MULCHING
Straw (crop stalks) is mixed with leaves, etc., to produce a dressing from which nutrients such as potassium and phosphorus are leached into the soil. The latter is also protected from wind and water erosion.

D CROP ROTATIONS CENTRED ON LEGUMES
The inclusion of legumes is important because they, via symbiotic associations with nitrogen-fixing bacteria, enhance the nitrate content of the soil. Other crops in the rotation then benefit and their productivity is increased.

E INTERCROPPING
The cultivation of several different crops in juxaposed rows generally improves the productivity of each of the crops. This is because each crop type has specific nutritional and light requirements; choosing complementary crops in terms of their nutritional demands makes agronomic sense. It also evens out labour requirements but does not facilitate mechanisation.

F RELAY CROPPING
The seed of a second crop (in multicropping systems) can be sown prior to the first crop being harvested.

G LEVELLED FIELDS
This reduces erosion, curtails nutrient loss and encourages the even distribution of rainfall. It is, however, labour intensive.

H FERTILISING WITH RIVER/CANAL MUD
River/canal mud provides an appropriate milieu for mixing with organic refuse (see B above) to produce compost. Enriched with algal remains and organic matter, such muds add to the organic content of compost as well as providing a cohesive medium.

and other agriculturally related enterprises as well as light industries such as garment manufacture.

An analysis of nutrients and energy flows of Zhang Zhuang village, near the city of Su Zhou in Jiangsu province, by Guo and Bradshaw (1993) indicates that the productivity of even these intensive Chinese agricultural systems can be improved. Zhang Zhuang village has a population of 2353 at a density of 15 people ha^{-1} and its land area of 155 ha comprises 6.9 ha of flat fields with 46 ha of ponds, rivers and channels, the remainder being housing, roads, etc. To maintain productivity in the face of a population growth rate of 0.25 per cent per year, the use of nitrogen fertiliser has doubled in thc last decade leading to cultural eutrophication of the land

and watercourses. Guo and Bradshaw, however, on the basis of a number of field experiments have suggested several ways of improving productivity whilst curtailing eutrophication. These include the extension of fish production into embankments and channels, the deepening of fish ponds to accommodate more fish and the use of certain types of water plants and grasses for the channel and pond banks. Experimental data show that fish yields increased by a factor of 370 per cent and that the use of water plants to absorb nitrogen, phosphorus (the agents of eutrophication) and potassium not only curtailed eutrophication but also provided a source of feed for pigs. Although the productivity of the latter did not increase, the use of water plants as components of their feed meant that they required *c.* 19 per cent less commercially produced feed. This, plus the fact that pig excreta is returned to the arable fields as fertiliser, means that the use of water plants facilitated improved recycling of nutrients within the agricultural system as a whole. Similarly the wastes from the fish ponds, etc., were applied to the fields, illustrating the benefits of closely linking primary and secondary productivity. Guo and Bradshaw conclude that there was an overall increase of productivity of 25 per cent within the village of Zhang Zhuang which not only provided the villagers with plenty of food but also allowed them to increase the export of grain, pork and fish thereby increasing their incomes.

Another example of mixed farming is the agriculture of Nepal. This dominates the country's economy and employs 90 per cent of the population. Most of the agricultural activities are traditional, low energy input, subsistent types (Rijal *et al.*, 1993). The three physiographic zones of Nepal are illustrated in Figure 7.4. along with the dominant types of crops cultivated and livestock reared. Rijal *et al.* have examined the energy characteristics of three villages, each of which is considered to be typical of one of the three physiographic zones shown in Figure 7.4. Most of the energy inputs are from human and animal labour, as shown in Table 7.5, except for the village of Baijnathpur where there is an input of 903 GJ yr^{-1} from diesel. In all the villages the energy is used for farming activities, i.e. ploughing, weeding, planting, irrigation, harvesting, threshing and fertilising. Of concern, however, is the use of both animal dung and crop residues to provide domestic fuel. As Rijal *et al.* point out, some 28 per cent of the agricultural residue and 75 per cent of the dung in the terai is needed for domestic cooking and heating. This is mainly because of a lack of fuelwood. What it means, however, is that less manure is available for crop land and so the recycling of nutrients is inadequate. It also means that less crop residue is available as fodder for livestock, so curtailing another aspect of nutrient recycling. The problem is nearly as acute in the hill village of Lekhgaun whereas in the mountain village the availability of grassland for grazing and forests for fuelwood reduces the pressure on crop residues and manure which can then be recycled on the land. Similar problems are apparent in some Indian villages, e.g. Bhabinarayanpur, north-east India, where mixed farming is also practised (Nisanka and Misra, 1990). The use of dung for domestic energy production, in place of fuelwood, is leading to nutrient depletion which may ultimately lead to a reduction in agricultural productivity. Nisanka and Misra make the point that the problem could be overcome by increasing local fuelwood production and introducing biogas plants similar to those of China (see above).

Whilst small, subsistent farms characterise much of India they also dominate the agricultural sector throughout South-East Asia (Indo-China, Brunei Darussalam,

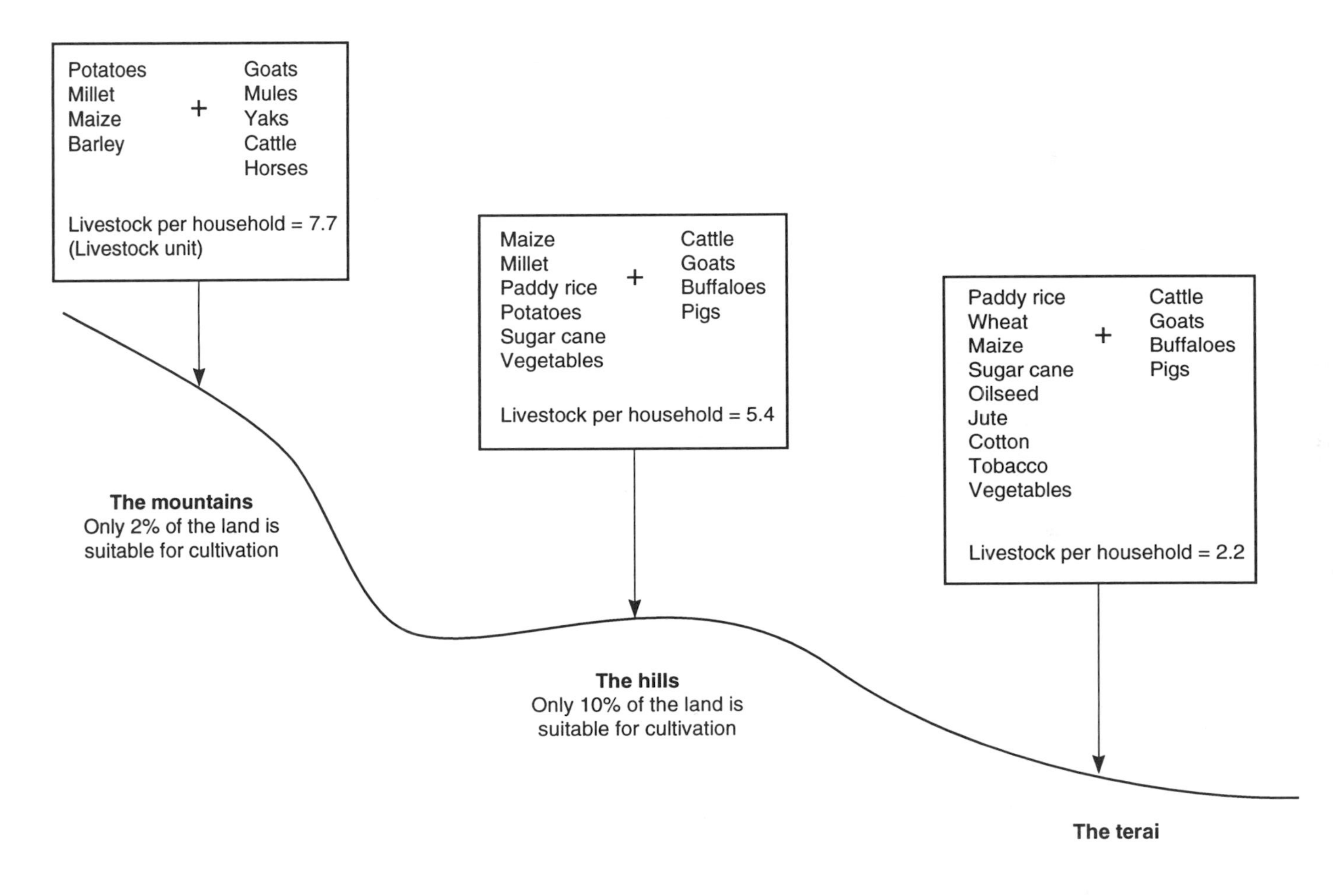

Figure 7.4 Nepal's physiographic regions and their dominant agriculture (based on Rijal *et al.*, 1991). Livestock types are listed in order of importance

Table 7.5 The energy characteristics of three Nepalese village agricultural systems (based on Rijal *et al.*, 1991)

	Animal labour (GJ yr^{-1})	Human labour (GJ yr^{-1})	Farmyard manure (GJ yr^{-1})	Chemical fertiliser (GJ yr^{-1})	Electricity diesel (GJ yr^{-1})
Baijnathpur (terai)	952	841	363	351	911
Lekhgaun (hills)	1246	1382	641	29	0
Marpha (mountain)	357	552	281	153	0

Cambodia, Indonesia, Laos, Malaysia, Myanmar, Philippines, Singapore, Thailand and Vietnam). As Devendra (1993) points out, the average size of farms in this area is 1–2 ha and they can be classified into three categories: farms in rainfed agriculture, farms in irrigated agriculture and farms in plantation agriculture. Livestock are rarely included in the last of these three but are usually an important component in the first two. In comparison with mixed farming systems in Africa and Latin America (see Section 7.3), those of South-East Asia have higher densities of ruminants, pigs, poultry and ducks. Apart from crops, pigs and poultry make the most significant contribution to food supply. Figure 7.5 illustrates the characteristics of these types of farms in South-East Asia which comprise integrated crop–livestock production units. Whether rainfed or irrigated, such enterprises usually combine crop production with ruminants or non-ruminants, e.g. pigs, as well as fish. Often livestock will be components of agroforestry systems, as discussed in Section 7.4. Moreover, the livestock usually consist of a variety of species, each type playing a multipurpose role. Most important is the recycling of nutrients between the crop and livestock components (this is a characteristic feature of 'organic' agricultural systems, as discussed in Section 7.1). Such practices are similar to those described above for Chinese village farms. In Java, for example, goats, sheep, chickens and ducks are reared in combination with mixed cropping. The variety of products from such a system provides a degree of security because at least a reasonable proportion of the products will find a ready market. In Indonesia in general there are 3.5×10^6 buffaloes, 103.5×10^6 cattle, 11.3×10^6 goats, 5.75×10^6 sheep, 590×10^6 poultry, 29.5×10^6 ducks and 6.8×10^6 pigs, of which some 90 per cent are owned by small farmers practising mixed farming (Devendra, 1993). Buffaloes and cattle are particularly important in rice-producing regions because they provide draught as well as milk, etc.

The primary production underpinning these small farms provides four types of feed: forages, crop residues, agro-industrial by-products and non-conventional feed resources. Forages comprise grasses and certain types of tree leaves. Crop residues, notably cereal straws, are particularly important. Agro-industrial by-products comprise crop remains after processing, e.g. molasses, rice bran, pineapple waste and coconut cake. Non-conventional feeds are products never traditionally included in livestock husbandry but which are nutritious. Examples include wheat straw and palm kernel meal. According to Devendra (1993), a major principle of Asian integral systems is to raise livestock so that they can contribute to crop production

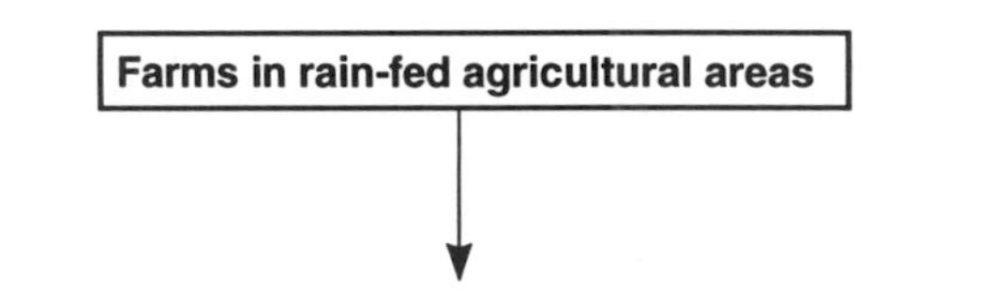

Upland and lowland

1000 -1500 mm rainfall per year

Sedentary and permanent

Traditional, often subsistent agricultural systems

Ruminants, especially in upland regions

Non-ruminants

Livestock provide food, draught, manure, a means of wealth accumulation

Crops: Varied and may include rice, other cereals, tree crops (agroforestry), e.g. coconuts, oil palms, rubber and fruit

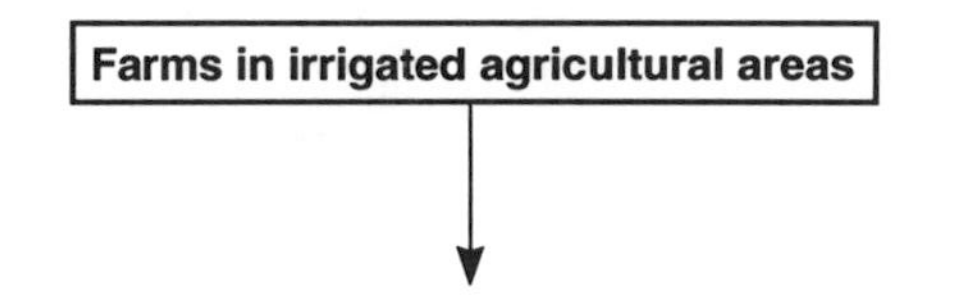

Inadequate rainfall means that irrigation is essential

Crop cultivation is intensive

Sedentary and permanent agricultural systems

Large ruminants predominate, especially swamp buffaloes and cattle

Small non-ruminants play a minor role, e.g. pigs, poultry and ducks

Livestock provide food, draught, manure

Crops: Mainly rice but could be sugar cane

Figure 7.5 The characteristics of mixed farms in South-East Asia (based on Devandra, 1993)

(i.e. manuring and traction) without competing for land. Ruminant animals are, consequently, reared by roadside, communal and stubble grazing, by tethering in confined areas, e.g. on rice fields after harvesting to ensure even stubble grazing, and by 'cut-and-carry' feeding which involves stall feeding with fodder that may be purchased outside the farm. All three systems may be used on any given farm over a few years. Alternatively, livestock may be reared as components of agroforestry systems (Section 7.4). Non-ruminants, especially pigs, ducks and poultry, are also fed on crop residues, etc., as well as organic kitchen refuse. Some may scavenge the residues from rice and cash crop fields or processing plants. Ducks will graze on rice crop residues and the weeds of rice fields and their banks. Not only do they provide meat and eggs but they also provide manure for fish ponds. In some South-East Asian small farms, livestock may be confined semipermanently or permanently. The latter is uncommon but where any form of confinement is employed, root vegetables (e.g. cassava or sweet potatoes) are a major source of fodder.

7.3 MIXED FARMING IN LOW-TECHNOLOGY AGRICULTURAL SYSTEMS: AFRICA

As in the case of Asia, the majority of farming enterprises that characterise Africa are subsistent mixed farms which also support the majority of the population. In many cases, it is a struggle for some farmers to produce sufficient food for their own families. Apart from environmental constraints on productivity the difficulties of food production may be aggravated by war and civil unrest. Nevertheless, a substantial proportion of small farmers in Africa and Latin America, just as they do in Asia, produce sufficient surplus food for sale in urban markets. They are not subsistent in the sense that they produce only enough food for their families' basic needs. In all types of subsistent agriculture in Africa there is a heavy reliance on human, and sometimes animal, labour rather than on mechanisation or oil-based products (Binns, 1994).

Mixed farming may be rainfed or reliant on irrigation. This is probably the only major distinction that can be applied to a diverse array of agricultural systems. Gahukar (1993) has examined the agricultural systems of Sub-Saharan Africa where savanna vegetation zones vary according to the amount of rainfall and number of wet seasons. In arid and semi-arid zones, field crop cultivation is integrated with livestock production, both of which vary according to local soil and rainfall conditions. The diversification is a buffer against the failure of a proportion of the arable crops due to drought. Goats (with some cattle and sheep) are often the preferred livestock and *Acacia* trees are maintained in fields to provide fodder for goats as well as gum arabic. The major crops are sorghum, millet, maize, groundnuts, beans, rice (where there is no shortage of water), cassava and plantains. This is a region of Africa (Figure 7.6) that is regularly adversely affected by drought, often leading to famine. It is thus imperative for appropriate agricultural strategies to be developed, especially in the light of a population of 950×10^6 and one which has the highest growth rate in the world at *c.* 3 per cent per year (Gahukar, 1993). This contrasts with an increase of only 1–2 per cent in food production (Okigbo, 1990). Innovative methods of crop production will involve the use of more productive hybrids, the use of crop plants resistant to various diseases which requires active crop-breeding policies and

Figure 7.6 Location of sites/regions referred to in Section 7.3

institutions, improved irrigation and efficient use of artificial fertilisers. In relation to artificial fertilisers, it would be most beneficial if their use could be avoided, or at least kept to a minimum, whilst adopting, if possible, intensive green manuring, i.e. the use of nitrogen-enhancing legumes as components of the cropping system.

McCown *et al.* (1992) have described various strategies that are applicable to the semi-arid Machakos and Kitui Districts of eastern Kenya where crops and livestock are combined in integrated farming systems. Here, traditionally, the production of both maize and pulses, which is precarious because of the unreliability of rainfall, is combined with the rearing of cattle, sheep and goats. The latter are corralled and their faeces collected to manure crop-growing terraces. Although farmers have access to artificial fertilisers they rarely use them, preferring instead to graze cattle, goats and sheep on uncultivated areas on and off farm and to feed them crop residues. McCown *et al.* used computer modelling programs to establish alternatives to the use of bush fallows (see Section 4.2) which are being curtailed because of population pressure. Without adequate fallow periods the land would decline in fertility rapidly, leading to considerably reduced crop productivity. The various possibilities involve the inclusion of forage legumes in the arable component and increasing crop, especially maize, productivity. Although the use of legumes would increase nitrogen availability in the soil, neither found favour with local farmers because there was no perceived need to improve the diet of animals. Moreover, the use of artificial fertilisers to maintain soil was considered inappropriate because they would not be retained in the poorly buffered soils with depleted organic matter. They could, however, contribute to increased productivity in combination with the addition of manure. The latter not only adds nutrients to the soil but adds cohesion, thereby improving the structure of the soil and its water-and-mineral-retention capacity. Crop residues provide a similar service but they are also required for animal feed, for fencing and as a biomass fuel. Nevertheless, the application of artificial fertilisers in limited amounts would, according to McCown *et al.*, increase crop production, leading to a surfeit of crop residues. Even after demands for animal feed, fencing and fuel are taken into account, there would be sufficient crop residue remaining to plough back into the soil and thereby improve its structure, etc. Indeed, McCown *et al.* suggest, in common with several other researchers in this field, e.g. Ruthenberg (1980) and Okigbo (1991), that the use of artificial fertiliser in agricultural systems like those of Machakos and Kitui Districts is essential to the development of sustainable agricultural systems. Apart from the problem of poor soils there is the added challenge of increasing population pressure, both of which can only be counteracted in relation to the food supply, if soil fertility is substantially increased. In the longer term, however, the purchase of fertiliser may prove to be cheaper than buying food and certainly better than relying on foreign aid. Moreover, the environmental problems (Chapter 8) caused by fertiliser use could be avoided by careful management; at the very least, land degradation through continued cropping without adequate nutrient addition would be avoided.

The central plateau of Burkina Faso (Figure 7.6) is occupied by two major ethnic groups. Of these, the Mossi, are sedentary and combine livestock, notably sheep, goats and poultry and occasionally cattle and donkeys, with crop production. They occupy the centre of the plateau whilst the other ethnic group, the Peul-Rimaibe, are nomadic or semi-nomadic livestock herders (Eger, 1990). Approximately 15 per cent of the land on the central plateau is cultivated; each extended family (*c.* 10 people) has about 6 ha under cultivation. The most important crops are sorghum and millet, though, as shown in Table 7.6, a variety of other crops are also cultivated in order to meet the needs of individual families. The livestock (see Table 7.6) are fed on crop stalks though this means that there is little organic matter to form a mulch on the soil. In addition, there is some overstocking of animals which is causing land degradation. On the positive side, however, the animals from the village, plus those of the nomadic Peul-Rimaibe people, are herded onto arable land to manure it. Long fallow periods can no longer be employed because of land and food scarcity, nor is crop rotation practised. This system of agriculture is unsustainable. Eger has described various measures which are essential to correcting this situation. The basis of the strategy is the adoption of soil and water conservation measures which include the construction of bunds and small permeable dams as well as the prevention of gully erosion. The bunds, which are *c.* 30 cm high and consist of compacted impermeable boulders/soil, are constructed along contour lines between 10 and 20 m apart so as to trap water. The dams are 50–300 cm high and 50–300 m long. They trap water, though it will percolate through slowly, replenishing soil moisture levels, and can be used to stabilise gullies. Both bunds and dams help to trap sediment which also increases fertility, and they can be shielded from wind and water erosion by planting them with grasses, etc. Experiments to determine the value of such measures

Table 7.6 The characteristics of crops/livestock systems in the central plateau of Burkina Faso (based on Eger, 1990)

Organisation of agriculture: small extended family landholdings of *c.* 10 people

Production type: subsistence for individual families mainly; there is a small surplus for urban markets

The characteristics of mixed agriculture

A Arable agriculture	
Cultivatable area:	6 ha
Cropping pattern:	5 ha of cereals; mainly sorghum and/or millet which are intercropped with legumes. 1 ha of other crops, e.g. groundnut, sesame and cotton
Fallow:	Ranges from extensive to virtually nil
Technology:	Hoe, cutlass—no machinery
Control of soil fertility:	Mulching with crop residues to increase the organic content of soils. Addition of manure, i.e. dung via the employment of transhumant animal herds
B Livestock husbandry	
Animals per holding:	
Cattle	0.75
Sheep	1.0
Goats	1.0
Donkeys	0.5
Poultry	4.0

N.B. Crop residues are the major source of animal feed; there are other inputs to Mossi households, notably the use of fuelwood for domestic consumption. Such fuelwood derives from forest preserves.

indicated that they could lead to increases in crop productivity of as much as 100 per cent.

Food security is also a problem in Botswana (Hesselberg, 1993) where 67 per cent of the total population of 1.4×10^6 live in rural areas. Traditionally, mixed farming dominates the country's agriculture in so far as it is practised by 40 per cent of farmers. Cattle are the dominant livestock along with smallstock (e.g. poultry), and crops. The crops, however, have increased in importance since the drought years of the late 1960s and 1980s which left many households without animal traction. The majority of farms are smallholdings of less than 5 ha, which is below the 7 ha estimated to be necessary to support the average family. Hesselberg states that 'Crop cultivation in Botswana can be characterised as "high risk/high-reward", owing to a low-cost input type of production.' The dominant crop is sorghum though productivity, even in years of high rainfall, is low at 300 kg ha^{-1}. The other major crops are millet and maize. In combination, about 50 000 t are produced annually, representing only 25 per cent of Botswana's requirements. This contrasts with the fact that the country was 90 per cent self-sufficient in the 1930s. Despite being subsidised, i.e. receiving grants from the government via the Arable Lands Development Plan (ALDEP) for donkeys and implements, the farmers' returns are so low that arable cultivation does not provide a living. Thus many farmers are actually sub-subsistence producers; serious crop failures may occur in three out of ten years. The growth of a large-scale cattle-rearing industry, instead of the bolstering of small farmers' livestock herds, is only exacerbating Botswana's food supply problem which is likely to deteriorate further in the future.

The problems are not as severe in Africa's humid zone (Figure 7.6) because there is much more water available than in the semi-arid Sahel. As Jabbar (1993) has discussed, mixed farming is actually increasing in West Africa's humid zone. Formerly, the high incidence of the disease trypanosomiasis, which is carried by the tsetse fly, restricted cattle rearing. Successful control measures, however, have reduced the risk and these, together with population pressure and forest clearance, have led many transhumant cattle herders (see Section 4.1) to settle permanently. Moreover, existing crop farmers are beginning to incorporate livestock. Jabber states that in West Africa approximately 80 per cent of cattle are reared in crop–livestock systems to fully integrated systems. Farm size is, on average, 3.3 ha with 2 ha under a crop and 1 ha left fallow. The dominant crops arc cassava, yams, maize and sorghum which are grown singly or in mixes. Crop residues are harvested and fed to animals or the animals graze crop residues directly as well as browsing from local trees. Grazing land is also utilised. The arable land bcnefits from the animals' manure. It is likely that this increase in mixed farming will continue because ready urban markets are increasing as the population grows. This is causing environmental change as forests are cut down to expand the area of crop production into which livestock are then introduced. As Jabber points out, it is imperative that these farming enterprises should benefit from advances in agricultural research, e.g. new and improved crop types and soil conservation techniques, so that agriculture can develop in a sustainable manner.

7.4 MIXED FARMING IN LOW-TECHNOLOGY AGRICULTURAL SYSTEMS: LATIN AMERICA

Grigg (1993c) asserts that there have been few attempts at classifying Latin American agricultural systems though, broadly, a distinction can be made between smallholders and large-scale agricultural enterprises. The former are more likely to practise mixed farming, whilst the latter generally comprise monocultural crop production or livestock production. In several Latin American countries traditional forms of agriculture have declined as land-holdings have become amalgamated and smallholders have migrated to the cities. This is particularly well exemplified by the changes that have occurred in Mexico's agriculture since World War II (Sonnenfeld, 1992). Nevertheless, livestock and crop production have rarely been combined in traditional smallholding enterprises. This is because farmers could afford to employ long fallows to allow the fertility of the soil to be replenished. The shortening of fallows, as has occurred in Africa (see Sections 4.2.1 and 7.3) has, however, created land and food supply pressures, some of which could be ameliorated by adopting crop–livestock complexes which do characterise certain types of smallholdings.

Several traditional fallow systems which incorporate crop and livestock production have been discussed by Kass *et al.* (1993). Of the six fallow systems, which are given in Table 7.7, the babassu palm (*Orbignya phalerata*) forests of central and northern Brazil, carbon negro (*Mimosa tenuiflora* Willd.) fallow of the wet–dry Pacific zone of

Table 7.7 The characteristics of some traditional fallow systems of Latin America (based on Kass *et al.*, 1993)

A ENRICHED FALLOWS OF THE AMAZON
Certain valuable trees are left standing when the forest is cleared by burning or they are planted during the cropping period. The latter leads to enrichment of the subsequent fallow. Overall, there is an increase in the population of the favoured species (e.g. fruit trees) which also help protect the soil and improve its organic and nutrient content as well as suppressing weeds.

B BABASSU PALM (*Orbignya phalerata*) OF CENTRAL AND NORTHERN BRAZIL
The babassu palm tends to dominate cleared land because young palms can survive cycles of cutting and burning. Leaves from mature stands will be used for burning before cropping; their organic and nutrient content enriches the soil. The trees require four years to recover. Annual crops and grazing for animals are produced during the cropping phase. The trees also provide a wide range of products for the farming household.

C BRACATINGA (*Mimosa scabrella* Benth) IN SOUTH-EASTERN BRAZIL
This is found in areas above 1000 m. 2 ha are cut and burned annually and planted with maize and beans. The burn allows the bracatinea seeds, produced during the preceding fallow, to germinate. The seedlings, after thinning, are allowed to remain in the field. Within six years there are different-aged stands of bracatinga and the cycle begins again with clearance of the oldest field.

D CARBON NEGRO (*Mimosa tenuiflora* Willd.) IN THE PACIFIC ZONE OF CENTRAL AMERICA
Carbon negro occurs 0–1200 m above sea-level in areas with a dry season of up to eight months. Fallows last 12 to 15 years but can also be as short as 4 to 7 years. After 12 years the species is overtaken as the dominant tree by other, longer-lived, species. During the fallow carbon negro is used for fuelwood, fence posts, etc. Cattle are also grazed whilst in the cropping period maize and sorghum are produced.

E FALLOWS OF FRIJOLILLO (*Senna guatemalensis* Donn Smith) IN SOUTHERN HONDURAS
At 1300–1500 m above sea-level, the traditional fallow–cropping cycle takes 22 years, of which only three are cropped for maize. The frijolillo fallow is managed, with weeding out of seedlings of other species to produce dense stands, i.e. 2400 trees ha^{-1} before burning to begin the cycle again.

F CARAGRA (*Lippia torresii*) FALLOW IN COSTA RICA
This occurs in the humid forest zone 1000–1200 m above sea-level. The species invades maize fields during the cropping stage. After harvest, cattle may graze the fields and the caragra though the latter is not cut until a substantial stand has formed. Then the trees will be cut for fuelwood; the branches and leaves provide a mulch for the soil.

Central America and the Caragra (*Lippia torresii*) fallow of Costa Rica incorporate crop–livestock complexes (see Figure 7.7 for locations). The babassu palm dominates areas that are repeatedly cleared by shifting cultivators. In established stands the leaves are cut for burning prior to land cultivation. During the cropping period, which is four years long at most, annual food crops are produced and animals may be grazed. This diversity provides the farmer with a wide variety of products: twine, baskets and roofing material from the leaves of the babassu as well as food and oil from its seeds, animal products and a range of annual crops, e.g. maize. In southern Honduras, which is typical of the carbon negro fallow zone of Pacific Central America, there is a dry period of up to eight months. Where possible, fallows last

Figure 7.7 Location of sites/regions referred to in Section 7.4

from 12 to 15 years, though four to seven years is not uncommon in areas of high population pressure. The tree is used to provide fuelwood, fencing and charcoal whilst the fallow is grazed by cattle. These help to raise the nutrient and organic status of the soils prior to maize and/or sorghum production. The improved fallow of caragra in Costa Rica occurs at altitudes of 1000–1200 m above sea level in the humid forest zone. Maize is cultivated on land cleared of forest though the caragra is allowed to form stands. Cattle graze the maize residues after harvest. Eventually the caragra is cut for fuelwood whilst the leaves and branches are retained in the cultivated area as a mulch for the subsequent maize crop. Kass *et al.* suggest that some or all of these fallow systems could be used to expand and improve agriculture in much of Latin America. Ecologically, the practices appear to be successful but there are no economic data on which to determine or predict financial success.

Home gardens, another form of traditional farming in Latin America, are also a type of mixed farming combining crops and livestock. Indeed Gliessman (1990) goes so far as to say that such enterprises 'incorporate most of the criteria for sustainability'. The dominant characteristic of home gardens, which are sometimes referred to as kitchen gardens, is their diversity. There is usually a canopy of trees (and so home gardens could be considered as a form of agroforestry; see Section 7.5) which provide various commodities, e.g. oils, fruit and shade. Shrubs, herbs and vines occupy the space from ground to canopy, thus mimicking the architecture of the forest and making the maximum use of light and nutrients. Apart from the production of a wide variety of crops, including some for medicine, fuelwood, ornamentals and spices to meet the family's needs, harvesting occurs all year round so that there is always work available. The diversity of plant species is complemented by a range of domesticated animals that also satisfy a range of purposes. Pigs, goats and chickens may be penned and fed on kitchen refuse or crop residues or they may be allowed to graze freely. Gliessman states that a typical home garden on the outskirts of Canas in Guanacaste Province, Costa Rica, included 71 plant species in an area of just 1240 m^2. The most important tree species was mango, the other major crops being maize, squash, beans, papaya, bananas and cassava. The home garden was not, however, the only source of income as the male of the household had a job in the nearby town. Other examples of home gardens in Latin America have been discussed by Gillespie *et al.* (1993). They point out that those of the Petén region of northern Guatemala are the modern versions of the home gardens that sustained ancient Maya populations of between 400 and 500 people km^{-2}, as compared to current population densities for the Petén region in general of only 5 people km^{-2}.

In the next decade or so it is likely that mixed farming will expand substantially into Latin American savanna lands. The changes taking place in these areas, where cattle have traditionally been the mainstay of agriculture, have been discussed by Madeley (1993). The soils of these areas are acidic, well drained, nutrient-poor and aluminium-rich. Such characteristics are not conducive to crop production, and in Venezuela only about 19 per cent of the savanna lands are used to produce crops of maize, beans, cassava and cotton (Silva and Moreno, 1993). This may change if current experiments in Colombia's savannas prove successful. These are the efforts of the International Centre for Tropical Agriculture (CIAT), 300 km south-west of Bogota. According to Madeley, CIAT's efforts in its Tropical Pastures Programme have focused on three possibilities: new varieties of pasture grasses to improve grazing quality which should boost carrying capacity and livestock production; legumes to increase the fertility of the soil; and crop plants suitable for the relatively inhospitable soils. Experimental success with new grasses, namely species of *Brachiaria* from Africa, coupled with the tropical legume *Arachis pintoi* (known as tropical white clover), to produce pastures suggests that both meat and milk productivity in these savanna areas could be increased.

Nevertheless, in order to encourage diversification and so reduce the risk factors in cattle rearing, the Inter-American Development Bank have given CIAT funds to develop a crop–pasture system. Rice, it appears, after the screening of a variety of crops, is the most suitable crop. The chosen species is, however, the result of a number of plant breeding experiments that produced *Oryzica sabana* 6 in 1991. This will grow in acid soil, and has resistance to a range of diseases and some pests. In eastern Colombia, farmers have grown 4000 ha of rice-pastures. They plant rice and the grass–legume pastures simultaneously. The rice is planted in rows and the grass–legume seeds are scattered throughout. The rice is harvested after three to four months and then cattle are grazed on the remaining pasture. The cattle gained weight more quickly than formerly and can be marketed within 16 months instead of the usual two years. Overall, Madeley suggests that carrying capacity increased from one animal per 10 ha to two animals per ha. Approximately 3 t ha^{-1} of rice were also produced, providing another source of income. Although its quality is not as good as that of irrigated rice, which commands a better price, an improved strain is already available to counteract this. However, the system will probably only sustain a rice crop for three or four years. Consequently, farmers are being encouraged by CIAT to maintain half of their land under rice-pasture and the other half under pasture. Further improvements may materialise if *Brachiaria* spp. with resistance to leafcutter ants and froghoppers are discovered or bred. Research is also underway to find other varieties of crops that will thrive under these conditions. The major worry, however, is that owners of large land-holdings will see rice as a way to 'get rich quick' and ignore the balance necessary to ensure that agriculture is sustainable. Even the continued production of rice, meat and milk on a sustainable basis may not benefit the poor of Colombia if this produce is exported. This new system will, however, benefit farmers with smallholdings through diversification and its associated increased productivity.

7.5 AGROFORESTRY

The International Council for Research into Agroforestry in Nairobi, Kenya, (MacDicken and Vergara, 1990) have defined agroforestry as, 'A land use that involves deliberate retention, introduction, or mixture of trees or other woody perennials in crop/animal production fields to benefit from the resultant ecological and economic interactions.' Agroforestry is not new. As Nair (1991) points out, it is a practice that farmers, especially those in the tropics, have been employing for centuries to produce a diverse range of products/crops from the rearing of livestock, crop husbandry and the maintenance of woodlands and hedges. In common with many of the agricultural systems described in Chapter 4 and Sections 7.1, 7.2 and 7.3, some of which may include an arboreal component, agroforestry systems are considered to mimic the ecosystems they replaced and thus to be more in tune with the environment than monocultural systems. A major characteristic is diversity which contrasts with the tendency to monoculture that characterises high-technology agricultural systems (see Chapters 5 and 6) and facilitates improved resource use, i.e. nutrient uptake, carbon dioxide assimilation and water. In this context it resembles intercropping (see discussion in Ong and Black, 1994, and Section 5.1).

Agroforestry thus represents a form of integrated land use which has received a great deal of attention in the last two decades as a possible means of achieving sustainable agricultural practices (see Section 10.4). Nair (1991) states, 'agroforestry is not all that new. What is new is the science of agroforestry; the art is old, indeed very old.' Certainly, there is abundant evidence for the integration of farming, forestry and hunting in pre-industrial societies (see Chapter 3). This is hardly surprising since, in many parts of the world, agriculture was developed on land originally forest-covered, prior to which the woodland itself and its fauna were the primary resources. The demise of forest cover, the increasing sophistication of agricultural techniques and, eventually, research specialisation caused agriculture and forestry to develop into separate disciplines. Now the two are coming together again in an attempt to find solutions to both food supply and environmental problems. Originally, agroforestry research concentrated on tropical regions where the need to increase productivity and protect the environment was greatest. Today, its value worldwide has been recognised and agroforestry has become almost synonymous with 'trees on farms'.

Several different systems for the classification of agroforestry types have been proposed, as reviewed by Carne (1993). That presented by Nair (1985, 1990, 1991) is one of the most comprehensive; an abridged version of it is given in Table 7.8. This was devised for tropical agroforestry systems but is also relevant to agroforestry systems elsewhere. All agroforestry systems represent a form of mixed farming; at the very least, crops and trees are combined. However, here mixed farming, as defined at the beginning of this chapter, refers to crop–livestock combinations. Consequently, only the silvopastoral systems and agrosilvopastoral systems are, strictly speaking, relevant to this chapter. This is not intended to understate the importance of agrosilvicultural systems, some examples of which are referred to in Table 7.8. The many categories of agroforestry, in combination with the varied examples given in Table 7.8, testify to the versatility of agroforestry in relation to its provision of goods (e.g. fodder and fruit) and services, such as the stabilisation of soils, shade, etc.

In relation to silvopastoral agroforestry systems, the use of protein banks and the combination of pastures, animals and tree crops provide a wide variety of goods and services. The use of trees to produce protein-rich fodder is, in effect, the production of a crop that can then be fed to stall-reared or free-range livestock. There are many examples of such agroforestry systems and yet other examples of potential gains if such systems were to be adopted. Gutteridge and Shelton (1994), for example, have reviewed the use of the tree legumes in both cropping and grazing systems. The principle involved is, of course, the availability of nitrogen and the role of leguminous species as a means of overcoming its scarcity as a limiting factor to crop growth. As Gutteridge and Shelton observe, there are approximately 17 600 species of legumes, of which about 6000 are trees or shrubs. Not only do they augment the available nitrogen content of the soil but they also help stabilise soils, even in saline or arid environments. In addition they provide a source of timber and fuelwood, shade and, sometimes, fruit and vegetables, as well as forage.

In areas where there is a land shortage and farmers occupy only smallholdings, tree legumes can provide a range of useful products. They may be grown on unused land, e.g. along field borders or fence lines, on rice paddy bunds or in home gardens. The leaves can be harvested, as part of a 'cut-and-carry' system, to provide high-protein forage for livestock. This supplements low-quality fodder/forage from crop residues. As an example, Guttridge and Shelton (1994) quote the case of the Batangas region of the Philippines which was initially reported by Moog (1985). Here, a 2-ha plot was used to produce *Leucaena leucoephala*, a major tree legume, in combination with the fruit tree *Anona squamosa*. The *Leucaena* provided sufficient forage for 20 growing cattle for six months. At the same time the tree legume helps to replenish the nitrogen in the soil and hence produce 'green manure'. In terms of dry weight, tree legumes contain between 2.5 per cent and 5.5 per cent nitrogen. Apart from improving soil fertility, tree legumes also improve moisture and nutrient retention by adding organic matter to the soil and so improving its structure. Their extensive root systems allow tree legumes to thrive on slopes that are too steep for

Table 7.8 Types of agroforestry systems (based on Nair, 1991, 1993)

	Agroforestry type	Description	Example	Reference
A	AGROSILVICULTURE	(Crops—may be shrub, vine or tree crops—with trees)		
(i)	Improved fallow	Woody species allowed to grow during the fallow	Amazon Basin	Kass *et al.* (1993)
(ii)	Taungya	Woody and agricultural species during the early stage of plantation establishment	Sri Lanka	Ranasinghe and Newman (1993)
(iii)	Alley cropping (hedgerow intercropping)	Woody species in hedges; crops in alleys between hedges	SW Nigeria	Kang (1993)
(iv)	Multilayer tree gardens	Trees dominate in a multilayer arrangement	Java	Jensen (1993)
(v)	Multipurpose trees on croplands	Trees scattered randomly; trees arranged in patterns	Australia	Scanlan *et al.* (1992)
(vi)	Plantation crop combinations	Shade trees; intercropping; multistorey arrangements	Smallholder plantations	Nair (1991)
(vii)	Home gardens	Multistorey combinations of shrubs, trees, ground cover	Petén, Guatemala	Gillespie *et al.* (1993)
(viii)	Trees in soil conservation and reclamation	Trees on bunds, terraces	Africa, Asia	Gutteridge and Shelton (1994)
(ix)	Shelter belts, windbreaks	Trees on or around farms	Widespread	Carne (1993)
(x)	Fuelwood production	Planting firewood species in or around farms	Karnataka, India	Garforth (1993)
B	SILVOPASTURE	(Pasture for animals with trees)		
(i)	Trees on ranges or pasture	Irregular or regular spacing of trees	Queensland, Australia	Wildin (1994)
(ii)	Protein banks	Production of protein-rich fodder for stalled animals	Wide use	Gutteridge and Shelton (1994)
(iii)	Plantation crops with pastures and animals	Pasture occurs between rows of crop plants	SE Asia	Reynolds (1988)
C	AGROSILVOPASTURE	(Crops plus pasture/animals with trees)		
(i)	Home gardens with animals	As (vii) but with animals	Kerala, India	Kumar *et al.* (1993)
(ii)	Multipurpose woody hedgerows	Wide variety of combinations and products	Widespread	Nair (1993)
(iii)	Agriculture with trees	Trees for honey production plus livestock		Nair (1993)
(iv)	Aquaforestry	Trees lining fish ponds; they provide food for fish	China	Zhaohua, *et al.* (1991)
(v)	Multipurpose woodlots (social forestry)	Woodlots to provide a variety of goods as well as grazing	Widespread	Shepherd (1993)

cropping or even grazing. Such land thus becomes productive and the trees provide protection against erosion.

This strategy has been particularly effective in the Philippines where it is known as Sloping Agricultural Land Technology or SALT (Laquihon and Pagbilao, 1994). The development of SALT began because of declining crop production in several upland areas due to the depletion of soil nutrients, a consequence of erosion. To combat this the Mindanao Baptist Rural Life Centre (MBRLC) eventually obtained seeds of *Leucaena leucocephala* and planted them in double rows along contours 4–5 metres apart. The *Leucaena* grew to form double hedges and between each double hedge a variety of crops and fruit trees were grown. This strategy proved so successful that further variations were developed, including a goat-based agroforestry project comprising 40 per cent livestock. Today, a variety of legumes are used to produce hedges which are allowed to grow to about 2 m in height and then cut back to 0.4 m. The clippings are used to produce a mulch and source of nutrients in the alleys where perennial crops, e.g. coffee, banana or cacao, are grown with cereals, e.g. maize and sorghum, and other crops, e.g. roots, legumes, fruit bushes. The goats not only produce milk, from which the farmer can derive an income, but also manure for return to the alleys. This is an example of a successful agrosilvopastural system (see Table 7.8).

Another form of an agrosilvopastoral system is that which predominates on the archipelago of Vanuatu in the south-west Pacific Ocean. According to Weightman (1989), Vanuatu's farming systems are family-based concerns that produce cash crops, including coconuts, which are often grazed by cattle if not interplanted with cocoa. This practice is combined with a multi-crop annual food garden which incorporates pigs that are penned and free-range poultry. On average each family has 2 ha of permanent cash crops and between $\frac{1}{2}$ ha and $\frac{1}{3}$ ha occupied by a food garden which is cultivated for three years. Thereafter the land is left fallow for an average of seven years but from which some crops, e.g. banana, are still obtained. Yams and taro are the major crops. Where fallows are less than five years, because of pressure on the land, *Leucaena* is being employed to improve soil fertility. In some parts of Vanuatu it has replaced forest species as the dominant vegetation type during the fallow. Plantation agriculture is also important, having been introduced to Vanuatu by Europeans in the 1860s. Historically, coconuts, maize, cocoa and arabica coffee were produced; cattle were grazed under the coconuts, cattle having been introduced in 1845 from Australia. The cattle in coconut plantations were of secondary importance to the collection of copra. To make collection of the copra efficient, plantations were divided into paddocks and the cattle moved from one paddock to another to trample and graze the grass and weeds. The cattle were not selectively bred or well-cared for and many escaped to the wild. Today, however, there are breeding programmes and animal health programmes, though these are limited to large-scale enterprises. Smallholder herds are still not always properly managed; livestock numbers are sometimes allowed to increase beyond the carrying capacity of the under-tree pastures which become weed infested.

Whilst the examples given above attest to the value of agroforestry systems, it is essential that they are properly managed. Moreover, it is ironic that in some parts of the world agroforestry is being heralded as a means of improving the productivity of the land and reducing environmental degradation (e.g. Le Houérou, 1990), whilst in

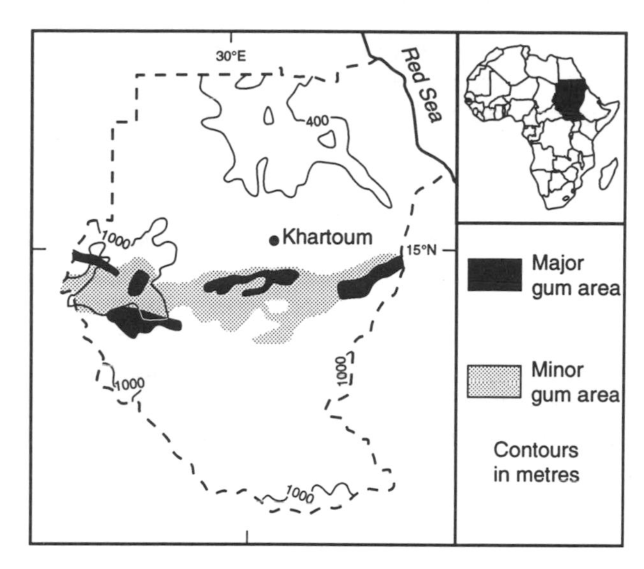

Figure 7.8 The gum-arabic producing region (gum belt) of Sudan (based on Jamal and Huntsinger, 1993)

others there are instances of declines in traditional agroforestry systems. Jamal and Huntsinger (1993), for example, have examined the case of deteriorating gum arabic gardens in the Kardofan region of Sudan. Here there are two major gum-producing species: *Acacia senegal* and *A. seyal*, which are commonly known as hashab and talh respectively. Both are drought-resistant and are important components of an agrosilvopastoral system that has been in existence for at least 200 years. Apart from the fact that these acacias produce a gum that has a wide variety of uses, e.g. textiles, confectionery, paper, ink and glue, they also contribute to soil stability and nitrogen availability in a similar manner to *Leucaena leucocephala* (see above). As such they help support the agricultural systems that produce a substantial proportion of Sudan's grain and livestock. The major gum-producing region of Sudan is known as the 'gum belt', the location of which is shown in Figure 7.8. The trees occur in pure stands and can be both wild and cultivated; the gum they produce is of the highest quality. As well as producing gum on a commercial basis, the acacias produce fruits that are used for medicinal purposes and the mast and leaves provide a good-quality fodder. The livestock comprise camels, goats, sheep and cattle which graze in the gum gardens during fallow periods when understorey crops are not being grown. The animals consume grass, weeds and acacia pods. Nomadic camel caravans may also be allowed to graze. The animals' manure enhances soil fertility whilst acacia seedlings are protected until they grow above the browsing level. This ensures their regeneration. Grazing is managed so that a good grass cover persists. This is important to protect the soil and to provide an indication of when the gum trees should be tapped. Only when the grass dries out should gum collection commence; this takes place during the period of winter dormancy, by which time the tree will have accumulated a large carbohydrate reserve from the preceding wet summer season.

In their examination of the decline in this system, Jamal and Huntsinger refer to the overall decline in gum production that has taken place during the last decade.

This was partly due to the occurrence of drought between 1979 and 1986, though, as the acacias are drought-tolerant, there are other factors involved as well. Despite high demand for the gum, prices were held at low levels by a marketing monopoly; consequently, farmers felt that a return of $10 per ton was insufficient to justify the large amount of labour involved. This resulted in farmers cutting down gum trees for fuelwood; others moved to the cities. Additional factors causing the breakdown of the system included insect pest attack, over-tapping and weakened trees. Not only did gum production decline but livestock herds were also decimated. Lack of water and pasture, degraded pasture and necessity forced farmers to sell livestock at low prices. Thus neither gum nor livestock from the gum gardens were providing a livelihood. Desertification (Chapter 9) set in as gardens were abandoned. As Jamal and Huntsinger state, 'The low gum price paid to farmers removed any motivation to keep the trees through the drought. . . . Logging was one of the few ways of obtaining the cash they have become dependent on.' Here, then, one of the major factors causing the breakdown of what, for at least two centuries, had been a viable agroforestry system was concerned with economics rather than ecology. In the longer term there may well be more such breakdowns as biotechnology facilitates the manufacture of speciality goods like gum arabic in the laboratory (see Section 10.2).

Shortages of food, declining productivity and environmental degradation in Africa, especially in the semi-arid region, have caused governments and aid agencies to seek alternative agricultural practices. Agroforestry features high on such agendas, as has been discussed by Kessler (1993). He points out that the majority of traditional practices derive from the humid and sub-humid zones rather than the semi-arid zone, one notable exception being the gum gardens of Kardofan described above. In consequence, Kessler cautions that simply extrapolating such practices, especially if they are based on only limited experimental data, may lead to failure and further environmental degradation. He states, 'Although isolated successful experiences exist, the introduction of agroforestry techniques generally has not had noticeable impacts in semi-arid regions.' Instead of transposing practices from areas with very different environmental conditions, Kessler suggests that indigenous agroforestry systems should be investigated, along with an assessment of the value of native tree and shrub species for increasing soil fertility and enhancing crop productivity, etc. Le Houérou (1990) also draws attention to the difficulties of extending or introducing such systems into new areas that have no tradition of agroforestry.

Notwithstanding these potential drawbacks, Loker (1993) advocates the value of agroforestry systems incorporating cattle as alternative and sustainable means of food production for small- and medium-scale land-holders in the Amazon Basin. He recognises the need for such systems to be compatible with the local ecology and to be amenable to adoption by local farmers. As well as increasing productivity, Loker suggests that the development of agroforestry could reduce pressure on the land and, most importantly, reduce the rate of destruction of primary forest. The Amazon Basin has been the focus of settlement for thousands of families, many with no tradition of agriculture, from heavily populated parts of Latin America. They practise slash-and-burn cultivation in combination with cattle raising. The cattle are raised on pastures which are cultivated on cleared land after it has produced an annual crop. The pastures extend the life of cleared plots, though fertility rapidly declines, and signifies ownership through 'land improvement'. The crops provide basic needs whilst the cattle represent wealth via dairy products and/or beef. Such agriculture is, however, unsustainable because of rapidly declining soil fertility related to the prevention of a return to forest, i.e. the fallow. As an alternative, Loker suggests that an agroforestry system based on grass–legume pastures, rotational grazing and the management of natural forest regeneration would be most suitable. After burning, the land could be planted with an annual crop, e.g. rice or maize. Then a grass–legume pasture could be sown, the species to be compatible with local conditions (examples include the *Brachiaria* spp. referred to in Section 7.4) and with the aim of promoting species diversity. In addition, during the first few years selected native trees would be allowed to regenerate to densities of between 50 and 100 trees per hectare. Some may produce a crop but their main purpose would be to recycle nutrients to be released when the land is burnt again. The grass–legume sward beneath these trees would be a type of grazed fallow. After six years of grazing, the livestock are moved to another grass–legume pasture while the vegetation is allowed to regenerate. This may be accelerated by seeding with a tree or shrub legume. Loker estimates that a further three years of fallow are necessary to restore the pre-clearance levels of fertility, allowing the land to be burnt again and the cycle to recommence. Such practices are indeed preferable to the continued removal of primary forest, though it is unlikely that they will significantly halt the rate of forest destruction when the spread of logging and mining activities, as well as land-hungry settlers, is considered.

7.6 CONCLUSION

Mixed farming, involving crop–livestock complexes, dominates smallholder agriculture worldwide. Of all the agricultural systems discussed in this text, those concerned with mixed farming are the most diverse. Mixed farming is not as common in Latin America as it is in Africa and Asia but in all these regions it supports large populations of subsistent farmers. Some of these farmers may be operating at levels that barely supply the family unit with sufficient food for survival; others produce a surplus for sale in nearby towns and cities. Such commercialisation is an important component of development processes not only because it supports urban populations but also because it provides raw materials for agriculture-related industries. Many of the examples presented in the foregoing sections reflect the true integration of crop and livestock production. This is especially so in the developing world where the integration leads to the efficient recycling of nutrients between the environment, livestock and crop components of the agricultural system. In some cases, notably China, artificial fertilisers supplement nutrients available from the soil and biomass. Despite China's efficient agriculture there is evidence that even its highly organised dual systems can be made more efficient than they are at present. The key to this appears to be diversification, often through an aquaculture component to produce fish which supplement protein available from livestock and livestock products. Another important component of successful mixed farming enterprises is the inclusion of legumes in the cropping sequence. Through their association with nitrogen-fixing bacteria in their root nodules, legumes enhance nitrate availability in the soil.

These subsistent small-scale enterprises contrast markedly with the mixed farming systems of the developed world. The latter fall into two categories: those which are heavily subsidised by fossil fuel and those where the subsidy is comparatively small.

Where the subsidy is high there is little integration between crops and livestock. This is well illustrated in Figure 7.1 which shows the energy flow characteristics of a mixed farm in central England. The benefits, in terms of enhancing productivity, are minor in such agricultural systems when compared with the mixed farming systems of China and Asia, for example. Mixed farms with a low fossil-fuel subsidy in the developed world are generally referred to as organic farms. The integration between crops and livestock is much more pronounced than in the highly subsidised systems. To overcome the lack of fossil-fuel subsidy in the form of pesticides and artificial fertilisers, organic farms rely on crop rotations, animal manure and fallow periods to maintain soil fertility and overall productivity. In this respect they have more in common with the systems of agriculture that prevailed in the period before the Industrial Revolution than with the modern fossil-fuel-intensive systems which, even if they are in name mixed farms, produce only a limited range of crops and livestock. They also utilise the same principles as those of the mixed farming systems as the developing world in relation to nutrient recycling but they are usually much larger enterprises and rarely produce such a diverse array of crops. Organic enterprises represent only a very small proportion of agricultural systems in the developed world though they are increasing in response to public demands for pesticide-free foods.

Agroforestry systems are another form of mixed farming; all involve trees which fulfill a variety of purposes. They contribute to the overall productivity of a mixed farm in a number of ways, one of the most important of which is the enhancement of nitrogen availability. To this end many of the species involved in agroforestry are legumes. There are many types of agroforestry (see Table 7.8), some of which involve livestock. Most research on agroforestry systems has been undertaken to identify sustainable agricultural systems for the developing world, especially for those regions experiencing high rates of population growth and where environments are fragile. Much success has been achieved and the practice is now spreading to the developed world. Caution, however, needs to be exercised because the euphoria about agroforestry that abounds in the literature may not materialise if the species chosen turn out to be unsuitable for a specific environment, or if they do not meet the needs of local people. The various examples given above, like those of other mixed farming enterprises, simply illustrate the many and varied ways that biotic resources can be manipulated to provide sustenance for humankind.

7.7 FURTHER READING

Gutteridge, R.C. and Shelton, H.M. (eds) (1994) *Forage Tree Legumes in Tropical Agriculture*. CAB International, Wallingford.

Jarvis, P.G. (ed.) (1991) *Agroforestry: Principles and Practice*. Elsevier, Amsterdam.

Kotschi, J. (ed.) (1990) *Ecofarming Practices for Tropical Smallholdings*. Verlag Josef Margraf, Weikersheim, Germany.

McIntire, J., Bourzat, D. and Pingali, P. (1992) *Crop Livestock Interactions in Sub-Saharan Africa*. World Bank, Washington, DC.

Nair, P.K.R. (1993) *An Introduction to Agroforestry*. Kluwer Academic Publishers, Dordrecht, in co-operation with the International Centre for Research in Agroforestry (ICRAF), Nairobi.

Soulé, J.D. and Piper, J.K. (1992) *Farming in Nature's Image*. Island Press, Washington, DC.

CHAPTER 8

The Environmental Impact of Agriculture in Middle and High Latitudes

Agriculture, to state the obvious, has had a profound influence on the Earth's surface and the processes that operate thereon. There are few parts of the globe that remain unaffected by agriculture. Even where there has been no direct modification of landscapes the indirect consequences of agriculture, e.g. contamination with pesticide residues and water pollution, are often manifest. This is true of high and low latitudes alike but because it is such a vast topic this discussion on the impact of agriculture has been subdivided into two. This chapter concerns middle and high latitudes, which correspond roughly with the developed world. The notable exception is the tropical region of Australia which is also considered in this chapter. The environmental impact of agriculture in low latitudes, which correspond mainly with the developing world, is examined in Chapter 9. This does not mean that the two are entirely separate. In fact they are intricately linked by trade; markets in the developed world, for example, are often the stimulus for certain types of agriculture, e.g. plantation agriculture and cash cropping, in the developing world. This situation is summarised succinctly by Goodman and Redclift's (1991) question: 'When the North [the developed world] eats does it eat the South [the developing world]?' Moreover both are linked through their impact on biogeochemical cycles which operate at all scales and so contribute to global environmental change (see Mannion, 1991, for a review).

The environmental impact of agriculture began with the initial domestication of plants and animals and the inception of the first agricultural systems 10 000 years ago (Chapter 3). Thereafter, as agriculture spread from its centres of origin (see Figures 2.4–2.6) throughout prehistory and history (Chapter 3), its impact intensified. Currently, there are quite different trends occurring in the developed and developing worlds. In the former, excess food production in many nations has led to the formulation of policies to encourage set-aside. Thus, in the European Union and the USA, for example, decisions taken by politicians and shaped by economic considerations are manipulating, as they have throughout history, the character of the landscape as it is influenced by the practice of agriculture. In the developing world the trend is in the opposite direction; as a consequence of high population growth and the existence of a large population of landless poor, agriculture is expanding into lands hitherto undisturbed. Here, then, social and economic pressures, sometimes effected through government resettlement policies, are the major stimuli for the spread of agriculture. This will be discussed in Chapter 9.

Globally, agriculture affects the atmosphere, the land and the hydrosphere, though its impact on the atmosphere is not discussed separately; rather it is included in the sections which discuss its impact on the land and the hydrosphere. In relation to the land, the most obvious impact of agriculture is its replacement and/or modification of the natural vegetation. Thus agroecosystems replace natural ecosystems (see Sections 1.2 and 1.3). In middle and high latitudes agriculture replaced forests, grasslands and wetlands, causing wildlife habitats to become reduced in extent and/or fragmented. Inevitably, this has caused plant and animal extinctions, so reducing global biodiversity. The alteration of the natural vegetation cover may also prompt soil erosion by wind and/or water, so accelerating the natural processes of weathering and land denudation. Nutrient depletion has ensued with adverse affects on crop productivity. The eroded soil may itself create environmental problems. In middle and high latitudes the large-scale use of crop-protection chemicals, i.e. pesticides, has also had a significant environmental impact; even substances now banned by regulatory authorities continue to exert an influence on local and global ecology. In addition, the wide-ranging use of artificial fertilisers, which in some respects has masked the problem of soil erosion, has caused pollution in the hydrosphere by altering the nitrogen biogeochemical cycle. In particular, drainage water from agricultural land is rich in phosphates and nitrates which stimulate algal growth in watercourses and lakes. This process, known as cultural eutrophication, reduces floral and faunal diversity and thus impairs the value of rivers, lakes and wetlands as wildlife habitats. Nitrate-rich drainage can also adversely affect marine ecosystems in the coastal zone and cause problems for the extraction of drinking water from aquifers. Other impacts of agriculture, such as desertification, salinisation and waterlogging, are considered in Chapter 9 because they are more widespread in low latitudes than in middle and high latitudes.

8.1 LANDSCAPE CHANGE: LOSS OF BIODIVERSITY AND NATURAL HABITATS

Landscape change is a vast topic and so can only be discussed here in brief. It has, in common with the subtitle of this book, temporal and spatial dimensions that, to do the subject matter justice, require a separate text. The temporal dimension of landscape change due to agriculture concerns the entire duration of the current interglacial period. The inception of agriculture, as rapid environmental change occurred at the end of the last ice age (see Section 2.2), was not only a momentous period in the history of modern humans but also the beginning of a process that has, on the one hand, accommodated human population increase and, on the other hand, transformed ecosystems from biotically diverse habitats to species-poor, highly controlled agroecosystems. Indeed, Hannah *et al.* (1994) report that some 75 per cent of Earth's habitable land has been disturbed in some way. Only 15.6 per cent of Europe's land area remains undisturbed whilst 31.8 per cent is partially disturbed and 56.6 per cent is heavily disturbed, producing what Hannah *et al.* describe as a human-dominated environment. This generally bears little resemblance to the natural ecosystems it replaced. In addition, early agriculturalists and/or innovative ideas relating to agriculture spread from spatially disparate centres of domestication (see Figures 2.4–2.6) into distant lands within a temporal framework. The transformation of natural ecosystems has, in many parts of the developed world, but especially in Europe, been so overwhelming that it is impossible to find existing woodland habitats that can be considered as analogues of the primeval forests characteristic of the climatic climax vegetation of 7000 to 5000 years ago. The numerous pollen diagrams that have been constructed from various parts of Europe (e.g. Huntley and Birks, 1983) and the Near East (see Section 2.2) attest to the impact of agriculture throughout the current interglacial but they cannot provide the fine resolution needed to define precisely the extent (or the detailed character) of the natural vegetation sacrificed. Palaeoecology remains an imprecise but nevertheless invaluable tool for assessing the temporal impact of early agriculturalists. As Williams (1989a) states, 'Deforestation implies a diminution from some previous original stock that existed in immediate postglacial times, but clearly, the task of reconstructing that stock is difficult.'

However, data for the last 300 years are more accessible than those for pre-1700 and have been tabulated by Richards (1990) whose results are given in Table 1.3. It should be noted that these data are only estimates of global land use, the difficulties of assembling which have been discussed by Meyer and Turner (1992) and Turner and Meyer (1994). Whilst Williams (1989a, 1990) states that the world's forests have decreased by 15.15 per cent and the world's woodlands by a further 13.8 per cent, Richards (1990) observes that 18.7 per cent of this 28.95 per cent overall decrease has occurred since 1700. This reflects the significance of the Industrial Revolution, and the earlier agricultural innovations, for facilitating continued population growth as well as the migrations of Europeans into the Americas and, later, to Australia. The 466 per cent increase in cultivated land that occurred in the 1700–1980 period involved an area of *c.* 12×10^6 km^2 (Meyer and Turner, 1992). Since 1980, Lal (1994a) has suggested that the amount of cultivated land has increased to 14.75×10^6 km^2, representing an addition of 23 per cent on 1980 values. However, as Table 1.3 shows, the extension of cultivated land did not occur uniformly across the globe. The largest increase occurred in North America (a massive 6666.7 per cent), followed by Latin America (1928.6 per cent), South-East Asia (1275.0 per cent) and developed countries in the Pacific, i.e. Australia, New Zealand and the Pacific Islands (1060.0 per cent). In North America, for example, the major expansion of agriculture began in the late 1700s, reaching a peak in the late 1800s as frontiers to the west were opened up (Williams, 1989b). In the former USSR (Richards, 1990) the largest expansion of arable agriculture occurred between 1850 and 1920. Much of this occurred in the Russian Plain (Alayev *et al.*, 1990) where the amount of arable land doubled between 1877 and 1966. During this time many changes in agriculture occurred, including a huge increase in crop productivity from 450 kg ha^{-1} to 900 kg ha^{-1}, the development of large-scale collective and state farms as opposed to a feudal system based on small farms, and agricultural industrialisation, i.e. mechanisation, etc. Even if the data quoted above are accurate only to plus or minus 20 per cent they still reflect a vast amount of land transformation due to arable agriculture. Other transformations involving the spread of pasture have also occurred. These are given in Table 1.3 and are particularly significant in Latin America and tropical Africa.

Forests and woodlands are not the only ecosystems and habitats that have

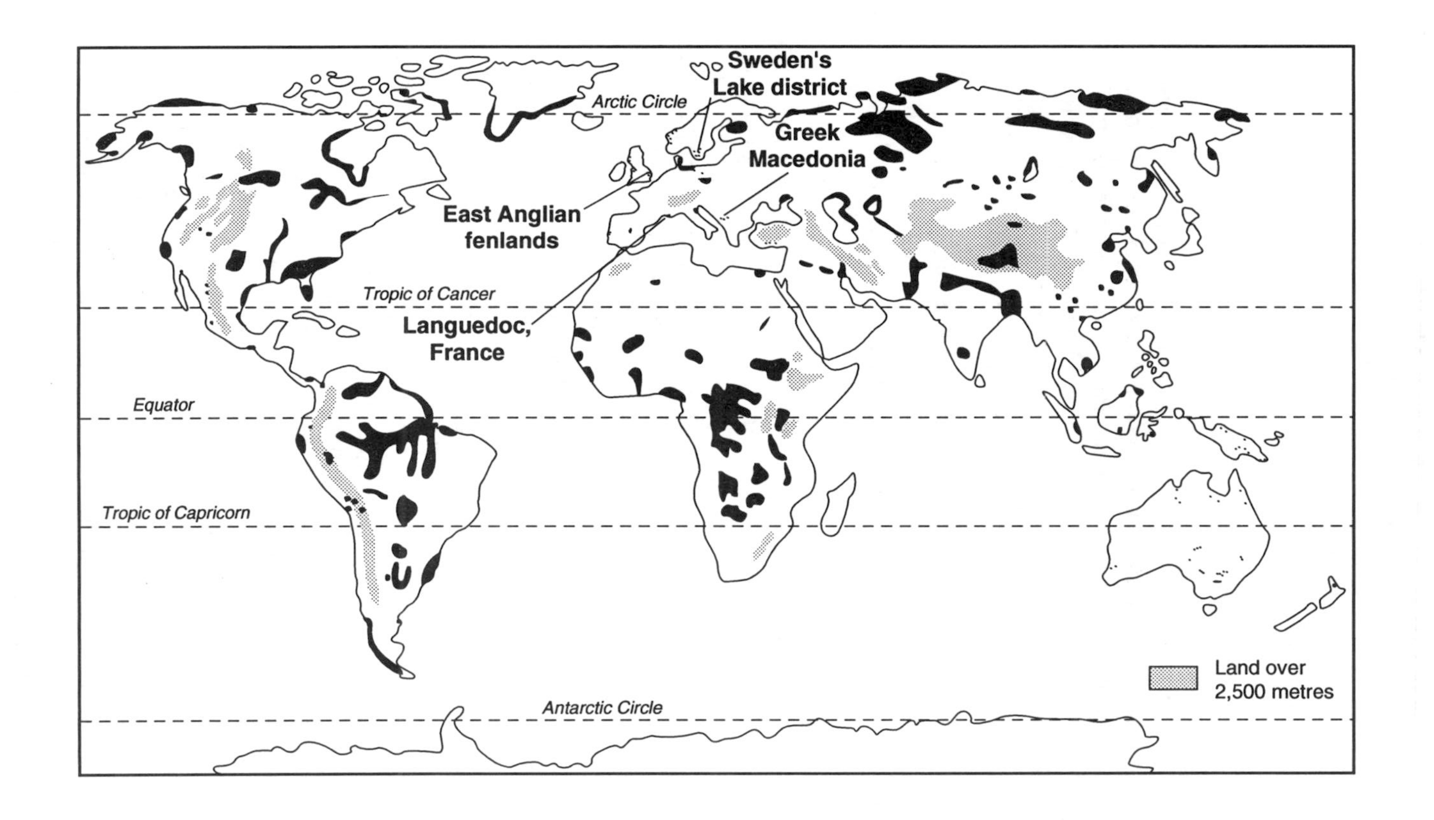

Figure 8.1 World distribution of wetlands and the location of sites mentioned in Section 8.1 (based on World Conservation Monitoring Centre, 1992)

Table 8.1 Data on wetland losses in selected states of the USA (abstracted from Mitsch and Gosselink, 1993)

State	Extent of original wetlands *c.* 1780* 1×10^3 ha	1954 Survey* 1×10^3 ha	Inventory mid-1980s* 1×10^3 ha	% change 1780–mid-1980s
Arkansas	3 986	1 532	1 119	−72
California	2 024	226	184	−91
Florida	8 225	6 955	4 467	−46
Georgia	2 769	2 396	2 144	−23
Illinois	3 323	173	508	−85
Indiana	2 266	115	304	−87
Iowa	1 620	56	171	−89
Louisiana	6 554	3 904	3 555	−46
Michigan	4 533	1 302	2 259	−50
Minnesota	6 100	2 042	3 521	−42
Mississippi	3 995	1 048	1 646	−59
Missouri	1 960	153	260	−87
N. Carolina	4 488	1 641	2 300	−44
N. Dakota	1 994	616	1 008	−49
Ohio	2 024	40	195	−90
Oklahoma	1 150	113	384	−67
Texas	6 475	1 514	3 080	−52
Wisconsin	3 966	1 129	2 157	−46
Total wetlands (all US wetlands)	158 395	–	111 060	−30

* Surveys used different methods so they cannot be directly compared.

diminished in the wake of agricultural development. Figure 8.1 shows the distribution of the world's wetlands. Comparing this with Figure 1.5, which illustrates the world's major agricultural regions, it is clear that there is considerable overlap. As a result wetlands have been gradually reduced in extent to accommodate the expansion of agriculture. There are no data for wetland demise which are comparable with those given above and in Table 1.3 for forests and woodlands. Thus it is impossible to provide a global overview of wetland loss. However, there is evidence that even in Roman times (approximately 300 BC to AD 480) wetlands were being drained in many parts of Europe in order to use them for cattle grazing and crop production. Rackham (1986), for example, describes the efforts of the Romans to drain parts of East Anglia's fens. As discussed by the World Conservation Monitoring Centre (1992), wetlands currently occupy some 5 per cent of the Earth's land surface; they can be found in all the major climatic zones (see Figure 1.3). The largest areas of wetlands occur in the former Soviet Union, South America and Canada.

According to Larson (1991), 24 per cent of the world's wetlands are in Canada where they comprise mainly bogs and fens occupying 1.27×10^6 km^2. In addition, Dahl (1990) has estimated that the USA, excluding Alaska and Hawaii, probably contained 89.0×10^4 km^2, of which only 47 per cent are still in existence. The largest losses (Table 8.1) have occurred in California and the states of the Mid-West and most have been due to the expansion of agriculture. In parts of Australasia, and

Table 8.2 The characteristics and reasons for landscape fragmentation in the Netherlands (based on Opdam *et al.*, 1993; Vos and Zonneveld, 1993)

A	An increase in woodland, mainly because of the afforestation of heathlands and peatlands
B	A decrease in small woodlands and hedgerows within the agricultural landscapes. This has occurred in order to increase field size and hence improve the efficiency of mechanisation
C	Wetland and grassland habitats have declined in agricultural landscapes, often for the same reasons as in B
D	A major decrease in heathlands and peatlands due to reclamation for agriculture and reafforestation

especially in the lower reaches of the Murray River basin, wetlands have been flooded with saline water derived from upstream irrigation systems (Finlayson, 1991). This has altered the ecology of rivers, swamps and lakes. Many of the wetlands in the Mediterranean basin have suffered a long history of human impact, mainly via agriculture. Coastal lagoons, for example, were once much more widespread than they are today (Lepart and Debussche, 1992). Delta regions and river floodplains such as those of the Camargue in France, the Po in Italy and the Danube in Romania, have also diminished significantly in size (Dugan, 1993). Even in the less populated region of Scandinavia, large areas of wetland have been lost since 1800. In Sweden's lake district, for example, an ambitious series of drainage schemes had, by the mid-1900s, affected a large proportion of lakes. A typical such lake is that of Hornborga where water levels declined by 2 m due to drainage schemes designed to provide arable land. However, shrinkage of soils caused major problems and since 1989 a restoration scheme has been introduced to reinstate the lake and its status as a bird habitat (Dugan, 1993). In other parts of Europe the situation is irredeemable. Greek Macedonia, for example, has lost 94 per cent of its marshland since 1930. According to Hollis and Bedding (1994), only 56 km^2 of these once extensive wetlands now survive.

These data on forest, woodland and wetland demise illustrate the magnitude of change due to the expansion of agriculture. All other major ecosystem types have also been affected by this process. Not only have these ecosystems diminished in extent but many have become so fragmented that maintenance of biotic diversity is threatened. Although it is difficult to define accurately the minimum size of an ecosystem or habitat in any given area that it is necessary to conserve in order to maintain diversity, there are concerns that habitat fragmentation has caused, and will continue to cause, the extinction of plant and animal species (see discussion in Tilman *et al.*, 1994). This occurs because the isolation of 'small' conservation areas leads to a reduction in the genetic resource base of the biota. The fragmentation of natural habitats has been most widespread in Europe due to the long history of agriculture in this continent and the other pressures of population growth, and urban and industrial spread. An example of landscape fragmentation (or disintegration) is that of the Netherlands which has been described by Opdam *et al.* (1993). The characteristics and reasons for this fragmentation are given in Table 8.2 which shows that agriculture is a major cause of landscape change. As Vos and Zonneveld (1993) report, the use of crop-protection chemicals has also caused a loss of biodiversity. This pattern is repeated throughout western Europe and has occurred by degrees rather than through large-scale clearance. The resulting mosaic of vegetation communities, as Newman (1993) observes, is not necessarily disadvantageous to species survival. Much depends on whether there are corridors linking fragments of a similar nature so that interbreeding and replenishment can occur. For any species there is a minimum viable population. What makes management and conservation difficult, however, is determining what this is before the fragments become too small. It is also related to genetic diversity and whether or not invasion from outside the fragment is possible.

The loss of wetlands in the UK since 1900 illustrates the problem of habitat fragmentation. This, Fisher (1993) attributes to agricultural change, especially the reclamation by drainage and ploughing that was deemed necessary during World War II to increase food production. This, and the post-war intensification of agriculture, has in Fisher's words, 'reduced East Anglian fenland to a few isolated and endangered relicts'. The Norfolk Broads, also part of the UK fenland, have become even more impoverished since the early 1970s. Not all of this is due to agricultural practices as the impacts of partially treated sewage and recreation have taken their toll. Not only has habitat fragmentation occurred; so has cultural eutrophication, a process that is discussed in Section 8.4.

The issue of habitat fragmentation and its impact on bird populations has been discussed by Jarvis (1993). In illustration, he gives the example of the demise of lowland heath communities in southern Britain. This habitat has diminished in extent due to soil improvement for agriculture and a host of other factors such as afforestation, mineral extraction and urban spread. The heathlands of east Dorset, for example, have decreased in the last 200 years from *c.* 39 960 ha to just *c.* 7900 ha. In addition to this 80 per cent decline in the habitat, the remaining 7900 ha has become severely fragmented. What was once an almost continuous expanse of heathland has disintegrated into more than a hundred parcels. According to Chapman *et al.* (1989), only about 18 of these areas are more than 100 ha in extent. Research on the invertebrate populations of these areas (Webb, 1989; Webb and Thomas, 1994) suggests that there is a negative relationship between the diversity of the invertebrate community and the area of the heathland and its degree of isolation, i.e. heaths of small area and/or which are isolated have a reduced diversity when compared with large and/or non-isolated areas. The same relationship is true for birds. In the case of lowland heath fragmentation, the major threat is to the Dartford warbler. Jarvis (1993) also reports that the drainage of wetlands for agriculture in the UK has rendered several already rare species, e.g. the bittern, water rail and black-tailed godwit, susceptible to extinction.

Changes in invertebrate populations due to habitat destruction and fragmentation in the UK have been examined by Thomas and Morris (1994). Their examination of a variety of data has allowed the construction of Table 8.3 which details the numbers of extinct and vulnerable or endangered invertebrate species (these are categories recommended by the IUCN—the International Union for Conservation). These data, however, do not take account of pre-1900 extinctions due to lack of records. Butterflies, in particular, have been especially vulnerable to extinction, while Orthoptera (grasshoppers and crickets) are amongst the currently most vulnerable/endangered species. These data on the plight of butterflies nationwide has been reinforced by further study by Thomas and Morris in Suffolk. Here, recording has been undertaken for the last 150 years and shows that between 3 and 12 per cent of

Table 8.3 The percentage of invertebrate species classified in UK Red Data Books and other reviews as extinct or endangered (from Thomas and Morris, 1994)

Order		No. of British species	% presumed extinct	% vulnerable or endangered	% in Red Data Book
Insects	Odonata (dragonflies)	41	7	7	22
	Orthoptera (grasshoppers, crickets)	30	0	17	20
	Heteroptera (bugs)	540	1	4	15
	Trichoptera (caddis flies)	199	2	6	17
	Lepidoptera (butterflies)	59	8	5	20
	Lepidotera (macro-moths)	*c.* 900	2	3	11
	Coleoptera (beetles)	*c.* 3900	2	6	14
	Aculeate Hymenoptera (ants, bees, wasps)	580	4	7	28
	Diptera (flies)	*c.* 6000	0.05	8	14
Others	Mollusca (slugs, snails)	*c.* 200	0	9	17
	Annelida (leeches)	16	0	0	19
	Myriapoda (centipedes, millipedes)	*c.* 88	0	0	6
	Crustacea (woodlice, amphipods)	64	0	0	9
	Arachnida (spiders, pseudoscorpians)	*c.* 665	0	8	13

invertebrates have been lost, with butterflies topping the list at 42 per cent species decline. Moreover, the species considered to be 'vulnerable' or 'endangered' in Table 8.3 may become extinct in the next few decades. Thomas and Morris (1994) also discuss the factors that cause species declines. They discovered that most extinct and threatened species are those characteristic of the early or late stages of succession within woodlands, grasslands and heathlands. The diminution of these stages has been caused by changes in land-use practices, notably the abandonment of coppicing in woodlands and the conversion of heaths and grasslands to agriculture. It is likely that this is the case elsewhere in Europe (e.g. Thomas, 1991).

Apart from its impact on natural and semi-natural habitats (an example of the latter is the lowland heathland of Dorset referred to above), agriculture has caused the demise of another, though somewhat artificial, habitat: the hedgerow. Although this is not a feature that is unique to the British countryside, its history in the UK is well documented (e.g. Rackham, 1986) and it is clear that the histories of the hedgerow and agriculture are intimately associated. The present distribution of hedgerows reflects many centuries, if not millennia, of agricultural change. Used to confine animals as well as to denote the boundaries of land-holdings, the distribution and location of hedgerows until the twentieth century reflected changes in both of these factors. The enclosures of open fields and arable land which occurred throughout England between the fifteenth and eighteenth centuries, for example, caused many new hedgerows to be planted. Hoskins (1988) remarks, 'The thousands of miles of new hedgerows in the Midland countryside, when they came to full growth after a generation, added enormously to the bird population.' This was probably the 'golden age' of the hedgerow. However, a major change occurred during and after World War II. Not only were hedgerows removed to facilitate the construction of aerodromes but the intensification of agriculture, which came in the wake of mechanisation, led to the extensive grubbing up of hedgerows. Though not universal, the removal of hedgerows to amalgamate fields and to provide additional crop-growing land has altered a large part of the British countryside. To quote Hoskins (from the original 1955 edition) again: 'in some parts of England such as East Anglia, the bulldozer rams at the old hedges, blots them out to make fields big and vacant enough for the machines of the new ranch-farming and the business-men farmers of five to ten thousand acres.'

Overall, approximately 200 000 km of hedgerows have been removed since World War II (Body, 1987). Although some 30 000 km of new hedgerows have been constructed, there was a net loss of hedgerows which amounted to 22 per cent of the total for pre-1940 England and Wales. The removal of hedgerows on such a large scale has not only altered the character of large parts of the countryside but also destroyed an important wildlife habitat. Hedges are particularly important for game birds; up to 90 per cent of Britain's partridge population have their breeding grounds in hedgerows. Indeed, Southerton and Rands (1987) have suggested that the decline in hedgerows has been the major cause of the decline in numbers of both grey- and red-legged partridges in the last few decades. In addition, the use of crop-protection chemicals, notably insecticides and pesticides, on fields adjacent to surviving hedgerows has caused the diminution of the habitat's biodiversity. The decline in area of many other natural and semi-natural habitats, e.g. peatlands, moorlands, grasslands, woodlands, etc. (see reviews in Mannion, 1991; Tivy, 1990), has also contributed to the declining biodiversity of the British countryside.

The spread and intensification of agriculture has unquestionably caused the extinction of species. In view of the fact that the biota of the Earth has, to date, only been partially identified and classified, it is impossible to estimate rates of extinction accurately. Wilson (1988), for example, suggests that there may be more than 4×10^6 species whilst May (1989, 1994) estimates that there are in excess of 100×10^6 species (see also discussion in Colwell and Coddington, 1994). Moreover, the Earth's biota has a long history of expansion punctuated by periods of extinction due to natural causes (reviewed in Wilson, 1992). Of major concern, however, is the speed at which humans are causing extinction; the rapidity with which this is occurring appears to have no precedent in the past. Smith *et al.* (1993a, b), for example, have discussed global plant and animal extinctions since 1600. They estimate that 600 plants have become extinct, representing 0.25 per cent of an estimated total, and that 486 animal species have become extinct, i.e. approximately 0.04 per cent of the total. In particular, Smith *et al.* report that a marked increase in extinction rate occurred between 1850 and 1950, corresponding with the annexation of Europe's colonies throughout the world. The extinctions represent the impact of European agriculture and resource exploitation on the biota of the colonies.

One of these colonies was Australia. The impact of European settlers was considerable and enduring; much of it was due to the introduction and expansion of agricultural systems suited, by dint of many centuries of practice and innovation, to Europe. In some areas there was complete removal of the native vegetation cover; in

Table 8.4 The impact of European settlement on Australia's fauna and flora (based on data in Hobbs and Hopkins, 1990; Humphries and Fisher, 1994)

A THE IMPACT OF EUROPEAN SETTLEMENT ON AUSTRALIA'S FLORA
- (i) 70% of the floral communities have been altered
- (ii) 65% of the original tree cover has been removed
- (iii) 75% of the rain forest has been cleared for grazing and agriculture
- (iv) 165 species of plants (out of 20 000) are now extinct
- (v) 209 species of plants are considered endangered
- (vi) 784 species of plants are considered vulnerable
- (vii) This means that 5% of the flora is extinct or under pressure

B THE IMPACT OF EUROPEAN SETTLEMENT ON AUSTRALIA'S FAUNA
- (i) 20 species of mammal (out of 263) are now extinct
- (ii) Examples of these include the thylacine, the Alice Springs mouse, four species of wallaby, four species of hopping mice and two species of bandicoot
- (iii) At least five species of birds have become extinct (out of 522)
- (iv) The flightless birds have been most acutely affected, e.g. the Tasmanian emu

others, modification occurred (Hobbs and Hopkins, 1990). These changes, which were particularly intense during the major expansion of population between 1850 and 1950, also brought about many extinctions. Overall, since the initial colonisation of Australia in 1788, 70 per cent of the floral and faunal communities have been altered (Humphries and Fisher, 1994). Some data relating to these changes, including extinctions, are given in Table 8.4, though the introduction of carnivores, notably cats and foxes, has played an important role in causing extinctions. Other animals, e.g. the Tasmanian wolf, have been hunted to extinction. The advent of Europeans also caused plant extinctions; the World Conservation Monitoring Centre (1992) states that 165 species of plants are now considered to be extinct. This represents 0.83 per cent out of an estimated total flora of *c.* 20 000 species. The environmental impact of Europeans in both Australia and New Zealand has also precipitated accelerated land degradation, as is discussed in Section 8.2.

8.2 SOIL EROSION

Soil erosion is the removal of soil particles at a rate that exceeds their replacement by weathering. It is a natural process that is part of terrain deflation. However, agricultural practices have accelerated rates of soil erosion throughout the world since the inception of agriculture. Until recently, soil erosion has been particularly associated with the developing world (see Section 9.2), especially parts of Africa where its occurrence may bring crop failure and famine. Soil erosion is a much more widespread and usually more insidious process than the occasional, but sometimes life-threatening, much-publicised incident. Accelerated soil erosion is one of the most significant ways in which agriculture brings about environmental change, though not all of the world's soil erosion is caused by agricultural practices. Deforestation, logging, mining and construction activities all contribute to the 75×10^9 t of soil that *The Gaia Atlas of Planet Management* (Myers, 1993a) states are lost from the Earth's land surface annually. Agriculture is, however, the major cause of this substantial alteration to an important Earth surface process. Approximately 11×10^6 ha of the world's arable land are lost annually as a result of soil erosion. The areas most severely affected are shown in Figure 8.2.

How and why soils are eroded have been discussed by Morgan (1986), Barrow (1991) and Lal (1994a). Soils are eroded by the agencies of wind and/or water, the significance of which depends on the prevailing climate and the density of vegetation providing protection. This, in turn, depends on land use. Why soil is eroded hinges on rainfall/wind erosivity, i.e. how effective the rainfall/wind is at entraining soil particles, and soil erodibility which is the susceptibility of the soil to removal (see definitions in Lal and Elliot, 1994). Both of these factors are influenced by the degree of pressure exerted on the land which, in turn, is reflected in the nature of the vegetation/crop cover and how it is managed, i.e. tillage practices, soil conservation practices and water management. One of the most important considerations is whether or not a crop or fallow cover is maintained throughout the year. Moreover, soil erosion only becomes significant if the rate of soil removal exceeds the rate of replacement. Clearly, the rate of replacement will vary spatially since it depends on the bedrock type, climate and vegetation cover. Estimates also vary. For example, *The Gaia Atlas of Planet Management* (Myers, 1993a) reports that 2.5 cm of topsoil can take between 100 and 2500 years to form; it can, however, be lost in as little as 10 years. Such variability in soil replacement and the long timescales involved mean that it is difficult, if not impossible, to generalise about rates of soil replacement through weathering. The severity of soil removal is equally difficult to quantify. Monitoring programmes at sites throughout the world nevertheless indicate that current rates of soil erosion worldwide are reaching unacceptable proportions, especially in the context of developing sustainable agricultural systems (see Section 10.3). It is also unequivocal that the impact of soil erosion on food production globally, but especially in the developed world, has been, and continues to be, masked by the increasing use of artificial fertilisers and improved crop varieties.

Only since the mid-1970s has soil erosion been acknowledged as a problem worthy of address in the developed world. In part, this has been due to the application of science and technology to improving crop productivity. Mechanisation, the use of crop-protection chemicals and artificial fertilisers as well as irrigation, improved crop varieties and no-tillage practices, often within a political framework geared to protectionist policies for agriculture, have all contributed to yield increases per unit land area. However, the land degradation on the loam-rich plains and plateaux of Europe (Wisherek, 1993) has become too expensive to ignore and has stimulated a considerable amount of research on Europe's soil erosion problem. Generally, this is considered to be a consequence of changing agricultural practices since World War II. Some examples are discussed below. Elsewhere in the developed world, notably Australia and the USA (see Figure 8.2), soil erosion issues have been a major concern for a far longer period than they have in Europe. In the USA and Australia, large-scale erosion was precipitated by the introduction of European agricultural practices into regions that were unsuited to them. Indeed, in the USA the infamous Dust Bowl of the 1930s, which caused the impairment of some 40×10^6 ha of the Great Plains, provided an admonition about injudicious agricultural practices that reached the world but which largely went unheeded until the 1980s.

There are other lessons from the past that reflect the significance of soil erosion. The palaeoecological record from lake sediments, for example, reflects periods of

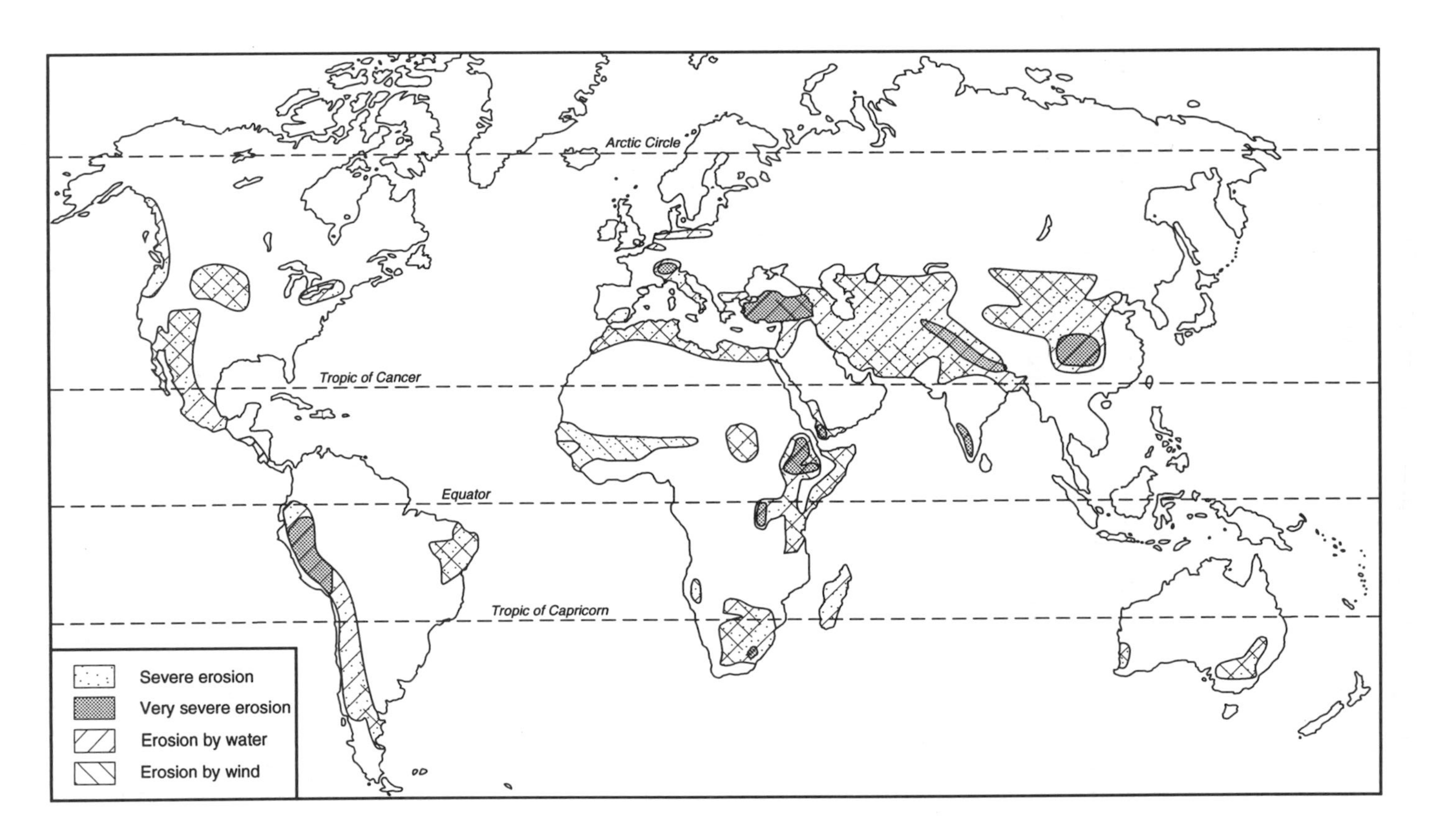

Figure 8.2 Areas of the world most severely affected by soil erosion (based on sources quoted in the text)

erosion in the UK, e.g. in the Lake District (Haworth, 1985) and in the River Severn catchment (Brown and Barber, 1985) due to Neolithic and Bronze Age farming respectively. Bronze Age, Iron Age and Romano-British colluvial deposits, produced following clearance for agriculture of the chalk downlands have been reviewed in Bell and Walker (1992). Soil erosion is, thus, not a phenomenon confined to the post World War II era (see also Evans, 1990a, 1992). Nevertheless, it is only in the last few years that efforts have been made to quantify rates of erosion in the UK (reviewed in Arden-Clarke and Evans, 1993; Evans, 1993a, 1995). This comes in the wake of Morgan's (1985) estimate that 37 per cent of the arable land of England and Wales is at risk from soil erosion (see also Evans, 1990b). Patterns of soil erosion have been investigated using a variety of methods including caesium-137 measurements on arable soils from a range of sites in England and Wales (Walling and Quine, 1990; Quine and Walling, 1991). The method involves the measurement of caesium-137, a by-product of the testing in the atmosphere of thermonuclear weapons between thc 1950s and 1970s, in soil profiles. The radionuclide reaches soils via rainout and washout and is absorbed by clays. The horizontal distribution in a given soil is the result of this deposition plus or minus any subsequent removal or erosion of soil particles. Thus, the establishment of a caesium profile from an undisturbed site, e.g. from beneath pasture or woodland, within an area characterised by a uniform soil type, provides a reference stratigraphy. Against this, caesium-137 profiles from adjacent sites on arable land can be compared to determine how much soil has been deposited or removed in the last 30 years or so. Some of Quine and Walling's (1991) results are given in Figure 8.3. The data, which have been compared with monitored or measured rates of soil erosion to verify the accuracy of the methodology, show a large variation in erosion rates from 0.6 t ha^{-1} yr^{-1} to 10.5 t ha^{-1} yr^{-1}. The lowest rates occur on clay soils while the highest rates occur on sandy and sandy-loam soils. It must, however, be borne in mind that there are many difficulties associated with the use of caesium-137 to measure rates of soil erosion. Moreover, the data illustrate that variations can occur in relation to the erodibility of soils. For example, sandy and sandy-loam soils, because of their structure and composition are naturally more susceptible to erosion than are clay soils. The problem can then often be exacerbated by land-use type.

There is also a considerable amount of data for soil erosion on the South Downs (Boardman, 1990). Much of this is attributed to changing agricultural practices, notably the switch to autumn-planted winter cereals which now occupy *c.* 55 per cent of agricultural land in the area. After harvesting in July and August there is often a gap before the drilling of winter wheat is completed, usually in early November. This is followed by rolling of the seed-bcd which produces a smooth and compact soil surface. These operations, when the soil has little to bind it, coincide with the autumn peak of rainfall that characterises this region. Initially, rill systems become established, through which soil particles are transported to be deposited as fans on valley floors or at the bases of slopes. The distribution of rills is also influenced by tractor-wheel tracks, especially those which originate on steep valley sides. Boardman's (1990) data show that over a six-year period erosion rates varied from 0.7 m^3 ha^{-1} to 5.0 m^3 ha^{-1}. Not only are these rates cause for concern in relation to long-term crop productivity, though Boardman reports that on-farm repercussions have been minimal during the short-time interval concerned, but there have been off-farm

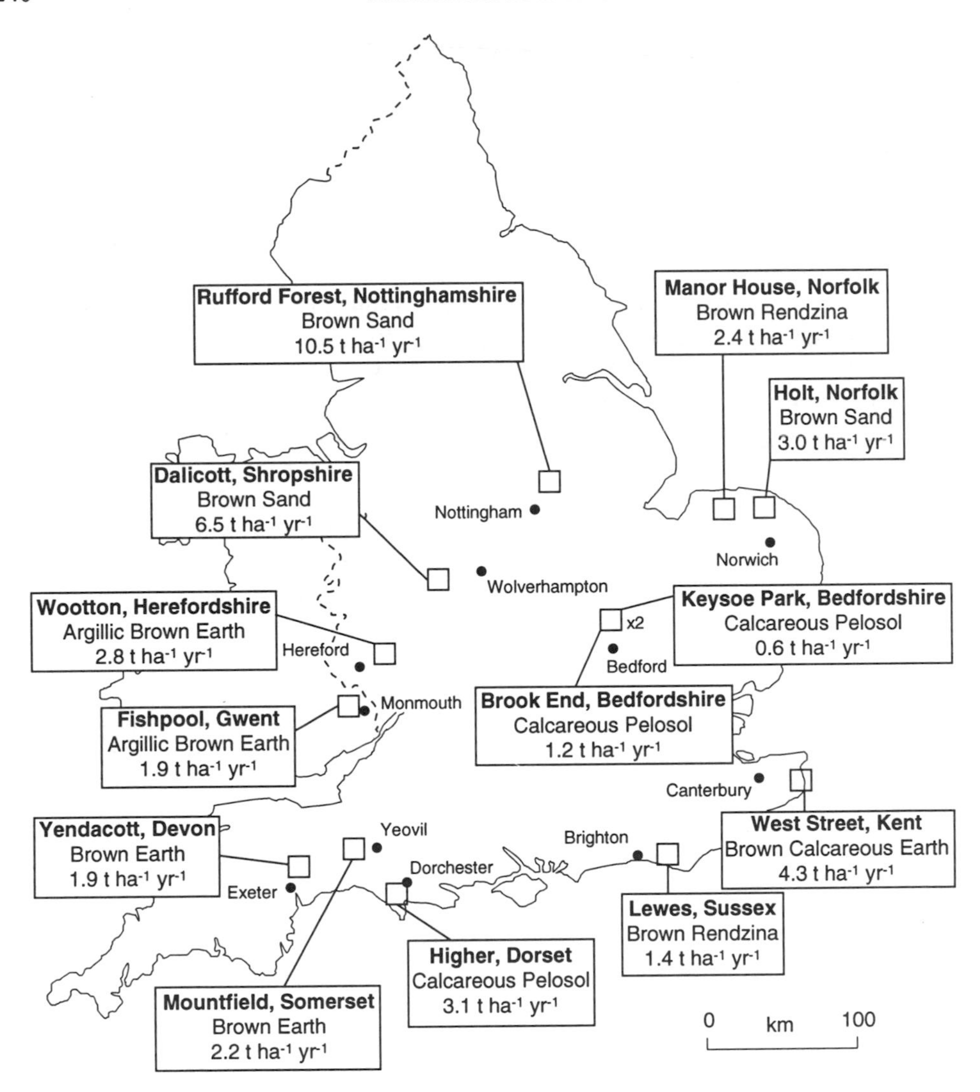

Figure 8.3 Soil erosion data from selected sites in England and Wales (based on Quine and Walling, 1991). Data are net

impacts with significant financial consequences. For example, nearby housing has suffered flooding with runoff carrying silt. The effects of water and silt can be very damaging to property, though, as Boardman points out, it is local authorities and insurance companies who shoulder these costs, not the farming community. The relationship between rainfall and erosion, albeit a complicated one in relation to rainfall intensity and frequency, in this region was further illustrated in 1989–1990 (Boardman, 1991). In contrast to the data referred to above, the majority of erosion occurred on land prepared for or cultivated with spring cereals. This was because of an unusual distribution of rainfall: considerably above-average amounts were received in December, January and February. By this time winter cereal crops had become well established so they acted as a stabilising factor whilst the converse was true for land allocated to spring cereals.

Evans (1993b) has also surveyed the extent, frequency and rates of erosion at 24 sites on arable land in England and Wales, many of which correspond with the locations given on Figure 8.3. On average, erosion recurred frequently, often more than once in three years, even on soil types not particularly susceptible to erosion. Moreover, in areas prone to erosion an average of 5 per cent of the arable land was affected, though in the worst case, in Nottinghamshire, 13.9 per cent of the arable land was affected. This area has a sandy soil and the comparatively large extent of erosion is paralleled by high rates of erosion (see Figure 8.3 and data for Rufford Forest). Evans also suggests that these data are likely to represent 'best case scenarios' in so far as no exceptional rainfall events occurred during the monitoring period. Of particular note is the fact that the recurrence of erosion was most frequent on irrigated land in Kent. The implication of this is the possible impact of global warming if it warrants an increase in irrigation extent and frequency. Evidence that relatively long periods of low-intensity rainfall can cause substantial soil erosion is produced by Kirkbride and Reeves' (1993) report on the impact of 50 mm of rainfall over 24 hours on recently seeded arable fields in Angus, Scotland, in 1992. Although rainfall intensities rarely exceeded 4 mm h^{-1}, rill erosion affected 30 per cent of newly sown fields and a further 28 per cent were affected by sheet wash. The impact was considerably less on fields with growing crops indicating the considerable protection afforded by even young crops. The examples referred to above concern the water erosion of soils; undoubtedly the UK suffers from this process at least on a local scale. For example, Walling and Quine (1990) suggest that some aspects of the caesium-137 profile at Rufford Forest Farm, Nottinghamshire (Figure 8.3), are due to the deposition of locally derived wind-borne soil. Soil erosion as a result of wind erosion is, however, considered to be relatively insignificant in the UK when compared with the extent of that due to water erosion.

As stated in the introduction to this section (p. 236), soil erosion has become apparent in parts of Europe other than the UK. The most severely affected areas are illustrated in Figure 8.2; these are south-eastern Spain, the north European plain and much of Italy, particularly the alpine region. This does not mean that soil erosion is confined to these areas but that available data reflect the severity of erosion; for many areas data are not available whilst for others historical rates of erosion are documented. For example, Alayev *et al.* (1990) report that soil erosion was occurring on agricultural lands in the broad-leaved and coniferous forests zones of the Russian Plain by 1700. Today, an average of 6.6 per cent of the arable land of this region is significantly eroded. Much of this is due to water, especially as a result of snowmelt in the spring. The steppes, however, also suffer wind erosion causing 'black storms'. As Figure 8.4 illustrates, the percentage of eroded land in the regions of the Ukraine varies from less than 2 per cent to more than 50 per cent. Alayev *et al.* also state that 'A storm in 1960, with wind speeds of up to 28 m per second, completely blew away a 40 cm arable layer from a chernozem field not far from Donbass. Depressions are sometimes formed 10 m in diameter and up to 1.5 m in depth.' Problems such as these have resulted from agricultural practices that leave the soil bare and susceptible to gullying, the distribution of which is given in Figure 8.4. In addition to the impact of arable agriculture, Ray *et al.* (1993) report that the elimination of elements of the

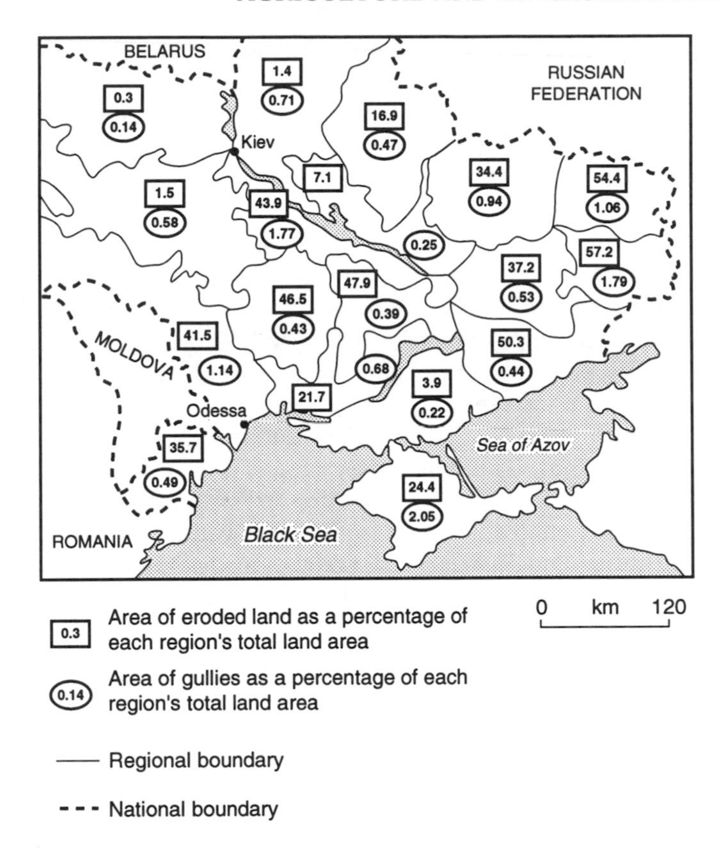

Figure 8.4 Soil erosion and gullying characteristics in administrative regions of the Ukraine (based on Alayev *et al.*, 1990)

flora of steppelands due to overgrazing also causes accelerated soil erosion with extensive gullying.

In the loess belt, an intensively cultivated region of Europe (see Figure 8.2), soil erosion is also widespread. Information on this region has been widely published, e.g. De Ploey (1986). There are reports of annual erosion rates of between 3 and 100 t ha^{-1} occurring, with the highest erosion rates being in areas with steep slopes. This incidence of soil erosion is ascribed to changing agricultural practices, notably the demise of traditional intercropping and its replacement by mechanised monoculture. The reduction in organic matter, a characteristic of the Ukraine chernozem soils discussed above, is also considered to be an underlying cause of soil erosion as grass leys have been replaced by arable land. The problems associated with gullying, and its prevention, in this region have been addressed by Poesen (1993) and Vandaele (1993). Further north still, in southern Sweden, changes in agricultural practices have increased the erodibility of soils by water. Not only is the loss of soil becoming a problem but eroded particles entering water bodies are thought to be contributing to cultural eutrophication (Section 8.4) because of the phosphorous bound to the soil particles (Alström and Åkerman, 1992). A wide range of values for erosion rates, due to rill, interill and ephemeral gully erosion, is characteristic of the studied area in

Scania, southern Sweden. Over a three-year period, rates of erosion varied temporally and spatially within the range 1 kg ha^{-1} yr^{-1} to as much as 120 t ha^{-1} yr^{-1}. Much of this variation can be explained by differences in precipitation rates, variations in the rate of increase of spring temperatures that cause slow or rapid soil thaw, and topography. Overall, some 7 per cent of the land monitored experienced an average rate of erosion of 825 kg ha^{-1} yr^{-1}, with much of the soil being removed by rill and ephemeral gully erosion.

There are numerous examples of soil erosion problems in other parts of the developed world. In both North America and Australia, there is much evidence for the impact of early colonists who had little knowledge of their new-found soils. In Georgia, USA, for example, Trimble (1992) has demonstrated that the swamp lands of the Alcovy River are the result of the deposition of soil eroded from the upstream Piedmont region. Here, the soils are coarse with 30–75 per cent sand (see the results of Quine and Walling (1991), referred to above, which indicate the increased erodibility of sandy soils as compared with clay-rich soils), slopes are steep and rainfall is abundant. The removal of natural vegetation and subsequent injudicious cultivation of these soils caused rapid soil-particle transfer through gullies and streams, as was noted by the famous British geologist Sir Charles Lyell on a visit in 1845. The deposition downstream of this bed load caused streams to break their banks and so flood the bottomlands. Later changes in agricultural practices in the USA have also been reflected in the record of soil erosion. Earle (1992) has reviewed the impact of agricultural innovations, especially the adoption of artificial fertilisers (mainly mineral phosphates) in the US cotton belt between 1870 and 1930. The increased yields of cotton that ensued with initial fertiliser applications encouraged sweeping increases in the amount of land being cultivated. Some of this was expansion into marginal land; other changes included the replacement of the cotton–corn–cowpea rotation with monocultural cotton and an increase in row width from 20 to 40 inches, and eventually to 60 inches. Earle states that these changes generated sediment yield increases of 80 per cent and 136.5 per cent respectively. Overall, Earle observes that soil erosion in the cotton belt has a tripartite history as is described in Table 8.5. These changes in land use and agricultural practice are apparently reflected in the geomorphological history of the region recorded in 1931 by Happ, Rittenhouse and Dobson (quoted in Earle, 1992).

Another historical perspective, on soil erosion in southern Minnesota, has been presented by Beach (1994) and is based on studies of alluvial deposits in three drainage basins in an agrarian landscape. Over a period of 137 years, much of the soil eroded from the upper reaches of these basins (87, 65 and 63.5 per cent) is still present in the respective watersheds; indeed between 47 and 65 per cent of the eroded material has travelled less than 3 or 4 km. Sources and sinks in this landscape are thus close together. Nevertheless such deposits are less cohesive than the original soils, and may well be liable to remobilisation, should vegetation and/or land use change as might be the case if global warming ensues. Thus, the resulting sediment patterns of past and present land use, in this region at least, may be the source of major erosion problems in the future. Currently, about one-third of US croplands, consisting of some 60×10^6 ha, are experiencing soil erosion (Myers, 1993a). The level of concern is reflected in the government-promoted incentives that are available to farmers to institute soil conservation measures (Napier, 1990). The

Table 8.5 Changing agricultural practices and their relationship with soil erosion in the cotton belt of the USA, 1600–1930 (based on Earle, 1992)

A	INITIAL SETTLEMENT AND CONSOLIDATION 1600–1840 Accelerated soil erosion after initial colonisation and then modest soil erosion as a crop complex based on cotton/tobacco maize and small grains. Abandonment of cotton occurred with the depressions of the 1830s and 1840s
B	INTRODUCTION OF COTTON–MAIZE–COWPEA ROTATION 1840s–1870s Decelerated soil erosion as the rotation ensured an improved soil cover. This was a short-lived period
C	INTRODUCTION OF COMMERCIAL FERTILISERS 1870s Agricultural innovations which included the introduction of commercial fertilisers, faster cultivation methods, à move towards monoculture as the cotton–maize–cowpea rotation was abandoned in favour of cotton monoculture and a massive expansion of the cotton acreage, caused a huge increase in soil erosion. This lasted until the 1930s

costs of soil erosion are not only reflected in the loss of a valuable resource but also in the off-site costs associated with soil deposition. The 2.7×10^9 tonnes of soil removed annually from US croplands cause silting in reservoirs, lakes and rivers which then require dredging. In the USA alone this costs some $300 million annually (Moran *et al.*, 1986). Remedial measures, to combat the problem at source, are therefore essential in order to safeguard the soil resource itself and habitats likely to be affected by soil deposition. In consequence, soil conservation measures, which may include suitable tillage practices as well as crop residue management practices (e.g. Bower, 1994), are essential components of sustainable agriculture. This is discussed in Section 10.4.

The problem of soil erosion is even more acute in Australia than it is in Europe or the USA. According to K. Edwards (1993), more than half of Australia's land area used for agriculture has been subject to some form of degradation. This amounts to *c.* 38×10^6 ha and much of it is due to soil erosion. Another form of degradation is salinisation due to irrigation which is discussed in Section 8.4. Soil erosion occurs in both the arid and non-arid zones and more than half of the damaged land requires treatment involving earthwork construction rather than management through appropriate agricultural practices. According to Boucher and Powell (1994), some of this erosion began soon after the initial settlement of Australia in 1788, particularly on sodic soils (i.e. soils with high concentrations of sodium salts) as exemplified by those in the state of Victoria. K. Edwards (1993) states that the most severe soil erosion problems occur in the cropping areas of northern New South Wales (see also Graham, 1992) and southern and central Queensland. Some of the heaviest soil losses, in the 70–150 ha^{-1} yr^{-1} range, have been recorded in Queensland, though losses of up to 500 t ha^{-1} yr^{-1} are not unheard of. Much of Australia's soil erosion problem stems from the introduction of hard-hooved grazing animals (i.e. sheep, cattle and camels), the advent of the rabbit and arable agriculture. Although efforts have been made to improve agricultural practices (e.g. long bare fallow periods are now avoided), crop production is still a major cause of soil erosion. For example, there is still a summer fallow period between annual winter crops. In both cropping regions this fallow period coincides with the occurrence of erosive rains. A crop which leaves soil particularly susceptible to erosion is sugar cane. This is because seed-bed preparation, exposing the soil, may take up to a year and because the crop is planted in rows. K. Edwards (1993) also reports that changes in commodity prices influence rates of soil erosion. When grain prices are high, for example, crop production may be extended into lands that, in terms of their rainfall, are marginal. Poor crop growth exposes the soil and increased erosion ensues. Even when crop production ceases, the land remains susceptible to erosion because there are rarely suitable schemes available for pasture establishment. Plant succession then takes several years before a vegetation cover develops; even then degradation may occur. Other factors that exacerbate soil erosion are drought and subsidies for land clearance which existed up until 1983. This situation illustrates that the treatment of the symptoms of soil erosion is relatively ineffective; agricultural practices need to change substantially or cease entirely.

Soil erosion problems have also developed in New Zealand, most of which are due to the adoption of pastoral agriculture on lowland steeplands since the advent of Europeans 150 years ago. Blaschke *et al.* (1992) have reviewed the historical context for this problem which stems primarily from the removal of the natural forest cover on both North Island and South Island. The location of New Zealand's steeplands, which occupy 40 per cent of the country's land area, are given in Figure 8.5. Here the problem of soil erosion and management is exacerbated by the fact that these steeplands, as part of New Zealand's mountain region lying on the boundary between the Indian and Pacific lithospheric plates, are tectonically active. Continued uplift coupled with a regular rainfall, which is particularly high on South Island, provide streams and runoff of sufficient strength, even in the rain shadow region east of the high mountains, to cause serious water erosion. The variation in environmental conditions along the north-west to south-east axis of New Zealand's steeplands varies considerably and the data of Blaschke *et al.* data relate to monitored sites on North Island, as shown in Figure 8.5. In particular, their data on landsliding, which is the most ubiquitous form of erosion in New Zealand, show that it has increased substantially since forest removal because of the loss of soil cohesion provided by extensive tree roots. Not only has landslide density increased but so has landslide frequency. In the Taranaki district, for example, landslide scar density in forests is only 2.5 km^{-1} while it is 3.5 km^{-1} in scrub and a huge 7.3 km^{-1} in pasture. A similar pattern occurs in the Gisborne–East Coast area where high-intensity rainstorms are a major cause of landslides in Wairarapa and west coast sites (see Figure 8.5). Estimating erosion rates since deforestation, Blaschke *et al.* calculate a soil depletion rate of 2000 $m^3\ km^{-2}\ yr^{-1}$ as deep forest soils are replaced by shallower pasture soil; up to 20 per cent of the soil profile may be lost depending on the steepness of slope. Equally important is the fact that pasture production on landslide scars rarely exceeds 70–80 per cent of the production on the remnant forest soil in its first 20–40 years. These data point to the fact that pastoralism is not sustainable on the steeper slopes and should be limited to gentler slopes. Moreover, agroforestry practices (see Section 7.5) may contribute to conservation-based land-use practices in these areas. In a country as small as New Zealand and one that derives considerable income from the export of its animal products, soil conservation measures are essential to economic stability in the long term. Similar but less acute

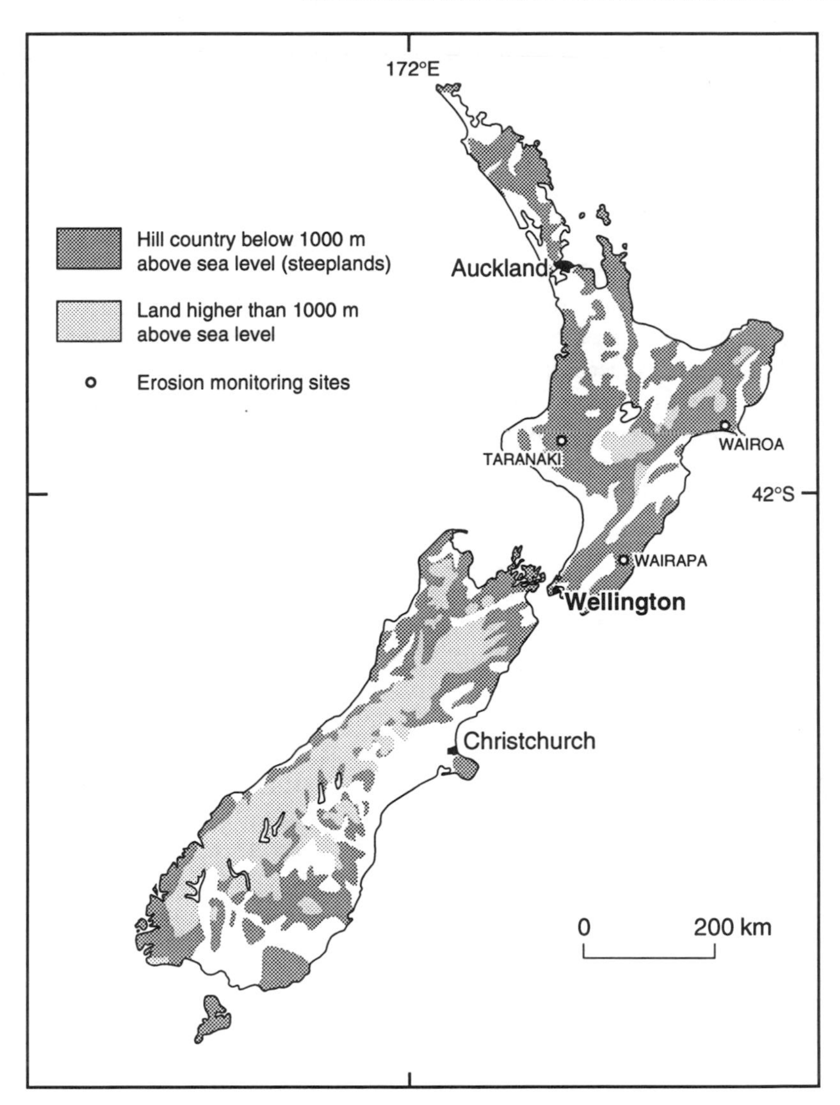

Figure 8.5 The location of New Zealand's steeplands and erosion monitoring sites (based on Blaschke *et al.*, 1992)

soil erosion is occurring on the Falkland Islands. Settled about 150 years ago, large-scale sheep farming began in the mid-1800s. Until then the islands, like Australia and New Zealand, had no large herbivores. Today, erosion is most extensive in sandy soils in coastal areas and is due not only to physical environmental characteristics but also to land management practices, notably grazing management, stocking rates and the frequency and intensity of pasture burning (Wilson *et al.*, 1993). From the information presented in this section, soil conservation measures should be a priority for most middle and high latitude nations.

8.3 THE ENVIRONMENTAL IMPACT OF CROP-PROTECTION CHEMICALS

Crop-protection chemicals are compounds that are produced industrially on a large scale for use in agriculture, horticulture and silviculture to improve the efficiency of crop production. They do this by providing protection from disease, from competitors such as non-crop plants and insects and by enhancing or controlling growth. There are other crop-protection agents which are biopesticides, i.e. they comprise populations of organisms such as bacteria, fungi and viruses which are pathogenic to specific pests. A discussion on the use of these agents is included in Section 10.1 which is concerned with biotechnology. The majority of crop-protection chemicals are pesticides, the general categories of which are given in Table 8.6. About 4.4×10^6 tonnes of pesticides are used every year (quoted in C. A. Edwards, 1993). In 1993 the global agrochemical market was worth US $\$25.28 \times 10^9$ (Wood Mackenzie Consultants Ltd, 1994). Many pesticides are active against a wide range of organisms, i.e. plants or animals, and consequently are known as broad-spectrum herbicides or insecticides. Others are target-specific, having been designed to kill particular pests.

However, defining a pest is more problematic than defining a pesticide because there are no pests in natural ecosystems; all plants, animals (including insects) and micro-organisms are interconnected and essential constituents of the biota. Such organisms are only termed pests when they compete with humans for resources. Insects, for example, reduce crop biomass by consuming it. Weeds prevent the realisation of the optimum biomass by competing with crop plants for light, water and nutrients. Other pests may consume biomass during storage of the harvest prior to processing. Animal health products are also considered under the heading of crop-protection chemicals. These are designed to improve animal (secondary) productivity by reducing the impact of disease. Plant growth regulators (PGRs), another component of the crop-protection armoury, are chemicals produced to influence the way crops grow so that harvesting efficiency can be improved or to manipulate the architecture of crops so that they maximise the light, water and nutrients available to them.

All of these substances have an impact, if only to reduce or eliminate pest populations in a specific arable field. Others have had far-reaching effects, influencing regions of the Earth where they have never been directly applied. Much of this was not anticipated, especially during the production of the early chemical pesticides immediately following World War II. The ecological and human health implications of widespread pesticide use was brought to the attention of the public by Rachel Carson's classic book *Silent Spring*, published in 1962. Although many of her criticisms were dismissed at the time, many of her fears have been realised subsequently. The response to these concerns politically has at least led to the establishment of regulatory authorities whose requirements for the registration of a pesticide are now quite stringent (reviewed in Mannion, 1991). Regulatory bodies include the Environmental Protection Agency (EPA) in the USA and the Ministry of Agriculture, Fisheries and Food (MAFF) in the UK. A summary of the tests required by the EPA is given in Figure 8.6. Despite these exacting requirements, there are currently about 1400 approved pesticide products available for use in agriculture and horticulture in the UK alone (Ivens, 1993).

Table 8.6 The major pesticide categories

	Herbicides	Fungicides	Insecticides	Acaricides	Nematicides	Molluscides	Rodenticides	Plant Growth Regulators
Target organisms	Weeds	Fungi	Insects	Mites	Nematode worms	Snails, slugs	Rodents	Crops
Examples of types of pesticides	Phenoxyacids Bipyriliums Phosphonates Pyrazoliums Aryloxy-Phenoxyacetic acids Nitrodiphenyl-ethers Acetanilides	Pyrimidines Alanines Triazoles	Organochlorines Organophosphates Carbamates Pyrethroids Avermectins	Organochlorines Organotins Tetrazines	Fumigants Carbamates Organophosphates	Aldehydes Carbamates	Coumarins Inorganic compounds, e.g. zinc phosphide	Hormones Triazoles

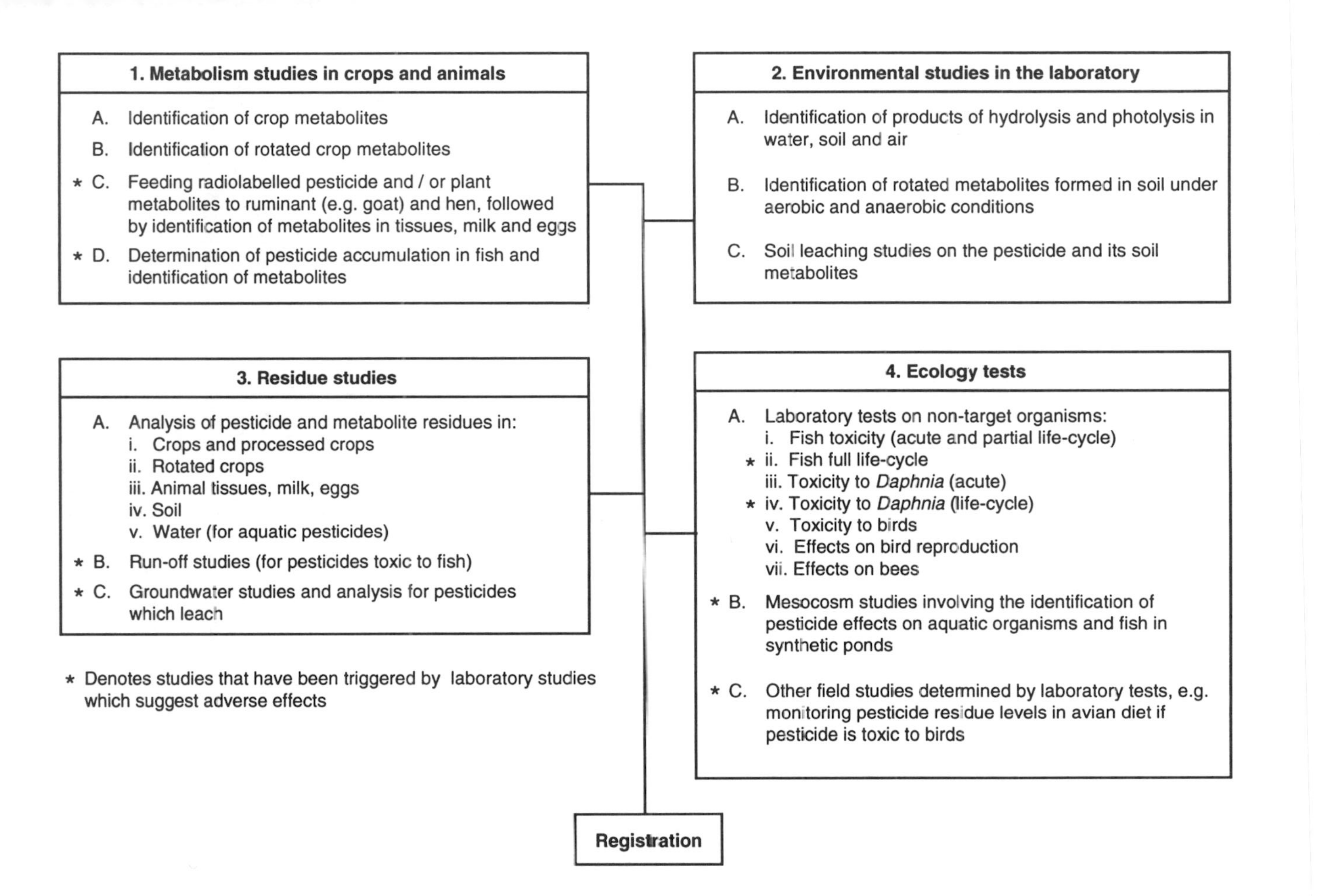

Figure 8.6 Environmental tests required for the registration of a new pesticide by the US Environmental Protection Agency (based on Mannion, 1991)

Apart from the promotion of controls and legislation on pesticide registration, which is essential before a product can be marketed, the problems associated with the early chemical pesticides have resulted in agrochemical companies changing their approach to finding and developing new pesticides. Brief histories of insecticides and fungicides have been given in Mannion (1995a, b). Essentially, the traditional methods of crop protection employed prior to World War II involved crop rotations, small fields of different crops (see Sections 3.2.1 and 3.3) and the integration of crop and livestock. Thus seasonality, crop susceptibility to disease, predation and competition, the use of fallow and crop organisation provided farmers with a number of permutations that could assist in pest control as well as the maintenance of soil fertility. With the advent of artificial fertilisers in the late eighteenth century and the onset of the mechanisation of agriculture in the 1920s and 1930s, agriculturalists began to look for alternative means of crop protection that would provide adequate safeguards for the increasing monoculture of crops in Europe and North America. For two to three decades the emphasis was on synthetic chemicals, i.e. compounds with toxophores (toxin-carrying parts of the molecule) assembled in the laboratory. Included in this group are the organochlorines, the most well known of which is DDT (see Table 8.6). These pesticides were effective but persistent in the environment and capable of accumulating, in a process known as biological magnification, as they were transferred through food chains. In consequence such compounds had a substantial impact on insect and faunal associations (see below).

The lessons provided by the organochlorines did not go unheeded (Lehman, 1993; Perkins and Holochuk, 1993) although, even today, DDT is still used in many developing countries because it is cheap and effective. As a result, attention turned, in the 1960s and 1970s, to natural pesticides. In particular, attempts were made to isolate the active ingredient of pyrethrum, a chemical present in the flowers of *Chrysanthemum cinerariaefolium* that had been used in powdered form ('insect' powder) on a small scale since the late nineteenth century (Davies, 1985). The active chemical, pyrethrin, was isolated and its structure determined. Since then it has provided a template for a range of synthetic analogues known as pyrethroids. The first pyrethroid product was marketed in 1977; subsequently there have been many others, most of which are broad-spectrum contact insecticides. Unlike the earlier organochlorines, the pyrethroids are not persistent in the environment and there is little evidence that they have any significant impact on the wider environment. One reason for this is that they are hydrophobic, i.e. they do not readily dissolve in water, and so they do not easily diffuse beyond the area of point of application. Most pyrethroids are, however, toxic to fish and aquatic crustaceans. This successful development of pesticides based on naturally occurring templates is now one of the preferred routes for the generation of new pesticides. Many agrochemical companies, in common with many pharmaceutical companies exploring the same avenues for the generation of new drugs, are now probing botanical gardens, plant collections and natural vegetation communities for leads that may eventually produce new crop-protection chemicals with little or no adverse environmental or ecological impacts. This is an aspect of biotechnology discussed in Mannion (1994a) and in Section 10.1. In addition, completely new ways of pest control have been developed, e.g. biopesticides based on fungi, bacteria and viruses (Section 10.1).

In relation to the environmental impact of pesticides, most has been written about dichlorodiphenyltrichloroethane (DDT). This was developed in the early 1940s by Paul Muller, a swiss chemist who was later awarded the Nobel Prize for Medicine and Physiology in 1948. Used initially in public health and military hygiene to control body lice, it was, and still is in some parts of the world, used to control malaria-carrying mosquitoes. According to Horn (1988), DDT was first used in crop protection in 1941 to halt an outbreak of the Colorado potato beetle in Switzerland. Its versatility, ease of production, low cost, stability and low acute toxicity to mammals made DDT a particularly attractive pesticide. Even in the 1940s, however, there were reports of adverse ecological effects but it was not until the early 1970s that DDT was withdrawn in some countries. The ecological effects of DDT have been reviewed in Mannion (1991) and Mellanby (1992). Classic experiments in the 1960s on *Brassica* crops, e.g. Brussels sprouts, which suffer infestations of larvae of the cabbage white butterfly, indicated that initial spraying with DDT caused a major reduction in larvae populations. Pest control was, thus, effective. However, populations soon revived; over a three-year study the population growth of the larvae was in fact more rapid on sprayed crops than on unsprayed control crops. This is explained by the fact that the initial spraying kills not only the cabbage white larvae but also its natural predators which live in the soil. Moreover, the persistence of DDT in the soil prevented populations of these predators from recovering whilst leaf growth and the removal of the DDT by rainwater from the crop provided an unprotected niche for the pest. There are other instances of DDT use which actually caused outbreaks of new pests for similar reasons. For example, the red spider mite, although present in small numbers in European and North American orchards did not become a significant pest until DDT was used. Similarly, the control of mosquitoes by DDT in California orchards gave rise to an outbreak of the cottony cushion scale due to substantial declines of the populations of its predator, the ladybird.

More notorious than the examples given above are the effects that DDT has had, and continues to have, on populations of birds of prey. Potts (1986) considers that eggshell thinning due to DDT and its residues was the prime reason for declines in populations of the peregrine falcon and sparrow-hawk. Thus, although the birds are not killed directly their reproduction rates are reduced through lack of survival of the eggs. Similar problems occur with fish-eating birds. DDT is not very soluble in water but even concentrations as low as 0.00005 ppm may eventually give rise to a build-up of DDT and its residues that is lethal. This occurs because of the way in which fish and other aquatic organisms obtain oxygen. They take in water to pass over their gills. Just as oxygen is absorbed so too are other substances. As Mellanby (1992) reports, a fish of 1 kg in weight passes some 700 litres of water per day through its body. This could result in the absorption of some 700 mg of DDT if the pesticide is being actively applied. Some types of fish, e.g. trout, will concentrate DDT to lethal levels. Aquatic insects and crustaceans may also be adversely affected. Ingestion of fish, insects, etc., by birds of prey can cause death directly or population decline due to the impairment of their reproductive capacity. DDT is not the only pesticide to have such an impact. The cyclodienes, e.g. aldrin, dieldrin and endrin, are cases in point. Dieldrin in particular has been widely used as a sheep dip and as a seed dressing; it is considered responsible for declines in populations of birds of prey. The processes involved are similar to those operative in the case of DDT.

Organophosphate insecticides largely replaced organochlorines. Many organophosphates, which are related to carbamates and work in a similar way by attacking the nervous systems of insects, were initially developed as nerve gases during World War II. These substances are not as persistent in the environment as the organochlorines (C. A. Edwards, 1993) but many are particularly toxic to mammals and birds. Their environmental impact has generally been localised. For example, Dempster (1987) reports that the treatment of cereal seeds with carbophenothion caused several instances of geese poisoning. Five hundred greylag geese died after feeding on recently sown fields in 1971. Concerns about pollution of the Mediterranean Sea with organophosphate pesticides resulted in a meeting of the Barcelona Convention in 1991. All of the Mediterranean countries, except Albania, agreed to phase out the use of these substances by the year 2005 (Dinham, 1993).

The most recently developed and now widely available insecticides are the pyrethroids, the first of which was permethrin marketed in 1977. Unlike the organochlorines, the pyrethroids are not persistent in the environment, nor are they susceptible to biological magnification. They are effective at only low dosages so costs are acceptable and their toxicity to mammals is very low. However, as they are broad-spectrum contact insecticides their application will in some cases adversely affect beneficial species. For example, Pynosect (a pyrethrin containing α-resmethrin) is harmful to bees (Ivens, 1993). The pyrethroids are also toxic to fish and aquatic organisms. Overuse, however, is leading to the development of resistance to the insecticides. This is a major problem associated with insecticide use in general; resistance is exacerbated by insecticide overuse and acts as a stimulant for the pesticide industry. According to Cremlyn (1991), the next generation of insecticides currently being developed are the avermectins, a group of natural compounds present in the actinomycete *Streptomycetes avermitilis*. These compounds are related to the milbemycins, a group of natural products derived from another soil micro-organism. The avermectins and milbemycins are, Cremlyn states, amongst the most powerful insecticidal compounds so far discovered. In addition, they complement the pyrethroids in that they are active against nematodes as well as insects, ticks, lice and mites. Other new synthetic pesticides (not based on natural products) being marketed are fipronil and imidacloprid to combat lepidoptera and sucking pests respectively (Moffat, 1993).

Herbicides are an even more important component of the market in crop-protection chemicals than insecticides. Herbicides may be systemic, i.e. effective when they are absorbed into the plant via the root system; others are contact herbicides which kill weeds when they come into contact with the foliage. Examples of systemic herbicides, the so-called hormone herbicides, include 2,4-dichlorophenoxyacetic acid (2,4-D) and 4-chloro-2-methylphenoxyacetic acid (MCPA). According to Cobb (1992), these herbicides mimic the plant hormone, auxin, which controls cell elongation and root development. This mode of action involves growth stimulation in certain plant tissues that eventually causes death. These herbicides were amongst the first to be developed in the post World War II period. 2,4-D was first marketed by the American Chemical Paint Company as 'Weedone' in 1945 and in 1946 MCPA was marketed by ICI as 'Agroxone'. They were the first chemical herbicides that replaced hoeing and weeding by hand. Apart from their efficiency as broad-leaf weedkillers, these herbicides were cheap to produce and are not very toxic to mammals.

C. A. Edwards (1993) states that these systemic herbicides cause few serious direct environmental problems. This is because they are non-persistent though some systemic herbicides, especially those used to control aquatic weeds, are toxic to fish. The lack of persistence also means that resistance has not developed on a large scale. Nevertheless, the widespread use of these chemicals in cereal crops had inadvertent effects due to their impact on broad-leaved species. Not only have populations of such species been reduced or eliminated in cereal fields but they have also been reduced in the hedgerows due to spraying at field margins. Moreover, the use of MCPA in the UK has led to an increase in populations of wild oats. This is a monocotyledon like the cereals and so is not affected by auxin-type herbicides. Consequently wild oats replaced many broad-leaved species as a weed of cereal fields. Potts (1986) also regards the use of such herbicides as an additional reason for the decline in UK partridge populations (see above). This is due to the elimination of certain weeds on which insects and their larvae feed leading to a dearth of these insects for the partridge chicks.

The next category of herbicides to be developed were the bipyridylium herbicides. These were designed to eliminate all weed species so they are true broad-spectrum herbicides. Examples include paraquat and diquat which are also contact herbicides. They can be used to clear fields completely of weeds prior to planting and thus obviate the necessity of seed-bed preparation. Most importantly, these herbicides expedited the development of direct seed drilling, allowing the simultaneous application of herbicide, crop seed and fertiliser. This 'no tillage' practice is considered to be an important aspect of sustainable agriculture (see Section 10.4). In relation to the environmental impact of these herbicides, there appear to be few problems as they are inactivated in the soil. They are, however, highly toxic to mammals, including humans. Cobb (1992) states that their use in some countries is now prohibited. Glyphosate is another broad-spectrum herbicide with similar properties to the bipyridyliums, though it is an organophosphorus compound. Glyphosate is slow-acting but has a longer lasting effect than paraquat; it works by inhibiting the production of essential amino acids in plants. Its adverse environmental impact is considered to be minimal and its value as a broad-spectrum herbicide, including its potential for future use, is witnessed by efforts to genetically engineer crop plants with in-built resistance to glyphosate (Section 10.2). This will allow the continued use of the herbicide to reduce weed populations in fields where crop plants are actually growing. Glyphosate is, however, harmful to fish (Ivens, 1993) and should not, therefore, be used where it may enter watercourses.

Fungicides comprise approximately 20 per cent of the pesticides market. Amongst the problems caused by fungi are the downy and powdery mildews, potato late blight, *Botrytis* grey mould and a variety of rots, smuts and scabs (Briggs *et al.*, 1992). As Chrispeels and Sadava (1994) point out, most fungicides are applied either as powders (dusts) to field crops or as seed dressings. Fungicides are only really effective in a preventative capacity, i.e. they need to be present prior to fungal attack in order to prevent, or at least limit, the initial growth of fungi. The earliest fungicides were based on inorganic substances, e.g. copper and mercury compounds, which were used as seed dressings for cereals and as sprays for fruit crops. However, such substances are toxic to mammals and can contaminate the wider environment. The triazoles were developed in the 1970s and 1980s as alternatives. These are inhibitors of fungal hormone biosynthesis and are generally considered to be environmentally benign; no harmful residues or adverse effects have been reported. Like many of the insecticides discussed above there is, unfortunately, growing fungal resistance to the triazoles. One of the most recent groups of fungicides to be developed is the phenylamides, e.g. Metalaxyl®. These are effective because they inhibit the synthesis of ribonucleic acid (RNA), an essential ingredient of fungal cell nuclei. In addition, a new group of fungicides, the methoxy acrylates, based on strobilurin which is a natural product derived from a specific fungus, are entering the market-place.

Plant growth regulators (PGRs) are not, strictly speaking, crop-protection chemicals although they are a component of the chemical armoury available to farmers. They do not offer protection but a means of altering the growth of a crop to improve its structure or use of light, to improve harvesting. PGRs account for about 6 per cent of the agrochemical market. One example of a PGR is Alar®. This improves firmness and colour in apples but is no longer available in the USA for use on food. According to Hathaway (1993), Uniroyal, the company that produced it, ceased its domestic sales in June 1989. This transpired because of concerns about exposing children to pesticides via food products. Alar® was first registered for use in food in 1968 and by the mid-1970s there was some evidence to link Alar® and its breakdown product with cancer. Even so it was not banned. A subsequent study on the general issue of children's risk of contracting cancer in 1987 indicated that Alar's® breakdown product was the biggest single risk. The fact that this study reached television as well as newspapers resulted in a public outcry. Eventually, and in the wake of hugely declining sales, the EPA decided to reduce the legal limit for Alar® residues in food and to make any residue illegal by 1991.

The Alar® story highlights one of the costs of the use of crop-protection chemicals. One of the objectives of organic farming (Section 7.1) is to eliminate the need for such commodities and thus eliminate any risk associated with the use of chemicals. It is also interesting to note that in the USA, despite the wide use of pesticides, pests destroy 37 per cent of all potential food and fibre crops (Pimentel *et al.*, 1993a). Estimates suggest that if no pesticides were used losses would increase by 10 per cent. Thus, pesticides do improve food supplies globally and increase profits from food production. However, Pimentel *et al.* (1991) have demonstrated that agricultural practices also contribute to the pest problem. Between 1945 and 1989, despite a tenfold increase in insecticide use in the USA, crop losses to insects increased from 7 per cent to 13 per cent. Much of this Pimentel *et al.* (1993a) attribute to the development of monoculture and the abandonment of crop rotations, as has occurred in the corn belt (Section 5.2). This is an important point, especially in relation to agriculture in the developing world. Here, the desired increased productivity could perhaps be more effectively achieved by improving traditional agricultural systems (cf. organic farming, Section 7.1) with integrated pest management strategies rather than complete reliance on crop-protection chemicals.

8.4 THE IMPACT OF AGRICULTURE ON WATER QUALITY

Agricultural practices can have a variety of impacts on drainage and groundwater quality. As discussed in Section 8.2, for example, injudicious cultivation cycles can give rise to accelerated soil erosion, the particles from which may be deposited in rivers, lakes and reservoirs. In extreme cases, e.g. downstream of the Piedmont region of the USA (Section 8.2), deposition may result in the creation of new habitats such as swamplands. In addition, the silting of waterways and reservoirs can be expensive to remedy. However, the two most important issues caused by the impact of agriculture on water resources are cultural eutrophication and salinisation. Both occur extensively in middle to high latitudes though they also occur elsewhere. The salinisation problem in low latitudes, for example, is discussed in Section 9.4.

Just as soil erosion occurs as a natural process so does the eutrophication of lakes, rivers, coastal areas and groundwater. Eutrophication is the nutrient enrichment of these waterbodies and watercourses. Cultural eutrophication is the acceleration of these processes by direct or inadvertent human activity. The most significant nutrients are nitrate and phosphate, both of which may be provided by artificial fertilisers applied to arable land. Such agricultural practices are not the only suppliers of these nutrients in excess. Untreated or partially treated sewage, runoff from urban areas and removal of natural vegetation cover and the subsequent release of soil nutrients can also contribute to cultural eutrophication. Agriculture, however, is considered to be the major cause of what is now a widespread problem though there is no available global survey. As a rough guide, the problem is most acute in the developed world as is illustrated by the Global Environmental Monitoring System (GEMS) Water Programme (reported in Tolba *et al.*, 1992). This records average values for nitrate concentrations in European rivers as 4500 mg l^{-1} as compared with 100 mg l^{-1} outside Europe. Although crude, these figures rather reinforce the view of Heathwaite *et al.* (1993) that in the last two decades the nitrate issue has developed from a local pollution problem to one of regional-scale, if not continent-scale, proportions. This is especially so if marine as well as freshwater pollution by excessive nitrates is considered. Moreover, this environmental problem has resulted primarily from increasing agricultural productivity, especially cereal production, since the 1950s when the expansion of cropland at the expense of natural ecosystems began to diminish. According to Gilland (1993), expansion was replaced with a trend toward intensification which was facilitated by the increased use of artificial fertilisers, particularly nitrates. Even between 1980 and 1990, nitrogen consumption (in artificial fertilisers) increased from 57.25 Mt to 79.08 Mt. This accompanied an increase in cereal production from 1567 Mt to 1767 Mt (figures quoted in Gilland, 1993). Almost all of this available nitrogen is now produced by the Haber-Bosch process, which was commercialised in 1913; it now uses 1.3 per cent of world energy production. This inevitably contributes to atmospheric pollution via the production of greenhouse gases as well as the nitrate pollution of water resources. The nitrate issue is, however, only part of the cultural eutrophication problem as phosphate is also important.

Agricultural practices can contribute to cultural eutrophication in several ways. First, artificial fertilisers provide nitrate and phosphate. Nitrate, because of its high solubility, is easily washed out of the soil profile into drainage water and/or groundwater. This is illustrated in Figure 8.7. Phosphate is not readily soluble but is removed from the soil profile attached to soil particles which may also enter drainage systems (Figure 8.7). This process is, thus, intimately related to soil erosion (Section 8.2). Secondly, farm animals produce excreta rich in nitrates and phosphates. Where animals are housed indoors the excreta is concentrated and may be disposed of into

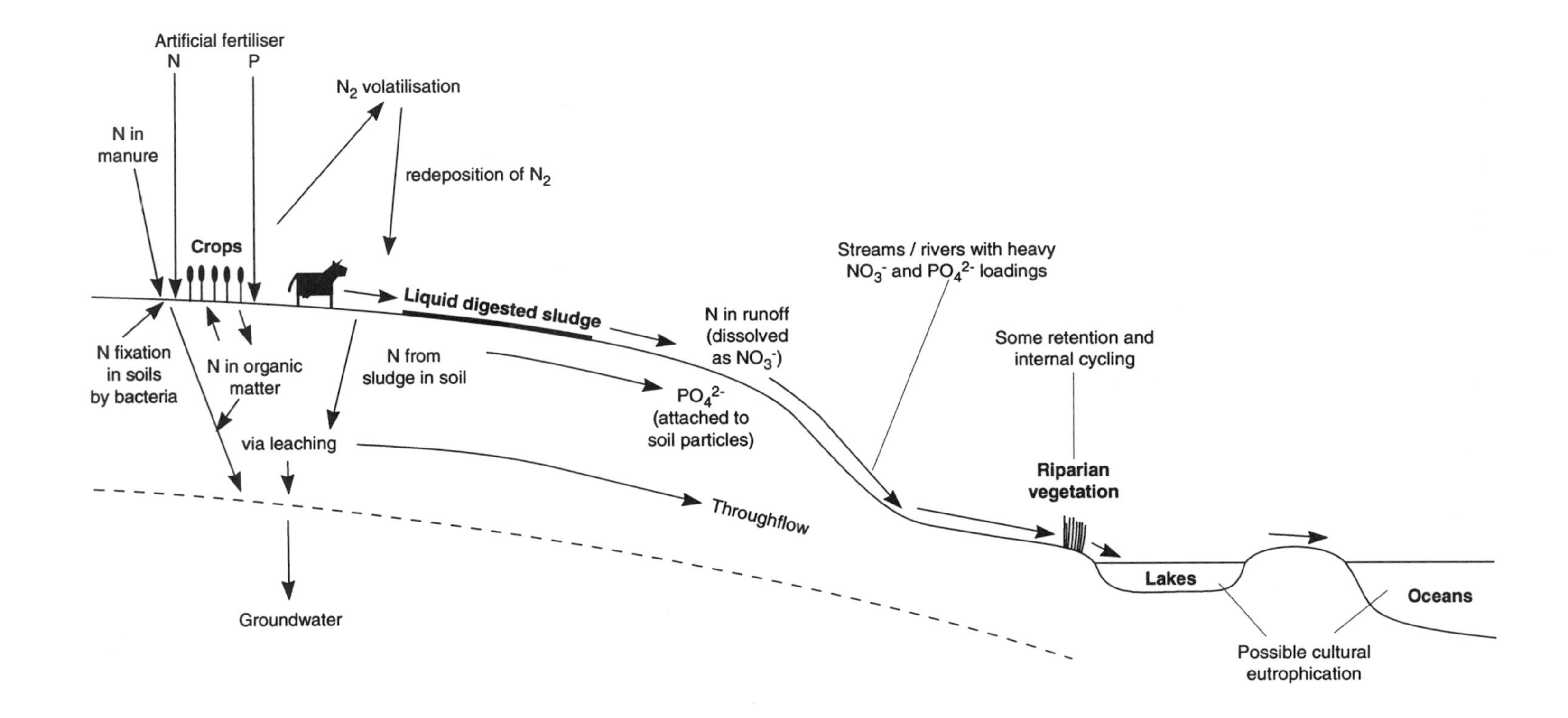

Figure 8.7 The processes involved in the transfer of nitrate and phosphate from agricultural land into drainage systems, groundwater and the oceans

watercourses. Moreover, the addition of organic manures, i.e. slurry or liquid digested sludge, adds to the nitrate and phosphate content of arable/pasture soils and is subject to the same removal processes as mineral fertilisers. Similarly the ploughing in of crop residues adds nitrate to the soil which can be subsequently washed out. The rate and frequency of flushing will depend, as it does for mineral fertilisers, on the time of application in relation to rainfall incidence, temperatures, land use (i.e. crop type, which determines how much and when the nutrient is absorbed) and management practices. In general, most of these factors give rise to non-point sources of pollution, except for the discharge of slurry from animal houses.

As stated above, the nitrate problem is particularly acute in Europe where it has prompted EU legislation to both curb the pollution and safeguard human health. In Denmark, for example, the input of nitrogen to agricultural land has doubled since the early 1960s and produces a total nitrogen discharge to the aquatic environment of 290 000 t (Kronvang *et al.*, 1993). Similarly, the discharge of phosphorus has increased considerably. Although much of this is due to discharge from sewage plants, non-point agricultural sources and discharge from fish farms, especially into the Kattegat and Baltic Sea, accounted for *c.* 8 per cent. The consequence of these excess nutrients in inland waters is poor water quality. In coastal regions, the nitrate enrichment has caused biotic changes in marine ecosystems. For example, eelgrass communities have diminished in extent and the depth of water in which they survive has decreased by 3–4 m due to turbidity. Similarly, the depth distribution of micro-algae has decreased. These problems prompted the Danish Government to establish an Action Plan on the Aquatic Environment in 1987, the objective being to reduce nitrogen loading by 50 per cent and phosphorus loading by 8 per cent. Action of a similar kind is required by countries surrounding the Baltic Sea (Rosenberg *et al.*, 1990) in order to curtail algal growth and prevent the occurrence of anoxic (oxygen poor) waters that reduce secondary productivity, i.e. fish and crustaceans cannot obtain sufficient oxygen and therefore high mortality occurs.

In relation to the cultural eutrophication of freshwaters the characteristics of lakes experiencing cultural eutrophication are given in Table 8.7. This shows that the biological, chemical and physical components are all adversely affected, culminating in declining amenity use and the need to treat water for domestic consumption. Whilst Table 8.7 gives generalised characteristics, Moss (1988) has discussed in detail the specific case of the Norfolk Broads, UK. Here nutrient enrichment has altered both water quality and lake ecosystem components. Instead of clear water with a high degree of light penetration, turbid water predominates with a high population of algae. The rapid growth of algae, as well as the decomposition of dead algae, reduces oxygen concentrations in the water. Consequently, invertebrate and fish communities have been adversely affected causing an overall decline in biodiversity. Submerged plant communities have diminished in extent and the loss of their protection along banks has accelerated erosion. High nitrate concentrations in UK rivers also have adverse effects. National patterns of nitrate concentration in UK rivers have been detailed by Betton *et al.* (1991) and Johnes and Burt (1993). In upland Britain mean concentrations are usually below 2.5 mg NO_3-N l^{-1}; in lowland rivers mean concentrations are usually above 5.0 mg NO_3-N l^{-1}, especially in eastern England but generally in areas of intensive agriculture, including intensively managed pasture. Moreover there is evidence that nitrate concentrations have increased substantially

Table 8.7 The characteristics of lakes experiencing cultural eutrophication (from Mannion, 1992a)

BIOLOGICAL FACTORS:
(a) Primary productivity: usually much higher than in unpolluted water and is manifest as extensive algal blooms
(b) Diversity of primary producers: initially green algae increase, but blue-green algae rapidly become dominant and produce toxins. Similarly, macrophytes (e.g. reed maces) respond well initially but due to increased turbidity and anoxia (see below) they decline in diversity as eutrophication proceeds
(c) Higher trophic level productivity: overall decrease in response to factors given in this table
(d) Higher trophic level diversity: decreases due to factors given in this table. The species of macro- and micro-invertebrates which tolerate more extreme conditions increase in numbers. Fish are also adversely affected and populations are dominated by surface-dwelling coarse fish such as pike and perch

CHEMICAL FACTORS:
(a) Oxygen content of bottom waters (hypolimnion): this is usually low due to algal blooms restricting oxygen exchange between the water and atmosphere. Oxygen-deficient (anoxia) conditions develop, especially at night when algae are not photosynthesising. Thus seasonal and diurnal patterns of oxygen availability occur. The decay of algal blooms also produces anoxia
(b) Salt content of water: this can be very high and a further restriction on floral and faunal diversity

PHYSICAL FACTORS:
(a) Mean depth of water body: as infill occurs the depth decreases
(b) Volume of hypolimnion: varies
(c) Turbidity: this increases, as sediment input increases, and restricts the depth of light penetration which can become a limiting factor for photosynthesis. It is also increased if boating is a significant activity

WATER USES:
(a) Water quality for domestic and industrial uses: this is usually poor
(b) Amenity use: this can be severely impaired due to the production of noxious odours and loss of floral and faunal attractions

since the 1970s. Johnes and Burt's analysis of data for the River Windrush, a tributary of the River Thames, shows that concentrations of 5.12 mg NO_3-N l^{-1} occurred in 1973–1974. By 1989–1990, however, the concentration had risen to 9.37 mg NO_3-N l^{-1}. At the very least, this increase indicates that a proportion of the nitrate fertiliser being applied is wasted in relation to crop productivity. This also means that the fossil-fuel energy expended to produce the fertiliser is contributing towards environmental problems, i.e. global warming and acid rain, other than the ones with which it is usually associated. This is clearly not a wise use of resources, nor is it in the spirit of sustainable agriculture (Section 10.4). Ecologically, increasing concentrations of nitrate (and phosphate) in rivers contribute to the cultural eutrophication of lakes, wetlands and coastal regions. In rivers themselves high nutrient levels can lead to infestations of aquatic plants which clog waterways and hinder their use as transport media. In the lower Mississippi and the Nile, for example, the water hyacinth is now such a nuisance. It not only clogs waterways, but reproduces rapidly to produce a dense mat, rather like an algal bloom (see Table 8.7), which inhibits the diffusion of oxygen to below-surface water which can cause the mortality of fish and other aquatic species.

As Figure 8.7 shows, nitrate can leach through the soil profile below the root zone and enter groundwater in porous sedimentary rocks. The occurrence of high nitrate concentrations in aquifers, notably those of chalk and sandstone, is causing problems for the provision of domestic water in developed countries. This is because both the World Health Organisation (WHO) and the EU have recommended that nitrate concentrations in drinking water should not exceed 50 mg l^{-1}. Concerns about the role of nitrates in causing cancer, for example (although the case for their promotion of stomach cancer is not proven), and the unequivocal role of nitrates in causing methaemoglobinaemia ('blue baby' syndrome), have prompted these directives (Cartwright *et al.*, 1991). In the UK, for example, complying with this directive, particularly in the post-privatisation era of the UK's regional water authorities, may prove problematic. Water issues such as this exemplify the relationship between a vital resource and its control and safety, both of which involve legislation, as well as its economics (see O'Riordan and Bentham, 1993, for a review). According to Dudley (1990), Anglian and Severn Trent water authorities are having to deal with water supplies with nitrate concentrations in excess of WHO and EU recommendations.

Solving this pollution problem is not just a question of legislating about acceptable nitrate concentrations in domestic water supplies. In fact there are several, but costly, ways in which treatment processes can reduce nitrate concentrations, as Hall and Croll (1993) have discussed. Such measures simply treat the symptoms of pollution problems rather than their underpinning causes and just add costs that could be avoided. The control of land-use practices that use artificial fertilisers would be a more appropriate approach than trying to remedy the causes. Burt and Haycock (1992), for example, have discussed the possibilities associated with the management of nitrate fertiliser application. Factors that need to be considered include the timing and amount of fertiliser dressing, and cultivation practices. All can be adjusted to minimise nitrate loss. In addition, Haycock *et al.* (1993) have suggested that riparian buffer zones could be employed to reduce the amount of nitrate entering water-courses. The mechanisms involved include the retention of the nitrate through uptake by the vegetation and denitrification by bacteria which emit gaseous nitrogen to the atmosphere. This too addresses symptoms rather than causes but it could be included as part of a package of catchment-level management strategies. Obviously, the adoption of organic farming practices discussed in Section 7.1, which do not involve artificial nitrate fertiliser use, would reduce nitrate losses substantially. Changes in the EU's Common Agricultural Policy in 1988, and subsequently, have also provided possibilities for reducing the nitrate problem. The set-aside programme began in the UK in 1988, the aim being to take land out of production in order to reduce the production of surplus food. Measures to reduce the level of nitrate in water were introduced in 1990. Two types of areas were designated: Nitrate Sensitive Areas (NSAs) of which there are ten, and Nitrate Advisory Areas (NAAs) of which there are nine. In the NSAs, restrictions on the use of artificial fertilisers, slurry, etc., have been introduced on a voluntary basis and involve financial compensation. In the NAAs, advice but not compensation is available. The effects of these measures have yet to be determined.

Cultural eutrophication is not the only cause of poor water quality. In some regions, especially arid and semi-arid regions, where irrigation is practised the salinisation and/or alkalinisation of soils, soil water and drainage systems can pose problems for the provision of domestic water and irrigation water. The processes involved in salinisation/alkalinisation are discussed in Section 9.4 since the problem is most acute in developing countries. In brief, the use of irrigation systems leads to an increase in the exposure of water to the atmosphere when compared with its confinement in a river channel and/or in an aquifer. Increased evaporation causes an increase in salt concentrations in the irrigation water itself and in the irrigated soils which may also become waterlogged. Some 17 per cent of the world's cultivated land, i.e. 2.35×10^6 ha (Postel, 1993) is irrigated, of which 4.3×10^6 ha are affected by soil deterioration, especially salinisation (Tolba *et al.*, 1992). Irrigation worldwide is responsible for 73 per cent of the global use of freshwater and is responsible for the irretrievable loss of a vast proportion of this resource. In the developed world, the areas most severely affected by salinisation, etc., are parts of Australia, especially the Murray-Darling Basin, the USA and the region around the Aral Sea in the Republics of Turkmenistan, Uzbekistan and Kazakhstan (Figure 8.8). According to Postel (1992), irrigation is now a cornerstone of global food security; it has contributed substantially to supporting the rapid growth of world population (from 1.6×10^9 to 5×10^9) that has occurred since 1900. Although it fulfils a vital role, irrigation is also the cause of much land degradation. Either way it makes a significant environmental impact.

In Australia there are approximately 1.88×10^6 irrigated hectares, much of which is in the Murray-Darling River Basin in south-east Australia. Not only does this river system provide water for irrigation, it also provides domestic water for South Australia. Apart from the fact that more than 80 per cent of the basin's annual runoff is required for irrigation, and is thus subject to some degree of salinisation, the bedrock comprises consolidated marine and floodplain sediments which add to the high salt concentrations in soils through weathering. Pigram (1986) reports that typical salinity concentrations of 40–50 ppm occur in the upper Murray River but in the lower reaches of the basin near Adelaide concentrations of 400–500 ppm are not uncommon. Lawrence and Vanclay (1992) report that salinisation caused by irrigation affects 1200 km^2 though a further extensive area is affected by dryland farming, mainly due to tree removal. Overall, salinisation, through its adverse effects on soil chemistry and soil structure, causes losses in productivity that amount to more than $100 million annually. To combat these problems various management practices have been introduced since the 1970s. For example, the use of tiled drains and piped water supplies have reduced evaporation and the demand for irrigation water. Evaporation basins have been established into which water is pumped and the salt recovered for industrial use.

The region most severely affected by the adverse impact of irrigation in the USA is the arid and semi-arid states of the south and west, including parts of California, where the majority of the USA's 20.162×10^6 ha irrigated area is located (Postel, 1992). For example, water is extracted from the Colorado River and transported via a 714-km aqueduct system into California where it is used to irrigate fruit and olive crops in the Imperial and Coachella valleys. The aqueduct journey takes the water through arid and semi-arid regions where approximately 10 per cent of the water

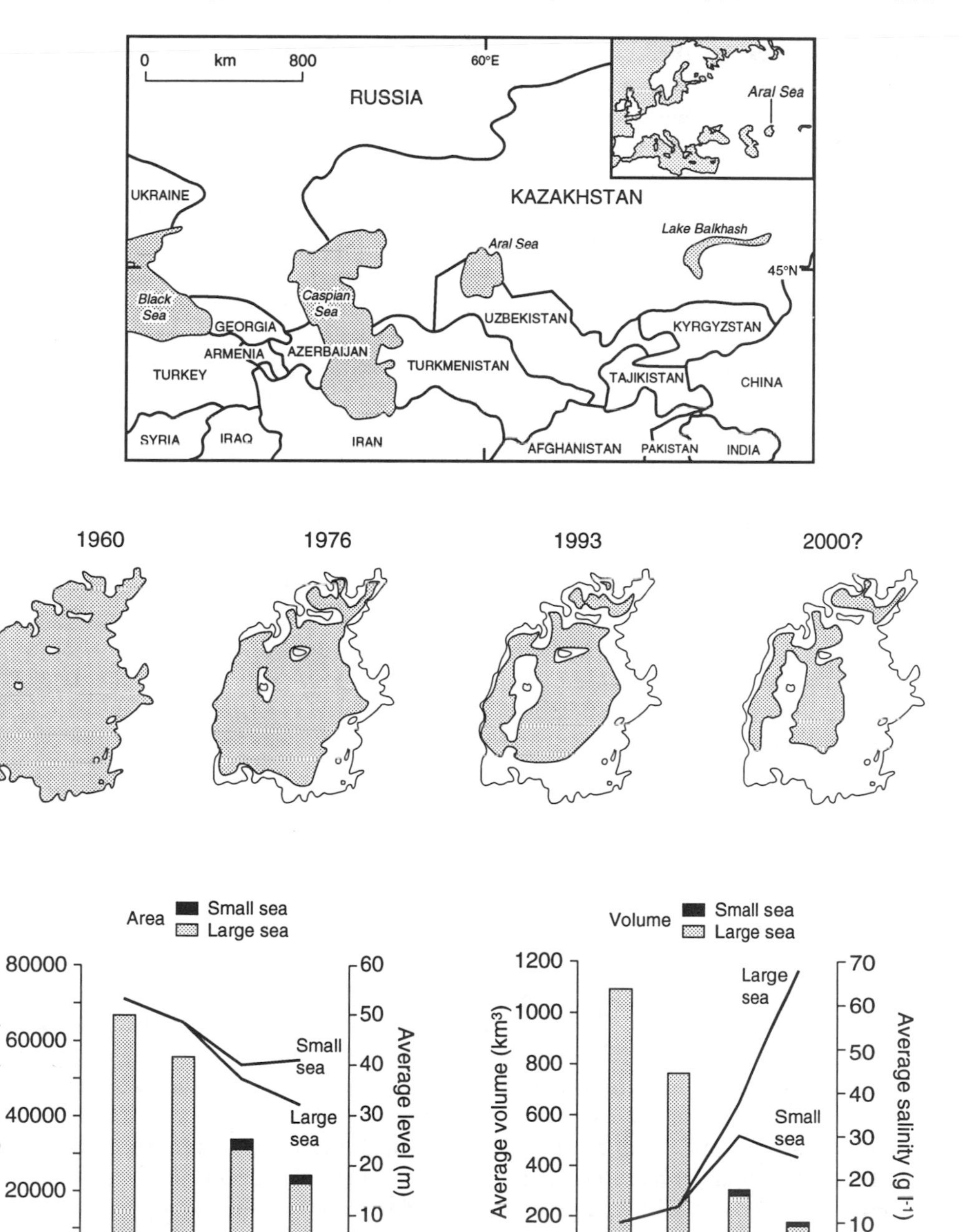

Figure 8.8 Changes in the Aral Sea, 1960–2000 (based on Micklin's update in Glantz *et al.*, 1993)

evaporates. Even at this stage the water is too saline to be used for domestic drinking water; other sources of water have to be used to dilute it. After use in irrigation, the water rejoins the Lower Colorado River whence it enters Mexico. Complaints about high salinity concentrations by Mexico have resulted in the USA funding a $\$350 \times 10^6$ desalinisation project at Yuma in Arizona. There is also a price to pay in terms of declining crop productivity where there is a build up of salt. Postel (1993) reports that crop losses of between 25 and 30 per cent can occur. In extreme cases land might actually be abandoned. Various practices that can be employed to minimise water loss, aquifer depletion, salinisation and waterlogging have been discussed by Suarez (1992) who suggests the possible use of urban waste water and brackish waters as well as the more conventional methods of improved water delivery, e.g. sprinkler and drip irrigation systems.

The mismanagement of irrigation and the environmental price that it exacts is nowhere better exemplified than it is by the Aral Sea (Figure 8.8). Not only does the current plight of the Aral Sea illustrate the problems associated with injudicious agricultural practices but it also reflects environmental mismanagement at regional and national levels within a centrally planned economy. The political background to what is one of the world's most pressing ecological problems is discussed by Glantz *et al.* (1993) whilst Micklin (1988, 1992) has examined the environmental impact. The region has a long history of irrigation use. In 1950, for example, 2.9×10^6 ha of land in Central Asia were irrigated. This increased to 7.6×10^6 ha by the late 1980s, requiring 104 km^3 of water, much of which was extracted from the two major rivers, the Amu Dar'ya and Syr Dar'ya, supplying water to the closed basin of the Aral Sea. This occurred as cotton production spread rapidly in the wake of what Glantz *et al.* (1993) describe as a 'cotton independence policy' that emanated from Moscow in the 1950s. The environmental repercussions of this policy have been enormous and are probably irreversible. The level of the Aral Sea declined by 15 m (from 53.4 to 38 m) causing the area of water to split into two as illustrated in Figure 8.8. The exposure of 27×10^3 km^2 of bottom sediments has resulted in a massive accumulation of salt which inhibits vegetation colonisation. The lack of cohesion in the substrate has also given rise to active sand dunes (an indication of active desertification; see Section 9.3) and serious dust storms. Some 43×10^6 t of salt are removed from the basin annually and are deposited over a *c.* 200×10^3 km^2 area. The deposited calcium sulphate, and sodium sulphate and chloride, contribute to the salinisation of the surrounding area. The salinity in the basin has increased from 10 g l^{-1} in the pre-1960 period to a current 30 g l^{-1}. This has inevitably altered faunal and floral communities in the basin itself; in particular, commercial fishing and hunting have declined sharply. There are problems associated with the provision of drinking water, harbours no longer at the water's edge, and loss of employment as well as health problems such as high infant mortality, a high incidence of congenital deformity and of thyroid cancer. This is attributed to the heavy use of fertilisers and crop-protection chemicals for cotton production. Various schemes to mitigate these problems have been proposed. Until recently, the statement of Glantz *et al.* that 'The Aral Sea will most likely be allowed to continue to decline as a result of inaction (which is also a form of action in favour of "business as usual") by governments' held true. Moreover, some of the mitigation schemes proposed, notably the diversion of rivers flowing north, could in all likelihood create a different set of problems with equally significant environmental repercussions. A recent report by Pearce (1994a), however, suggests that all may not yet be lost as interested republics have just sealed a deal for improving irrigation in the region.

8.5 CONCLUSION

The above issues and examples unequivocally attest to the very significant impact of agriculture in middle and high latitudes. Landscapes have been transformed in both aesthetic and ecological ways. Overall, biodiversity has declined as agriculture has contributed substantially to the demise and fragmentation of natural habitats as well as to increasing rates of extinction. Mosaics of natural habitats, reflecting nature's chaos, have been replaced by ordered arable fields and pastures. This epitomises society's attitude of attempting to tame nature for service to a growing world population. In the developed world, agriculture has been highly successful in providing a more than adequate food supply but not without environmental cost. The ecological factors referred to above are part of this cost. Another component is the accelerating loss of soil resource. Much of this is caused by cropping patterns, the adoption of monoculture rather than polyculture and a decline in soil organic matter. However, the impact of soil erosion on crop productivity has been masked somewhat by the maintenance, or even the improvement, of crop yields due to improved crop varieties and intensive fertiliser use.

In the years following World War II much of the agricultural land of middle and high latitudes has been subject to the application of crop-protection chemicals designed to eliminate competition from so-called pests. These chemicals have undoubtedly increased crop yields by decreasing losses due to insects, weeds, fungi and virus diseases but they have not been without ecological impact. Those due to DDT, for example, are still being felt but the more recent generations of crop-protection chemicals appear, generally, to be environmentally benign. The major drawbacks of such substances are that they represent an input of fossil-fuel energy into agriculture and their use may be comparatively short lived as 'pests' develop resistance. There are also concerns, despite stringent registration requirements, that some crop-protection chemicals leave residues in food that are health hazards. The years since World War II have also witnessed a huge increase in artificial fertiliser use and irrigation. Both have assuredly contributed to the support of the Earth's increasing human population during this time but impaired water and soil quality is the price paid for these innovations. High nitrate and/or high salinity concentrations in rivers and aquifers used for domestic water supplies now require treatment, though the most appropriate action requires changes in agriculture and water management. All of these problems are also occurring in the developing world as is discussed in Chapter 9.

8.6 FURTHER READING

Barrow, C.J. (1991) *Land Degradation: Development and Breakdown of Terrestrial Environments*. Cambridge University Press, Cambridge.

Chadwick, D.J. and Marsh, J. (eds) (1993) *Crop Protection and Sustainable Agriculture*. John Wiley and Sons, Chichester.

Gleick, P.H. (ed.) (1993) *Water in Crisis: A Guide to the World's Fresh Water Resources*. Oxford University Press, New York.

Goudie, A. (1994) *The Human Impact on the Natural Environment*, 4th edn. Blackwell, Oxford.

Mannion, A.M. and Bowlby, S.R. (eds) (1992) *Environmental Issues in the 1990s*. John Wiley and Sons, Chichester.

Pimentel, D. (ed.) (1993) *World Soil Erosion and Conservation*. Cambridge University Press, Cambridge.

CHAPTER 9

The Environmental Impact of Agriculture in Low Latitudes

The low latitudes, the tropics and subtropics, are occupied mainly by developing countries. The problems they face in relation to agriculture are very different to those of developed countries. Inevitably, the nature of the physical environment provides the overall constraints in which cultural factors operate. However, the latter are stimuli for agricultural and land-use changes that contrast markedly with those of the developed world (Chapter 8). For example, rates of population increase ensure that there is an ever-growing demand to increase food production. In the developed world quite the opposite trend is apparent; land is being taken out of production. Efforts to improve general health via improved nutrition in the developing world, coupled with the need to produce crops for export, also enhance demands for an increase in the area of cultivation and an intensification of production from existing agricultural systems.

The environmental impact of agriculture in the developing world bears many similarities to its impact in the developed world. In terms of general issues, the loss and fragmentation of natural habitats, as well as the decline in biotic diversity, are serious problems. Most importantly, the rapid demise of tropical forests is cause for concern not only because of the loss of genetic resources, as biotic communities are impoverished and reduced in extent, but also because of the climatic implications. Forests are efficient assimilators of atmospheric carbon dioxide; decreasing their extent and impairing their efficiencies as carbon sinks means that carbon dioxide remains in the atmosphere where it contributes to the enhanced greenhouse effect. Furthermore, the removal of trees, especially by burning, as is often the case prior to cultivation of the land, releases the store of carbon from the woody biomass, the litter and the soil. This too adds to the carbon dioxide reservoir in the atmosphere. Wetlands, savanna woodlands and coastal mangrove ecosystems are also being transposed into agricultural systems.

Soil erosion is a much more extensive and acute problem in the cultivated areas of the low latitudes than it is in the middle and high latitudes (see Figure 8.2). The semi-arid and arid lands are particularly vulnerable, especially where unsuitable or over-intensive agriculture is practised. This may be overgrazing and/or imprudent arable cultivation, or it may be the result of fallows of too short a duration. Soil erosion may be associated with desertification which involves a number of processes including salinisation and loss of biological productivity. As will be discussed below, desertification is a controversial issue that may or may not be associated with natural and/or cultural climatic change. In common with the middle and high latitudes, agriculture in the low latitudes can also alter water quality. In many arid and semi-arid regions there are major problems with irrigation systems and the salinisation of both water and land.

Curtailing and mitigating the effects of agriculture in low latitudes is difficult and complicated. As discussed in Chapters 5, 6 and 7 there are ways to improve productivity and reduce environmental degradation (see, for example, Section 7.5 in which the value of agroforestry is examined). However, tackling the primary causes of inappropriate land tenure, debts to the developed world that promote cash cropping to earn foreign currency, lack of investment, poverty, problems of technology transfer and high rates of population increase, is quite another matter. Once again, it is a question of addressing the causes and not the symptoms though the latter are usually easier to identify and alleviate in the short term than are the former.

9.1 LANDSCAPE CHANGE: LOSS OF BIODIVERSITY AND NATURAL HABITATS

The destruction of natural ecosystems, particularly the demise of tropical forests, must rank along with potential global warming as one of the most important environmental issues of the 1990s. As discussed above, the two are in any case closely related. Overall, Hannah *et al.* (1994) estimate that 43.5 per cent, 48.9 per cent and 62.5 per cent of Asia, Africa and South America, respectively, remains undisturbed whilst 29.5 per cent, 15.4 per cent and 15.1 per cent, respectively, can now be classified as 'human dominated', i.e. has been changed considerably by human activity. However, the reduction in the extent of low latitude ecosystems is not entirely due to agricultural encroachment. In the forests, for example, logging companies take their toll as the market for tropical hardwoods continues to be buoyant in the developed world. In savanna lands and mangrove ecosystems a major threat is presented by the scarcity of fuelwood. Neither the monitoring of habitat destruction nor determination of the proportion due to agriculture are easy matters. The difficulties of precisely defining the extent and causes of land-use change have been referred to in Section 8.1 (see also Meyer and Turner, 1992; Turner *et al.*, 1994). However, in this particular context, the major problem is defining forest and different types of forest. The destruction of natural habitats reduces global biodiversity and increases rates of extinction. The loss of organisms from the Earth's biota, many of which may never have even been described or named, amounts to a loss of opportunities for the future. The preservation of such opportunities is a fundamental principle of sustainable development (see Section 10.4). They include the use of biota as a food source or the constituents of the biota being developed into useful substances such as pharmaceuticals or crop-protection chemicals. The risk of extinction is also increased by the fragmentation of habitats, as has been discussed in relation to UK habitats in Section 9.1.

There are many different types of tropical forests. According to Vanclay (1993), 40 per cent of the tropics is forested, the remainder being too dry to support forests. Of the forested portion, approximately 50 per cent is rain forest whilst the rest comprises seasonal forests, savanna woodland and other types of open forests. Tropical moist forests include rain forest, seasonal or monsoon forest and mangroves. These are the forests about which there is grave concern. They are most extensive in the Americas, followed by Asia and then Africa, as shown in Figure 9.1. Apart from their role in influencing atmospheric composition, tropical moist forests are amongst the most biodiverse ecosystems on the Earth, containing more than 50 per cent of the world's plant and animal species. Maintaining biodiversity may be a key factor in ensuring the efficiency of tropical moist forests as carbon assimilators. Recent experiments using an Ecotron, a series of chambers with controlled conditions, Naeem *et al.* (1994) have demonstrated that of the plant communities they monitored, the most biodiverse was the most productive. In addition, forest clearance by burning diminishes the carbon pool in the biosphere. The carbon sequestered in trees and forest soils is released into the atmosphere where it contributes to the enhanced greenhouse effect (see Section 10.4). For example, Dixon *et al.* (1994) report that in 1990 deforestation in low latitudes alone emitted 1.6±0.4 petagrams of carbon.

Just as estimates of forest cover vary, so too do estimates of deforestation and the causes thereof. These issues have been discussed by WCMC (1992) and Whitmore and Sayer (1992). First, there are the problems associated with precisely defining forest and the associated problem of defining various categories of forest as referred to above. Secondly, there are different definitions of what exactly is meant by deforestation. For example, this may mean complete removal of trees so that the land can be used for alternative purposes, or it may refer to disturbance involving only partial removal of the tree cover. The latter is particularly imprecise because there are so many levels of disturbance possible. Moreover, examining temporal trends over more than 20 years or so is notoriously difficult. As Whitmore and Sayer (1992) point out, it has only been the advent of remote sensing that has allowed the determination of global forest cover at a given moment in time. Even then there are discrepancies due to the varying interpretation of satellite imagery. However, the Food and Agriculture Organisation in Rome undertook a survey in 1980 which has since been revised. This survey is the basis of several commentaries (e.g. Lanly *et al.* (1991)) and is particularly useful because it provides country-level statistics, all of which have been collected using identical criteria.

Table 9.1 has been compiled from FAO/UNEP (1981) and FAO (1988) sources. This shows that for closed forests, i.e. tropical moist forests, the rate and extent of deforestation are highest in tropical America. For all regions, rates of deforestation increased between the two recording periods. Later FAO reports in 1990 and 1991 indicate that rates of deforestation increased to 0.8, 0.9 and 1.4 per cent in Africa, Latin America and Asia, respectively. By 1990 deforestation rates were, on average, 0.9 per cent as compared with 0.6 per cent in 1980 for tropical regions. Table 9.2, compiled from a range of data sources quoted in WCMC (1992) and Sayer and Whitmore (1991), gives data on rates of losses of tropical moist forests in selected countries. This reflects not only the considerable spatial variability but also the discrepancies that exist between various estimates. The most important conclusion that such data indicate, even allowing for a 20 per cent error in calculating the extent and rate of deforestation, is that low latitude nations are depleting their forests at an alarming and accelerating rate, with all the implications that this brings for global climatic change and loss of biodiversity.

In relation to biodiversity, Wilson (1992) estimates that in tropical rain forests alone deforestation causes some 27 000 species to become extinct annually. Moreover, of the 18 so-called 'hotspot' areas identified by Myers (1988, 1990) as containing

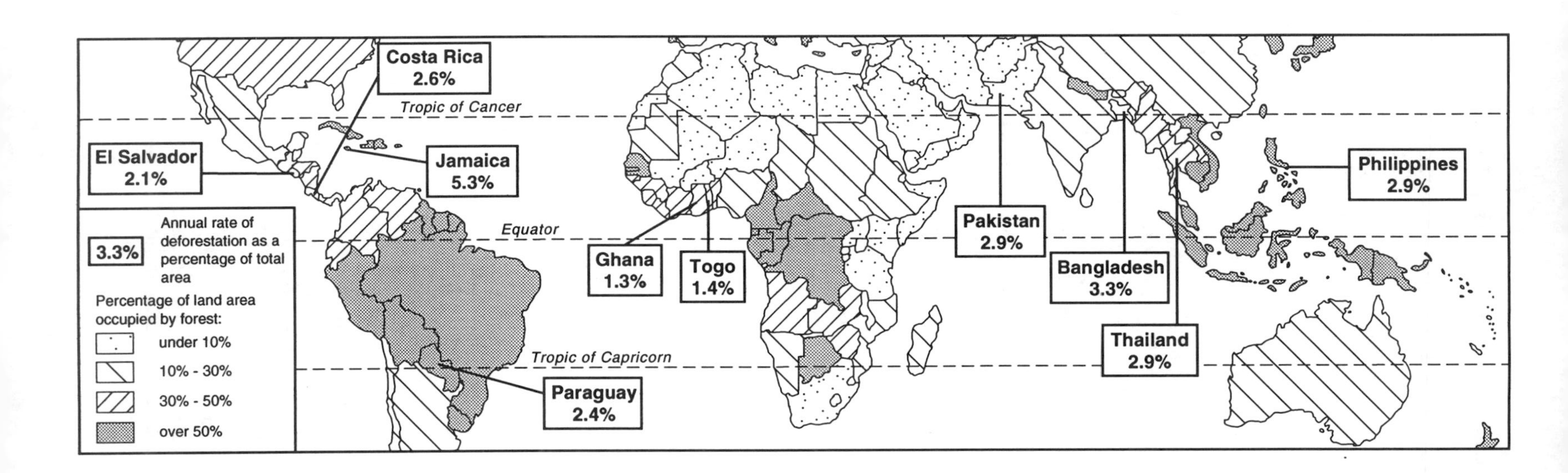

Figure 9.1 Percentage of land area in various countries occupied by forest and woodland and rates of deforestation between 1981 and 1990 as a percentage of closed forest area in selected countries (based on sources quoted in the text)

Table 9.1 Estimates of tropical forest extent and rates of deforestation (based on data in FAO/UNEP, 1981; FAO, 1988)

	Forest extent (km^2)	Area deforested (km^2)		Annual rate of deforestation (%)	
		1976–1980	1981–1985	1976–1980	1981–1985
Closed forest*					
Tropical America	6.79×10^6	41 190	43 390	0.60	0.63
Tropical Africa	2.17×10^6	13 330	13 310	0.61	0.61
Tropical Asia	3.06×10^6	18 150	18 260	0.59	0.60
World	12.01×10^6	72 670	74 960	0.60	0.62
Open Forest†					
Tropical America	2.17×10^6		12 720		0.59
Tropical Africa	4.86×10^6		23 450		0.48
Tropical Asia	0.31×10^6		1 900		0.61
World	7.34×10^6		38 070		0.52

* Closed forests: closed broad-leaf forests (tropical moist forests).
† Open forests: mixed forest–grassland formations. Tree cover exceeds 10% of the ground, e.g. savanna woodland.

Table 9.2 Annual rates of loss of tropical moist forests[1] in selected countries

	Annual rates of loss				
	FAO (1988) Extent (km^2)	%	Sayer & Whitmore (1991) 10^3 km/year	Other (quoted in WCMC, 1992) Extent (km^2)	%
Tropical America					
Brazil	14.8×10^3	0.4	35.0×10^3	$35–80\times10^3$	1.0–2.2
Costa Rica	0.65×10^3	4.0	1.24×10^3	1.24×10^3	7.6
Columbia	8.20×10^3	0.4	6.0×10^3	6.0×10^3	0.3
Ecuador	3.40×10^3	2.4	–		
Mexico	4.70×10^3	1.8	–		
Asia					
India	1.47×10^3	0.3	–	15.0×10^3	4.1
Thailand	3.79×10^3	3.0	–	3.97×10^3	3.1
Malaysia	2.55×10^3	1.2	3.1×10^3	3.10×10^3	1.5
Philippines	0.92×10^3	1.0	1.3×10^3	1.43×10^3	1.5
Indonesia	6.00×10^3	0.5	10.0×10^3	10.00×10^3	0.8
Africa					
Cameroon	0.80×10^3	0.4	1.0×10^3	1.00×10^3	0.6
Congo	0.22×10^3	0.1	–		
Nigeria	3.00×10^3	5.0	–		
Ivory Coast	2.90×10^3	6.5	–		
Madagascar	1.50×10^3	1.5	$1.5–3.0\times10^3$		

[1] includes open woodland

higher concentrations of species than any other ecosystems, and which are experiencing rapid rates of attenuation, 12 are in tropical regions. In view of the fact that human population is likely to double, from *c.* 5 billion people to *c.* 10 billion people, in the next 50 years (Ehrlich and Wilson, 1991) and much of this is going to occur in tropical regions, the prognosis for the survival of tropical forests is not very optimistic. As has been discussed elsewhere in this book, it is not only the loss of biodiversity that is disquieting but also the repercussions for the atmosphere (e.g. Keller *et al.*, 1991; Penner, 1994), the hydrosphere (e.g. Rogers, 1994) and the pedosphere (e.g. Pimentel, 1993). However, the problem of establishing rates of extinction accurately is not unlike the problems relating to estimating rates of deforestation which are referred to above. Existing estimates for rates of extinction vary from *c.* 2 per cent of the world's organisms per year to a massive 11 per cent (data are given in Reid, 1992); the majority of the species would be arthropods that have not yet been described and named (WCMC, 1992). Apart from the loss of the goods and services that deforestation and the associated loss of biodiversity are likely to cause there could be further repercussions. For example, Vanclay (1993) suggests that by the year 2000 about 50 per cent of people in the developing world will not have enough fuelwood for their requirements. In addition, a large source of wood and wood products destined for the developed world will have disappeared. Vanclay believes that only 10 out of the present 33 timber-exporting nations will continue to export timber. There will thus be economic repercussions for those nations who lose valuable export earnings.

There is no doubt about the main causes of deforestation: the spread of agriculture, logging and mineral extraction. The three are not necessarily independent since the opening up of inaccessible forest regions by logging and mining companies provides roads, albeit crude, for shifting cultivators to follow in their wake. Moreover, there are further threats to tropical forest survival from dam construction, especially in the Amazon region, and from fragmentation (see also Section 8.1). McCloskey (1993), for example, reports that only 33 per cent of primary rain forests in the tropics are to be found in large wilderness blocks whilst the other 66 per cent are fragmented by roads and river access making them especially vulnerable to destruction. An example of the threat of fragmentation to wildlife is that of the panda which was formerly more widespread in China than it is now. Currently, it is limited to six separate mountain areas each of which comprises a dissected landscape. The populations are thus isolated and sometimes as small as 20 adults (O'Brien and Knight, 1987). Despite efforts by the World Wide Fund for Nature (WWF) and the Chinese Government to increase populations via breeding in captivity, and legislation to prevent poaching, the panda may not survive into the twenty-first century. Habitat fragmentation is also a major threat to the survival of many species of birds and invertebrates. Fragmentation is acute in Africa where only 20 per cent of tropical rain forest occurs as large wilderness tracts; there are 15 blocks, comprising 33×10^6 ha. The situation is bleakest in South-East Asia where only 12 per cent of the remaining tropical rain forests occur in large wilderness tracts. Even in South and Central America, the region of the world's least fragmented rain forests (41 per cent in 257×10^6 ha), development schemes are increasing the forests' vulnerability to destructive forces. In consequence, the annual rates of deforestation quoted in Tables 9.1 and 9.2 may well increase substantially in the future.

In relation to agriculture, the two most important stimuli to deforestation in low latitudes are the spread of cattle ranching and shifting cultivation. According to Myers (1993b), the latter, largely through the effects of what he refers to as 'shifted cultivators', accounts for more than 50 per cent of tropical moist forest destruction. The reasons why this occurs are complex and difficult to resolve when compared with those associated with logging, mining and cattle ranching (see also the discussion by Dove, 1993a, b, who suggests that if the value of goods, and possibly services, derived from tropical forests provided an adequate income the problem of shifting cultivators would cease). The factors underpinning this form of deforestation are just some of the numerous driving forces that operate to bring about global land-use/land cover change (see discussion in Turner *et al.*, 1994). Shifted cultivators, as discussed in Sections 4.2.1 and 4.2.2, are not traditional shifting cultivators; rather they are 'displaced peasants', often from the cities, who have little alternative but to seek a subsistent existence in new lands opened up by mining and logging companies. In addition, families with a tradition of shifting cultivation are growing rapidly and in some parts of the world (Sections 4.2.1 to 4.2.3) this population pressure is reducing the length of the fallow period to such a degree that it no longer allows the soil to replenish its nutrient store. In many tropical countries there are government resettlement or transmigration programmes which have been devised to open up new lands for development (Whitten, 1991). Examples of such policies include the Repelita programme of Indonesia which involves the migration of people from the densely populated islands of Java, Lombok, Mandura and Bali to the sparsely populated outer islands of the archipelago. The total number of resettled people is already in excess of 5 million; the aim is to eventually resettle 65 million people. Clearly, this will have a major impact on Indonesia's forests which are currently experiencing deforestation at the rate of 0.5 per cent per year (Table 9.2). Other such projects have been sanctioned by Peruvian, Ecuadorian and Brazilian governments to accelerate the development of Amazonia. In Rondonia, for example, the population increased from *c.* 10 000 in 1965 to more than 1 000 000 in 1985. This has resulted in the destruction of 20 per cent of Rondonia's rain forests (Browder, 1988; Johns, 1988). A further stimulus, but one that is related to rapidly growing populations and resettlement policies, is land tenure. Land-owning elites and multinational corporations are powerful politically, often opposing rural and agrarian reform. In many tropical countries, but especially in Latin America, most of the best agricultural land lies in the hands of the few. Most of these landowners are engaged in the money-making enterprises of plantation agriculture and ranching. Sometimes referred to as agro-export agriculture or industrial agriculture, this type of food production has been favoured by governments wishing to promote private enterprise and boost export earnings. As Utting (1993) states, in relation to Central America, 'The profit opportunities associated with the cotton, beef and sugar booms of the 1950s, 1960s and 1970s, further intensified processes of land concentration and landlessness. By 1970, approximately half of all rural families were either landless or farmed sub-subsistence plots of less than a hectare.' This reflects the marginalisation of a substantial proportion of the population subsequently forced into the ghettos of the cities or into the forests to take their chance as shifted cultivators. Similar problems exist in the Ecuadorian Amazon (Rudel, 1993) and the Ecuadorian Andes (Stadel, 1986). The latter illustrates that forested lands of low latitude nations are not the only landscapes to come under threat by shifting and shifted cultivators.

The question of land tenure is thus linked to economic factors. These also include subsidies and tax incentives provided by governments to encourage foreign investments in developing nations. Such incentives, in combination with ready markets in North America and Europe, have led to a massive expansion of cattle ranching. Once again, this is concentrated in Latin America where it matches, and in some countries exceeds, the amount of deforestation due to shifting and shifted cultivators. Its role in changing the landscapes of Central America has been discussed by Utting (1993). In Honduras, Nicaragua and Costa Rica, for example, rates of deforestation increased dramatically in the 1960s. By the mid-1970s the area under pressure increased from 3.9×10^6 to 9.4×10^6 ha, comprising *c.* 66 per cent of all agricultural land. Much of this expansion of cattle ranching was a response to growing markets for beef in North America. It caused Nations and Komer (1987) to describe the relationship as the 'hamburger connection'. Since the early 1980s markets for this beef have also increased in Europe. Much of the land converted to pasture was purchased from shifting cultivators or peasant producers. This resulted not only in the establishment of large ranches but also pushed these shifting cultivators to move further beyond the agricultural frontier into the forests or into squatter settlements in cities. Again, this relationship highlights the impact, through several agencies, of a distant market on indigenous cultivators and Central American forests.

Similar developments have taken place in Brazil as has been discussed by Browder (1988) who suggests that in 1980 some 72 per cent of the clearance was for the creation of pasture alone. This reflects several factors not least of which was the government's development policies for specific Amazon regions associated with road construction providing hitherto unavailable access. A case in point is that of the Belém–Brasilia highway, which opened up the eastern part of the Amazon in the mid-1960s. Both shifting cultivation and cattle ranchers availed themselves of new-found opportunities. The latter, for example, assisted by tax incentives offered by Brazil's government to encourage local and foreign investment, established 470 cattle ranches with an average size of 23 000 ha between 1965 and 1983. Unfortunately, many of these enterprises have not succeeded. The high rate of subsidy and tax incentives have not helped to create the internal wealth that the government had anticipated. Indeed, what profits have been made have largely been accrued by multinational companies who answer to shareholders in the developed world. The ecological damage that such enterprises have created is enormous. The large size of ranches, even when abandoned as some have been, means that recolonisation by forest is very slow or impossible. The failure and non-profitability of these ranches coupled with their low carrying capacity of only 0.5 animals per hectare has not, however, stopped the development of new ranches or the expansion of existing ones. In the case of the latter, the subsidies, etc., make it more economic to clear new land when pastures become unproductive through weed infestation or nutrient depletion rather than employ a weeding and fertilising programme. Even in the period 1980–1990 cattle ranching and shifting cultivators caused 4×10^6 ha of deforestation in Amazonia (Salati *et al.*, 1990). This is despite a relative decline in the rate of deforestation since 1987 (Fearnside, 1993) which, as reflected by the data given in Table 9.3, declined from 22 000 km^2 yr^{-1} in the 1978–1988 period to half that amount in the 1990–1991 period. Fearnside

Table 9.3 The extent and rate of deforestation in Brazil's Legal Amazon (based on Fearnside, 1993)

	Deforested area (10^3 km^2)					Deforestation rate (10^3 km^2 yr^{-1})			
	1978	1988	1989	1990	1991	1978–1988	1988–1989	1989–1990	1990–1991
Due to all causes except HEP dams	152.1	372.8	396.6	410.4	421.6	21.6	18.1	13.8	11.1
Due to HEP dams	0.1	3.9	4.8	4.8	4.8	0.4	1.0	0.0	0.0
Total deforestation	152.2	376.7	401.4	415.2	426.4	22.0	19.0	13.8	11.1

suggests that the reasons for this are due to Brazil's economic recession in the late 1980s and early 1990s (along with the rest of the world) rather than to any changes in government policy. Moreover, Fearnside's assessment of the land-tenure characteristics of this deforestation indicates that *c.* 30 per cent is due to small farmers with holdings of less than 100 ha whilst 70 per cent is caused by medium- to large-size ranches. Thus, the problem of poverty as a stimulus, via shifted and shifting cultivators, is not quite as important as previously thought. The problem of deforestation in Brazil, if Fearnside's data are an accurate reflection of reality, is more to do with subsidies, development policies and speculation than with subsistent farmers though they remain significant.

Beyond Latin America the clearance of tropical forests on a large scale for the production of cash crops, e.g. sugar cane, tea, coffee, cacao and palm oil, is continuing. In peninsular Malaysia, for example, Kishokumar *et al.* (1991) report that in the late 1970s high rates of deforestation of 2500 km^2 yr^{-1} were experienced, as compared with an average of 956 km^2 yr^{-1} for the 1985–1990 period. They also point out that most of this deforestation has been and continues to be a consequence of cash cropping, in contrast with Latin America. Beginning on Malaysia's west coast, private enterprises established plantations for rubber, coconut and oil palm production. Later, government-sponsored development schemes encouraged the spread of plantation agriculture, much of this being achieved under the auspices of the Federal Land Development Authority (FELDA). For example, FELDA had developed 8355 km^2 of plantation, much of it in Pahang and Johor, by the end of the Fifth Malaysia Plan in 1990. As well as forest clearance, such schemes are associated with resettlement programmes aimed at providing employment for landless people whose labour is necessary for operating the plantations. The changes which have occurred in the Malaysian peninsula (excluding Singapore) and the neighbouring states on the island of Borneo have been discussed in detail by Brookfield *et al.* (1990). They state that since the onset of major forest clearance for agriculture in the early 1950s about half of the 73 per cent of the area considered to be forest covered will have been cleared by the mid-1990s. Beginning with the establishment of the first rubber plantations in Malaysia around the turn of the century, FELDA and other developments have since, as referred to above, led to many changes in land use, as shown in Figure 9.2. Similar changes have occurred in Sumatra (also illustrated in Figure 9.2). In Borneo, wetland reclamation has been undertaken to increase the amount of agricultural land (see below) but this is mainly by family units engaged in rice production. Plantation agriculture has, however, been expanding in parts of Sarawak and Kalimantan.

In Africa the removal of tropical moist forest has been most extensive in southern Africa where only 8.7 per cent of the original cover remains (Sayer, 1992). In Africa as a whole, 36.2 per cent of moist forest remained in 1988 (FAO, 1988), most of it being located in central Africa. For centuries the forests of Africa have been used by shifting cultivators but, as has been discussed in Sections 4.2.1, problems with maintaining adequate fallow periods are causing land degradation. This, coupled with high rates of population increase sometimes in excess of 3 per cent per year, is exerting considerable pressure on Africa's remaining forests, many of which have already been subject to intensive logging. According to Cleaver (1992), one of the prime causes of land degradation in Africa is rapid population growth. Whilst this is a simplistic view it nevertheless reflects a social factor which is related to a plethora of

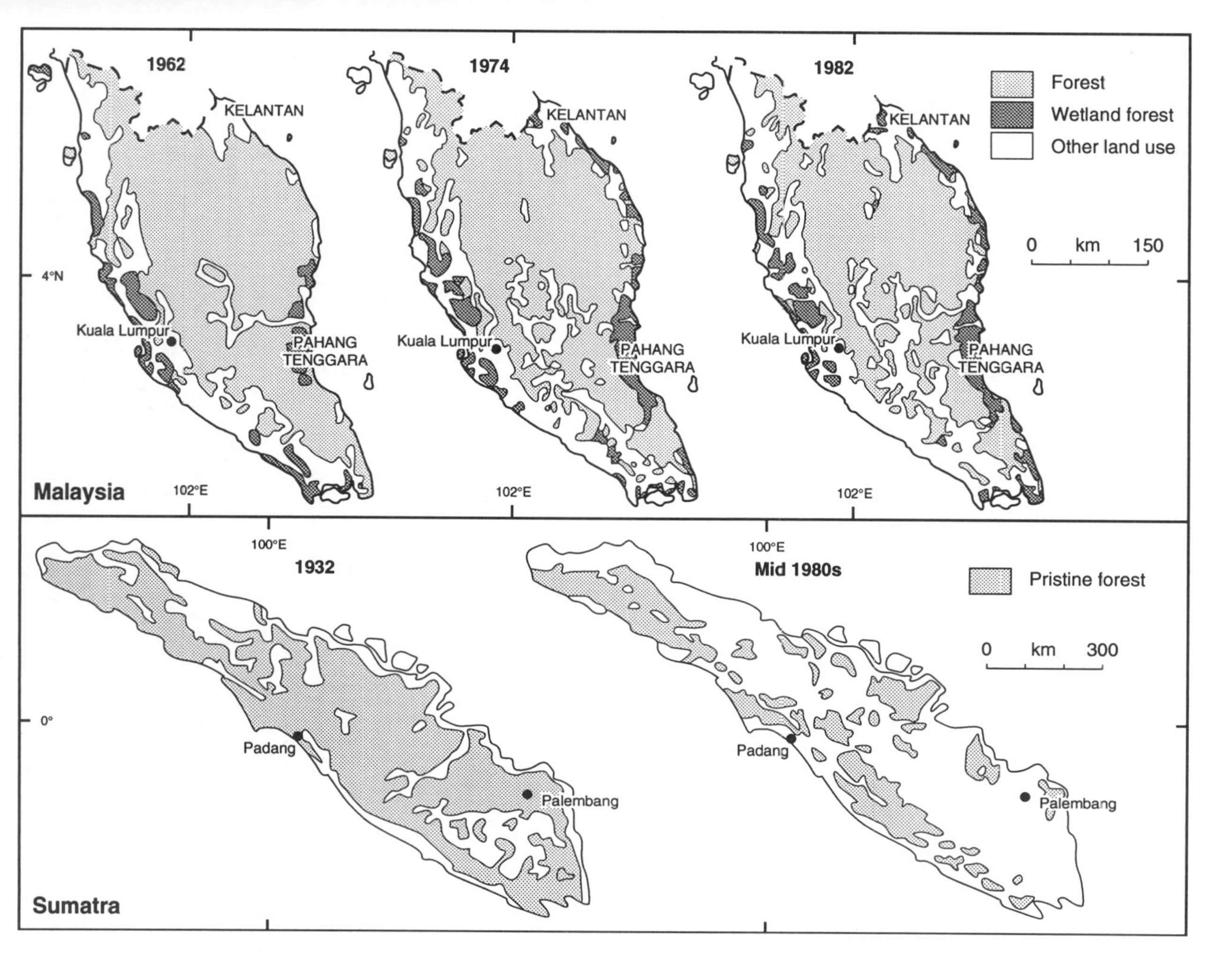

Figure 9.2 Changes in land use in peninsular Malaysia (based on Brookfield *et al.*, 1990) and Sumatra (based on Cox and Collins, 1991)

Table 9.4 The 14 major fronts of deforestation currently occurring as defined by Myers (1993b) with additional data from sources quoted in WCMC (1992) or in the text

Region or country	Original forest estimate (km^2)*	Remaining forest 1991 (km^2)	Rate of annual deforestation	
			km^2	%
Latin America				
Southern Mexico	–	50×10^3	5 000	10.0
Central America	*c.* 340×10^3	80×10^3	4 100	5.1
Colombian Choco		50×10^3	1 500	3.0
Western Amazonia }	4000×10^3 †	200×10^3	5 900	3.0
Southern & Eastern Amazonia }		612×10^3	10 000	1.6
Africa				
Eastern Nigeria and south-western Cameroon	*c.* 400×10^3	54×10^3	2 900	5.3
Madagascar	275.01×10^3	20×10^3	2 000	10.0
Asia				
Eastern Myanmar	*c.* 200.00×10^3	125×10^3	6 000	4.8
Northern and northeastern Thailand	*c.* 130.00×10^3	51×10^3	4 900	9.6
Vietnam	280×10^3	51×10^3	4 900	9.6
Eastern Malaysia	*c.* 15×10^3	53×10^3	3 500	6.6
Sumatra/East Kalimantan/ South Kalimantan (Indonesia)	*c.* 900×10^3	375×10^3	8 600	2.3
Philippines	295×10^3	45×10^3	3 000	6.7
Total	N/A	$1\,879.5\times10^3$	63 400	3.4

* Extent of original forest for the region as a whole, not just the sectors mentioned.
† Brazil's Legal Amazon Region.

economic factors. This again reinforces the opinion expressed in Chapter 1 that agriculture is a manifestation of a complex range of natural and cultural variables. The problem of deforestation is particularly severe in Madagascar (Harcourt, 1992) where rates of deforestation between 1950 and 1985 were 1110 $km^2\ yr^{-1}$ in the eastern moist forests. Myers (1993b) estimates the rate of deforestation in 1991 to have been 2000 $km^2\ yr^{-1}$ (Table 9.4). It has been estimated that even at the lower rate of deforestation only rain forests on the steepest slopes will survive the next 35 years (Green and Sussman, 1990). The major threat comes from slash-and-burn agriculture to grow dryland rice, maize and cassava. This reflects a rapidly growing population which is increasing at a rate of 3.2 per cent per year. Between 1960 and 1990 it increased from 5.4×10^6 to 72×10^6 and the highest concentrations of 100 inhabitants per km^2 occur on the east coast where rain forest is predominant (Harcourt, 1992). The threat to the survival of rain forest in Madagascar is thus substantial and it is considered by Myers (1993b) to be one of the world's main deforestation fronts in low latitudes. Other deforestation fronts and rates of deforestation are given in Table 9.4 and specific details of some of these regions are given in WCMC (1992).

Forests are not the only natural vegetation communities in low latitudes to come under threat from agriculture. For example, reference was made in Section 7.4 to the expansion of mixed farming, which includes rice production, into Latin American savanna lands. Wetlands are also threatened. In fact agriculture has, throughout historic time, been the biggest threat to the world's wetlands. Most of this is the result of reclamation schemes to increase the area available for crop production but a small proportion is due to the adverse impacts of agricultural activity in the catchment, e.g. cultural eutrophication, sediment deposition and the extraction of water for irrigation. Many of the regions discussed above in relation to the demise of tropical moist forests also have problems with wetland destruction. In the Amazon Basin, for example, development activities including agriculture have adversely affected aquatic and wetland communities. Some areas of Várzea, a riverine forest community, have been cleared for the production of jute, maize and rice as well as livestock (Dugan, 1993). Mangrove ecosystems can also be adversely affected by agriculture. Many such areas have traditionally been used for the cultivation of tidal rice, e.g. the Gambia River. Slash-and-burn agriculture is also practised in the mangrove ecosystems of Cameroon and Nigeria (European Communities Commission, 1992) where vegetables, tubers and fruit are produced on sandy, slightly elevated areas within the mangrove swamps and along their borders. In India there are 3150 m^2 of mangroves, mainly along the east coast and Andaman Islands (Jagtap *et al.*, 1993); this represents 70 per cent of the original cover, the other 30 per cent having been cleared for agriculture or urbanisation. The mangrove communities of the Indus Delta are also under threat mainly due to the withdrawal of water upstream for irrigation. The reduction in the flow of freshwater in the delta has increased salinities which are not conducive to mangrove growth and successful replacement (Dugan, 1993). Similar problems are occurring in the Sundarbans, the single most extensive mangrove forest in the world which is located where the Ganges, Brahmaputra and Megha rivers enter the Bay of Bengal (Broekhaven, 1991).

9.2 SOIL EROSION

Figure 8.2 illustrates the areas of the world where soil erosion is most severe and data on wind and water soil erosion are given in Table 9.5. As the table shows, at least 50 per cent of all soil erosion is due to agriculture, with overgrazing and cultivation being the major causes. In addition, soil erosion is one of the processes involved in desertification which is discussed in Section 9.3. Soil erosion is mainly brought about by water and wind, the processes associated with which are given in Table 9.6. Additional important factors which influence rates of soil erosion include the erodibility of soils, i.e. their degree of susceptibility to erosion. For example, sandy soils and soils with little organic matter to provide cohesion are more vulnerable to soil erosion than are soils with a high degree of cohesion. The erosivity of rainfall and/or wind is also significant and relates to rainfall and wind intensity and frequency. Such factors influence natural rates of soil erosion whilst land use may increase erodibility by affecting soil structure and it may effectively increase the erosivity of wind and/or rain by failing to provide an adequate protective cover for the soil. The land-use factor is just as significant, and possibly an even more significant stimulus to soil erosion, as the physical characteristics of the soil (see discussion in Stocking, 1994). As discussed in Section 8.2, there are several ways of

Table 9.5 The extent of water and wind erosion of soils on each continent, 1945 to mid-1980s (based on World Resources Institute, 1992). Note that these data do not include physical and chemical degradation and are approximate

	Water erosion		Wind erosion			
	Total (10^6 ha)	% of all degraded land	Total (10^6 ha)	% of all degraded land	Main cause(s)	% due to main cause
Africa	227.4	46	186.5	38	Overgrazing and cultivation	73
North and Central America	106.1	67	39.2	25	Overgrazing and cultivation	96–60
South America	123.2	51	41.9	17	Overgrazing and cultivation	54
Asia	440.6	59	222.2	30	Overgrazing and cultivation	53
Europe	114.5	52	42.2	19	Overgrazing, cultivation and industry	52
Oceania	82.8	81	16.4	16	Overgrazing	88
World	1093.7	56	548.3	28		63

Table 9.6 The processes involved in soil erosion

Process	Description
A WATER EROSION	
Surface processes	
1. Rainsplash	This may result in the formation of a crust that reduces infiltration capacity and so promotes increased surface runoff which is itself erosive. Soil particles may move downslope as they are displaced.
2. Overland flow	When soil is saturated further rainfall causes sheet flow (sheetwash) or braided channel flow, both of which carry particles downslope.
3. Rill erosion	Flowing water concentrates in small channels i.e. rills through which soil particles are carried. Rills may be ephemeral.
4. Gully erosion	Gullies are enlarged rills which become permanent features of the landscape. Their size increases through headwall and bank corrosion.
5. Stream bank collapse and removal	Soils are eroded as stream banks collapse and are added to the sediment load as it is carried downstream.
6. Mass movements	These comprise slides, rockfalls and mudflows which cause material to be moved downslope.
Subsurface processes	
1. Piping	Channels form underground and carry soil material away. Channel roofs may collapse to form gullies.
2. Tunnel erosion	This is common in loess soils in which tunnels are washed out just above a less permeable layer or bedrock. Tunnel roofs may collapse to form gullies.
B WIND EROSION	
1. Suspension	Fine particles are transported high into the air and may be carried long distances.
2. Surface creep	Coarse particles of soil, etc., are rolled along the ground.
3. Saltation	Particles of soil are moved along the ground surface in a series of jumps.

measuring soil erosion, none of which are ideal. The direct measurement, i.e. recording the amount of soil removed per unit land area, relies on measurements from plots, and extrapolation of the data to a wider area must be undertaken with caution. Similar limitations apply to the use of caesium-137 data, whilst the monitoring of stream sediment loads cannot always take into account the redeposition of eroded material within the catchment. It is thus difficult to identify the temporal and spatial contexts of soil erosion. Nevertheless, there is no doubt that soil erosion is one of the most important environmental issues of the 1990s.

The most serious consequence of enhanced soil erosion on agricultural land is loss of productivity, though in regions where artificial fertilisers are used increasingly intensively the detrimental impact of soil erosion on productivity may be masked (Section 8.2). In many parts of the developing world, however, this is not the case. Productivity is reduced because the water retention capacity of the soil is impaired and because nutrients and organic matter become depleted (see discussion in Pimentel *et al.*, 1993b). The latter may also occur if there is excess leaching or if there is little replacement of nutrients lost from the system in the harvest. The decline in productivity varies enormously within and between soil types but it can be as much as

100 per cent. Moreover, in some circumstances it is impossible to reverse the process and restore productivity. According to Döös (1994), between 5 and 10×10^6 ha yr^{-1} of croplands are lost as a result of soil erosion. Even taking the lower figure, which is an FAO estimate, and assuming that 50 per cent of the land is producing grain, the annual loss of grain to the world's food supply is conservatively in the order of 5×10^6 tons. Much of this is in areas that are marginal for food production but which have been brought into cultivation through expediency and the need to feed a growing population. Consequently, areas that can least afford it are losing a valuable resource that is essential for food production. Whilst the world population continues to grow, the issue of soil erosion is crucial and must be included in strategies for achieving sustainable agriculture (see Section 10.4).

In Asia some of the most severe soil erosion problems occur in China, where agriculture on croplands is particularly intensive (see Sections 5.1, 5.3 and 7.2) in order to support a population of *c.* 1.12×10^9, and in mountainous regions, e.g. the Himalayas. Smil (1993a) reports that between 1957 and 1980 China's farmland loss due to soil erosion was, on average, 1×10^6 ha per year and that a further 5×10^6 ha will be lost in the decade of the 1990s. This represents a halving in the rate of erosion as a result of conservation measures but it is still unacceptable in the face of such a large population. As Dazhong (1993) notes, the average land area per person in China is only 33 per cent of that in the world as a whole. Apart from desertified land and land vulnerable to desertification (Section 9.3) which accounts for 3.5 per cent of China's land, and 28 per cent comprising unusable land (i.e. deserts, tundra, etc.), a further 15.6 per cent is water-eroded land. Overall, approximately 42×10^6 ha of land under cultivation are experiencing serious wind and water erosion; this is more than one-third of the total 130×10^6 ha which are cultivated.

The major soil erosion regions in China and their annual rates of soil loss are given in Figure 9.3. As this shows, the highest rates of soil loss occur in the loess plateau and the southern region. The former region is the world's largest loess plateau, occupying 53×10^6 ha at 1000–1400 m above sea level (Shengxiu and Ling, 1992). It is drained by the Yellow River and its tributaries, the former being so named because of the silt that it carries. This is derived from the yellow loess which is 50–100 m deep, having been blown in from the deserts of Mongolia during the cold episodes of the Quaternary period. The yellow soil is a loose, loamy deposit whose natural vegetation is steppe grassland. Most of this, however, has been removed in the conversion to cropland. Currently 13.3×10^6, or 25 per cent, is cultivated, 15 per cent of which is sloping land. The average erosion rate is 60 t ha^{-1} though only about 33 per cent of the loess plateau's soil erosion occurs on cultivated land. At worst, the annual rate of loss can be up to 100 t ha^{-1} (Figure 9.3). According to Zhang *et al.* (1994), on cultivated land most of this erosion is due to water; it occurs because intense and concentrated rainfall can, at times, exceed the rate at which water percolates into the soil. The resulting runoff causes sheet, splash and rill/gully erosion of the poorly cohesive soils (Table 9.6). According to Shengxiu and Ling (1992), the Yellow River now transports, on average an annual sediment load of 1.6×10^9 t which it collects from 35 tributaries. This represents a 25 per cent increase over the last 35 years (Smil, 1987) and is 80 times that of the Yangtze River (Dazhong, 1993). Some 400×10^6 t are deposited in the river bed in the lower reaches whilst the remainder enters the sea. The deposition on the river bed results in a rise of 0.1 m yr^{-1}, requiring that the river

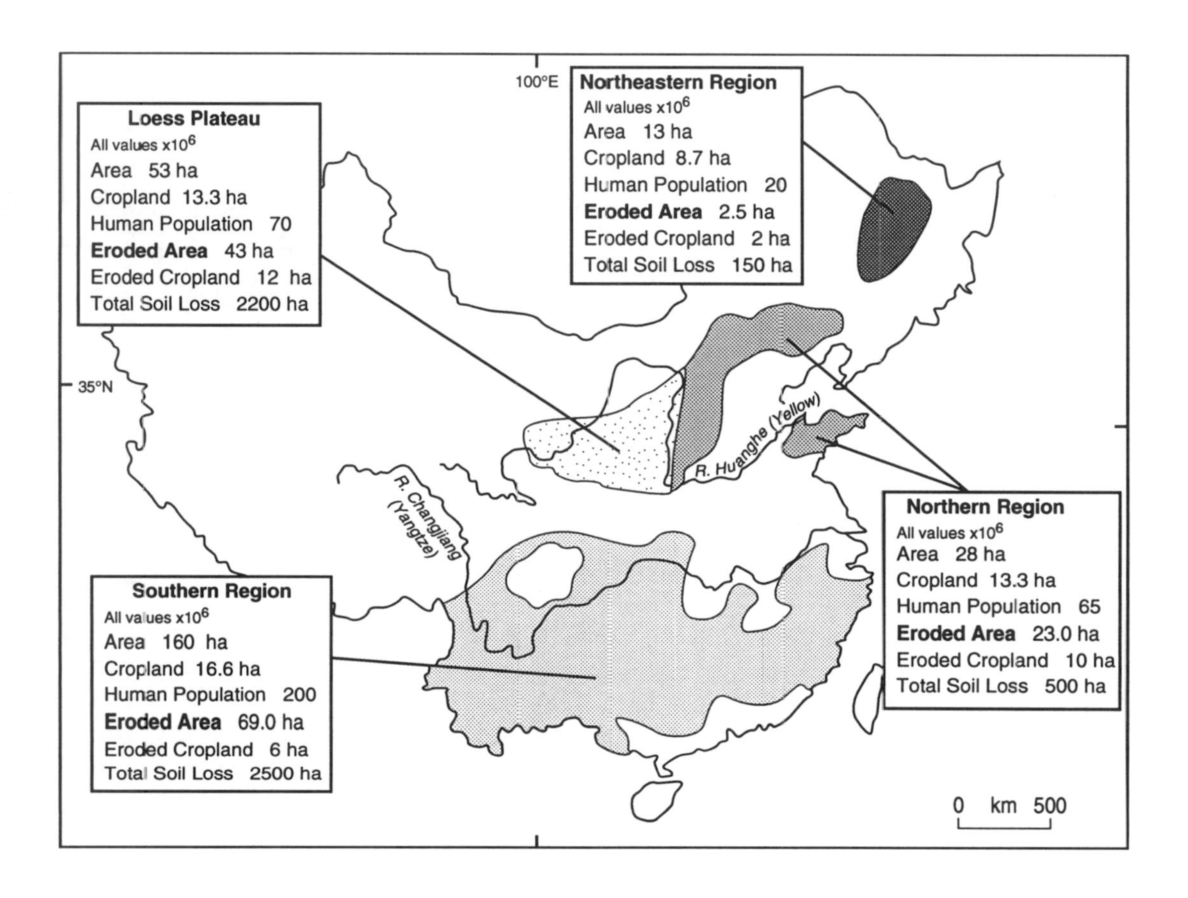

Figure 9.3 The regions of China most severely affected by soil erosion (based on Dazhong, 1993)

bank be raised every 10 years. In some places, e.g. Kaifeng City and the surrounding region in Hunan Province, the bed is actually 8 m higher than the city, giving rise to a 'suspended' river. This increases liability to flooding. As Dazhong states, 'The large amount of sediment on the riverbed is seriously threatening the security of agricultural production, five oil fields, seven railway lines, and 100 million people in the North China Plain.' Sediment deposition has also resulted in a decrease of *c.* 30 per cent of the total water-storage capacity in the loess plateau along with 48 per cent of total irrigation capacity of the reservoirs. Not only does this prove costly, it also means that vast amounts of nutrients are lost to the China Sea. All of these factors conspire to diminish agricultural productivity so the erosion problem becomes a vicious circle. This is despite the long history of erosion control that characterises this region (reviewed in Shengxiu and Ling, 1992). Such measures include strip-furrow cultivation, levelling the land, terracing, diversion of flood waters, the construction of silt traps and soil dams, afforestation and grass planting. Agricultural measures to curtail soil erosion include the cultivation of crops in pits, deep ploughing combined with soil dams, contour ploughing, crop rotations, intercropping and overcropping (i.e. layered cropping). The combination of engineering and improved agricultural practices, within a government programme of erosion control, are beginning to reduce erosion but there is still much to be achieved.

China's other region which experiences severe soil erosion is the southern region (Figure 9.3) of the tropical and subtropical zone. Dazhong (1993) states that this region comprises 160×10^6 ha of which 16.6×10^6 ha is cultivated, and contributes 100×10^6 t of grain annually to China's food production. Much of it is paddy rice. Some 69×10^6 ha are eroded lands with an average annual erosion rate of 36 t ha^{-1}. As in the case of the loess plateau, the soils of the southern region have become depleted of essential nutrients and organic matter. The problems of silt deposition are similar in character to those of the Yellow River though are not quite so widespread. Much of the erosion is again due to water rather than wind erosion but the sediments so derived have reduced the transportation capacity of the Yangtze River system by about 40 per cent since the 1960s. This has economic as well as environmental repercussions. There are many other parts of Asia suffering from soil erosion (Dregne, 1992) including India where 113.3×10^6 ha and 38.7×10^6 ha are subject to water and wind erosion respectively (Khoshoo and Tejwani, 1993). Erosion is widespread but is particularly acute in the hill regions of the Outer Himalayas, the Punjab and south-east India. Overall, 5.33×10^9 t of soil are eroded annually of which *c.* 50 per cent enters India's rivers and 10 per cent is deposited in reservoirs causing a decline in their water-storage capacity. Most importantly, this erosion removes an estimated 5.37×10^6 t of plant nutrients. Khoshoo and Tejwani also provide details of land use and land degradation in two small watersheds in Uttar Pradesh in the Outer Himalayas. Here, communal lands are used for grazing but poor management is causing degradation. In addition, much land is cultivated that should be forested or used for pasture. The absence of soil and water conservation measures such as bench terracing accentuate erosion. Erosion is also a significant problem in the Punjab especially in the north-east where soils and sediments are particularly erodible (Kukal *et al.*, 1991) due to slope angle, soil character and rainfall distribution. Parts of north-eastern Pakistan, also in the Outer Himalayas, and where grazing and cultivation occur below 2000 m, are subject to erosion rates in the region of 30–150 t ha^{-1} yr^{-1}

(Ellis *et al.*, 1993). The management of the forest cover appears to be a crucial factor in determining erosion rates. In some Himalayan regions, however, tectonic uplift is likely to be contributing to sediment removal. There is, for example, evidence for an increase in the forest cover of the region, due to the institution of environmental management practices, between 1972 and 1990 (Schreier *et al.*, 1994) which goes some way to dispelling the view that deforestation is directly responsible for accelerated erosion and flooding downstream (see also discussion in Ives and Messerli, 1989). Following successful establishment in India, many Nepalese farmers are using vetiver grass (*Vetiveria zizaniodes*) to help combat soil erosion. According to the National Research Council (1993), this practice originated in the 1950s in Fiji as sugar-cane growing encroached on the hill regions. Since then it has been widely adopted with much success. The stabilisation of soils increases crop yields because nutrients are retained within the fields. In Nepal, mostly in the terai region, vetiver is now being used to protect the front lip of terraces and even to stabilise roadsides.

Of all the world's soil erosion problems, however, it is that of Africa which attracts most attention in the general media. This is not because the problem is more acute in Africa than elsewhere in the developing world (Figure 8.2) but because it is often associated with drought, famine and loss of life. In fact these phenomena are associated more with the complex process of desertification than with soil erosion *per se* (see Section 9.3) and occasionally there are additional cultural stimuli such as warfare. Nevertheless, soil erosion is a pressing concern in many parts of a continent that has severe food production limitations. Inevitably, the rates of soil erosion that occur reflect a range of physical environmental factors, such as soil erodibility and rainfall/wind erosivity (see above) and a number of cultural factors which include land management, population growth and market stimuli. Soil erosion is prevalent in all of Africa's climatic zones which are illustrated in Figure 9.4. As Table 9.5 shows, water erosion affects 227.4×10^6 ha and wind erosion affects a further 186.5×10^6 ha. One of the most severely affected regions is the arid zone of Sub-Saharan Africa where wind erosion is prevalent. This, as a component of desertification, is examined in Section 9.3.

Lal (1993) has reviewed the soil erosion rates that occur in West Africa, some of which are given in Table 9.7. As might be expected, erosion rates are high when the natural vegetation cover has been disturbed or removed. In addition to the susceptibility of some soils to erosion, the problem is often exacerbated by inappropriate management practices. Lal states that in the forest zone these include mechanised land clearance and uncontrolled burning whilst in the savanna zone, as in the Sahel, they include uncontrolled grazing and overstocking. However, the implementation of such practices is often due to socio-economic factors (Blakie and Brookfield, 1987). For shifting cultivators these include shortening of the fallow, land scarcity and inadequate resources to institute soil conservation. The very fact that farmers shift their plots every two to three years can mean that they are not aware of the problem. Local perceptions of land degradation, including the view that it is inevitable, may be additional constraints on the implementation of conservation. The problem has been observed in Burkina Faso by Lindskog and Tengberg (1994) who suggest that failure to address it may be an important factor in the ineffectiveness of many development projects. If land is communally owned, soil erosion is often more severe than on privately owned land though the latter, as Lal (1993) states, may suffer

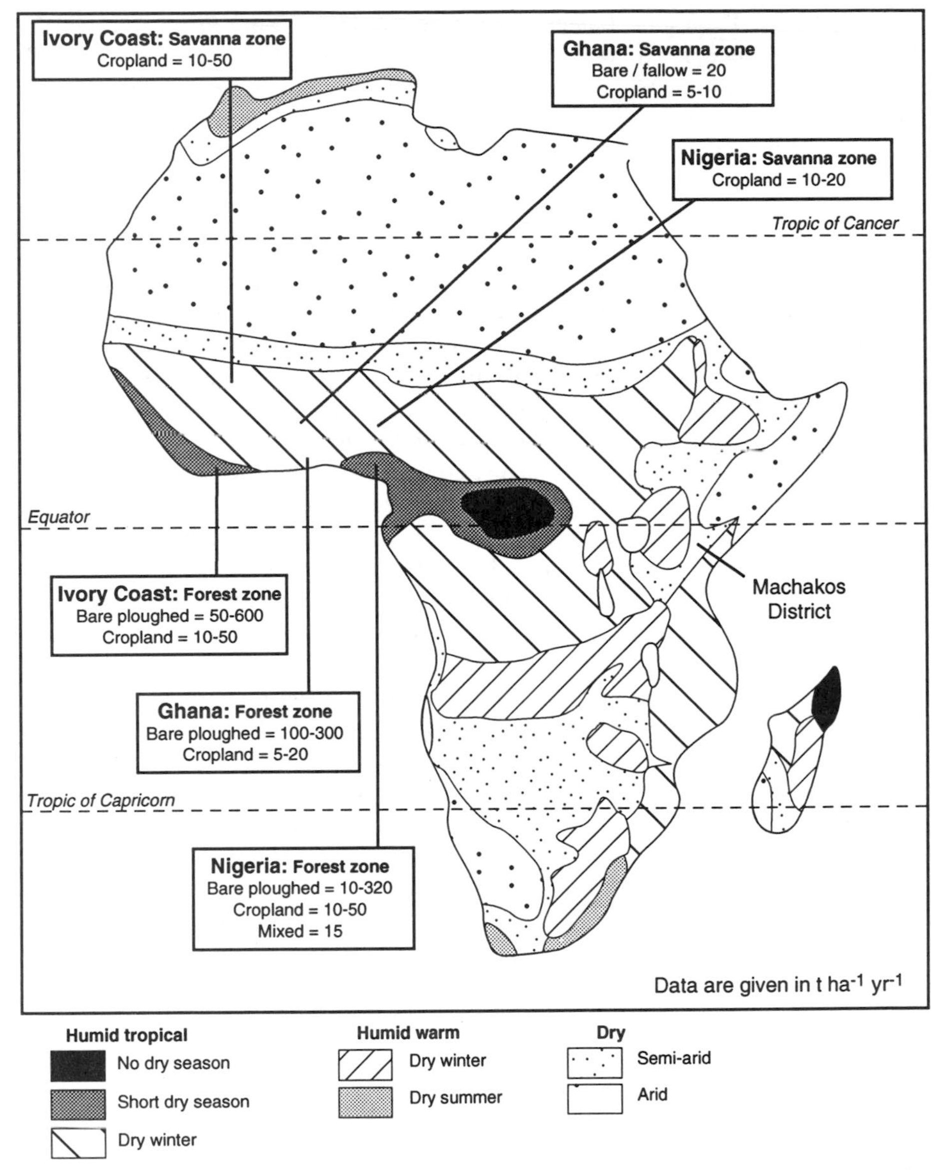

Figure 9.4 Rates of soil erosion in selected regions of West Africa in relation to the climatic zones (see Table 9.7)

from fragmentation when shared amongst large families. Zimbabwe's communal lands represent another example of badly eroded soils due to inappropriate cropping and stock management. According to Vogel (1992), existing tillage practices involve inversion tillage using a single-furrow mouldboard plough. This produces a soil of fine tilth. With little vegetation cover to protect it the soil is particularly susceptible to sheetwash erosion (Table 9.6). Rates of up to 9.5 t ha^{-1} yr^{-1} have been recorded. The same is true if clean ripping of the soil and hand hoeing are undertaken. Both of these

Table 9.7 Rates of soil erosion in the savanna and forested zones of West Africa (abstracted from Lal, 1993)

Country	Bioclimatic* zone	Land use	Rate of soil erosion (t ha^{-1} yr^{-1})
Ghana	Savanna	bare-fallow	20
Ghana	Savanna	cropland	5–10
Nigeria	Savanna	cropland	10–20
Ivory Coast	Savanna	cropland	10–50
Ivory Coast	Forest	cropland	10–50
Ivory Coast	Forest	cropland	10–50
Ivory Coast	Forest	bare-ploughed	50–600
Nigeria	Forest	bare-ploughed	10–320
Nigeria	Forest	cropland	10–50
Nigeria	Forest	mixed	15
Ghana	Forest	bare-ploughed	100–300
Ghana	Forest	cropland	5–20

* See Figure 9.4.

methods of land preparation produce erosion rates of up to 5 t ha^{-1} yr^{-1}. With no-till tied ridging (i.e. the creation of ridges rather than furrows) soil erosion was reduced to less than 0.5 t ha^{-1} yr^{-1}. Ripping into crop residues, which offer some protection for the soil, was nearly as conservational, producing an erosion rate of 0.6 t ha^{-1} yr^{-1}. Thus, land preparation practices can alter significantly the rate of soil loss. Where soil erosion is particularly severe, as has occurred on the Irangi Hills of the Kondoa region of Tanzania (Payton *et al.*, 1992), land may have to be abandoned. Once heavily cultivated, the pediment midslopes of the Irangi Hills now have only a sparse cover of *Brachystegia* and the terrain is dissected by deep gullies.

Changing the type of crop or combination of crops grown can also alter the rate of soil erosion. Experiments conducted at the University of Benin's farm in Nigeria's cassava-growing region show that erosion rates increase when traditional polycultural cropping systems are abandoned in favour of monocropping (Odemerho and Avwunudiogba, 1993). Moreover, ridging of the soil across the slope prior to planting rather than using flat land reduced soil erosion rates from 29.5 t ha^{-1} to 16.8 t ha^{-1}. These experiments show that if changes in the type of crop cultivation occur it may be necessary to alter land-preparation practices as well in order to control soil erosion. However, there is little doubt that the initial removal of a natural vegetation cover is followed by high rates of soil erosion which eventually diminish as pasture or crops provide protection. In Algeria, for example, erosion is particularly acute in the Jijel region which is located in the north along the Mediterranean coast (Zaimeche, 1994). Much of this area is mountainous and receives on average 1000 mm of rainfall per year. The removal of forest for crop cultivation and the burning of forest to create grazing lands is causing massive soil erosion; in some areas entire soil profiles have been lost. Zaimeche states, 'In the Sidi Abdel Aziz commune which, not long ago, contained some of the densest forests in the country, lands have been so degraded that the area resembles land close to the Sahara rather than a Mediterranean region.'

In the East African highlands (including the countries of Kenya, Ethiopia, Uganda and Tanzania) soil erosion is a major problem. According to Ståhl (1993), population pressure is causing an expansion in the numbers of impoverished smallholders. The resulting land scarcity is causing a reduction in fallow periods so the land is becoming nutrient deficient. In addition cultivation has been extended into lands traditionally used by pastoralists who are forced to move into unsuitable areas or to overgraze their herds. Ståhl reports that efforts by Kenya's government to halt and reverse land degradation have reaped some rewards but as many as 66 per cent of the small farms in need of conservation measures have not yet been contacted. The most effective soil conservation measures include bench terracing, bunds, and trash lines consisting of branches, twigs, etc. These structures not only trap soil as it moves downslope in sheetwash but also conserve water. Other measures that are recommended include agroforestry (Section 7.5), the trees and shrubs of which provide a permanent protective cover for the soil as well as organic matter to help bind the soil. Livestock management can also be improved. Land shortages are already leading to the penning of animals, thus reducing damage to terraces, etc., on which forage grasses can be grown. Most importantly, Ståhl suggests that major improvements would ensue if the system of land tenure was changed. Government ownership of land with individuals being granted land rights is not conducive to good custodianship, especially if governments are not trusted by their tenants or if they are liable to change land-use policies too frequently. Despite acute degradation problems there are measures that can be taken to improve the lot of peasant farmers.

Indeed, considerable success has been reported by Tiffen *et al.* (1994) in the Machakos District of southern Kenya (see Figure 9.4). Here the main types of erosion are inter-rill, rill and gully erosion, the former two combining to produce sheet erosion. Erosion affects cropland and grazing land. In the Iiuni catchment, for example, degraded grazing lands (37 per cent of the area) have a soil loss of *c.* 53 t ha^{-1} yr^{-1} whilst on cultivated land (43 per cent of the area) it is 16.0 t ha^{-1} yr^{-1}. Averages for the Machakos District overall are, however, in the range 5–15 t ha^{-1} yr^{-1} for whole catchments. This, Tiffen *et al.* state, represents an improvement on the rates of the 1930s and 1940s, when deforestation and a series of droughts occurred and soil erosion affected *c.* 75 per cent of the area. Improvements have been brought about by terrace construction on cropland and improved land rights and land tenure on grazing lands. The latter in particular illustrates that the individual ownership of land encourages conservation and forward planning. In the Machakos District communal grazing lands have almost disappeared. The achievements, and there is still room for further improvements, also highlight the significance of cultural factors in causing and mitigating a severe environmental problem.

The mountain zone of South America is another area of the world that is prone to high rates of soil erosion. As in the case of the examples discussed above, land-use types and practices influence erosion rates substantially. In Andean Argentina between 2000 m and 3500 m, for example, Molinillo (1993) has examined the environmental impact of traditional pastoralism. This involves transhumance to take advantage of altitudinal variations in forage availability. There appears to be little relationship between grazing and erosion in the forested zone of the lower altitudes but a strong relationship in the grassland zones of the higher altitudes. Consequently, the latter require particularly careful management. Sarmiento *et al.* (1993) have also discussed erosion on land under traditional agriculture in the Venezuelan high Andes. Here, between 3000 and 4000 m above sea-level, plots of land are cultivated for two or three years producing potatoes initially and then a cereal crop. The fallow can be anything between 7 and 20 years. The vegetation cut down to make way for cropping is used as a mulch and a green manure which not only adds nutrients and organic matter to the soil but also protects it from water erosion. In consequence, erosion rates are low. Sarmiento *et al.* report that overall erosion rates are less than 0.6 t ha^{-1} yr^{-1}. In addition, in crop fields the erosion rate is only 0.08 t ha^{-1} yr^{-1} more than in fields left fallow for 25 years, which is attributed to the water-retaining capacity of the organic fraction of the soil. This is considered to be a sustainable type of agriculture (see also Section 10.4). A further study, undertaken by Harden (1993) in Ecuador, also indicates that land-use practices need not lead to excessive erosion. Although there is sufficient sedimentation in the Amaluza Dam (on the Paute River), which produces more than 50 per cent of Ecuador's electricity, to warrant dredging, the sediment is produced mainly from abandoned farmlands. This abandonment is caused by 5 per cent of the rural population moving to the USA every year. The abandoned land may be grazed by unpenned animals which restrict and slow the rate of vegetation colonisation. Once again this case study highlights the significance of cultural rather than environmental factors in soil erosion regimes.

9.3 DESERTIFICATION

Desertification is a controversial issue. This is partly due to the many definitions that exist, which confuse the issue, and partly due to the difficulties of measuring its extent and rate of development. Kovda (1980) uses the term 'desertification' to describe land aridisation. This involves several processes that reduce the effective moisture content of soils and so impair its production capacity. Kovda does not distinguish between natural and cultural agents that may bring about these processes or accelerate existing processes. This is part of the controversy surrounding desertification which is a term often used to imply a cultural cause of land degradation. It contrasts with the term 'desertisation' which is considered to refer only to natural agents of environmental change. Rapp (1987), however, defines desertification as follows:

> Desertification is the long term degradation of drylands, resulting from either over-use by man [*sic*] and his animals, or from natural causes such as climatic fluctuations. It leads to loss of vegetation cover, loss of topsoil by wind or water erosion, or loss of useful plant production as a result of salinisation or excessive sedimentation associated with sand dunes, sand sheets or torrents.

This definition involves both natural and cultural factors and reflects the variety of processes that may be involved. Desertification is, thus, not just a simple matter of mobile desert boundaries. Moreover, agricultural practices are often one of the root causes.

The global extent of the world's arid land areas at risk from desertification are illustrated in Figure 9.5. As this shows, desertification is not a problem confined to the developing world; it is also significant in the central Asian republics, the USA and Australia. The problem of desertification is, however, more acute in the developing world than in the developed world because of rapidly increasing populations and the lack of resources to combat it. As stated above and discussed by Grainger (1990) and

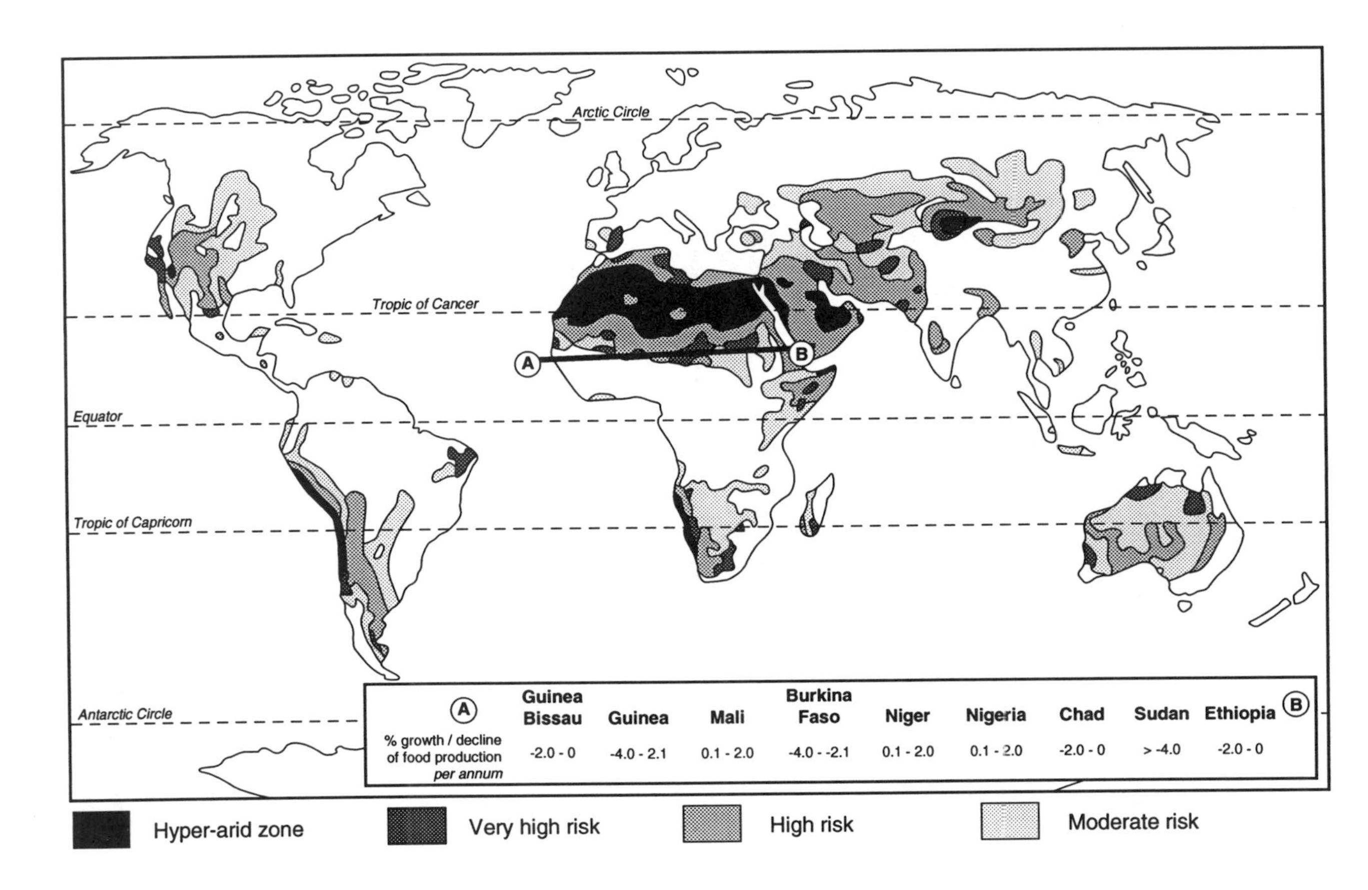

Figure 9.5 World map of the risk of desertification (based on Myers, 1993a) and the percentage growth or decline of food production in the Sahelian nations (based on Morgan and Solarz, 1994)

Table 9.8 UNEP estimates of types of drylands deemed susceptible to desertification, the proportion affected and actual extent. (Reproduced by permission from Thomas and Middleton, 1994)

	1977 UNCOD	1984 GAP	1992 GAP II
Climatic zones susceptible to desertification	Arid, semi-arid and sub-humid	Arid, semi-arid and sub-humid	Arid, semi-arid and dry sub-humid
Total dryland area susceptible to desertification (million hectares)	5281	4409	5172
Percentage of susceptible drylands affected by desertification	75	79	70
Total area of susceptible drylands affected by desertification (million hectares)	3970	3475	3592

Hellden (1991), estimates of the extent of desertification are unreliable. This is because of the varied definitions (see above) and because of different interpretations of field evidence and satellite imagery (these are similar problems to those discussed in Section 9.1 in relation to deforestation). Such problems also make estimates of rates of increase or decrease speculative.

Consequently, these and subsequent data prepared by UNEP in 1976–1977, 1983–1984 and 1992 must be viewed with caution. As Thomas and Middleton (1994) have discussed, and as is reflected in Figure 9.5, the original data were compiled not on the basis of actual measured desertification but on the basis of estimates of desertification hazard. The second attempt at compiling data comprised a questionnaire (see also the discussion in Grainger, 1992) sent to governments and further estimates by consultants whose definitions of degraded land did not coincide. The most recent assessment (UNEP, 1991, 1992) used a different approach again, namely a soil degradation database formerly developed to assess soil degradation due to human activity globally and not just in drylands. Soil degradation, however, is generally considered to be only one of several components of desertification (see quote by Rapp, 1987, given above). To the values for soil degradation, UNEP added estimates of vegetation degradation. A summary of these estimates is given in Table 9.8. Even if these data are in error by *c.* 25 per cent they still indicate that desertification is an environmental problem affecting many nations and many millions of people. Nevertheless, the unreliable and, to some extent, incomparable data given in Table 9.8 do not facilitate the accurate assessment of rates of desertification. In view of Thomas and Middleton's (1994) comments on the UNEP efforts to monitor desertification, it would appear that there is no accurate assessment of its worldwide extent or any current means of determining its rate of increase and/or decrease.

Table 9.9 The spatial expressions of desertification and their causes (based on Grainger, 1992)

A	EXPANSION Instead of increasing the productivity of existing farmland, the expansion of agriculture may occur onto marginal land which becomes degraded. In years of sparse rainfall this land may be abandoned.
B	CONFINEMENT Land use becomes more intensive, e.g. around urban areas where deforestation for fuelwood is common, around boreholes and villages where overgrazing or overcultivation may occur. Changes, especially restrictions on mobility, to the movements of nomadic peoples or their enforced settlement may also contribute to land mismanagement.
C	DISPLACEMENT Increasing cultivation of cash cropping may displace subsistence rainfed cropping which, in turn, displaces nomadic pastoralism. This creates a disequilibrium between land use and the land's productive capacity which is particularly fragile during periods of low rainfall. Fallows may also be shortened so that soil fertility never has an opportunity to recover. The reduced extent of rangelands may then cause desertification through overgrazing. Thus the cause of desertification need not be proximal but distant and outside the control of pastoralists.
D	THE TRAGEDY OF THE COMMONS Overexploitation of common grazing lands may occur leading to desertification. Some of this is due to C above but some is also due to the breakdown of traditional social structures that controlled land allocation (see Section 4.1.2).

According to Grainger (1992), there are four spatial expressions of desertification. These are given in Table 9.9 which also describes some of the many social and economic factors that underpin the physical environmental processes involved in desertification. Some of these factors have been discussed in Section 4.1.2 and expressed in Figure 1.1. Moreover, the annual loss of land from agricultural production, much of which is in the developing world, exerts yet more pressure on the remaining productive area in drylands. Nor is it conducive to the development of sustainable agricultural systems (see Section 10.4). Any factors or practices that give rise to, or contribute to, desertification, including those that are economic and extrinsic, even if they give rise to short-term gains, are going to destabilise food production in many countries that already suffer shortages. Drought and famine are thus associated with desertification in some nations, especially those in Sub-Saharan Africa.

On a global basis, desertification is most acute in Africa on two counts. First, the largest extent of desertified land occurs in the Sudano-Sahel zone, a large area of land at great risk of desertification, as illustrated in Figure 9.5. Secondly, this region has a large and rapidly growing population which can ill afford to lose productive land. All of the factors referred to in Table 9.9 are relevant to the situation in this region. To these must be added the possibility that climatic change, whether human-induced or not, is also occurring. There is no doubt that the boundary of the Sahara Desert has altered in the last two decades, as has been demonstrated by Tucker *et al.* (1991) who have monitored the expansion and contraction of the Sahara Desert in the 1980–1990 period using remotely sensed imagery. Between 1980 and 1984 there was a southward movement of the boundary but in 1984–1985, and again in 1985–1986, the boundary retreated northward. Overall, the mean position of the boundary (the 200 mm yr^{-1} precipitation isoline) was 130 km south of its position in 1980. Precisely what the causes were of these desert-boundary movements is impossible to determine. What is equally evident, however, is that most of this region, as well as areas to the south, has been experiencing a low rate of growth in food production or even a decline in the rate of food production, e.g. the Sudan (Morgan and Solarz, 1994). Indeed, few Sahelian countries have achieved more than a 1 per cent increase in food production in the 1986–1990 period for which data have been examined. Indeed, for the last 30 years or so the high incidence of drought, notably in the 1968–1973 and 1982–1985 periods, has caused considerable hardship. This has led to famines in some regions, namely Ethiopia, but as Hurni (1993), Mortimore (1989) and others have discussed, there is more to famine than drought. Such factors include the degree of rural development, political stability, standards of health and education and systems of land tenure.

Amongst the most important contributions that farmers can make to combatting desertification are the adoption of soil and water conservation measures. These have been discussed, in the context of Sahelian nations, by Critchley *et al.* (1992). They examine traditional systems of cultivation and the way these are used to avoid or mitigate land degradation. In mountainous regions, for example, earth and stone terrace and bunds are widely used, some of which are also designed to improve water harvesting. Interestingly, the findings of Critchley *et al.* (1992) contrast with those of Tiffen *et al.* (1994) on the Machakos District of Kenya (Section 9.2) that individually-owned land is less degraded than communally-owned land. (Solving many of the other problems that are associated with desertification, such as poverty, rural development and sustainable resource use are the remits of politicians at the regional, national and international levels.) However, most land that becomes desertified is rangeland which is used for grazing rather than croplands. As Grainger (1990, 1992) discusses, the wise and sustainable use of rangelands must be a cardinal element in policies or strategies to counteract and control desertification. Since this may be influenced by distant as well as proximate factors, the regulation of, and implementation of management practices in, grazing lands is particularly difficult. Determining the rate of stocking that is appropriate for a given rangeland is thus vital. Formulating and implementing such policies is, however, difficult.

Degradation problems that give rise to desertification are illustrated by recent land-use changes that have occurred in Goudoumaria, eastern Niger (Reenburg, 1994). Here, land use varies with the geomorphology. The dune topography, having originated during a period drier than today and then inundated by an expanded Lake Chad, has coarse sandy soils separated by depressions with calcareous soils. In the depressions rainfed cropping of millet, sometimes intercropped with cowpeas, is undertaken whilst the dunes are used for pastures for nomadic pastoralists. However, Reenberg's analysis of satellite imagery indicates that in the years 1936–1992 cultivation had expanded into the dune system. Apparently this is a response to the low amounts of precipitation received in this period. This precipitation is more efficiently used by crops on the sandy soils of the dunes than on the calcareous soils of the depressions. Unfortunately, the dune soils are more susceptible to erosion than the calcareous soils so this change in land use may give rise to accelerated wind erosion. This is a response to a shift in an environmental factor. Elsewhere in the Sahel, land-use changes have occurred in response to economic factors. For example, the need to earn foreign currency has encouraged some governments to stimulate the spread of cash cropping in marginal areas. Groundnut production in Senegal and Niger has caused soil exhaustion and the onset of desertification.

Overgrazing also causes desertification. This has occurred in the Sahel but it is also a problem elsewhere in Africa. Dean and Macdonald (1994) have examined the degradation of rangelands in the Cape Province, South Africa, and its impact on stocking rates. Their analysis of records of stocking rate for the 1911–1981 period in ranches and in the semi-arid districts shows that a decline in stock numbers occurred. The average stocking rate in the 1911–1930 period was 12.5 ± 8.6 large stock units per 100 ha; by the 1971–1981 period the average rate had fallen to 8.4 ± 7.7. This trend, Dean and Macdonald believe, is a direct result of degradation which has caused a reduction in primary productivity. The sinking of boreholes is also acknowledged as a catalyst to overgrazing, as are foreign markets. In Botswana, for example, the commercialisation of cattle raising for European markets has encouraged the abandonment of traditional transhumance and the increased use of the Kalahari sandveldt which is itself subject to accelerated erosion. Kenya, Tanzania, Tunisia, Morocco and Algeria also have land degradation problems caused by overgrazing.

Asia's largest extent of desertified land is located in China where the major impetus is increasing land-use intensity due to China's need to feed a large population from a relatively small land area suitable for agriculture. According to Smil (1993b), there was a nationwide survey of land degradation in the late 1980s. Soil erosion affected 31 per cent, desertification (desert encroachment) affected 5 per cent, salinisation and alkalinisation occurred on 6 per cent and waterlogging on 9 per cent of the 45×10^6 ha surveyed. In addition, desertification is considered to be occurring at the rate of 150×10^3 yr^{-1} ha, 25 per cent of which is due to inappropriate cultivation. Desertification is occurring on rangelands and croplands as has been examined by Zhenda *et al.* (1992). The current status of the problem is illustrated in Figure 9.6, which shows that most desertified land occurs in the semi-arid zone. One such area is the Mu Us region. Fullen and Mitchell's (1994) observations on this area (Figure 9.6) indicate that desert-like conditions develop in localised areas of rangelands which are not necessarily directly at a desert margin. These so-called 'blisters' enlarge and eventually merge. Reclamation efforts were begun in the late 1950s, the aim being to stabilise mobile dunes and eventually to recreate rangelands and croplands. The six techniques employed are described in Table 9.10. In addition, the scale of the problem in China, coupled with that of soil erosion (see Section 9.2), has prompted the establishment of the 'Project of Protective Forest System' (Kebin and Kaiguo, 1989). This was initiated in 1978; by 1985, 5.3×10^6 ha had been planted with trees, shrubs and herbs. It included 700 000 ha of forest planted specifically as a sand-fixation belt, another 700 000 ha of shelter belts to protect arable land and 170 000 ha of shelter belts for pasture. The aim is to plant a massive 8×10^6 ha to provide a 'green great wall'. Such efforts, whilst costly, are not always successful but the paucity of agricultural land in China means that there is little alternative but to invest in reclamation schemes.

Elsewhere in Asia, desertification is a problem in parts of India and Pakistan. As Figure 9.5 illustrates, parts of north-west India and much of Pakistan (as well as

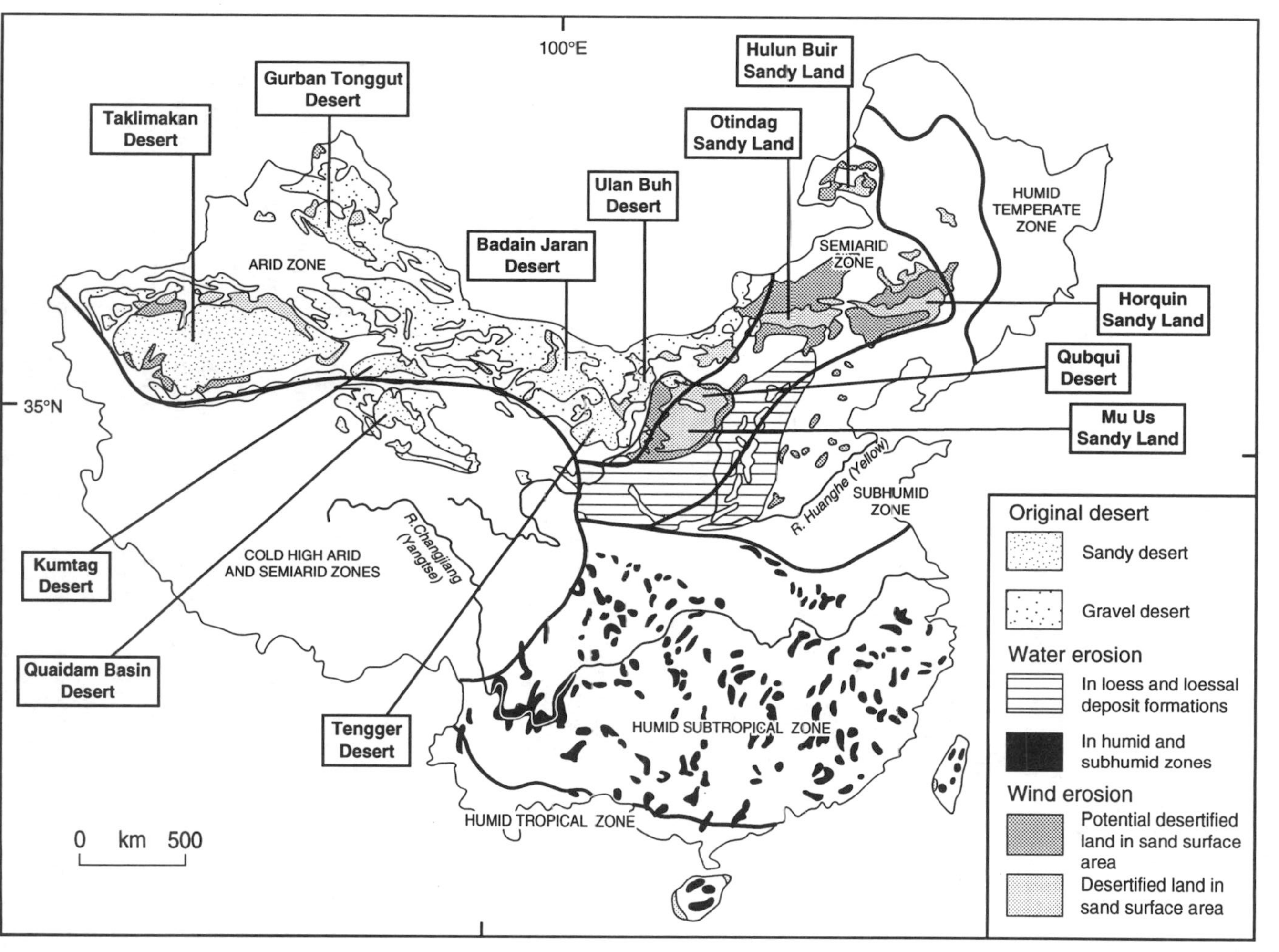

Figure 9.6 The distribution of desertified lands in China (based on United Nations Environment Programme, 1992)

Table 9.10 Techniques for the reclamation of desertified land (based on Goudie, 1990; Fullen and Mitchell, 1994)

A WINDBREAKS
Trees help to stabilise soils by adding organic matter to improve the soil texture. Lines of trees also trap moving sand and silt. Pines, poplars and willows are the most common tree species used.

B IRRIGATION
This improves the water content of the soil and so increases primary productivity. Water from rivers, especially the Yellow River, supply silt as well. This improves the nutrient status of the reclaimed soils.

C STRAW/CLAY CHECKERBOARDS
Artificial checkerboards of straw or clay are located in strategic places to increase surface roughness. This traps sand/silt and encourages colonisation by vegetation. It can increase the vegetation cover of sand dunes from 5 per cent to 30–50 per cent. The litter produced by the colonising plants enhances the organic content of the soil.

D LAND ENCLOSURE
The enclosure of land with fences allows stocking rates to be managed so as not to exceed the carrying capacity of the land.

E EXTRACTING PALAEOSOLS
This involves the recovery of buried palaeosols which may be fertile silts and loams. Mixing these with dune sand and adding sheep manure provides a soil capable of producing a grain or vegetable crop.

F CHEMICAL TREATMENT
The treatment of sodic soils, for example, to replace sodium with calcium or magnesium salts creates an improved soil structure.

much of Iran) are at considerable risk from desertification. The Rajasthan 'Desert', a semi-arid area surrounding the Thar Desert in north-west India and Pakistan, is heavily desertified. Moreover, parts of this region have been subject to an ambitious irrigation programme associated with the movement of water from the Punjab via the Indira Gandhi Canal Project. As is discussed in Section 9.4, there are now problems of salinisation in this region which can be sufficiently severe to prompt land abandonment. The problems of Rajasthan and the neighbouring sub-humid zone are also illustrated by Scott's (1994) account of environmental degradation and conservation measures in the southern Aravalli Hills. Lying as they do between the Thar Desert to the west and the sub-humid Gangetic plains to the east, these hills are a buffer zone. In the post-1900 period there has been an extension of both grazing and cultivation on sloping land which has accelerated rates of erosion. This, Scott states, is a consequence of population increase (*c.* 285 per cent for Rajasthan as a whole since 1900) not only intensifying agricultural practices but also exerting more pressure than hitherto on local woodlands for domestic fuel supplies. These changes are compromising the Aravalli region's capacity to withstand the eastward advance of the Thar Desert. This is evidenced by an expansion of vegetation types characteristic of the arid zone. However, the employment of soil and water conservation measures, as well as reforestation, is now being used to redress environmental degradation. Such measures include the construction of low-lying stone walls and bunds to reduce soil loss and conserve water. This ecotonal area is a fragile environmental zone that may constitute what Mather and Sdasyuk (1991) describe as a red zone area, i.e. an area at risk of rapid environmental change. Similar problems occur elsewhere in India. Ghosh (1994), for example, reports that in the Gulbarga region of Karnataka State (*c*. 400 km south-west of Bombay) overgrazing and forest removal caused an extension of desertified land between 1975 and 1991.

The above examples illustrate the significance of desertification in developing nations. Lest there should be any doubt that it is a problem in the developed world, Goudie (1994) has examined its extent, causes and reclamation procedures in the USA. Here, overgrazing is one of the most important causes, especially in New Mexico, Colorado and Idaho. In the Central Valley of California, irrigation systems have caused both soil and water salinisation (see Section 9.4). Reference has already been made to the problems of land degradation in the vicinity of the Aral Sea (Section 8.4) whilst desertification is considered to be a major land-degradation issue in Australia. Barrow (1991) states that most of Australia's drylands, which are currently used for livestock rearing, are moderately desertified. Areas in New South Wales, South Australia and Victoria have been severely affected though the problem has been exacerbated by rabbits which were introduced by early European colonists.

9.4 SOIL AND WATER SALINISATION AND RELATED PROBLEMS

As discussed in Section 9.4, salinisation, alkalinisation and waterlogging in soils are often considered to be components of desertification. Whilst this is clearly the case, these processes are also associated with irrigation which is employed solely to allow or enhance crop production in areas where water shortage is a limiting factor. As well as influencing soil characteristics and environmental quality, irrigation schemes affect water quality. The direct and indirect impacts of irrigation in the Aral Sea region (Section 9.4) illustrate these relationships and the vicious circle that can operate if irrigation management is poor. Highly saline water, for example, is not suitable for irrigation and if aquifers become saline there may be problems for the provision of water for domestic consumption. Some of these issues have been referred to in Section 8.4. Postel (1992) has stated that irrigation is the cornerstone of agricultural production in many regions. This is particularly true in the arid and semi-arid zones which are occupied mainly by developing countries. The heavy reliance of developing countries on irrigation schemes means that such systems must be adequately maintained so as to create as little soil and water degradation as possible. Indeed, the importance of irrigation and its management must make it a core issue in sustainable agriculture policies. Moreover, and in view of its role as a manipulator of water resources, irrigation may need to be substantially altered if global warming occurs (Section 10.5).

Samad *et al.* (1992) and Postel (1993), briefly reviewing the recent history of irrigation, state that the 1950s to the late 1970s witnessed a rapid establishment of irrigation schemes. By the mid-1970s the number of irrigation schemes had increased substantially. By 1989 there were *c*. 233×10^6 ha of irrigated land worldwide (FAO, 1990), of which 73 per cent was located in the developing world. The distribution of this is given in Figure 9.7. Irrigation, thus, supports some 21 per cent of the cropland

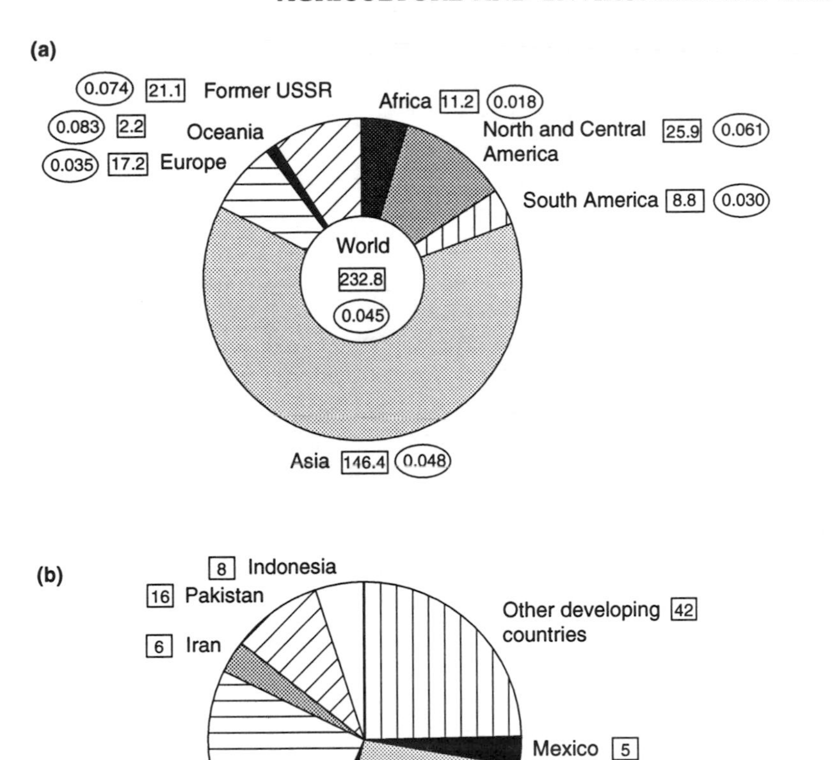

Figure 9.7 The distribution of irrigated land: (a) world; (b) developing countries (based on Food and Agriculture Organisation, 1990; Postel, 1992, 1993)

in the developing world. China and India together contain about 50 per cent of the developing world's irrigated land. Although there has been limited development of irrigation in Africa this situation in changing as the need to improve food supplies is increasing. In addition, the establishment, extension and improvement of irrigation systems may be one way to increase the world's food supplies from existing agricultural land. Such intensification, along with other improvements to productivity, will limit the spread of agriculture into marginal lands and so conserve natural ecosystems. Currently, approximately 33 per cent of the world's food derives from irrigated land and, as Samad *et al.* (1992) suggest, the development and improvement of irrigated land, especially in Asia, made a major contribution to the so-called 'second green revolution' of the 1960s and 1970s which was spearheaded by the breeding of improved crop varieties. However, costs are high. Postel (1993) reports that capital costs for new irrigation capacity arc about \$US1500 ha^{-1} in China, and between \$US1500 and \$US4000 ha^{-1} for large projects in India, Indonesia, Pakistan, and the Philippines and Thailand. Costs may be as much as \$US10 000 ha^{-1} in Mexico and even higher in parts of Africa where the infrastructure is less well developed.

The environmental costs of irrigation may also be high. These depend on the efficiency and management of the irrigation system. For example, Postel (1993) observes that less than 50 per cent of the water diverted from the reservoir into the field may enhance crop productivity. In Asia efficiencies in large irrigation schemes are usually in the order of 30 per cent. Where water use is inefficient environmental problems ensue, not least of which is loss of the water resource itself or its chemical modification so that the water is no longer useable. The three major problems that cause impairment of crop production and even, in the worst cases, land abandonment, are salinisation, alkalinisation and waterlogging of irrigated soils. The relationships between these processes and irrigation are given in Figure 9.8. The water itself may become saline as well as contaminated with silt, pesticides and fertilisers. Not only does this have implications for the further supply of irrigation water but it may give rise to aquifer contamination which, in turn, creates problems for domestic water supply.

Salinisation (reviewed in Thomas and Middleton, 1993) is a process whereby the concentrations of salts, particularly sulphates, chlorides and carbonates, increase in soils, surface water and groundwater. This is a process that occurs naturally in arid and semi-arid regions because the rate of evaporation is high. It brings saline water to the soil surface by capillary action; the water evaporates leaving the salt in the soil. In extreme circumstances salt flats will develop which cover large areas of land, as occurs in central Australia. Even moderate amounts of salt will hinder vegetation growth, the lack of which leaves soils susceptible to wind and water erosion, i.e. some of the processes involved in desertification (Section 9.3). Obviously, salt accumulation will restrict crop growth in irrigated crop lands where inadequate flushing of the soil has allowed salts to accumulate. The very nature of irrigation, i.e. the exposure of water to the atmosphere as it is transferred from concentrated to diffuse sources, is inevitably conducive to salt accumulation in the water and soils in any area where evapotranspiration exceeds precipitation. Only where rates of water application are in excess of the rates necessary to promote leaching are salts flushed out of the soil and back into the drainage water. Alkalinisation is a similar problem; instead of calcium and magnesium salts, sodium compounds which are mainly carbonates and bicarbonates, accumulate. This can cause the loss of soil structure and the development of a crust. Where irrigation is practised without a fallow or rest period, or where there is seepage, waterlogging may ensue because of a rise in the water-table. Apart from creating an anaerobic root zone, which is not conducive to crop growth, waterlogging encourages salinisation and alkalinisation and is a misuse of the water resource itself.

These adverse environmental impacts of irrigation cause a reduction in crop productivity, the very commodity that irrigation was designed to enhance. Although estimates vary as to how much reduction in crop productivity and land abandonment actually occur globally, Postel (1990) suggests that approximately 65 per cent of the world's irrigated land requires upgrading in order to remain productive. Rhoades

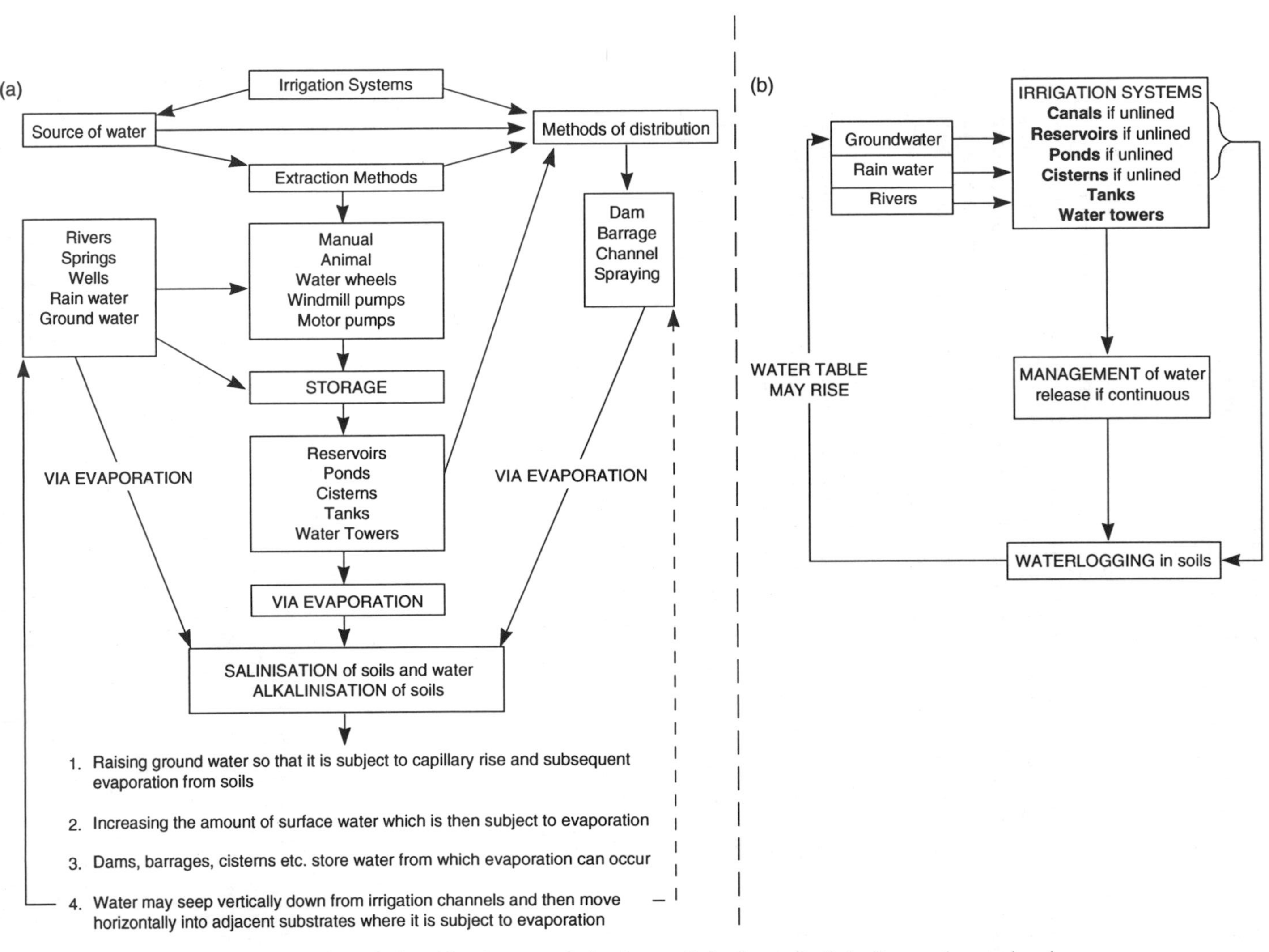

Figure 9.8 The relationships between irrigation, salinisation, alkalinisation and waterlogging

(1990) reports that crop productivity on *c*. 30 per cent of all irrigated land (data are given in Figure 9.7) is reduced because of salinity problems. In Egypt and Pakistan, for example, environmental impairment may be responsible for a 30 per cent decline in crop yields (quoted in Postel, 1993). Even in developed nations there are serious problems. In the USA, 5×10^6 ha are adversely affected by salt accumulation (Rhoades, 1990) and reference has already been made to the problems in the region surrounding the Aral Sea (Section 8.4) where as much as 2.5×10^6 ha may be salinised. There may also be adverse effects on aquifers which receive drainage from irrigated land. High nitrate levels are a case in point (see Section 8.4) and may create problems for the provision of domestic water, as may contamination with pesticides and heavy metals.

Aquifers that provide irrigation water may also become depleted causing irrigation systems to become ineffective and eventually abandoned. The excessive pumping of water from aquifers occurs in many countries. In the USA the most well-known example is that of the Ogallal aquifer in the high plains of Texas, Oklahoma and New Mexico (Johnson, 1992). Here, water removal has been greatly in excess of the recharge rate of 0.5–1.3 cm yr^{-1} from precipitation. Groundwater irrigation of 18 583 km^2 occurred between the 1930s and the 1980s; the sustained use of water to continue to irrigate this area is likely to leave only 49 per cent of the original water content by 2020. The overexploitation of aquifers in coastal regions can lead to the intrusion of salt water which reduces its use for irrigation and domestic consumption. This has occurred in the Lebanon (Khair *et al*., 1992), where the near-coastal aquifers have been heavily exploited for a variety of purposes including irrigation. The haphazard and unregulated drilling of wells around Beirut has reduced freshwater levels to such an extent that saline water has intruded, via a heavily faulted bedrock, into the aquifer. Salinity increased from 340 mg l^{-1} in the early 1970s to 22 000 mg l^{-1} in 1985. A similar problem has occurred in Argentina near Mar del Plata city on the Atlantic coast (Hernández *et al*., 1992). Several other nations have already developed, or have plans to develop, systems for exploiting aquifers containing 'fossil' water. This is so called because the water accumulated in the aquifer millennia ago and there is little, if any, recharge. This is a form of water mining so it represents the exploitation of a non-renewable resource. According to Ali-Ibrahim (1991), 75 per cent of Saudia Arabia's water is so derived. This nation intends to increase its food production and thus it is likely that the reduction in the groundwater resource is set to continue. Libya has a similar scheme underway. From an aquifer beneath the desert in the south of Libya, fossil water is extracted and pumped through underground pipes to irrigate land near the Mediterranean coast. The aim is to irrigate 240×10^3 ha and so improve Libya's food production.

In the 1950–1980 period the global expansion of irrigation was implemented mainly via the construction of large dams. Many of these were built not only to provide water for irrigation systems but also to provide hydroelectric power. The most obvious example of such a project is that of the Aswan High Dam which was completed in 1964. As Pearce (1992, 1994b) has discussed, this dam is 4 km wide as it extends across the Nile at Aswan. It now provides Egypt with more than 70 per cent of its electricity needs and has allowed substantial development to accommodate Egypt's growing population. However, there have been some serious disadvantages; for example, the dam's impoundment of Lake Nasser not only required the displacement of *c.* 100 000 people (a problem not uncommon in such enterprises, e.g. the Narmada Dam project in India and that of the Three Gorges in China) but also increased the surface area of water exposed to evaporation. Moreover, this upstream control on the Nile's waters is now threatening the survival of the delta region where, throughout its occupancy since ancient times (see Section 3.1), agriculture has been widely practised. There is also a problem with nutrient availability and general soil fertility in the Nile Valley and its irrigated corridors. The Aswan Dam ensures that silt from the Nile's upper reaches is trapped, denying the lower reaches of the fertilising silt that sustained the civilisations of ancient Egypt. Now Egypt is obliged to use artificial fertilisers to sustain its agricultural output. The gains from eliminating the insecurity associated with flood years are thus counteracted to a certain extent by the need to supplement soil fertility. There are also problems with salinisation. Approximately 1 tonne of salt is deposited per hectare per year. Mitigation measures require the expenditure of \$US2×10^9, annually. This is an expensive and ongoing programme. In addition, Pearce suggests that the extension of irrigation beyond the Nile Valley into the adjacent deserts is unlikely to be sustained in the long term due to salinisation and waterlogging. Apart from the environmental issues that the Aswan Dam has created for Egypt, there are a number of political issues that epitomise the arena of hydropolitics. Egypt does not own the Nile. The headwaters arise in the highlands of East Africa whose nations include Ethiopia, Uganda, Zaire, Tanzania, Rwanda, Burundi and Eritrea (formerly part of Ethiopia). They all have a vested interest in the Nile; what they do in the long term will influence Egypt's diminishing advantages from its control of the Nile via the Aswan High Dam. As Biswas (1993) indicates, Egypt now has the lowest amount of arable land per capita in Africa despite these irrigation enterprises. In view of its rapid rate of population growth it is difficult to envisage how Egypt will continue to support its population into the next millennium without increasing its reliance on imported foodstuffs.

The Aswan Dam is not the only large-scale irrigation/hydroelectric scheme to be criticised on the grounds of its long-term environmental and economic impact. The extensive irrigation systems of Pakistan, for example, where approximately 75 per cent of the cropland is irrigated, have caused considerable environmental degradation. Today some 16×10^6 ha of the country is irrigated via canals, tubewells and a number of traditional practices that include lift irrigation (the use of a bucket on a scaffold) and *karez*, *ganat* or *fogarra* irrigation (the use of subterranean canals that begin at the base of hills to collect water and channel it as far as 10 km to distant fields). These practices have been described by Khan (1991), who also provides a brief history of irrigation which began in the 1630s. Most importantly, an extensive irrigation system has been developed in the Indus Basin and comprises a series of dams (e.g. the Mangla and Tarbela dams, which are also used to generate electricity), barrages and link canals.

The modern system is the product of the Indus Water Treaty between India and Pakistan which was established in 1947 when Pakistan was created. Well before this, however, in the late 1850s, waterlogging and salinisation had already become apparent in some areas, mainly due to perennial canal irrigation. By 1953, 6.5×10^6 ha had become poorly drained or waterlogged, with a further 0.4×10^6 ha being added to this total per year. Perennial irrigation, with no respite from water seepage, combined with unlined canals means that water is permanently saturating the soils at a rate that is far in excess of evapotranspiration. Consequently the water-table rises. Khan states that if this rises to less than 5 m below the surface the situation is compounded by the raising of water through capillary action. This water carries salts to the surface where they are deposited when the water evaporates. Thus salts are continually added to the soils, a process which accelerates as the water-table rises. When the water-table reaches 1.5 m below the surface, salinity becomes sufficiently acute that soils may have to be abandoned; even at 3 m depth crop productivity is adversely affected. Eventually, the water-table reaches the surface to cause permanent waterlogging in the soils. The situation has been exacerbated by the abandonment of plans for leaving fields fallow, mainly due to increasing pressure on the land from a rapidly growing population, and the rate of water application from the canals being inadequate from the outset for the thorough flushing of salts from the soils. Despite many attempts at improving this situation, notably through the Salinity Control Reclamation Project (SCARP) plans, large areas have been transformed into saline waterlogged swamps. More than 1×10^6 ha of land are currently classified as severely saline and are mostly unsuitable for crop production. A further 2.5×10^6 ha are seriously affected by salinity and waterlogging. Overall, these problems have reduced Pakistan's crop yields in the Punjab to amongst the lowest in the world.

Similar problems have occurred in China. Smil (1993b) estimates that between 1989 and 2000 China stands to lose $250–500 \times 10^3$ ha of farmland as a result of irrigation. Seepage and loss through evaporation may be as high as 80 per cent in the arid and semi-arid parts of north and central China (see Figure 9.6). Only large-scale upgrading of China's irrigation system will counteract these losses. It will involve the replacement of furrow irrigation by drip or sprinkler methods, for example. The latter, also referred to as micro-irrigation, are expensive but efficient. At the very least the efficiency of furrow irrigation can be improved by lining the canals, ditches and furrows and ensuring that the rate of water application is sufficient to flush out salts. Well-managed irrigation schemes can overcome many of these problems, as Dovrat (1993) has discussed in the context of irrigation schemes in Israel. He points out that water management must take account of soil types, salt content and crop water requirements. Flexibility of water application appears to be a key issue though in nations like Pakistan and China, where such systems have developed over hundreds and possibly even thousands of years, flexibility and upgrading are expensive and difficult factors to institute. Another approach is the breeding and cultivation of either salt-tolerant crops or crops with low water requirements. It is likely that biotechnology and genetic engineering (Sections 10.1 and 10.2) will be used to develop such crops. Bredero (1991) has discussed some of the possibilities associated with the use of saline water for irrigation in India. For example, barley will tolerate salinity concentrations when compared with maize or potato.

In other parts of the world, small-scale rather than large-scale irrigation systems have proved to be successful at a local level. There are success stories in the Sahel zone, for example, a range of which have been discussed by Brown and Nooter (1992). These systems comprise simple low-cost technologies which are implemented and controlled by local farmers on small plots. In Mali, for example, peri-urban irrigation has increased to supply food for urban populations. Many farmers are city-dwellers who invest in pump-based rather than gravity-based irrigation from rivers. Another example of low-technology irrigation is the hill-furrow system used by the

Sonjo people of Tanzania (Adams *et al.*, 1994) on the western side of the Rift Valley. The Sonjo build simple unlined canals to carry water diverted by stone and brushwood dams from spring-fed streams and rivers. The dams (or weirs) are built communally and comprise boulders which are sometimes strengthened with tree branches and mud. Occasionally dams of cement may be constructed but usually dams are repaired if not rebuilt every year. Thereafter, the water enters a complex system of canals which may carry it several kilometres before it enters the fields through breaks in the canal banks. Small basins, each of about 4 m^2, are irrigated in turn; how much water and how many inundations are allowed depends on the crop type and its water needs. The fields are cultivated by the Sonjo women whilst land preparation, i.e. burning and clearing, as well as the maintenance of the irrigation systems and the apportionment of water is carried out by the Sonjo men. The water allows a variety of crops to be grown, e.g. sorghum, sweet potato, beans, maize, cowpeas and cassava as well as fruit crops such as papaya, mango, lime and lemon. Schemes such as these appear to have been in operation for at least a century; they effectively enhance food productivity and appear to be more sustainable than many of the large-scale systems elsewhere in the world. Micro-irrigation in small-scale projects has also proved successful in North Africa (Van Tuijl, 1993), e.g. in Tunisia and Morocco, as well as in the Middle East, e.g. Israel and Jordan.

9.5 CONCLUSION

Agriculture has had, and continues to have, a major impact on low latitude lands. Inevitably, such land-use and land-cover changes bring about global environmental change. The topics discussed above represent important environmental issues that are likely to become more entrenched in the next few decades. This is because low latitude lands, most of which are occupied by developing countries, are experiencing high rates of population growth which, in turn, are putting increased pressure on often already beleaguered agricultural systems. The need to find land for growing crops and grazing cattle, some of which is to produce cash crops and meat for export rather than for the support of indigenous people, is leading to habitat destruction and fragmentation. Although the demise of tropical forests features prominently in the academic and popular literature as the most important type of habitat destruction, it is by no means the only habitat to be threatened. Savanna lands, mangroves and wetlands are also affected, sometimes indirectly through processes like salinisation or eutrophication. All of these impacts reduce biodiversity and the genetic resources contained within that biodiversity. In addition, it is likely that forest demise in particular will influence global climatic change via its impact on the global biogeochemical cycle of carbon.

The loss of the natural vegetation cover and its replacement with crops or pasture renders the soil susceptible to erosion. Low latitudes experience the highest rates of soil erosion in the world. This is a loss of a vital resource that developing nations can ill afford. Nor can they afford the silting problems that ensue as eroded soil is deposited in streams and reservoirs. Soil erosion is one of the components of the complex process of desertification, the other components being salinisation, alkalinization and waterlogging. Millions of hectares of cropland and pastures are either rendered unproductive annually or have their capacity to produce crops or animal products impaired. Both soil erosion and desertification can contribute to food scarcity, which at its most acute becomes famine. In many cases, these impacts are underpinned by social economic factors such as population increase, poverty, land tenure and ownership and political ideologies. These factors conspire with the fragility or robustness of the physical environment to bring about environmental change at local and/or regional scales. In addition, the need to generate cash crops to earn foreign currency and the ready markets that exist in the developed world encourage agricultural practices that are not necessarily ideal and which cause degradation. The gains are usually immediate but short lived.

Salinisation not only affects soils but also water resources, notably rivers and aquifers. This is caused by injudicious irrigation systems which create vicious circles. The system is instituted to improve agricultural production but waterlogging and salinisation can cause productivity to decline to such low levels that land may even be abandoned. It can also give rise to such high salinity concentrations in drainage systems and aquifers that the water is unsuitable for irrigation or, indeed, for the provision of domestic water supplies. Large-scale irrigation projects, possibly because of their inflexibility and sometimes poor design, appear to create the worst environmental problems, as is evidenced in Pakistan and China. Small-scale farmer-operated schemes, e.g. those in Tanzania, enjoy a high degree of success. Whatever the size of the scheme, good management is the key to success.

9.6 FURTHER READING

Agnew, C. and Anderson, E. (1992) *Water Resources in the Arid Realm*. Routledge, London.

Barrow, C.J. (1991) *Land Degradation*. Cambridge University Press, Cambridge.

Greping, Q. and Jinchang, L. (1994) *Population and the Environment in China*. Lynne Reinner Publishers, Boulder and Paul Chapman Publishing, London.

Hinman, C.W. and Hinman, J.W. (1992) *The Plight and Promise of Arid Land Agriculture*. Columbia University Press, New York.

Thomas, D.S.G. and Middleton, N.J. (1994) *Desertification: Exploding the Myth*. John Wiley and Sons, Chichester.

Whitmore, T.C. and Sayer, J.A. (1992) *Tropical Deforestation and Species Extinction*. Chapman and Hall, London.

CHAPTER 10

New Developments in Agriculture

Agriculture is entering a period of rapid innovation and adjustment if not major change. There are several factors deriving from quite different sources which are the agents of change in agriculture. First, there is the continued and intense application of science and technology, amongst the most important aspects of which are biotechnology, genetic engineering and information technology. Secondly, there is a growing awareness of the unsustainability of many existing agricultural systems and the need to remedy this situation. Finally, and the most unpredictable of these factors, is the impact of climate change if global warming does indeed occur.

As has been the case throughout history, science and technology develop to a certain level of sophistication whereupon they begin to generate what appears to be revolutionary progress. This has occurred in agriculture several times, as has been discussed in Sections 2.2 and 2.4 in relation to the early domestication of plants and animals, and in Section 3.3 in relation to the so-called agricultural revolution of the seventeenth and eighteenth centuries. Plant breeding programmes, based on the principles of heredity and genetics established in the mid-nineteenth century by Charles Darwin and Gregor Mendel, have led to the establishment of biotechnology and genetic engineering as major forces in agriculture. Modern biotechnology is the culmination of all the deliberate, and accidental, manipulations of plants and animals that have occurred throughout prehistory and history; indeed, agriculture itself is a form of biotechnology. However, the explicit breeding of desired traits in particular species has become a mainstay of twentieth century agriculture, giving rise to what is often referred to as the 'green revolution' of the post World War II period. The breeding of high-yielding varieties (HYVs) of the major crop plants in particular facilitated substantial improvements in crop yields throughout the world in the 1970s. The search continues for disease- and insect-resistant varieties as well as crop types that are tolerant of salt, drought and high and low temperatures. Genetic engineering, which involves the manipulation of chromosome constituents in plants and animals, is a particularly sophisticated means of obtaining such desired characteristics and it too is beginning to contribute to the production of improved species. Both biotechnology and genetic engineering, apart from the fact that they have many applications beyond agriculture, have been applied to aspects of agriculture other than plant and animal breeding. For example, biopesticides to control insects, fungi and weeds, are now widely used and certain bacteria are being harnessed to enhance nitrate availability in soils. Genetic engineering is being used to develop plants and animals that generate products other than food or fibre. For example, sheep and pigs can now be engineered to produce therapeutic substances for use in human health care, and rape can be engineered to produce a type of biodegradable plastic. The explosion in information technology that has occurred since the early 1980s has also brought about change in agricultural practices. The development of knowledge-based systems, remote sensing and geographical information systems represent yet another way in which technology has influenced agriculture. In particular, these developments have contributed to improved use of land and water resources and have provided a means of rapidly assessing the impact of agriculture.

There are other stimuli that generate change in agriculture. The growing awareness, publicly and at all political levels, of environmental issues and especially the realisation that many of society's activities, including agriculture, are not sustainable is prompting a change of attitudes. As Chapters 8 and 9 illustrate, the environmental impact of agriculture is so enormous that it is essential to find more efficient, less energy intensive and more enduring food production systems than many of those that exist today. In so doing the remaining natural and semi-natural ecosystems will be protected and valued as a source of novel materials and compounds. Sustainable agriculture is an important key to a sustainable future. The possibility that climatic change of some magnitude will occur as a response to global warming makes the possibility of sustainable agriculture particularly difficult to accommodate. Nevertheless, global warming is almost certain to occur. So far, little can be unequivocally predicted about its impact on soils, vegetation, crops and pests of crops and livestock. Indeed, as will be discussed below, much effort is being expended to develop predictive models because food security is such an important component of political stability.

Science, technology, environmental change stimulated by climatic change, as well as a world population that will continue to grow at least until the middle of the twenty-first century, are the major cultural and environmental factors that will change the nature of agricultural systems in the next few decades. The changes are likely to be profound.

10.1 BIOTECHNOLOGY

Biotechnology is a general term that pertains to the harnessing of organisms, living or dead cells and cell components to undertake specific processes with applications in fields as diverse as agriculture, medicine and pollution control (these have been examined in Mannion 1993a, b, 1995c). The manipulation of cell components can be undertaken at two levels: the subcellular level which involves whole cells without the cell walls, and the nuclear level which involves the manipulation of the genetic material only. This latter is genetic engineering and its applications in agriculture are considered in Section 10.2. Biotechnology is not just a product of the late twentieth century; its practice really began 10 000 years ago when the first plants and animals were domesticated and the earliest agricultural systems were inaugurated. The selection of plants and animals and their improvement that have occurred throughout prehistory and history all constitute a form of biotechnology. There is also evidence for the harnessing of other organisms, notably bacteria, in food and drink production. For example, bread, cheese, beer, wine and yoghurt production all rely on bacteria and all have their origins in prehistory. Some 8000 years ago the Sumerians were producing beer and 6000 years ago the Egyptians were using yeast to produce leavened bread. Today, modern biotechnology has a substantial and increasing role to play in the food industry. Indeed, the culture of the fungus *Fusarium graminearum* in a mixture of glucose and ammonia is the way in which the mycoprotein (a fungus-derived protein) commercially known as Quorn® is produced. This is now widely available in supermarkets and can be used as a meat substitute suitable for vegetarians. Pruteen®, first produced in the 1960s, is another biotechnologically produced protein. It comprises cultures of the bacterium *Methylophilous methylotrophus* and has been marketed as an animal feed. In the longer term it is possible that such products will provide alternative food sources to those produced conventionally.

However, the most important applications of modern biotechnology in agriculture fall into three main categories. First, there is its role in plant and animal breeding and, in relation to the former, biotechnology can facilitate the rapid production of large numbers of identical seedlings, i.e. clones, for marketing. Secondly, biotechnology is becoming increasingly important in the crop-protection industry. As discussed in Section 8.3, the adverse ecological impact of post World War II pesticides, coupled with the increasing resistance of pests, especially insects, to modern chemical pesticides, has led agrochemical companies to seek alternative means of pest control. In consequence, there is now a growing industry, as well as ready markets, for products known as biopesticides. These are commercially produced preparations containing micro-organisms that attack specific pests or groups of pests. The third category of biotechnological applications in agriculture concerns the enhancement of nutrient availability, particularly nitrate, in agroecosystems to which end bacteria can be manipulated and used to inoculate both seeds and soils. All of these applications centre on the improvement of the fundamental processes of energy flows and biogeochemical cycles which underpin all agricultural, and for that matter ecological, systems (see Section 1.2). In the context of the agroecosystem the improvements are intended to channel more energy, nutrients, light, etc., to the crop. All of these aspects of biotechnology are the focus of further improvements through genetic engineering, as is discussed in Section 10.2.

The improvement of existing crops is of prime importance in agricultural research. Until the 1970s, plant breeding programmes relied mainly on the manipulation of whole organisms. Such conventional methods, which involve the interbreeding of established crop species (i.e. landraces) and/or crossing with wild relatives, have been discussed by Stoskopf *et al.* (1993). Although it can take as long as 15 years to produce a species for marketing, many of the new crop varieties that fuelled the 'Green Revolution' of the post World War II period were produced in less than a decade. This was the result of concentrated international efforts that were brought to bear on the vital issue of improving the world's food output. In addition to the development of many improved crop plants, the efforts of plant breeders during this period led to the establishment of international research centres. These are now renowned for their excellence in research and for their role as gene banks, i.e. as repositories of seeds of cultivated species and their wild relatives. It is from these banks, in conjunction with additional acquisitions of seeds of landraces, that further improvements in crop species will derive as genetic engineering gathers pace (see Section 10.2). Corporately owned seed companies, however, are the major players on this lucrative stage. There are now concerns (Mackenzie, 1994) that the World Bank will take control over all these International Agricultural Research Centres (IARCs) leaving developing countries, who not only house most of these institutes but also contribute most of the seed collections, even less control over agricultural developments than they have currently. Conventional crop breeding programmes remain vital for the development of new and improved crops but modern biotechnology is providing a further range of techniques to accelerate and indeed revolutionise crop breeding. As Figure 10.1 shows, there are three ways in which cells or groups of cells can be manipulated to produce crop plants with propitious characteristics. Once a crop species has been identified after conventional breeding through crossing, etc., tissue culture (or micropropagation) can be employed to produce large numbers of clones. This ensures that crop quality and productivity are relatively uniform, an attribute that is advantageous for subsequent marketing. In addition, germplasm (chromosomal material) can be lodged in gene banks for future use. Tissue culture has been widely used to reproduce not only species used in agriculture but also those used in horticulture and as ornamentals, i.e. garden plants. Examples of crop plants so produced include strawberry, potato, tobacco, tomato and cauliflower. The use of tissue culture, as Figure 10.1 shows, allows the potential of their biotechnological methods to be realised. There are many examples of crops so produced. One such example is cassava (*Manihot esculenta* Crantz), which is a staple crop in tropical regions. Hershey (1993), reviewing the developments in cassava research, points out that much of it has been undertaken by the *Centro Internacional de Agricultura Tropical* (CIAT) in Colombia, South America and the International Institute of Tropical Agriculture in Ibadan, Nigeria. These institutes are government rather than privately owned and, in view of the monopolies that many agrochemical seed companies are beginning to develop in relation to the production and marketing of transgenic crop seeds (Section 10.2), they are particularly important for ensuring that developing countries benefit from biotechnology. The efforts of these institutes in screening cultivated and wild species of cassava have resulted in the discovery of species with resistance to bacterial blight and African cassava mosaic virus. Tissue culture can be used to generate large numbers of clones of these resistant types which can then be given to farmers. There are *c.* 4000 clones of cassava in CIAT's germplasm bank; 600 of these are elite clones which feature in 30 national cassava programmes in Latin America, Africa and Asia (Roca, 1989). The cultivation of one such clone, Nan-Zhi 188, on 30 000 ha in China has produced yield increases of between 30 and 70 per cent. This success story highlights the importance of germplasm collections and the need to preserve as large a gene bank as possible for crop species. Concerns that such biodiversity is diminishing are justified because it means that opportunities to increase yields, rather than expand the area used for agriculture, are lost.

In relation to animal biotechnology, the culture of mammalian cells has several applications that impinge on animal production as well as other applications such as human health care. One of the oldest applications of animal cell culture is that of viral vaccine production. Examples in human health care include measles, polio, rabies and rubella. One of the most important vaccines so produced for veterinary care includes that for foot and mouth disease. Cell cultures are also widely used to produce therapeutic substances for human health care. Examples include interferon and interleukin. The potential for this type of production is considerable, especially if genetic engineering is used to provide the cells with genetic codes for desired products (Section 10.2). The manipulation of the reproductive process in farm animals is yet another facet of biotechnology. Selective breeding, for example, is undertaken to

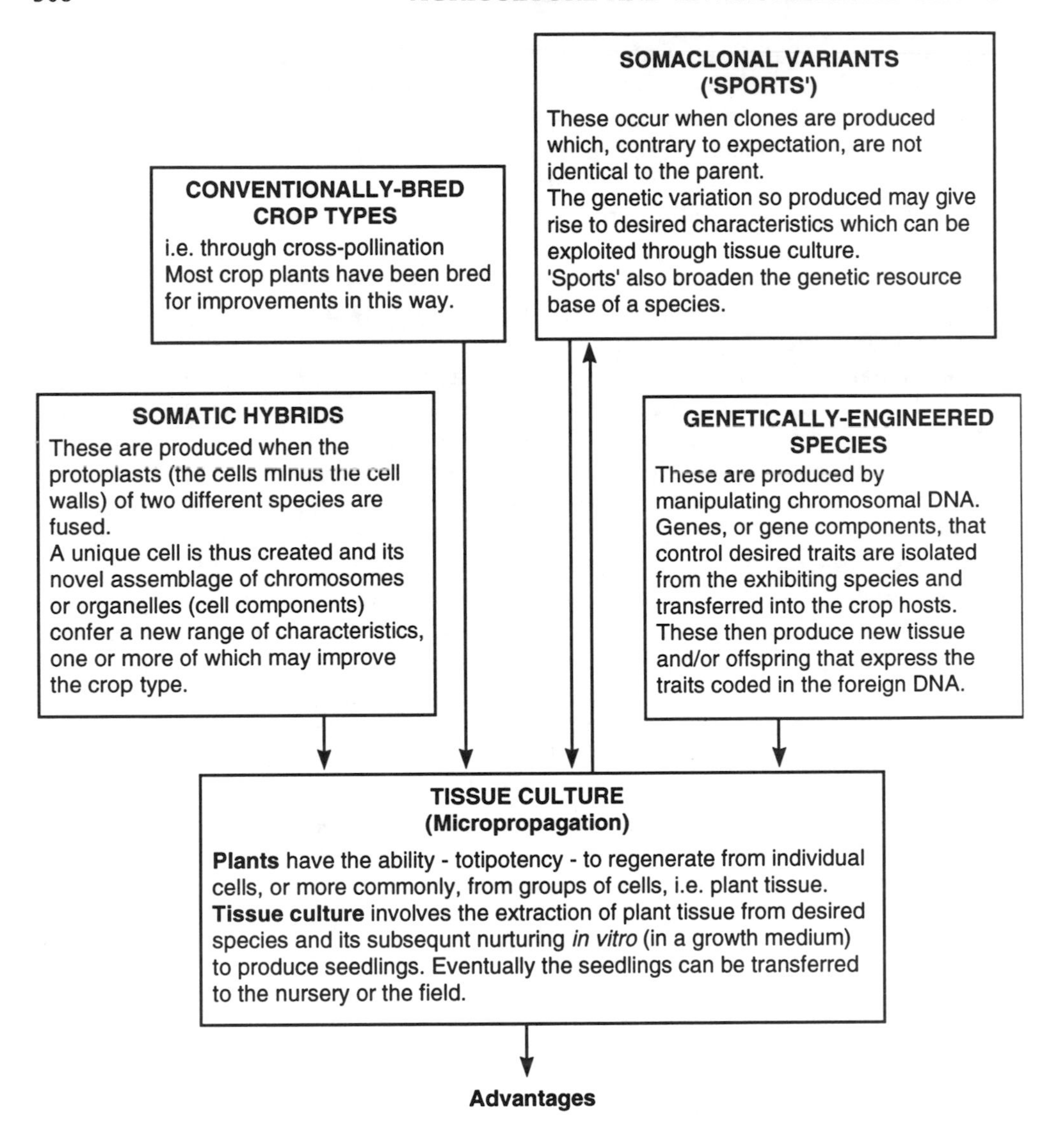

Figure 10.1 The various ways in which crop plants can be modified and duplicated

improve secondary productivity. Such manipulation is now easier than hitherto, and began with the development of artificial insemination techniques in the 1930s. Further developments are occuring via genetic engineering as discussed in Section 10.2.

Somatic hybridisation and somaclonal variation (Figure 10.1) can also be exploited to generate improved crop plants which can then be reproduced via tissue culture.

Somaclonal variation has been recognised in potato and sugar cane, some variants of which are resistant to certain diseases. Evans (1989) reports that there arc variants of the potato Maris Piper that exhibit resistance to virus attack and the fungal disease scab. In addition, Douches and Jastrzebski (1993) report that there are other variants with resistance to late blight fungus, the cause of the potato famine in Ireland in the 1840s (see Section 2.3.5). There are also somaclonal variants of sugar cane with resistance to eye-spot disease, Fiji disease and downy mildew (Evans, 1989). Somatic hybridisation (Table 10.1) has also been used to produce new varieties of tomato, potato and tobacco with resistance to a range of diseases. In relation to the potato, for example, variants have been produced with resistance to potato virus X, potato leaf roll virus and the herbicide atrazine (Evans, 1989). The advantage of herbicide resistance is that weed competitors can be eliminated whilst the crop remains unimpaired. This trait is being developed in wheat and other major crop plants through genetic engineering (Section 10.2). Tree crops in particular can benefit from micropropagation as has happened with oil palm. There is considerable variation in oil productivity between plants. Identifying elite species with oil production 30 per cent above average and then reproducing vast quantities of these plants via micropropagation is the way in which the oil palm industry replaces ageing trees and increases oil production.

The capacity of certain types of viruses, fungi and bacteria to cause diseases in crop pests has also been turned to advantage by biotechnology. There are now commercial preparations of these organisms, available for crop protection (Strobel, 1992). As discussed in Section 8.3, the rapidity with which insect pests, especially, become resistant to chemical pesticides has stimulated the development of these additional crop-protection agents which are known as biopesticides. Sometimes these are used as sole crop-protection agents but often they are components of integrated pest management strategies (for reviews, see Rappaport, 1992; Zalom and Fryi, 1992) which include chemical and cultural means of pest control. The latter include variations in planting times to avoid pest outbreaks, staggered harvesting, etc. There are many different types of biopesticides based on *Bacillus thuringiensis*—the Bt pesticides. This bacterium, in common with the other biopesticides discussed below, produces chemicals that are toxic to insects. According to Whitten and Oakeshott (1990), Bt produces two classes of insect toxins that work as stomach poisons when ingested by feeding insect larvae. The delta endotoxin is not harmful to plants or higher animals and is the active ingredient in three of the four commercial strains. The most widely used strain is *B. thuringiensis* var. *kurstaki* (Meadows, 1992). This is particularly effective against lepidopteran pests, i.e. butterflies and moths, and it is often employed in integrated pest management strategies for cotton. There is, however, evidence that some lepidoptera are developing resistance to Bt pesticides (e.g. Gibbons, 1991) which once again illustrates the capacity of insects to survive and the ongoing battle between insects (and other pests) and the farmer. This resistance is worrying in the light of the recently marketed (1995) transgenic cotton seeds with engineered resistance to the cotton boll worm (see Section 10.2).

Other types of biopesticides have been discussed by Chrispeels and Sadava (1994). They comprise myconematicides, mycoherbicides, mycofungicides and myco-insecticides, all of which employ fungi as the toxin-bearing carrier. According to Wainwright (1992), there are more than 400 species of fungi that attack insects and mites. There is, thus, considerable potential for the development of mycoinsecticides. Examples of mycoinsecticides include *Beauveria bassiana* which kills the Colorado potato beetle and *Metarhizium anisopliae* which is produced in Brazil to combat froghoppers in sugar cane. According to Dart (1990), *Beauvaria bassiana* is also used in China to combat the European cornborer in wheat crops. Examples of commercially available mycoherbicides are Collego®, Casst® and DeVine® (Ayres and Paul, 1990). Collego® is host-specific to the northern joint vetch which is a pest of soy bean and rice crops in the USA. Casst® is a pathogen of sickle pod and coffee senna which are pests of soy bean and peanut crops, whilst DeVine® combats milkweed infestations in Floridan citrus groves. Work on mycofungicides and myconematicides is experimental. Viral pesticides are also being developed which are species-specific. Viral pesticides are, therefore, target-specific rather than broad-spectrum but are harmless to plants and animals, including humans. Chrispeels and Sadava (1994) suggest that the use of baculoviruses, which have been most widely studied, could reduce the use of chemical insecticides by *c.* 60 per cent in California and *c.* 80 per cent in Central America. The development of viral pesticides is still in the early stages but there is much potential.

As stated above, biotechnology can be applied to enhance nutrient availability in the soils of agricultural systems. This involves the manipulation of whole organisms, in this case mainly bacteria and possibly some fungi. One of the most important constraints on crop productivity is the availability of nitrate in soils. This is why artificial nitrate fertilisers are so widely applied but, as discussed in Section 8.4, these can cause the cultural eutrophication of drainage systems, aquifers and coastal waters. As an alternative to artificial fertiliser, cultures of the bacterium *Rhizobium* spp. can be used. This species (*Rhizobium*) occurs naturally in a symbiotic relationship with leguminous plant species, including several important crops, e.g. alfalfa, peas and soy beans. Inoculating soils and/or crop seeds with *Rhizobia* can improve productivity substantially. However, as Catroux and Amarger (1992) have discussed, there must be certain conditions before *Rhizobia* inocula produce benefits: on the one hand, *Rhizobium* spp. are host-specific, e.g. *R. japonicum* forms a nitrogen-fixing association with soy beans and *R. meliloti* associates with alfalfa. This they do by entering the crop plant's roots where they multiply to produce root nodules. Usually only one type of *Rhizobium* can infect any given crop type and matching the two is essential if the symbiosis is to produce benefits. It is now possible for farmers to purchase the correct strains at the same time as the crop plants/seeds are purchased. The bacteria are packaged in a carrier such as powdered peat (this may not be the most appropriate or sustainable use for peat). On the other hand, the activity of *Rhizobia* will be curtailed if the soils are already rich in nitrate. Where a crop rotation includes a cereal and a legume, any artificial nitrate fertiliser applied for cereal production may inhibit nodulation of the legume in the following year. Using *Rhizobia* inocula and artificial nitrate fertilisers is, thus, counterproductive. According to Primrose (1991), much success has been achieved with *Rhizobia* inocula in Australia and New Zealand. Here forage legumes, such as clover and alfalfa, were introduced into a region with soils in which there were no indigenous *Rhizobia*. The inocula have to be repeated every year because the bacteria do not survive all year round. Much effort is being invested in determining the molecular basis of this symbiotic relationship in order to genetically engineer crops, other than the leguminous types, to develop a similar relationship.

Eventually, it may also become possible to engineer the capacity for nitrogen fixation in the major crop plants (Section 10.2). The use of *Rhizobia* inocula, in principle, is environmentally benign and provides an acceptable alternative to the use of artificial nitrate fertilisers which, as discussed in Section 5.1, are energy intensive. Apart from *Rhizobia*, fungi may be exploited to enhance nutrient availability in soils (Wainwright, 1992). For example, the fungus *Penicillium bilaji* releases zinc and phosphates from soils which can then be taken up by crops. Moreover, inocula of fungi that form mycorrhizal associations with crop plants give rise to an increased rate of nutrient uptake.

10.2 GENETIC ENGINEERING

Genetic engineering, which is also known as recombinant DNA (deoxyribonucleic acid) technology, gene cloning and *in vivo* genetic manipulation, involves the manipulation of DNA and the transfer of gene components between species in order to encourage the replication of desired traits. First, genes and/or their components which carry the genetic code for the desired characteristic are isolated. This DNA is then transferred via a vector, e.g. a bacterium, to a host organism. The host will then produce new tissue and/or offspring that exhibit the traits coded in the foreign DNA. The technique is thus another, but particularly sophisticated, means of plant and animal breeding. It is much quicker than conventional methods and more precise in so far as DNA only is manipulated. Once a transgenic species has been created it can, if it is a plant, be reproduced in large numbers by tissue culture (Figure 10.1).

There are many applications of genetic engineering in fields as diverse as medicine, waste disposal, pollution mitigation and resource recovery (reviewed in Mannion, 1991, 1993a). However, there is a vast potential for its use in agriculture. Indeed, the first genetically engineered crops are just appearing in the market-place. The so-called Flavr Savr® tomato is being marketed by Calgene and Zeneca plc and transgenic cotton seeds are to be marketed in 1995 by Monsanto. As Table 10.1 shows, the 1990 value of the biotechnology industry was *c.* $\$6.2 \times 10^9$, of which 47 per cent was accounted for by the agricultural sector. Projections for AD 2000 suggest that agriculture will remain the largest sector and will increase by a factor of 17. The industry is dominated by the USA, which accounts for 33 per cent, and is increasing, especially in Japan and Europe. According to the financial consultants Ernst and Young (1993), there are *c.* 1300 separate biotechnology companies in the USA. Of these the greatest concentrations are in the San Francisco Bay area, with 192, and New England, with 172. The agrochemical sector is dominated by Du Pont, Monsanto, Cyanamid, Dow Chemical and Pioneer Hi-Bred International which had an aggregate sales value in 1993 of US $\$71 \times 10^9$. The aggregate value for the top six pharmaceutical companies in the USA was US $\$40 \times 10^9$. Thus, biotechnology is a big business that is set to grow rapidly as genetic engineering brings valuable rewards.

Table 10.2 gives some of the many applications of biotechnology in agriculture (see also *Agrochemical Monitor*, 1994). In common with those applications discussed in Section 10.1, the basic aim is to improve crop and animal productivity. Improvements in crop productivity can be achieved by increasing the efficiency of resource use, i.e. energy and nutrients, by crop plants. As well as enhancing the existing attributes of crop plants new characteristics can be added. The introduction of pest-, disease- and

Table 10.1 The value (in US $\$\times10^9$) of world markets for biotechnology products in the early 1990s and projected values for AD 2000 (based on Kathuri *et al.*, 1992)

	Pharmaceuticals	Chemicals	Agriculture	Environment	Equipment	Total
Present World	1.50	0.12	2.90	0.50	1.20	6.20
World AD 2000	29.40	18.00	49.20	2.50	3.40	102.50

Table 10.2 Some applications of biotechnology in agriculture

Objective	Methods	Example
Crop breeding and Crop improvement	Tissue culture	Cassava
	Somaclonal variation	Potato
	Somatic hybridisation	Tomato
Disease resistance	Genetic engineering	Melon, tomato, tobacco
Insect resistance	Genetic engineering	Cotton, maize
Herbicide resistance	Genetic engineering	Tomato, oilseed rape, wheat, squash
Cold/frost tolerance	Genetic engineering	Ice minus bacteria
Salinity tolerance	Genetic engineering	Barley
Improved nitrogen availability	Fermenting	*Rhizobium* bacteria inocula
Flavour improvement	Genetic engineering	Tomato
Speed of ripening	Genetic engineering	Tomato
Production of speciality chemicals in plants	Genetic engineering	Oilseed rape
Production of chemicals in animals	Genetic engineering	Sheep, pigs
Production of organs for transplants	Genetic engineering	Pigs
Biopesticide production	Fermenting, genetic engineering	Bacteria, fungi

herbicide-resistance, for example, means that crop biomass is not consumed by pests, that photosynthetic ability is not impaired by disease and that weed competitors for light and nutrients can be eliminated. Genetic engineering also offers the possibility of designing crops to suit specific environments. This is a notable departure from traditional agriculture wherein the environment tends to be tailored to suit the crop. Crops that can withstand certain types of environmental stress, such as drought or high salinity concentrations, could improve productivity substantially. Improvements in nutrient availability, especially nitrogen, are also a target for genetic engineers, as are improvements in crop texture, flavour and chemistry. Biotechnology, and especially the subdiscipline of genetic engineering, offer much promise to improve animal productivity through the manipulation of animal reproductive processes. The technology is, however, not without disadvantages. Some of these are ecological and environmental; others are related to the economic and political organisation of the technology.

Since the mid-1980s the engineering of pest resistance in crop plants has been a major goal of the seeds/agrochemical industry (reviewed in Freyssinet and Derose, 1994). As in the case of biopesticides (Section 10.1), this is a response to the fast-developing resistance of many insects to conventional insecticides (see Section 8.3)

Table 10.3 The advantages of genetically engineered insect resistance in crop plants (based on Gatehouse *et al.* (1992) with additions)

A	Continuous protection is provided irrespective of insect life-cycles, season, or weather conditions. This is a major advantage over chemical pesticides.
B	Provided transgenic seeds are marketed at a reasonable price, the costs are lower than if chemical pesticides are used. There is no need for repeated applications which require labour.
C	The costs of bringing a transgenic crop to the market-place are much lower than those for developing a new chemical pesticide.
D	The entire crop plant is protected, including underground parts.
E	The protection is provided *in situ* and there is no possibility of contamination of the wider environment.
F	Protection is target-specific; beneficial insects remain unaffected. Consequently, there is less disruption of food chains and food webs than there is when conventional pesticides are used. There may, however, be other ecological disadvantages.
G	The active factor is biodegradable and there is virtually no possibility that it could become concentrated in the environment. There is, however, the possibility that the resulting gene products could be toxic to animals or humans.
H	Pesticide residues are absent so transgenic crops could become preferred by consumers. It should be a legal requirement for consumers to be informed, via labelling, of gene products.
I	Engineered pesticide resistance overcomes the problem of increasing resistance in insect pests to conventional pesticides. The advantage may be short-lived.
J	The reduction in the use of chemical pesticides will decrease the amount of fossil-fuel energy input into high-technology agricultural systems.

though there are many additional advantages as shown in Table 10.3. The genetic engineering of insect resistance in crop plants has focused on the bacterium *Bacillus thuringiensis*. As discussed in Section 10.1, several varieties of this are commercially available as biopesticides, so called because of the toxicity to various insects of the chemicals it produces. The gene(s) that code for toxin production have now been identified and transferred into several crop plants. Fraley (1992) reports success with tomato, maize, tobacco and cotton. The first of these crops to be marketed is cotton though a great deal of controversy has been generated by the patent granted to Agracetus, the biotechnology company that developed the transgenic seeds (Kidd and Dvorak, 1994; Mestel, 1994) and which is also working on the improvement of cotton fibre through genetic engineering (John, 1994). The patent relates to the entire species and thus provides the company with a monopoly. This is opposed on the basis that crop species, including improved varieties, are 'resources of common heritage' and therefore should not be under the exclusive control of any one individual or company. However Agracetus and Monsanto (one of the companies licensed by Agracetus to market the new seeds) believe that without the safeguard of a broad patent their investment in the research and development could be worthless. Such patents are important not only in the context of competitiveness between concerned companies in the developed world but also in relation to technology transfer to the developing world. This is discussed in more detail below. Initial trials with the transgenic cotton suggest that the crop is highly resistant to attack from the cotton bollworm. However, in view of the growing resistance in insect populations to Bt pesticides (Section 10.1) it is possible that the efficacy of these transgenic crop strains may be relatively short.

There are many other similar products nearing the market-place. Gasser and Fraley (1992) report that potato plants with resistance to the Colorado potato beetle have been produced. In addition, research on the engineering of insect resistance in wheat, maize and rice is underway. As these are the world's major cereal crops such developments represent not only big business but also a means of transforming the world's arable agricultural systems. Koziel *et al.* (1993) have discussed field trials of maize plants engineered with Bt genes to confer resistance to the European corn borer. Despite heavy infestations of the trial plots with the insects, the maize plants sustained little damage and so attest to the success of the inbred resistance. Ciba Geigy, in conjunction with Mycogen Corporation, plan to market transgenic maize within five years (*Agrochemical Monitor*, 1993). Rice is also the subject of research into engineering insect resistance. The company Mitsubishi Kasei, for example, is working on pest and virus resistance. Fujimoto *et al.* (1993) report the production of transgenic japonica rice plants (*Oryza japonica*) with Bt genes. Laboratory tests with two pests, the striped stemborer and leaf folder, indicate much more resistance in the transgenic plants than in non-engineered control plants. The potential for developing insect resistance in this way is considerable because of the wide variety of naturally occurring Bts. The case of sugar cane, for example, is discussed by Robinson (1993). The insecticidal properties of many of these Bts have yet to be investigated but there is a large gene resource. It may also be worthwhile improving already established Bt toxin genes by genetic engineering and then using these to produce transgenic crop plants. Such altered Bt genes could also provide a means of generating improved Bt insecticides directly (Feitelson *et al.*, 1992).

There are several other ways in which insect resistance can be engineered in crop plants though none are as far advanced experimentally as those described above. The use of viral preparations as biopesticides was referred to in Section 10.1. Leishy and Van Beek (1992) point out that the insecticidal properties of such viruses could be harnessed and transferred to crop plants. The efficacy of viruses could also be improved by genetic engineering, e.g. toxin-producing genes derived from insect predators and parasites could be introduced. In fact the first field trial of a genetically engineered baculovirus insecticide took place in Oxfordshire in the UK, in April 1994, amidst much controversy (Coghlan, 1994). The work has been reported by Cory *et al.* (1994) who modified a virus of the alfalfa looper, a pest of cabbage, with a toxin gene from the venom of the scorpion. The field trials indicated that the new virus was more effective than the unmodified virus. The controversy concerned opposition to the release of genetically modified organisms into the environment on the grounds that they may kill non-target caterpillars and spread uncontrollably. This issue is referred to again below. Other work concerned with the transfer of plant genes that encode for proteins with insecticidal properties is still experimental. Gatehouse *et al.* (1993) and Gatehouse and Hilder (1994), for example, have discussed the possibilities of engineering protection against the cotton budworm. This is a major pest of tobacco, cotton and maize. The cowpea, however, produces a natural insecticide (a larval growth inhibitor), the gene component for which has been successfully engineered into tobacco in laboratory trials. The screening of crops and their wild relatives for insecticidal properties could eventually produce a large gene bank for engineering insect resistance in a host of crop plants.

Another cause of substantial crop losses is disease. Diseases can be produced by viruses, bacteria and fungi. As a result, much research is underway to reduce such losses by engineering crop plants with inbuilt resistance (Gasser and Fraley, 1992; Brears and Ryals, 1994). Fungal diseases can be treated in the field but viruses are not so easy to treat. As reported in *Agrochemical Monitor* (1994), several companies are attempting to develop virus resistance in crop plants. The tomato, in particular, is the focus of such research and it is now possible to engineer the species to exhibit resistance to the tomato mosaic virus. In field trials of transgenic tomatoes, yield increases of 25–36 per cent have been achieved. Virus-resistant tomato plants could improve output by eliminating losses estimated to be as much as US$50 million per year (Primrose, 1991). Other crop plants which have been engineered with virus resistance include melon (Dong *et al.*, 1991), alfalfa (Hill *et al.*, 1991), cucumber (Gonsalves *et al.*, 1992) and potato (Truve *et al.*, 1993). In relation to the potato, the viruses that cause most losses are potato viruses X and Y and potato leaf-roll virus. Resistance to all of these can now be engineered though transgenic stains are not yet being marketed. Much research on virus resistance is being undertaken in China. For example, field trials on tomatoes with resistance to the cucumber mosaic virus have been undertaken and screening programmes to discover genes in other plants that confer virus resistance are underway (Chen and Gu, 1993). A prime target for engineering virus resistance is rice. The Chinese have already developed cultivars with resistance to red stripe and tungro viruses (Coghlan, 1993) and are attempting to develop resistance to bacterial blight. The latter can reduce rice yields by 10 per cent. Antifungal properties are another target for genetic engineers because fungal diseases cause crop losses that are as high as 15 per cent (Logeman and Schell, 1993). Chitinases, which are enzymes produced in certain plants, bacteria and fungi, are antifungal agents because they break down fungal cell walls. Identifying the genetic component responsible for the production of these enzymes and transferring it to crop plants can confer resistance to fungal attack. For example, Van den Elzen *et al.* (1993) report that engineering of resistance to the fungus *Fusarium oxysporum* in tomato is now possible, and Moffat (1992) has discussed the engineering of resistance to the late blight fungus (the cause of the potato famine in Ireland in the 1840s; Section 2.3.5).

According to *Agrochemical Monitor* (1994), the most extensive research on crop plants is concerned with the engineering of herbicide resistance. As Table 10.4 shows, the focus of interest is quite varied in relation to both crop plant and herbicide. This diversity reflects the market value of engineering this attribute. It also reflects the interest of agrochemical companies that manufacture the specific herbicides. The company itself, sometimes via a licensing agreement with a seed or biotechnology company, is aiming to produce and market both the transgenic seeds and the herbicide. Again, this has prompted critics to point out how prejudicial this could be in relation to farmers in developing countries (see discussion below). The advantage of herbicide resistance in crops, especially the major cereals, concerns the continued use of broad-spectrum, environmentally benign herbicides that allow weed competition to be removed without any damage to the crop. As discussed in Section 8.3, herbicides work in a variety of ways; many affect plant metabolic processes that are controlled genetically, e.g. enzyme production. These processes can be altered by genetic engineering so that continued herbicide use does not impair the crop. Most importantly, it is now possible to engineer resistance to glufosinate in wheat

Table 10.4 Efforts underway to engineer herbicide resistance in crop plants (based on *Agricultural Monitor*, 1994)

Crop	Objective	Status (if known)
Canola	Resistance to metribuzin	
Canola	Resistance to glufosinate	1995 release
Canola	Resistance to glyphosate	
Cotton	Resistance to bromoxynil	1994 release
Cotton	Resistance to glyphosate	
Soy beans	Resistance to glyphosate	
Soy beans	Resistance to glufosinate	
Maize	Resistance to imazaquin	Marketed
Maize	Resistance to imazethapyr	Marketed
Maize	Resistance to glufosinate	
Maize	Resistance to glyphosate	
Potatoes	Resistance to herbicide*	
Sugar beet	Resistance to herbicide*	
Sugar beet	Resistance to glufosinate	
Sugar beet	Resistance to glyphosate	
Tomatoes	Resistance to glufosinate	
Cucurbita	Resistance to herbicide*	
Squash	Resistance to herbicide*	
Cucumber	Resistance to herbicide*	
Cantaloupe	Resistance to herbicide*	
Wheat	Resistance to glufosinate	
Rice	Resistance to glufosinate	

* Unspecified.

(Vasil *et al.*, 1992) and as Table 10.4 shows, research is underway on other major crop plants (see also the discussion in Hoyle, 1993; Chrispeels and Sadava, 1994; Christou, 1994).

Changes to metabolic processes in crop plants can also alter the chemicals produced within the plants. Many crop plants are grown not for their food or fibre value but for speciality products such as oils (see Section 5.4). Genetic engineering can improve the productivity and quality of such substances (Somerville, 1993). Indeed, the cultivation of such crops which include biomass fuels, may well become big business in Europe if set-aside becomes widespread. Such commodities are renewable resources and could be used as substitutes for fossil fuels. Moreover, Zeneca plc has engineered oilseed rape to produce a substance called biopol® which is a type of biodegradable plastic (Selincourt, 1993). Although not directly relevant to agriculture, it is possible that plants and crop plants could eventually be engineered to produce therapeutic substances for use in human health care, as is occurring with some domesticated animals (see discussion below). Some of these possibilities have been examined by Spelman (1994). In addition, the genetic manipulation of crop characteristics can address traits such as ripening and composition. Indeed, the tomato has been the subject of genetic manipulation of both traits (listed in Beck and Ulrich, 1993). Grierson and Schuch (1993) have discussed the suppression of the softening enzyme. This trait would facilitate efficient marketing by curtailing waste due to over-ripeness. Composition affects flavour and as Miller reported in May 1994, the world's first genetically engineered food had just entered shops in the USA. This is the Flavr Savr® tomato developed by the US biotechnology company, Calgene. Much of this work has been prompted by the fact that the tomato is one of the most widely used fruits in the world because of its versatility.

Research on the genetic basis of tolerance to environmental stress is also in progress though it is not as far advanced as that described above on pesticide resistance. Nevertheless one of the first genetically engineered organisms to be tested in the field addressed the issue of frost damage. This was a so-called 'ice minus' bacterium developed in California to protect crops such as potato, strawberry and tomato from frost damage (Lindow, 1990). It was discovered that frost damage in certain crops occurs because bacteria in the foliage produce a protein which acts as a nucleus for ice formation in the temperature range 0–2 °C. The production of the protein, which occurs in the common bacterium *Pseudomonas syringae*, is genetically controlled. After removal of the responsible gene component, large numbers of the new bacterium were generated. Field trials, involving the spraying of the new bacterium onto crop foliage, showed a substantial reduction in frost damage. The commercial advantages of this are obvious: crop losses are reduced and profits increase. Further work on the genetic basis of resistance to chilling has been reported by Murata *et al.* (1992) who have linked it to the degree of unsaturation of fatty acids in chloroplast membranes. The generation of crops with frost resistance would not only reduce losses in frost-susceptible areas but also allow crop production to be extended to areas with a high incidence of frost.

Salamini and Moto (1993) state that although the genetic basis of environmental stress tolerance in crop plants has been little researched, it could provide numerous opportunities to overcome environmental limitations. This would be particularly advantageous in the developing world though it could also lead to the destruction of yet more natural ecosystems. Ramachandran (1992), for example, has drawn attention to the need to develop resistance to drought and salinity in crops such as wheat and millet in order to improve India's food supply. In addition, Qazi *et al.* (1992) stress the need to develop drought, salinity and acidity tolerance in crop plants if Pakistan is to continue to feed its population. This is particularly important in the context of Pakistan's problems with its irrigated lands (see Section 9.4). Other possibilities that are under investigation include heavy metal tolerance in soy bean and salinity tolerance in barley.

Genetic engineering may be able to contribute to the enhancement of nutrient availability in the soil. As discussed in Section 10.1, inocula of *Rhizobia* bacteria have been used successfully to promote nitrogen fixation in soils. It is possible that genetic engineering of the bacteria could improve this process. It is also possible that the responsible genes could be transferred into crop plants so that they could fix their own nitrogen. Yet another possibility is the engineering of the genetic basis for the symbiotic relationship that occurs between leguminous species and *Rhizobia* in other crop plants. These possibilities have been discussed by Mytton and Skøt (1993) who report that investigations are underway to determine the genetic basis for these functions. It is likely that non-nitrogen-fixing bacteria will be engineered first to extend the possibilities for using inocula. Eventually, the cereals may be engineered to fix their own nitrogen but this is a long way from becoming reality.

There are many other potential applications of genetic engineering in agriculture (Mannion, 1995c). For example, Pen *et al.* (1993) have engineered tobacco to produce large amounts of the enzyme phytase. When the seeds are fed to livestock the increased phosphate results in an increase in secondary productivity. Thwaites (1993) has reported on a transgenic clover for Australian sheep pastures that has increased wool production. This clover contains genes from sunflower which control the production of a protein rich in sulphur amino acids. These, in turn, enhance wool production. Genetic engineering also has a substantial role to play in animal production both directly and indirectly. There are possibilities for the manipulation of reproduction as Kruip and Spaan (1992a) have discussed. It is now possible to produce four embryos from one fertilised cell, for example, and to produce transgenic livestock (Kruip and Spaan, 1992b). Much of the work is experimental and much of it is ethically contentious (e.g. Fox, 1993; Mepham, 1993; Thompson, 1993). The obvious targets are improved milk production and disease resistance, especially for those diseases for which no vaccines are available. Animal productivity has already benefitted from genetically engineered substances, notably somatotropins and vaccines. The former are animal growth hormones which are now produced by transgenic bacteria. In dairy cattle these hormones are used to improve milk production though the practice is controversial as many consumers disapprove. Vaccines to improve animal (and human) health can also be produced using genetic engineering. Live vaccines, to combat viral diseases such as swine fever, are generated by engineering viruses like *Vaccinia*. The details have been discussed by Primrose (1991). Recently, Yilma (1993; see also Fitzgerald, 1994) has reported on the development of a vaccine for rinderpest which is a major problem for cattle in Central Africa, India and the Middle East. This too is a *Vaccinia* virus, which has been genetically engineered with the genes from the rinderpest virus itself.

Other developments in animal biotechnology involve many non-agricultural products though, in most cases, the animals can be kept under the same conditions as conventional livestock. Transgenic sheep that can produce the human protein α-1-Antitrypsin for the treatment of hereditary emphysema, have been successfully reared (Carver *et al.*, 1993). Human haemoglobin can also be produced by pigs (Sharma *et al.*, 1994). In addition, Concar (1994) has reported on the genetic engineering of pigs to provide organs that could be used for transplanting into humans. Already there are transgenic pigs with human DNA to make them compatible with the human immune system. It is anticipated that trials of such organs transplanted into humans will begin in 1996. This will undoubtedly generate opposition on ethical grounds.

As has been referred to above, there are many opponents of biotechnology and especially of genetic engineering. Whilst there are obvious advantages of biotechnology there are also a number of disadvantages and risks. The positive aspects of biotechnology in agriculture are illustrated in Figure 10.2. Not only are there benefits from increased food production but there could also be substantial benefits for the environment. In the latter context a major advantage concerns the increased productivity of agricultural systems which, in principle, should obviate the need to bring more land under cultivation. This should contribute to the conservation of ecosystems and habitats which provide essential services and house an important genetic resource. A second major advantage concerns the reduction of fossil-fuel inputs that would occur if crop plants were engineered to fix their own nitrogen and to exhibit pest resistance. All of the developments discussed above could contribute to making agricultural systems more sustainable than they are now and improving the

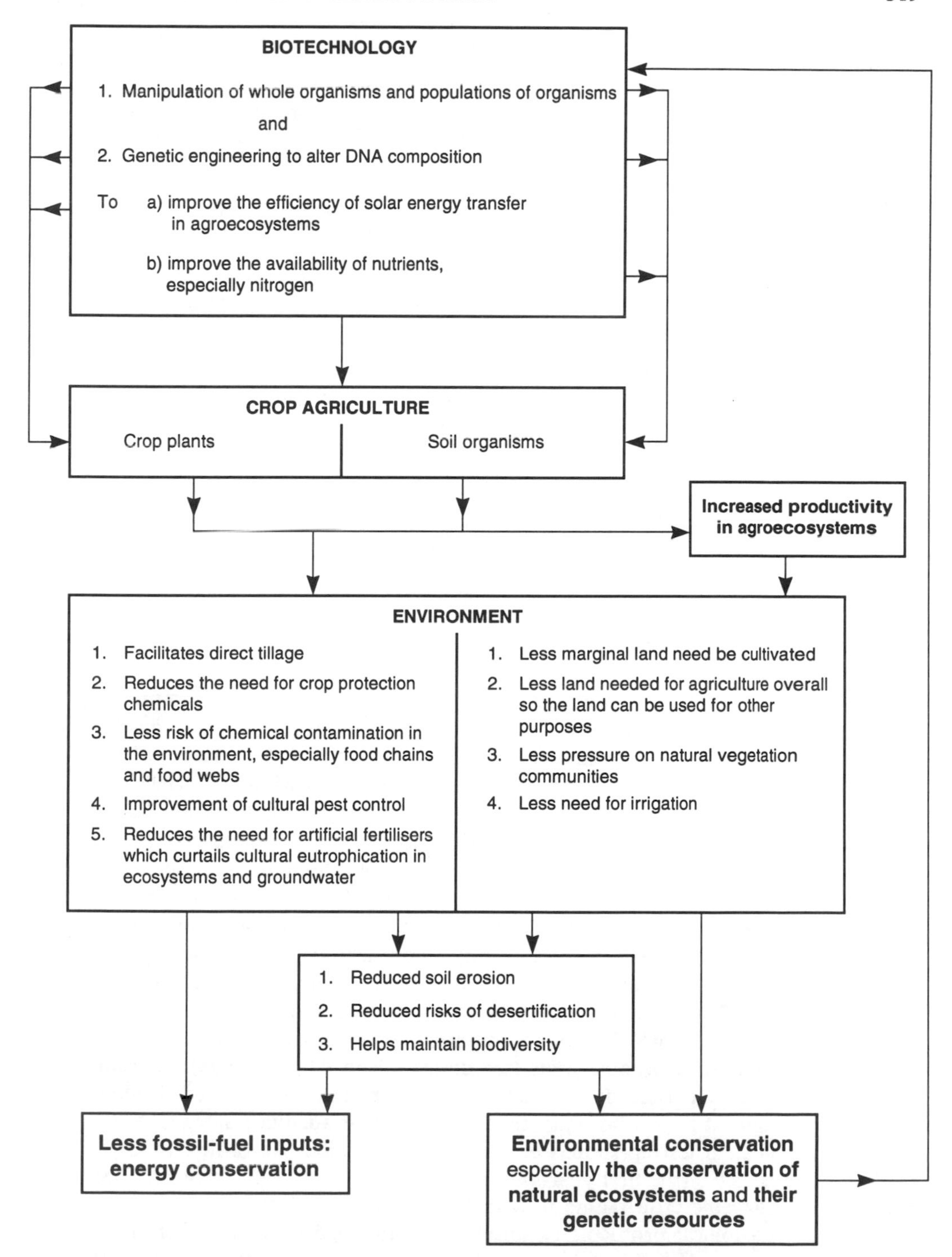

Figure 10.2 The positive relationships between agriculture, environment and biotechnology

Table 10.5 Some disadvantages of biotechnology

ECOLOGICAL/ENVIRONMENTAL DISADVANTAGES

A There is much potential to create ecologically damaging organisms, e.g. invasive plants that become pests, or plants whose engineered genes escape into wild relatives to create pests.

B There is much potential to create higher plants, animals or bacteria that are toxic to other organisms including humans.

C There is much potential to create organisms, especially bacteria, that may adversely affect biogeochemical cycles or which could be used in biological cycles or which could be used in biological warfare.

D Engineered species with resistance to environmental stress may encourage the spread of agriculture rather than its intensification. This would threaten the maintenance of biodiversity and ecosystem conservation.

E There are major difficulties in guaranteeing biosafety and establishing internationally-accepted protocols for assessing the risks of transgenic organisms.

F Countries without regulations for testing transgenic organisms may be exploited for field testing. These are likely to be developing countries.

CULTURAL/ECONOMIC DISADVANTAGES

A The biotechnology industry is dominated by transnational companies based in the developed world. This is disadvantageous to developing countries. TNCs have monopolies on transgenic seed production.

B Biotechnology in the food industry is reducing markets for some products, e.g. cocoa butter, that are produced traditionally in the developing world.

C Developing world resources, i.e. the genes of wild species and crop landraces, are being exploited without recompense.

world's food supplies (Mannion, 1992b). To be truly significant in improving the environment and the lot of all of humankind, the disadvantages of biotechnology must be recognised and controlled. The real and potential disadvantages of biotechnology are given in Table 10.5 and have been discussed in Mannion, 1995d.

It is possible that many of the advantages given in Figure 10.2 may not be realised. As Fox (1993) has pointed out, the production of crop plants to suit specific environments could lead to further conversion of natural habitats to agricultural land. This would deplete biodiversity, especially if it occurred in low latitude regions which house the greatest diversity of life. Thus, biotechnology could actually constitute a threat to the maintenance of biodiversity. Fox also suggests that the aim of biotechnology is not to improve food production but simply to reduce the costs of food production. The ecological concerns about biotechnology focus on the creation of transgenic organisms which may become pests or which may pass their genes on to wild relatives that then become pests (see discussion in Abbott, 1994). Consider, for example, the damage that deliberately and inadvertently introduced species have wrought throughout the world. It is, thus, essential that adequate safeguards through laboratory and field testing should be observed worldwide. According to Goy *et al.* (1994), between 1986 (the first year of such field trials) and 1992, there had been 675 releases of transgenic crop plants in field trials involving 31 species in 28 countries. The sorts of ecological protocols required are exemplified by Crawley *et al.*'s (1993) field trials to assess the invasiveness of transgenic oilseed rape. Each transgenic crop needs to be properly tested; no generalisations can be made about the way species will react in the environment. What a disaster it would be if, for example, resistance to herbicides developed in weed populations. Controls to ensure that gene products are non-toxic to animals or humans are also necessary. The engineering of bacteria and viruses could also lead to the creation of environmentally detrimental species and species that are harmful to human and animal health. In view of the difficulties of assessing risks such as these it is hardly surprising that there is much greater potential for producing transgenic organisms than for the initiation of safeguards. This situation needs to be redressed. Biosafety is all important and the various protocols that could be adopted to ensure this are examined in Persley *et al.* (1992). However, Persley *et al.*, in common with many other reviewers on the subject, tend to take the view that transgenic products such as crop plants, because they are similar to traditional products, pose the same risks which can be assessed in a conventional way. This is not necessarily the case.

As Table 10.5 shows, there are many other potential disadvantages of biotechnology which concern its cultural rather than ecological impact (see also the review by Perlas, 1994). First, the biotechnology industry is dominated by transnational companies (TNCs) which are based in the developed world and which are in business to generate profits. However, the major need for improved agricultural productivity is in the developing world which also houses many of the genetic resources required by the biotechnology industry. Apart from sharing the technology with developing countries through education and training programmes, they should also be given the opportunity to benefit from existing transgenic products. As stated above, there are concerns that TNC monopolies over transgenic seed production, and herbicide production, will disadvantage farmers in developing countries (Goodman and Redclift, 1991; Shiva, 1993). In addition, the application of biotechnology in the food industry has caused a marked decline in demands for products like cocoa butter, certain food oils and sugar. Such substances can be produced in laboratories often at a lower cost than if they were imported from farmers in the developing world. Reference has also been made above to the patenting of transgenic seeds which gives control over production to TNCs. It is not difficult to envisage the use of germplasm reserves (see reference above to international research centres in developing countries) and landraces to develop improved crop types which may even be sold to host countries. The latter need some safeguards if they are not to be exploited. This is also true of genetic resources used to produce new pharmaceuticals. The example of Costa Rica's National Biodiversity Institute (INbio) and its association with Merck and Company, the world's largest pharmaceutical firm, demonstrates that co-operation can be mutually beneficial and conflicts of interest avoided (Reid *et al.*, 1993). In this case Merck help fund INbio, and any products that ensue will generate a royalty for INbio which must be used to promote the conservation of Costa Rica's biodiversity. Similar arrangements for developing countries to share in profits from their crop genetic resources need to be instituted. This will require internationally accepted codes of practice to ensure mutual advantage. The payment of a royalty, for example, by TNCs to developing nations could be used to establish germplasm collections, to stimulate the compilation of inventories of plants and animals and to encourage the conservation of both habitats and landraces of crops. The question stated in the introduction to Chapter 8, i.e. 'When the North eats does it eat the South?' (Goodman and Redclift, 1991), could, in the context of biotechnology, be answered affirmatively but with some justification that there could be mutual benefits. As has been observed elsewhere (Mannion, 1995d), the advantages and disadvantages of biotechnology encapsulate the reciprocal relationships that exist between the developed and developing world and those between agriculture, ecosystems and the environment.

10.3 THE ROLE OF INFORMATION TECHNOLOGY IN AGRICULTURE

In Sections 10.1 and 10.2 the role of science in the modification of crop plants and livestock, as well as the harnessing of organisms such as bacteria, fungi and viruses, was examined in relation to crop and harvest improvement. Recent advances in biotechnology and genetic engineering, whilst they are opening up many new possibilities for increased food and fibre yields and improved crop protection and animal health, are not the only scientific advances of relevance to agriculture. The veritable explosion in information technology that has occurred since the 1970s has contributed, and will continue to contribute, to the increasing efficiency of agriculture. Whilst biotechnology and genetic engineering have focused mainly on improving the product, information technology is concerned with the use of resources, notably land and water resources. Its objectives are to optimise productivity by improving land-use practices, to monitor the environmental impact of agriculture and agricultural practices, and to reduce the environmental impact of some types of agricultural practices. Information technology can be employed in many aspects of food production, from the control of irrigation to the marketing of the produce. Amongst the many forms of information technology that exist, the types most relevant to agriculture are knowledged-based systems, i.e. computer programmes that assist the farmer to make management decisions, and remote sensing. The latter is the recording of information on environmental parameters, e.g. soil moisture and biomass, by satellites, cameras and radar systems. Knowledged-based systems and remotely sensed data may be combined, along with field survey and cartography, to produce geographical information systems.

The use of knowledge-based systems in agriculture has been addressed by Plant and Stone (1991) who stress the value of such systems as management devices. As Figure 10.3 shows, knowledge-based systems have four fundamental components: inference, knowledge representation, search and pattern matching. From these stem a wide range of applications. Of the most important and widely used are those related to fertiliser and pesticide treatments and irrigation management. In relation to fertiliser use, for example, many different types of knowledge-based systems have been developed. First, there are systems designed to determine the impact of on-farm best management practices (BMPs), in this case in relation to fertiliser use, on ecosystems and habitats beyond the source of the nutrients. Secondly, there are knowledge-based systems to assist farmers in the determination of the minimum amount of artificial fertiliser that must be applied at the most appropriate time to achieve optimum results. Both types of system are, if successful, conservational. Reductions in the amount of artificial fertilisers used reduce fossil-fuel supplements (i.e. in artificial nitrate fertilisers) and resource inputs (i.e. mined phosphate fertilisers) to agricultural systems. Financial costs are thus reduced, benefitting, at least in theory, both consumers and producers. The environmental costs are also curtailed as the wider environment is less polluted. Both types of system can, thus, be used to make agriculture more sustainable (see Section 10.4) than it is with traditional management.

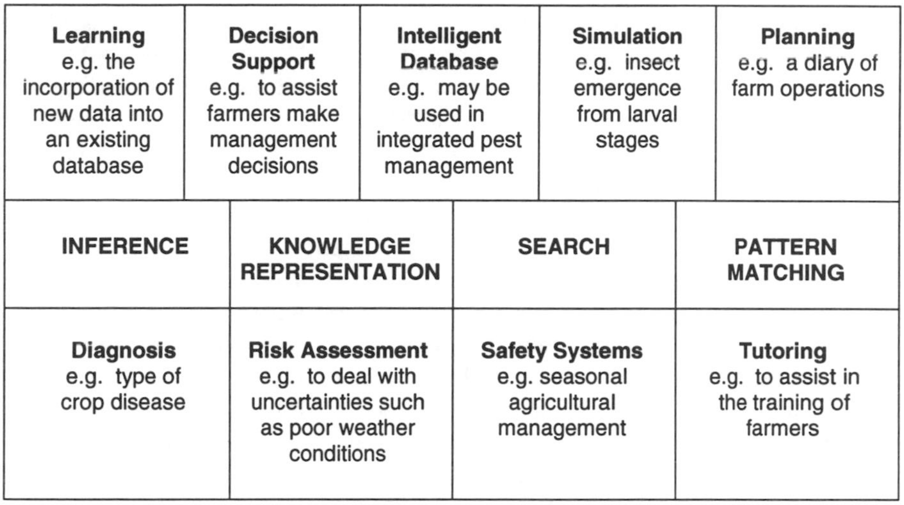

Learning e.g. the incorporation of new data into an existing database	**Decision Support** e.g. to assist farmers make management decisions	**Intelligent Database** e.g. may be used in integrated pest management	**Simulation** e.g. insect emergence from larval stages	**Planning** e.g. a diary of farm operations
INFERENCE	**KNOWLEDGE REPRESENTATION**	**SEARCH**	**PATTERN MATCHING**	
Diagnosis e.g. type of crop disease	**Risk Assessment** e.g. to deal with uncertainties such as poor weather conditions	**Safety Systems** e.g. seasonal agricultural management	**Tutoring** e.g. to assist in the training of farmers	

Figure 10.3 Types of knowledge-based systems in agriculture (adapted from Plant and Stone, 1991)

Willis *et al.* (1994) have described models devised to determine the effectiveness, or otherwise, of BMPs on the quality of water draining from the Everglades Agricultural Area (EAA) into the Everglades, a major freshwater marsh ecosystem in Florida, USA. Here some 700 000 acres of marsh were drained via a series of canals to produce prime agricultural land in the 1920s. The main products are currently sugar cane and winter vegetables; overall, the industry contributes more than $\$1 \times 10^9$ annually to the local economy. However, the intense land use since drainage, which has altered the quantity, quality and pattern of water movement, has had a detrimental impact on the flora and fauna of the Everglades. Much of this is due to changes in nutrient status and to help mitigate this problem the Everglades Surface Water Improvement and Management (SWIM) plan was established in 1992. This incorporated two objectives: to construct treatment systems to remove nutrients from drainage water before its release into the Everglades and to encourage, through the establishment of BMPs, a reduction in phosphorus loadings by a minimum of 25 per cent. Included in the BMPs were water management, fertility control and the provision of an aquatic crop cover. The models used to evaluate the feasibility of these practices had to take into account costs as well as impact. The models were manipulated to combine various possibilities for water management, fertility control and aquatic crop covers to achieve 25 per cent, 35 per cent and 45 per cent reductions in phosphorus loadings. The least-cost combination of BMPs to bring about a 25 per cent reduction in phosphorus loading on land used for sugar-cane production comprised a calibrated soil test (to optimise the relationship between soil phosphorus concentrations, the phosphorus added in the fertiliser and crop yields), rice production as an aquatic crop to store and ultimately remove phosphorus from the system, banding fertiliser (i.e. locating it near crop roots), the prevention of losses through misplaced fertiliser and an improved pump schedule to retain more water on farm. The cost of these measures per acre was $0.51 for sugar cane; for a 35 per cent reduction in phosphorus the cost rose to $1.07. These are feasible goals, especially when compared with the economically unfeasible $12.23 for a 45 per cent reduction. Clearly, such models can be useful as management adjuncts. It is also interesting to speculate what the 'real' costs of such ameliorative measures would be if the potential improvements in the Everglades environment were translated into dollars and included in the cost–benefit analysis.

Another example of a model to improve nitrogen management at the regional scale is described by Miller and Anderson (1994). The region concerned is Utah in the arid west of the USA where irrigation is essential for crop production. As discussed in Sections 8.4 and 9.4, the injudicious use of irrigation and nitrate fertilisers can give rise to major environmental problems. Thus, the availability of tools to effect improved management is to be welcomed. Miller and Anderson used NTRM, a soil–crop simulation model for nitrogen, tillage, and crop-residue management to model the growth of maize crops in relation to nitrate leaching. The input data into NTRM comprise daily weather data, tillage events, and the location and source of nitrogen in the soil profile. The model can be used to simulate crop responses to different soil types under conditions of changing soil characteristics. The results of this study are summarised in Figure 10.4 which shows that the best returns (in US $) derive from split rather than single applications of artificial nitrate fertiliser in combination with an irrigation schedule that provides water at a rate close to that of rates of evapotranspiration (i.e. controlled sprinkler irrigation) rather than in a biweekly application that takes place without due regard to rates of evapotranspiration.

In relation to pest management, knowledge-based systems can be used for a variety of purposes. For example Stone *et al.* (1990) have developed a model to predict the timing and magnitude of cotton bollworm emergence. This involved the compilation of data on past patterns of emergence in relation to weather conditions. The capacity to predict bollworm emergence is a valuable tool for advising farmers when to plant their cotton crop, especially in regions where cultural as opposed to chemical pest management is practised. It can be combined with a simulation model of cotton plant development in relation to likely weather conditions to determine the optimum date for planting. The aim is to plant the crop so that the cotton buds, the target of the weevils, develop after the weevils have emerged and died through the absence of buds. Knowledge-based systems can also be used to determine the optimum timing and optimum application rates of chemical pesticides. Plant and Stone (1991) have discussed the possibilities for cotton management. They list six factors that should be considered: the extent of bollworm infestation, the population of natural enemies, the stage of cotton plant growth, the moisture content of the soil, the anticipated weather and the future control measures that might be necessary depending on the action taken. Clearly, the decision-making process is complex and it is easy to envisage how an incorrect decision could be taken, particularly one involving spraying rather than not spraying as the farmer would judge such an approach to be an insurance measure. For example, if adequate predators of the bollworm are present, and if the latter is mainly present in the egg stage, spraying may not be necessary. Alternatively, a farmer may spray more often than is really necessary to control an insect outbreak and in so doing may eradicate the natural predators of the bollworm. Knowledge-based systems can remove much of the uncertainty in these kinds of situations and so reduce unnecessary pesticide applications.

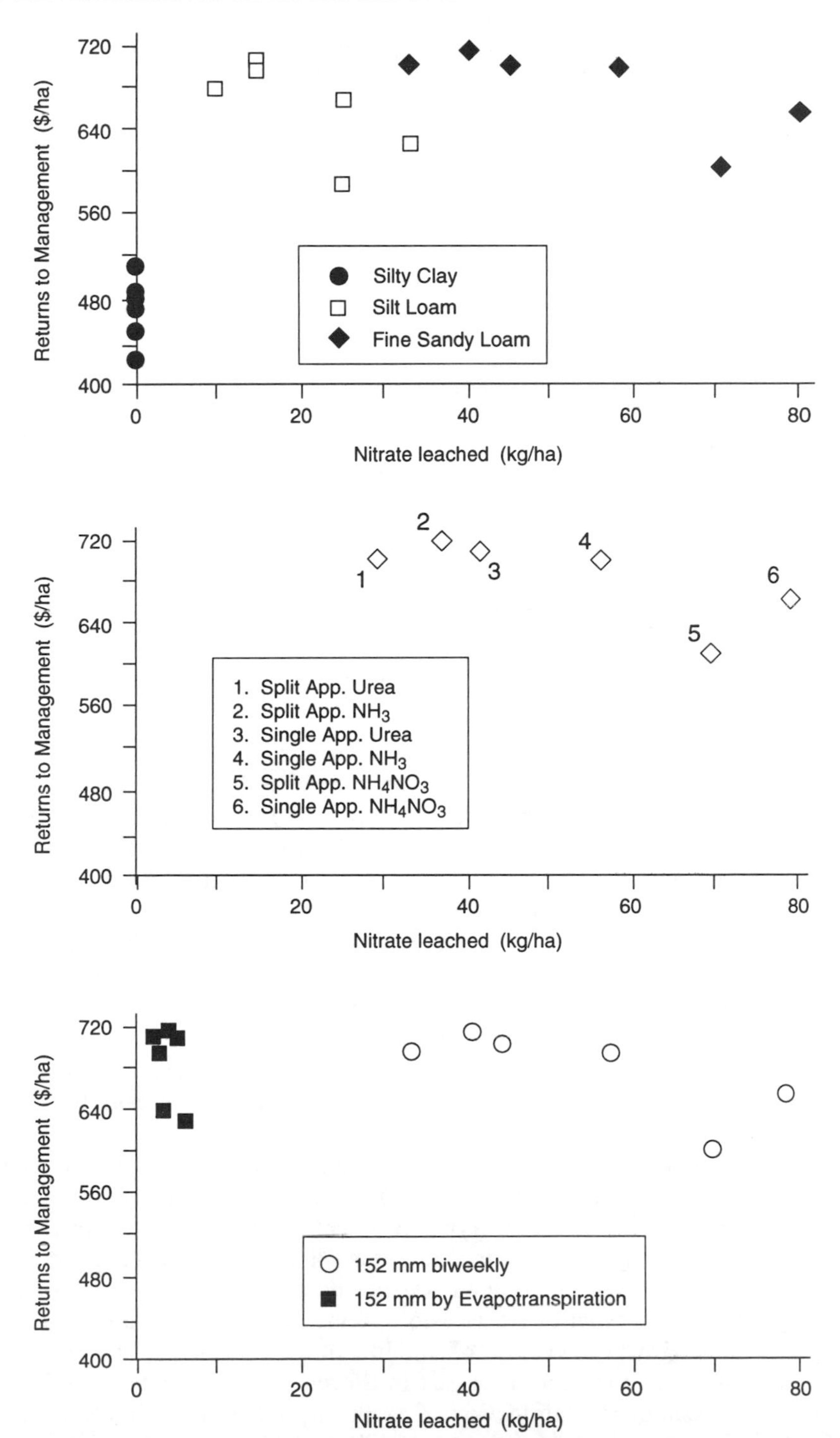

Figure 10.4 Interactions between soil type, artificial nitrate applications and their sources, irrigation schedules in relation to costs in US $ in Utah, USA (based on Miller and Anderson, 1994)

Table 10.6 The applications of remote sensing in agriculture and related environmental characteristics

A	**ENVIRONMENTAL/ECOLOGICAL/CROP APPLICATIONS**
1	Crop type
2	Biomass
3	Vegetation indices
4	Land-use change: enforcement of agricultural policies
5	Climate: at a variety of scales
6	Hydrology: at a variety of scales
7	Soil science: type and status
8	Agricultural statistics
B	**APPLICATIONS IN THE MANAGEMENT OF AGRICULTURE**
1	Crop condition
2	Crop damage
3	Crop productivity: actual and potential
4	Livestock inventory
5	Land degradation
6	Environmental impact
7	Irrigation: inventory, water use, salinisation and waterlogging
8	Soil: erosion, pollution, moisture and survey
9	Drainage
10	Rangeland: condition and carrying capacity

Other models have been developed to determine the rates of pesticide loss from sprayed areas with a view to improving rates and methods of application. For example, a model known as Groundwater Loading Effects of Agricultural Management Systems (GLEAMS) has been developed by Leonard *et al.* (1987) to assess and monitor the contamination of groundwater by pesticides. It has been used to make 50-year simulations to establish the likely annual losses of pesticide sprayed onto maize (Leonard *et al.*, 1992) under a variety of climatic conditions. The objective is to use such a simulation to advise farmers on appropriate spraying dates in order to minimise losses and so reduce the contamination of groundwater by pesticides. Several other simulation models designed to examine the environmental impact of non-point-source pollution on water quality have been discussed by Steyaert and Goodchild (1994) who also point out their role in geographical information systems (see below).

Another type of knowledge-based information technology with many applications in agriculture is remote sensing. According to Lillesand and Kiefer (1994), airphotographs can be used to produce a crop type classification for a given region, to assess the condition of a crop, e.g. the detection of disease, and to estimate crop yield. All of these crop parameters can be determined because different crops in varying states of health and growth development give rise to specific spectral response patterns and photograph textures. However, the use of satellite imagery has generated an increased range of applications, as has been discussed in Steven and Clark (1990) and detailed in Table 10.6. As a complete examination of all these possibilities is beyond the scope of this book, reference will only be made to limited examples. In relation to the forecasting of crop production, Sridhar *et al.* (1994) have described a method for such estimation based on data from the IRS-IB, an Indian remote-sensing satellite, for

1991–1992. The study area is Central Madhya Pradesh (in north central India), comprising 5.61×10^6 ha with a sub-humid climate and black clay soils. To derive a forecast, two parameters, i.e. the area occupied by the crop and yield estimates, were determined by sampling the imagery from the region. Using various procedures, and on the basis of ground truth, i.e. field data to identify wheat and distinguish it from other crops, Sridhar *et al.* were able to estimate the crop area with 80–90 per cent accuracy and the yield with 85–90 per cent accuracy. In a similar study, Rao and Mohankumar (1994) developed a cropland inventory methodology in the irrigated region of Krishnarajasagar, south-west India, in order to optimise water use from the Cauvery River. They used a combination of remotely sensed imagery from the IRS-IA and the Landsat satellites to identify irrigated croplands and crop types respectively. In addition, the use of various types of remotely sensed data to compile agricultural statistics, i.e. crop inventories on a national and international basis, can contribute to the formulation of agricultural policies. In the EU, for example, this type of data (see discussion in Vossen, 1992) could be used to determine crop quotas and the distribution of set-aside. Such monitoring could also be used to determine if set-aside policies are being correctly implemented and if payments to farmers for set-aside are justified. Some of these applications are part of the ongoing research at the Institute for Remote Sensing Applications at Ispra, Italy (IRSA, 1993).

Other applications of remote sensing in agriculture include the detection of plant stress. For example, Carter and Miller (1994) have demonstrated that it is possible to detect the early effects of herbicide treatments on soy bean crops using a camera system located 17 m above ground. Similarly, Taconet *et al.* (1994) have shown that microwave remote sensing, i.e. radar, can also be useful in an agricultural context. This is because radar can be used in all weathers whereas light-based remote sensing is hindered by cloudiness. Taconet *et al.* successfully used radar-derived data from ERASME, a helicopter-borne scatterometer, to monitor soil moisture conditions in Orgeval, a French watershed carrying a wheat crop. In addition to the monitoring of the characteristics of agricultural systems, remote sensing has been widely applied to monitoring the impact of agricultural practices. As discussed in Sections 8.1, 8.2 and 9.1–9.4, environmental change caused by agriculture is widespread. Habitat destruction—notably deforestation, soil erosion, desertification and salinisation—can, and have, all been monitored using satellite imagery. One such example is the work of Pickup *et al.* (1993) and Pickup and Chewings (1994) who have used multi-temporal remotely sensed satellite data, in conjunction with spatial models of grazing impact, to detect land degradation due to overgrazing in Australia. They make the point that such assays are the only possible means of rapidly assessing both the amount of degradation and its rate of occurrence in a land area as large as Australia's arid and semi-arid zones. The production of grazing gradients from these data allow the detection of grazing effects in rangelands despite the natural fluctuations that occur in such vegetation communities. Moreover, such data facilitate the assessment of degradation and/or erosion risk. This in turn relates to the stock carrying capacity of a given area.

A geographical information system (GIS) is a 'set of tools for collecting, storing, retrieving at will, transforming, and displaying spatial data from the real world for a particular set of purposes' (Burrough, 1986). Consequently a GIS may involve remotely sensed data and will almost always include a knowledge-based system.

Moreover, it is usually possible to manipulate such data in a temporal and/or spatial context and in a predictive role. There are many applications of GISs in agriculture and in the related, general field of land management. Robinette (1991) has described the applications of GIS to land management in Minnesota, USA, where, amongst other applications, a GIS was used to pinpoint areas of soil erosion and deposition. The objective was to inform landowners in particularly susceptible areas of the risks in order to encourage improved land-use practices. The data were also used in the allocation of state funding for conservation measures in critical areas. The GIS was developed on the basis of four categories of cropland (high to moderately productive) with twice the 'tolerable' loss of soil per hectare. The Universal Soil Loss Equation (USLE) and the Wind Erosion Equation (WEE) were then applied to the data set in order to model erosion for the entire state. The resulting map highlighted areas at risk in four categories which were determined according to priority need for remedial action. Planners and advisers could then act to offer advice on the best management practices and funding for mitigating measures.

Several GISs have been developed to assess the hazard, both actual and potential, of groundwater pollution. The nature of this type of hazard can be variable; in relation to agriculture the most important problems are pesticide and nitrate contamination and raised salinities due to irrigation. Spalding and Exner (1993) have reviewed the problem of nitrate contamination whilst Merchant (1994) has shown how this hazard, amongst others, can be assessed and mapped using a GIS known as DRASTIC. This model was developed by the US Environmental Protection Agency in conjunction with the US National Water Well Association. It has been widely applied, as is exemplified by Robinette's (1991) appraisal of its use in Minnesota. DRASTIC is based on geological and hydrogeological parameters, e.g. depth to water, net recharge, the rock type of the aquifer, the soil type, topography, the influence of the radose zone and hydraulic conductivity of the aquifer. The factors within these parameters are rated on a scale of 1 to 10 and the parameters themselves are weighted. These data facilitate the production of maps for the area being examined for each of the parameters. On each map, factors can be represented by their numerical score. The overall DRASTIC rating is derived by multiplying the factor rating by its parameter weights and then totalling the values. The data set can be manipulated to show the degree of vulnerability of designated landscape units to particular hazards such as fertiliser run-off and pesticide applications. Other types of GISs for assessing groundwater pollution hazards have been discussed by Khan and Liang (1989), Meeks and Dean (1990) and Mason *et al.* (1994). An example of a GIS incorporating remotely sensed data and of relevance to agriculture has been described by Wade *et al.* (1994). Vegetative indices derived from the Advanced Very High Resolution Radiometer (AVHRR) sensor on the National Oceanic and Atmospheric Administration (NOAA) satellite can be used to assess spatial patterns of crop conditions in the mid-west flood and south-east drought regions of the USA. Such information is used by policymakers in an advisory capacity and to predict crop yield. The GIS incorporates vegetation indices (derived from AVHRR) with data on topography, climate, ecological zones and land resource areas. The GIS is also being evaluated as a means of determining average dates for the first frost that may cause crop damage. This would require the inclusion of frost isolines into the GIS.

The rapid development of information technology in general, and particularly the innovations that are occurring in the recently emerged technology of GIS, will undoubtedly find many applications in agriculture and related fields in the future.

10.4 SUSTAINABLE AGRICULTURE

When the World Commission on Environment and Development's (1987) deliberations were published as *Our Common Future* one of the main themes was sustainable development. WCED's definition is now the most widely adopted: 'Sustainable development is development which meets the needs of the present without compromising the ability of future generations to meet their own needs.' It is, nevertheless, just one of many definitions (see discussion in Eden, 1994), all of which are rhetorically rather than factually useful. Why this phrase has become so popular in both the academic and popular media is probably because it engenders a semblance of hope and some degree of optimism for the future of planet Earth and its inhabitants. Indeed, O'Riordan (1991) suggests that the concept of sustainable development is a core theme in what he describes as the new environmentalism. Rhetorically, the idea appears somewhat positive when compared with the various doom and gloom scenarios that tend to emanate from many environmental activist groups. In this context the concept of sustainability is in danger of becoming a platitude. However, one of its strengths as a concept is the implication that there is a need to reconcile economic development, the quality of life, and the environment, within various political frameworks which interrelate at the international and global levels. This holistic view is one of the major reasons sustainable development is difficult to define in more than the broadest of terms. The numerous spatial and temporal scales inherent in its definition convey the complexity of the real world and epitomise the intricacy of people/environment relationships. In contrast to the dualism which has dominated the people/environment relationship, i.e. nature is there to be at the service of society and science is a way of achieving society's domination of nature, sustainable development entails a mutual, symbiotic relationship.

Although it may be in danger of becoming trite, the phrase 'still provides a useful point of entry in discussing development and the environment' (Redclift, 1992). WCED's definition may not overtly refer to the environment and the necessity of protecting it but it does so by inference. This is because the needs of the present population and those of future generations are the Earth's resources. The term 'resource' is used here in the widest possible sense; resources are not necessarily used directly to produce useful commodities as occurs with minerals, for example. Many resources provide services rather than goods (see Mannion 1994a) but the provision of these services is just as vital to the maintenance of society as the goods. Examples include the absorption of carbon dioxide from the atmosphere and the production of oxygen by green plants as they photosynthesise, water cycling, soil genesis and all types of biogeochemical cycling. These processes take place in the biosphere and they illustrate the significance of life in all its forms as a mediator in Earth surface and atmosphere processes (cf. Gaia hypothesis).

Apart from the provision of services for society's maintenance, the biosphere also provides the means for society's sustenance, the many and varied forms of which have been described in Chapters 5, 6 and 7. The production of food (also fibre, and now fuel) throughout history has not only provided food energy for society but it has also provided a means of wealth generation. As Chapters 8 and 9 illustrate, this generation of food, fibre and wealth has been achieved at considerable environmental cost. On a global basis, the practice of agriculture can hardly be considered as sustainable. Indeed, the unsustainability of many agricultural systems is somewhat masked by the application of science and technology. The increasing use of artificial fertilisers, for example, has negated the adverse impact of soil erosion on food production. It is possible that the improved crops promised by biotechnology (Setions 10.1 and 10.2) will continue to maintain productivity whilst the underpinning, unsustainable rates of soil erosion, salinisation, etc., remain unaddressed. Society appears eminently capable of ameliorating the symptoms of unsustainable practice but remarkably myopic when it comes to tackling the fundamental causes of environmental degradation. As discussed in Sections 1.2 and 1.3, agriculture is one of the most important, and widespread, ways in which society impacts on the Earth's surface. The fundamental aim is to channel energy into a harvestable and nutritious resource. However, a vicious circle can operate: in the process of channelling that energy, mainly through attempting to adapt the environment for efficient crop/animal production, the capacity to generate that energy may be impaired. The employment of irrigation, for example, can lead to salinisation and waterlogging (Section 9.4) which, in turn, can cause declines in crop productivity and the abandonment of land. Even where the impact of agriculture is not so drastic there is often a trade-off. For example, the deposition of eroded soil particles may cause the silting of dams and reservoirs (Section 8.2) and the intensive use of artificial fertilisers may lead to the cultural eutrophication of fresh waters and coastal regions (Section 8.4). Such practices are thus unsustainable; many are unsustainable even in the short term of a decade or two.

Another characteristic of many of the world's agricultural systems is their consumption of energy. Fossil-fuel energy is a non-renewable resource; its widespread use in high-technology agricultural systems (see examples in Chapters 5, 6 and 7) is aimed at increasing the efficiency of such systems to produce food energy. Consequently, fossil-fuel energy is used to manipulate the components of agricultural systems through artificial fertilisers, crop-protection chemicals and mechanisation. All of these have their environmental costs though it is also a fact of life that without such food production the world's population could not be supported. Moreover, it is fair to say that only a small proportion of the world's energy consumption occurs via agriculture. Nevertheless, the consumption of fossil-fuel energy means the generation of carbon dioxide and nitrous oxides. The former is a heat-trapping gas whilst the latter are not only heat-trapping gases but also contribute to acid rain. As is glaringly obvious from the wealth of literature, potential global warming is *the* environmental issue of the 1990s. Moreover, there is considerable effort being invested in predicting its likely impact on the world's agricultural systems. This is discussed in Section 10.4. The consumption of energy by agricultural systems, coupled with the release of carbon dioxide from biomass cleared to provide land for agriculture, is contributing to global warming. This, in turn, will impact on the agricultural systems themselves. As food security is so important for economic and political security the status and maintenance of the world's agricultural systems is all important. In an ideal world, agriculture should act as a counteragent to industry by absorbing the latter's production of carbon dioxide!

The issues of environmental degradation and global warming are factors intrinsic to the present and future practice of agriculture. This is because they are the consequence of, and impact on, agricultural systems. A further, equally important driving force in determining the nature and extent of agricultural systems is population growth (see discussion in Ehrlich and Ehrlich, 1990; Myers, 1993c). The relationship is complex. Does population growth stimulate agricultural expansion or does agricultural expansion stimulate population growth? This has been discussed briefly in the introduction to Chapter 2 and Section 2.2 and it is also examined in the endpiece. Whatever has been the case in the past, it is a matter of fact that the world's population is continuing to grow. The United Nations 1994 report on *The State of World Population* (UN, 1994) states that the world's population is currently 6.2×10^9. This is projected to grow to 8.5×10^9 by 2050. Agricultural systems need to be able to support these people without, as WCED states, jeopardising the needs of future generations. This is a tall order.

Much has been written on sustainable agricultural systems. For example, there are several publications directed at defining and explaining terms and defining the role of agriculture in sustainable development, e.g. Serageldin (1993). Anderson and Lockeretz (1992), whilst acknowledging that many definitions of sustainable agriculture exist, arrive at the crux of the matter when they state, 'sustainability must involve less reliance on non-renewable resources and less environmental degradation than the present agricultural system.' A more elaborate discussion has been presented by Brklacich *et al.* (1991) who suggest that six major factors underpin the concept of sustainable food production. These are given in Table 10.7. Such issues have been addressed separately by various authors in relation to both cultural and environmental factors. Carey (1993), for example, has examined the concept of carrying capacity. He points out that on a temporal basis the carrying capacity of an area changes as a result of technological innovations. Moreover, it is not a readily quantifiable commodity because of the numerous and disparate factors involved. These, Carey states, are physical, institutional, social and psychological. Again, this diversity reflects the multi-faceted people/environment relationship of which one manifestation is land use, and one component of land use is agriculture. It is also implicit in the systems approach to examining environmental processes and is holistic. Projections of carrying capacity for the next 50 years or so could assist in planning decisions. Other studies that adopt a similarly holistic approach include those of Runge (1992), Soulé and Piper (1992), Edwards and Wali (1993), Schaller (1993) and Otzen (1992, 1993). Otzen (1993) states, 'What is needed are agricultural policies that support the development of integrated land-use systems that are environmentally non-degrading, technically appropriate, economically viable and socially acceptable.' Otzen outlines a number of support factors that are needed for the institution of such systems and which must derive from governments in both the developing and developed worlds. These factors include trade agreements and support for the protection of habitats and ecosystems from encroachment by agriculturalists. As discussed in Mannion (1994a), the preservation of natural ecosystems could be achieved by valuing their products and services appropriately.

Whilst it is inevitable that governments should become involved in the delineation of such policies there are many arguments that the most effective policies for improving the sustainability, i.e. the non-degradational carrying capacity, of agricultural systems in the developing world should target individual farmers. In other

Table 10.7 The major factors underpinning the concept of sustainable food production systems (based on Brklacich *et al.*, 1991)

A	ENVIRONMENTAL ACCOUNTING: This identifies biophysical limits, i.e. types of degradation or pollution on and off farm.
B	SUSTAINED YIELD: This concerns the output per unit area (e.g. kilograms per hectare). It is a way of describing the impact of biophysical limits on crop yields; if degradation occurs, crop yields will decline, etc.
C	CARRYING CAPACITY: This is the maximum population size that the environment can support on a continuing basis. This is a complex issue with many interpretations. Carrying capacity will vary temporally if biophysical limits alter.
D	PRODUCTION VIABILITY: This concerns the long-term viability of agricultural production units, i.e. farms or groups of farms. This characteristic relates to economic and cultural factors, e.g. market forces and trade in commodities. Measures of how resilient to stress farm units are can be included in this category.
E	PRODUCT SUPPLY AND SECURITY: Self-sufficiency in food production is a key to political and economic stability (see text). For many of the world's population there is no such thing as food security despite the fact that sufficient food is produced in total to feed everyone. This reflects access to food.
F	EQUITY: This concerns intergenerational security (see text) which can only derive from food security. It is a complex issue: it concerns not only the capacity of land to produce food but also political organisation and the way it affects land tenure and land ownership. The equitable distribution of food (see **E**) is also important.

words, the directive may come from the top, i.e. the government, but the initiative must be left in the hands of local farmers. Such views have been expressed by Adams (1990) and in various papers in Sachs (1993). This is just one aspect of sustainable agriculture and it is clear from the examples of different agricultural systems given in Sections 4.1, 4.2, 5.1, 7.1 and 7.2 that indigenous systems based on small-scale activities can be successful over the long term, i.e. a century more. In addition, low-technology locally developed soil-conservation technologies can make a substantial contribution to achieving sustainable food production. The example of the Machakos District, Kenya, discussed in Section 9.2, is a case in point. Here, despite a rapidly growing population during the last 40 years, soil conservation measures have been so successful that food production is now sufficient to support the population (Tiffen *et al.*, 1994). This illustrates that the relationship between land degradation and population growth is complex. As English (1993) has discussed, population growth need not necessarily lead to land degradation provided other changes in society occur, e.g. the development of other resources and an infrastructure to facilitate development, as well as the introduction of 'good' land-use practices and conservation measures. However, at a time when there is a pressing need to reduce fossil-fuel consumption in all aspects of human endeavour, it is disquieting that in some parts of the developing world progress in food production will require increased use of fossil fuel. China is already in this position; the intensive use of widely available labour is insufficient to produce enough food to support a large population. As Zhao (1994) discusses, one of the few options that China has to enhance food production is to increase the yields from existing croplands, especially those classed as middle- and low-grade farmlands. Currently, such lands are characterised by grain yields of between 3 and 6 t ha^{-1}. Zhao's suggestion that these yields need to be doubled or tripled can hardly be achieved without recourse to considerable fossil-fuel inputs. He states that there is a need to 'Apply appropriate practices for use of irrigation, mechanization, chemical fertilization, electrification, and other modern equipment in all farmlands according to their specific conditions.' Other possibilities to increase China's agricultural output are discussed by Dazhong *et al.* (1992) and include intercropping, multi-cropping, no- or minimum-tillage, green manures, improved irrigation and agroforestry. In addition, reference to the necessity of so-called energy investments for the improved productivity of African farming systems has been referred to in Section 7.3. Similarly, Hall and Hall's (1993) analysis of agriculture in Costa Rica indicates that increasing inputs of fossil fuel will be necessary to enhance food production to support its population. Again a vicious circle is likely to develop since increasing food production (notably grain) to support a population growing at an annual rate of 2.6 per cent per year would require not only increased artificial fertiliser use (the result of fossil-fuel expenditure) but also the land to be occupied by perennial crops and pasture. A reduction in the latter would, ironically, reduce Costa Rica's exports of beef to North America and Europe and thereby curtail foreign earnings. These, in turn, are essential to purchase fertilisers because Costa Rica has no capacity, due to a lack of fossil-fuel resources, to produce artificial fertilisers itself. Apart from the intrinsic problem of Costa Rica's high rate of population growth which makes the achievement of sustainable agriculture very difficult, the situation is compounded by the country's heavy reliance on the export of agricultural products. This situation also exemplifies how the developed and developing worlds are interconnected. In this case, not only does the North eat the South (see quote in the introduction to Chapter 8) but it contributes to environmental degradation and complicates, if not makes impossible, the achievement of sustainable agriculture. Thus, markets and policies in the developed world can profoundly influence the sustainability or otherwise of agricultural systems in the developing world. As Ruttan (1991) states,

> The capacity to achieve sustainable growth in agricultural production and income will also depend on the changes that occur in the economic environment in which developing country farmers find themselves. The most favourable economic environment for progressing crop and animal productivity . . . is one characterised by slow growth of population and rapid growth of income and employment in the non-agricultural sector.

Approaches to sustainable development must inevitably focus on agriculture. As the above examples illustrate, the two are intricately related and, as stated above, the most appropriate approach to achieving the former is one based on holism since this identifies both symptoms and underlying causes of land degradation. Nevertheless, and to reiterate statements elsewhere in this book, such an approach is somewhat utopian. Attempts to improve or initiate sustainability by tackling localised issues or components of agricultural systems that create specific problems are also to be welcomed. Some examples have already been discussed in Section 7.1 which includes so-called organic farming types. These, as Tables 7.2 and 7.3 illustrate, attempt to operate, as far as possible, in harmony with natural systems rather than trying to dominate them and to recycle nutrients. In Section 7.1 the given examples of organic farming reflect the widespread changing attitudes to agriculture in the developed world. Such changes are to be welcomed and further examples are given in a number of papers that refer to organic farming in Canada (Hill and MacRae, 1992), California (Altieri, 1992a) and the north central and north-eastern USA (Liebman, 1992). Such systems are known as low input/sustainable agriculture (LISA). As discussed in Section 7.1, these agricultural systems attempt to minimise the expenditure of fossil-fuel energy and the input of synthetic chemicals. This they generally achieve by combining crop and animal productivity. Such goals are advocated by the Leopold Centre for Sustainable Agriculture, based in Ames, Iowa, USA, which is a particularly important influence in US sustainable agricultural practices and operates educational programmes. Altieri (1992b) has also examined the case of sustainable agricultural development in Latin America. He suggests an agroecological approach incorporating biological pest control or integrated pest management and crop diversification in combination with appropriate government policies such as the withdrawal of pesticide subsidies. To be a truly holistic approach to sustainable development these approaches to agriculture would need to be incorporated into economies geared to accepting the products of these systems as the main source, rather than a subsidiary source, of food, etc. This would mean changing the attitudes of consumers. To some extent this is occurring in the developed world where the market for organic produce is growing. Such comments, and others made elsewhere in this book, reflect those of Redclift (1994) who states that 'societies, not nature, are largely responsible for the absence of sustainable development.' Inherent in any policy for sustainable agriculture must be a focus on what Neher (1992) describes as socio-economic viability. This must also be reflected in agricultural institutes and their extension programmes (Dicks, 1992).

Within many agricultural systems the various practices can be made more sustainable than they are presently. One possibility is the use of 'green' manure and fertilisers instead of artificial fertilisers. Multicropping is more ecologically viable than monocropping, as is the use of crop rotations. As Caporali and Onnis (1992) have discussed, rotations which include polyannual legumes are particularly appropriate for sustainable agricultural systems because they maintain soil fertility and thus obviate the need to consume fossil-fuel energy via fertilisers, as well as avoiding cultural eutrophication (Section 8.4). Their data on a long-term rotation show that sunflower crop yields in central Italy are just as high when a legume is included in the rotation as when nitrate fertiliser is used. In addition the lucerne ley, the crop grown prior to sunflower, acted as a weed control so reducing the need for herbicide application. Callaway (1992) suggests that herbicide applications could be reduced if crop varieties tolerant to weeds were bred. His examination of 21 crop types, including the major cereals, indicates that varieties already exist with increased tolerance to non-parasitic weeds. Enhancing this characteristic through conventional breeding or genetic engineering could be yet another facet of sustainable agriculture. This, however, would not be in the interests of agrochemical companies. Where artificial fertilisers are essential, savings can still be made by judicious application. For example, Norris and Shabman (1992), referring to nitrogen management on the mid-Atlantic coastal plain, report that the replacement of a single application by split applications or the addition of organic nitrogen, reduce the amount of nitrogen loss from cropland and increase net returns. This issue has also been discussed by Logan (1993).

Soil conservation must be a prime target in sustainable agricultural systems. As discussed in Sections 8.2 and 9.1, soil loss is a major problem worldwide. Conservation can consist of appropriate agricultural practices and/or the construction of terraces, bunds, earthworks, etc., to limit soil loss and, in many instances, to conserve water as well as soil. Minimum- or no-tillage practices, facilitated by the direct drilling of seeds, reduce soil loss as has been demonstrated by van Vliet *et al.* (1993). Their comparison of soil losses on plots in the Peace River District of Canada's British Columbia show that zero tillage and reduced tillage curtail annual average soil losses by 81 and 55 per cent, respectively, when compared with conventional tillage. Amongst the practices that can reduce soil erosion are residue retention and the maintenance of a vegetation cover as has been discussed by Lal (1994b), the maintenance of biological diversity, especially in relation to soil microorganisms (Swift, 1994), as well as the retention of a good physical structure, e.g. porosity, aggregation and strength (Papendick, 1994). Land degradation, either through soil erosion and/or desertification, can be counteracted by multiple land use. In the Mediterranean Basin, for example, Le Houérou (1990) advocates the use of agroforestry and sylvopasture (wood pasture) to halt the decline in productivity and increase in soil loss. The significance of agroforestry as an inherently conservational agricultural system has been examined in Section 7.5 and the use of earthworks to combat soil erosion has been discussed in Section 9.2.

Clearly, there are many facets to sustainable development which is underpinned by sustainable agriculture. There are both intrinsic and extrinsic factors, which may be environmental or cultural, that require adjustment if sustainable agricultural systems are to become reality on a global basis.

10.5 AGRICULTURE IN A WARMER WORLD

There is no doubt that concentrations of heat-trapping gases, the so-called greenhouse gases, in the atmosphere have increased substantially since the Industrial Revolution of the mid-eighteenth century. What is as yet unclear, however, is whether or not the Earth is actually warming as a consequence. According to the International Panel on Climatic Change (IPPC; see Houghton *et al.*, 1990, 1992), over the past 150 years worldwide instrumental data indicate a global-mean warming of 0.45±0.15 °C. As Hulme (1994) comments, 'the overall warming of about 0.45 °C since the middle of the nineteenth century is suggestive of a long-term underlying cause of warming, for which the enhanced greenhouse effect is the primary candidate.' However, there have been many criticisms of the accuracy of instrumental records (see discussion in Jones, 1993; Hulme, 1994) which may underestimate or overestimate the real temperature change that has occurred during this period. Moreover, there is the possibility that the impact of global warming has been offset to a certain extent by anthropogenic sulphate aerosols (e.g. Wigley, 1991; Jones *et al.*, 1994) which are a product of fossil-fuel burning and one component of acid rain. Thus, one form of pollution may be counteracting the effects of another! This possibility is also suggested by the fact that the 0.45 °C increase in temperature falls within the lower limits of predictions for that period by model simulations, i.e. 0.5–1.1 °C. IPCC (Houghton *et al.*, 1990, 1992) also point out that an increase of 0.45 °C is the upper limit that could reasonably be considered as due to natural variability and that it will be another decade at least before this can be confirmed or refuted. Consequently, it is not yet certain that the Earth is experiencing enhanced greenhouse warming. However, concentrations of carbon dioxide (the most abundant greenhouse gas) have increased by 25 per cent, i.e. from *c.* 270 ppmv (parts per million by volume) to 350 ppmv. To the effects of carbon dioxide must be added those of nitrous oxide, CFCs and methane. These are all heat-trapping gases, as is water vapour which is more abundant in a warmer than in a cooler world due to increased evaporation. Such substantial changes to atmospheric chemistry in a relatively short space of time are unlikely to occur without some repercussions. Even if it is not yet certain that global warming is a matter of fact, curbing emissions of heat-trapping gases now is pre-emptive and at the very least conservational of a vital resource.

The uncertainty surrounding global warming makes the prediction of its impact difficult and equivocal, as do the inadequacies of the many numerical models that have been developed to make such predictions (see review by Henderson-Sellers, 1994). Nevertheless emissions of heat-trapping gases are continuing to increase, a trend that is likely to be maintained as many developing countries continue to industrialise and increase their consumption of fossil fuels. In China alone there has been a new era of modernisation since 1978 as the country has embarked on large-scale development. Zhao (1994) reports that the value of industrial production increased annually by 10.1 per cent with a major focus on textiles, engineering, chemical engineering and food industries. Simultaneously, China's coal production has increased from 66×10^6 t in 1952 and 872×10^6 t in 1985, to 1080×10^6 t in 1990. Despite this massive increase the percentage of coal in China's total energy consumption declined from 96.7 per cent in 1952 to 72.8 per cent in 1985. This difference reflects the increasing importance of oil and natural gas, the production of which has increased from 440×10^3 t and 8×10^6 m^3 respectively in 1952 to 125×10^6 t and 12.9×10^9 m^3 in 1985 (Zhao, 1994). Although China's economy is growing faster than any other nation in the world, its size and substantial reserves of fossil fuels mean that emissions of carbon dioxide (and other heat-trapping gases) are set to increase in the forseeable future. In view of such developments several general circulation models (GCMs) have been constructed to predict the impact of changing concentrations of greenhouse gases (excluding water vapour). As Henderson-Sellers (1994) has discussed, all such models have their limitations. There are, however, a number of similarities including the fact that warming is non-uniform globally, i.e. high latitudes are likely to experience more warming than lower latitudes but in neither zone is warming uniform. Of the two types of model so far developed those based on transient experiments, in which carbon dioxide increases gradually with time, indicate that global warming is not as likely to be as great as that predicted by simulations based on rapid increases in carbon dioxide. The latter are equilibrium models but are not considered to be as realistic as transient models because historical data reflect gradual rather than sudden increases in carbon dioxide emissions. (It could, however, be argued that a 25 per cent increase in carbon dioxide in the last 250 years is a sudden increase when considered in the context of geological time.) Presumably, gradual increases in heat-trapping gases allow the global carbon cycle to undergo readjustment involving absorption of the gas(es) in the oceans and/or the biosphere. Furthermore, the transient models suggest that the warming signal in the early stages may be masked by natural variability (see comments above), but as heat-trapping gas concentrations increase, the warming they induce overwhelms the natural variability. In the early stages of warming there may be negative temperature changes in some regions, making the detection of climate trends still more difficult and open to interpretation. Inevitably, global warming will influence global patterns of precipitation.

What has all this got to do with agriculture? Without definitive information on likely climate change it is impossible to evaluate with any degree of certainty the impact of global warming on crop/animal production. As discussed in Section 1.2, temperature and precipitation patterns provide the major constraints on primary productivity which, in turn, determines secondary productivity. The comments which follow must be considered with caution because most of the predictions and simulations of likely future patterns of global agriculture are based on the GCM predictions referred to above which themselves have a number of limitations. Thus, as the levels of abstraction increase, the further the model(s) are divorced from reality. Testing model accuracy, either GCMs or models of future agricultural patterns, is made particularly difficult because at no time in the short-term geological past (i.e. the last 10 000 years) have atmospheric concentrations of heat-trapping gases been as high as they are now. This means that there are no precedents on which to base or test predictions. Nevertheless, some attempt at prediction in relation to agriculture is essential because the lives of some 6.2×10^9 people depend on it and this number is increasing daily. In addition, much of the wealth and political security that characterises some areas of the world is reliant on adequate, even surplus, food production. Any major changes that may occur in agriculture as a result of global warming are thus highly politically sensitive and require both policies and planning for the future.

Despite these uncertainties there is a growing literature on the impact of global warming on agriculture at various scales from the global to the individual farm. There have been reports that the enhanced concentrations of carbon dioxide will act as a stimulant for increased primary productivity. For example, laboratory experiments under controlled conditions which include higher than ambient concentrations of carbon dioxide suggest that crop productivity could be increased substantially. Yields in grain crops show increases of as much as 36 per cent and those of cotton show increases of 100 per cent (quoted in Nilsson, 1992; see also the discussions in Parry and Swaminathan, 1992; Wolfe and Erikson, 1993). Such large increases in yields are, however, rarely observed in field crops in which decreases of between 33 and 25 per cent (Bazzaz and Fajer, 1992) have been recorded. This is because these experiments cannot take into account other facets of global warming such as its impact on soil moisture and pest populations. Nor can they allow for any possible effects of increased ultraviolet light reaching the Earth's surface because of stratospheric ozone depletion. As Kendall and Pimentel (1994) have discussed, this latter is an unknown factor in relation to pest distribution and crop productivity. Conversely, however, they draw attention to the already established impact on crop health of increased ozone levels at ground level. The effect is generally deleterious. As these comments illustrate, temperature increase, though it is the most important factor, is only one of many parameters that will cause change in agricultural systems.

Other approaches to the prediction of global warming on agricultural systems have involved regional or national assessments (e.g. Parry *et al.*, 1988; Parry, 1990; Tegart *et al.*, 1990). The results are summarised in Table 10.8 which shows that some regions will gain whilst others will lose. Clearly, the impact of global warming will provide opportunities for some nations and regions whilst it will exacerbate already existing problems of food supply in others. Scandinavia, other parts of northern Europe and Japan may be able to extend crop growing further north than occurs now. However, arid and semi-arid parts of Africa and Asia may well become even more marginal for crop and animal production than they are presently. In general, it is likely that there will be little change in the world's production of food overall but the problems of equitable food distribution, especially in the context of growing world populations from *c.* 6.2×10^9 to 10×10^9 by 2050 (see Section 10.4), will become even more severe and politically sensitive than they are now. Global food security in the twenty-first century is likely to be as elusive as it has been in the twentieth century. More recent assessments of regional impacts of global warming include those of Crosson (1993) who has examined the Missouri, Iowa, Nebraska and Kansas (MINK) region. This region is particularly important because it is part of one of the largest grain-producing areas of the world and one for which most GCMs predict significant warming with an accompanying decline in precipitation. The latter, coupled with increased evapotranspiration, can hardly fail to have an adverse impact on crop productivity. Using a crop-growth simulation model, Crosson has shown that for five crops—maize, wheat, soy bean, sorghum and alfalfa—production declines by 17.1 per cent (based on averages for the period 1984–1987) if there is no adjustment for carbon dioxide fertilisation (see above). If this is taken into account, yields still decline but by the reduced amount of 8.4 per cent. As Crosson points out, such changes would probably have been countered to some extent by changes in the types of crops grown and/or technological innovations. Globally, however, such a reduction in wheat and maize production in North America would most likely alter trade and food aid arrangements. These adjustments provide yet another set of variables which need to be included in the long-term (the next 50 years) prediction of global food production, as is discussed below. Crosson's view is rather optimistic, possibly reflecting a national rather than global perspective and, possibly, the USA's reluctance to introduce emission restrictions. A more recent commentary by Mount (1994) reflects a similar 'it'll be alright on the night' stance and focuses on economic rather than environmental readjustment. Rosenzweig and Parry (1993, 1994) and Parry and Rosenzweig (1993) have attempted to take the prediction of the impact of global warming on agricultural systems a stage further by combining crop-growth models for regions with the predictions of climate change models and a world food trade model. This approach, which is speculative, is at least holistic. As stated in Mannion (1994a), a most important unknown and unpredictable factor in the people/environment relationship is the human (cultural) response, which involves ingenuity, to external stimuli. The conclusion reached by Rosenzweig and Parry is that global food productivity will change little but that the major benefits will be to nations in the temperate zone. Developing countries will be disadvantaged. However, there are provisos. For example, the crop-growth models asume that the beneficial effect of fertilisation by carbon dioxide (see above) will occur. Moreover, as Reilly (1994) points out, the exercise relies heavily on the major cereal crops which are not

Table 10.8 The predicted regional impact of global warming on crop yields (based on sources quoted in the text)

	Temperature change	Soil moisture	Impact on yield
North America			
USA	Warming	Decrease	Decrease in maize, wheat, soy bean
Canada	Warming	Decrease	General decrease though northern limit of crop growth will move north
Central America	Limited warming	Decrease	May need more irrigation
South America			
Brazil	Limited warming	Varied	Increases overall
Andes	Limited warming	Increase	Some gains; cultivation possible at higher altitudes
Europe	Warming, especially in the north	Varied	Scandinavian wheat increases of 10–20% yield; losses in south Europe due to drought; increases in Alpine zone
Former Soviet Union	Warming	Varied	Increase in high latitudes, decrease in mid-latitudes aridity
Middle East	Limited warming	Decrease	Up to 40% reduction in wheat yields
Africa	Limited warming	Varied	Decrease in arid and semi-arid regions. Varied elsewhere
Eastern Asia	Limited warming	Increase; also flooding	Increase, e.g. rice; if monsoon in SE decreases, crop yields will fall
Japan	Warming	Increase	Increases of rice yield by 2–5%
Pacific Islands* Australasia	Limited warming	Varied	Some gains, e.g. New Zealand; some losses, e.g. Australia.

* Pacific islands will lose productive land to the sea as sea-levels rise. This is likely to have a greater impact on crop yields than warming.

necessarily the dominant crops in developing countries. Further investigations to include crops such as beans, root crops, sugar cane and various fruits would give a more comprehensive picture of the likely situation in tropical regions than that provided by Rosenzweig and Parry.

Another attempt at modelling the magnitude and impact of climatic change has been presented by Rotmans *et al.* (1994). This is the Evaluation of Strategies to Address Climate Change by Adapting to and Preventing Emissions (ESCAPE) model. It has been developed to provide an overview of the uncertainties surrounding likely global warming, which can be useful for politicians and planners. ESCAPE predicts that if carbon dioxide concentrations double from pre-industrial levels (a further 25 per cent than occurs at present), a mean global warming by 2050 of 1.51 °C will occur (based on 1990 temperatures). The EU will warm more than this. The winter temperature increase will be 1.85 °C whilst the mean summer change will be nearer to 1.51 °C. Precipitation changes will also occur involving values between a decrease of 13 per cent and an increase of 5 per cent, with drier conditions developing in the south and wetter conditions in the north when compared with those of 1990. These changes, especially in the north which is likely to be both warmer and wetter than now, concur with the suggestions of other models that agricultural productivity will increase. In relation to maize production, France, Germany and Belgium will experience an increase whilst Spain, Portugal and Italy will experience a decrease. For cold-climate grape production, Rotmans *et al.* indicate that potential cultivation zones will shift northward by 100–200 km. For at least a proportion of humanity global warming may not be such a calamity.

10.6 CONCLUSION

The future for agriculture is likely to involve considerable change and innovation. The application of biotechnology and its subdiscipline of genetic engineering hold much promise for improving crop productivity and contributing to sustainable agriculture, as does the use of information technology. On the positive side, it is possible that in another decade so-called designer crops will be available. Through genetic engineering crops will be developed to suit specific environments. This is in contrast to the conventional approach to agriculture which involves tailoring the environment to suit the crop, i.e. adding nutrients, increasing water supply and eliminating competitors where necessary. Similarly, it will soon be possible to engineer livestock to produce more fibre, milk or meat than they do now. It is even likely that animals will be used as bioreactors, to produce therapeutic substances, and possibly as organ donors. Like most technologies, biotechnology has its drawbacks. Designer crops could, for example, allow the expansion of agriculture into new areas and so destroy yet more of the Earth's remaining natural ecosystems. There is always the danger of creating organisms that turn out to be pests or which prove to be harmful to animal and human health. To avoid these and other problems, internationally accepted testing and registration procedures are essential. In addition, there are problems of technology transfer. Currently, most of the research in biotechnology is carried out by transnational companies in the developed world yet the biggest and most pressing need for improved crop types is in the developing countries. Since such nations are hosts to the largest share of the world's crop genetic resources, some form of recompense and technology transfer is necessary to ensure that the gap between the North and the South does not widen.

Biotechnology, applied wisely, could be a valuable tool for achieving sustainable agriculture. This is an almost impossible goal but attempts to achieve it must be improved in order to ensure that humanity's capacity for food production is not impaired needlessly. Soil and water conservation programmes are essential for sustainable agriculture. Soil erosion is a global problem that represents the loss of a vital resource, as does the contamination of water resources by nitrates and pesticides. However, many components of sustainable agriculture concern social and economic injustices such as inappropriate land tenure, lack of training and education and poverty. Population control is another very important component of sustainable agriculture and sustainable development in general. These are all actions that could be initiated immediately. They are measureable and tangible. To improve agricultural systems now would be a valuable hedge against the impact of global warming and all the uncertainties that surround it. Agriculture in a warmer world will need to be flexible and conservational. As Biswas (1994) states, 'In coming decades, preserving the environment will not be a choice, but an imperative, even in order to sustain present rates of food production. An even greater effort on behalf of the environment will have to be undertaken if we are to feed the global population in the year 2000.'

10.7 FURTHER READING

Barnett, V., Payne, R. and Steiner, R. (eds) (1995) *Agricultural Sustainability. Economic, Environmental and Statistical Considerations*. John Wiley and Sons, Chichester

Bud, R. (1994) *The Uses of Life: A History of Biotechnology*. Cambridge University Press, Cambridge.

Campbell, K.L., Graham, W.D. and Bottcher, A.B. (eds) (1994) *Environmentally Sound Agriculture*. American Society of Agricultural Engineers, St Joseph, Michigan.

Glantz, M.H. (1994) *Drought Follows the Plough*. Cambridge University Press, Cambridge.

Greenland, D.J. and Szabolcs, I. (eds) (1994) *Soil Resilience and Sustainable Land Use*. CAB International, Wallingford.

Kaiser, H.M. and Drennen, T.E. (eds) (1993) *Agricultural Dimensions of Global Climate Change*. St Lucie Press, Delray Beach, Florida.

Tudge, C. (1993) *The Engineer in the Garden*. Jonathan Cape, London.

CHAPTER 11

Endpiece

The role of agriculture as an agent of environmental change is intimately related to two major factors: technological innovation and population change, as illustrated in Figure 11.1. The relationship between these factors in an historical perspective is far from straightforward. This is because it is difficult to determine which have been forcing, reinforcing or reacting factors within the relationship. On the one hand, it could be argued that only the environment has been predominantly a respondent because it is being manipulated to produce food, though there have been times in the past when environmental change, notably climatic change, has been a forcing factor in agriculture. It may well become so again in the twenty-first century if global warming occurs. On the other hand, it could be argued that the environment has always been very much a forcing factor since it sets the broad physical limits in which technology can be applied to generate agriculture. As discussed in Section 2.2, there is substantial evidence that at least the domestication of plants may have been the result of diminishing wild food resources as climate fluctuated at the end of the last major ice advance *c.* 12 000 to 10 000 years ago. This possibility contrasts with many traditionally held views that the inception of permanent agriculture was a consequence of necessity created by population growth. There may also have been other stimuli: power (or greed) and security. These are related and are directly or indirectly linked with population and materialism. For example, the production of a food surplus, especially a surplus that could be planned for and calculated in advance, would have provided commodities with which to barter for other useful materials, e.g. stone tools, and thus provided an element of power and/or security. The trade-off for this power was hard work. Perhaps the power and/or security were worth it in a time when the success of hunting and gathering was unpredictable. Moreover, both the temporal and spatial interplay of these factors may well have changed. The relationship is not spatially uniform today; there is no reason to suppose that it was uniform in the past. Nor is there any reason to suppose that at any given time only one forcing factor was in operation. For example, climatic change and population pressure may both have led to the initiation of permanent agriculture.

However, this would not have occurred without the social infrastructure that characterised the earlier hunter–gatherer societies (see Section 2.1) within which modern humans (*Homo sapiens sapiens*) were organised. Going back further in time, it is difficult if not impossible to discern how or why modern humans or their ancestors, the hominids, grouped together as they did; or how and why so many characteristics developed that are peculiar to hominids rather than other mammals. Notable examples of such characteristics include bipedalism, increases in brain size

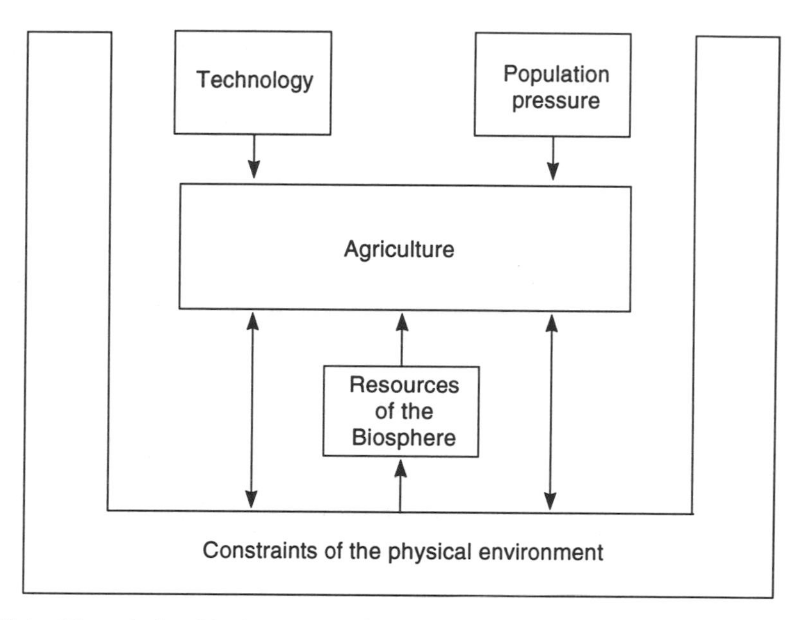

Figure 11.1 The relationship between agriculture, environment and the cultural factors of population pressure and technology

and tool making. The latter is particularly important since it is a capacity with which only primates, of all the animal kingdom, are endowed and which only the hominids have developed. Tool making is also technology, the development of which has profoundly affected and effected the relationship between people and environment in prehistory and history. Undoubtedly, the role of technology in this relationship is set to intensify.

11.1 TECHNOLOGICAL INNOVATIONS: TEMPORAL DIMENSIONS

Technology, from the most primitive to the most sophisticated, provides the means to transform a component of the Earth's surface into a resource. The first tools shaped by the hominid ancestors of modern humans some 2.5 million years ago represent the earliest attempts to dominate nature, thereby transforming humans into controllers of, rather than integral components of, ecosystems. For example, early stone tools allowed hominids to extract bone marrow from even the most denuded animal carcasses abandoned by other carnivores. This scavenging provided a supplementary food source to plant foods and made life in a seasonal biome such as the savanna, with a pronounced dry season, less precarious. The application of technology irrevocably altered the people/environment relationship. From simple beginnings the impact of technology has been considerable in relation not only to the environment but also to society. For example, social groupings, increases in brain size and the development of language are all likely to be interrelated with food procurement. In relation to the environment, technology provided, and continues to provide, the wherewithal to subjugate nature for society's ends and to create order from the chaos and complexity that characterise the natural world. It also introduced an element of predictability into food procurement. Equally important is the fact that as hunting and gathering became increasingly organised and planned, thus providing an increasing degree of food security, it became possible for a proportion of a social group to be freed from food procurement. This would have facilitated division of labour and allowed some people to concentrate on the exploitation of other resources. With the advent of permanent agriculture 10 000 years ago such specialisation intensified. Not long after this occurrence, for example, the archaeological record attests to the development of pottery and, by about 8000 years ago, metallurgy.

These developments not only broadened the resource base available to human groups but also provided additional means to improve food production. Examples include the development of pottery, which facilitated food storage and preparation, grinding implements with which to produce flour, as well as metal tools to speed up the removal of natural vegetation and to prepare the land for cultivation. In combination with human labour and animal traction, society's ability to manipulate nature was considerably enhanced. Throughout prehistory and history human ingenuity and inventiveness have allowed society to manipulate nature increasingly intensively. In space and time, such dexterity and control have also generated changing patterns of wealth and power within and between communities and nations. Variety may be the spice of life but technology provides the means of procuring advantage for one group over another and, possibly, a means of appropriating the latter's resources. Table 11.1 gives a list, with dates, of the major technological advances that have enhanced food procurement and production.

No one of these elements can be taken in isolation as a single major advance because all the developments build on earlier experience. Nevertheless there are certain stages in the history of agriculture that can be considered as turning points in the context of both culture and environment. For example, the inception of permanent agriculture 10 000 years ago allowed societies to develop in a way hitherto impossible. The division of labour that it facilitated allowed new resources to be exploited and new talents to be developed with further cultural and environmental ramifications. The 'expansion of Europe' in the fifteenth and sixteenth centuries was another important stage as crops were exchanged between the Old and New Worlds. This was facilitated by a form of technology unrelated directly to crop agriculture, i.e. navigation and shipbuilding. Then came improved crop rotations and agricultural intensification in Europe with associated cottage industries based on food processing. The advent of fossil-fuel use in Europe had many ramifications. Ready markets in expanding urban centres provided a stimulus to agriculture, as did improved transport and increasing global trade. The invention of artificial fertilisers, followed by mechanisation, began the so-called industrialisation of agriculture that characterised the early twentieth century in the developed world and which has since intensified. By the 1950s, however, a quite different approach to food production, to feed an ever increasing global population, was required because most land suitable for cultivation was already in use. Instead of continuing to expand the amount of cultivated land by clearing natural vegetation communities, science and technology were brought to bear increasingly on the problem of how to increase the productivity of land already under cultivation. New crops and improved varieties of existing crops were sought. The latter, known as high-yielding varieties (HYVs), have helped stave off widespread hunger and famine in many parts of the world. In conjunction with

Table 11.1 Some of the major technical developments that have occurred in relation to food procurement and production

Date	Stage/event/invention
1995	First genetically engineered crops marketed
1970s	First high-yielding varieties of crops introduced
1960s	Direct drilling
1950s	Expansion of agriculture into new areas declines
1940s	Chemical pest control introduced
1920s	Widespread use of combine harvester/tractors in use
1913	Discovery of the Haber process to produce nitrate fertilisers
1889	Murchland milking machine
1885	Stripper harvester invented in Australia
1880s	Refrigerated transport of meat from the Americas to Europe
1865	Mendel published his definitive work on heredity and genetics
1838	First combine harvester produced in the USA
1830s	Sodium nitrate from Chile for fertiliser
1786	Meikle's threshing machine invented
1750s	Beginning of Industrial Revolution
1730	Rotherham or Dutch plough invented
1700s	A period of significant innovation, i.e. new rotations, marls and seaweed, etc., used for fertiliser
1500–1800	Exchange of crops between Old and New Worlds
1492	Discovery of Americas
1500 BC–AD 1500	Deforestation of Europe
AD 200	Wetland drainage by the Romans
200 BC	Mouldboard plough/scythe invented
c. 3×10^3 years BP	Earliest known irrigation
c. 5×10^3 years BP	Plough; use of animal traction
c. 8×10^3 years BP	Metallurgy (bronze and, a little later, iron)
c. 10×10^3 years BP	First domestications of plants and animals; inception of settled agriculture
c. 20×10^3 years BP	Highly organised hunting and gathering
c. 2.5×10^6 years BP	The first stone tools made by hominids

artificial fertilisers, crop yields were increased as plant (and animal) breeding programmes exploited the principles of heredity in combination with micropropagation (see Section 10.1) to produce large numbers of identical high-yielding species. The current application of modern biotechnology, including genetic engineering, in agriculture represents another milestone in the history of agriculture.

However, no technology has ever been applied without direct and inadvertent repercussions. As Chapters 8 and 9 illustrate, there is abundant evidence for the degradation and impaired productivity of agricultural land as a consequence of technology out of control. Such problems are not confined to modern agriculture though they have undoubtedly intensified in the last century or so. Consequently, the application of human ingenuity to food production can, under some circumstances, be counterproductive. This is amply illustrated by badly managed irrigation whereby the enhancement of water availability may lead to salinisation and waterlogging. As discussed in Section 9.4, areas in which this is acute experience a rapid decrease in crop productivity, and agricultural land may, in extreme cases, have to be abandoned.

In addition, the inadvertent impacts of technology may be considerable as is demonstrated by the issue of cultural eutrophication and its impairment of aquatic ecosystems and water resources (Section 8.4). In some circumstances the application of one form of technology may simply mask the adverse impact of another technology. For example, in Europe the impact of soil erosion (Section 8.2), itself a consequence of mechanisation, on crop productivity is often masked by the increased application of artificial fertilisers (Section 8.4).

Tackling the symptoms of the injudicious use of technology is generally effective only in the short term and has no place in the achievement of sustainable agriculture. In view of the many environmental issues created by agriculture and the variety of agricultural systems described in this book, it is difficult to envisage quite what constitutes a truly sustainable type of agriculture. Furthermore, there are two other factors that militate against the establishment of sustainable agricultural systems. One of these is the problem of adequately predicting the degree and impact of global climatic change. The environmental parameters within which agriculture is constrained are set to change in directions that cannot now, if ever, be predicted with reliability. This uncertainty is itself a product of technology. The second factor is the apparent reluctance of society to learn from the lessons of the past; not least of these lessons is that virtually any technology has an environmental cost. Both technology and the environmental costs that it exacts give rise to social costs. Adequate safeguards are therefore essential in order to ensure that technology is applied in the least damaging, most environmentally sustainable and most socially equitable way. Such safeguards will include risk assessment, international legislation, safety for living organisms, monitoring within an internationally recognised framework, as well as technology transfer and education. The price of these safeguards will be high. Above all, the relinquishment of the power that technology endows on its originators will be necessary. Such grand altruism has few precedents in history. The development of technology requires investment which means that it has a value; its application will generate a product with a market price but there will be environmental and cultural costs. In some forms of agriculture these costs are unacceptably high and cannot be sustained for any length of time; the practice thus becomes self-defeating.

11.2 POPULATION INCREASE: TEMPORAL DIMENSIONS

As illustrated in Figure 1.1, one of the most important social or cultural factors that influences agriculture is population growth and its changes over time. Unfortunately, for a large part of human history there are no reliable population data on which to assess the relationship between population and the expansion of agriculture with its concomitant impact on the environment. True census data, for example, were first obtained in the 1750s in Scandinavia and the Austro-Hungarian Empire. In Britain the first census was taken in 1801 but for many African and Asian countries such data were not collected systematically until the late 1940s and 1950s. There are several reasonably reliable estimates of global population from about 1700 but for the earlier historic period estimates are speculative. For prehistory even less is known about population numbers except in the case of certain civilisations, such as ancient Egypt, for which there is archaeological evidence (reference has been made to the relationship between population and agriculture in ancient Egypt in Section 1.4). It is thus impossible to assess the precise relationship between global population change and the expansion of agriculture on a temporal basis. It is, however, possible to establish some generalisations.

Although entirely speculative, it has been suggested that there were some 10×10^6 humans scattered worldwide about 20 000 years ago, a time when the last major ice sheets reached their maximum extent. These were members of the species *Homo sapiens sapiens*, i.e. modern humans, whose ancestors had originated in Africa about 120 000 years earlier. At this stage, the Late Palaeolithic period, food procurement focused on organised and planned hunting and gathering (see Section 2.1). Why these people should have moved out of Africa (a hypothesis which is itself controversial but more likely than the alternative hypothesis of a multiregional or polycentric origin in various parts of the world from *Homo erectus*) is a matter for as much debate as the reasons why their descendants took to permanent agriculture. Obviously, if food resources became scarce then people would move to new areas or innovate *in situ* to enhance food availability. Either population pressure or environmental change could cause food scarcity. Either could cause migrations and/or innovation. These possibilities have been discussed in Chapter 2 where they are referred to as materialism and environmentalism. Conversely, food abundance can be a stimulus to population growth. It is thus impossible to determine unequivocally which were the forcing factors and which were the responding factors.

Why agriculture spread beyond the centres of origin (Figures 2.3–2.6; Chapter 3) is equally difficult to establish though the availability of a proximal and reliable food source is advantageous since it allows forward planning and so endows an element of power. If, as is suggested in Section 3.1, the practice of permanent cultivation and domesticated species were disseminated mainly through demic diffusion, i.e. migration, as well as through the diffusion of ideas, then migration must have been widespread in the four or five millennia following the initiation of permanent agriculture. Was this, too, caused by population pressure? In the Near East, for example, the abundance of archaeological sites indicates that population densities were relatively high. Perhaps the carrying capacity of the land within the social constructs of, for example, systems of land tenure, etc., was exceeded, forcing the landless to seek their subsistence elsewhere. There are parallels in the world of the 1990s (see Section 9.1). Alternatively, a food surplus may have provided leisure and learning time, prompting the curious to explore for riches elsewhere. The expansion of Europe in the sixteenth to eighteenth centuries was more to do with mineral resources than with food supply. It also reflected scientific advancement and technological achievement. Hence, whilst there is an indisputable relationship between agriculture and population change in the distant past, it does not lend itself to simplistic analysis.

Throughout the eight millennia between the inception of permanent agriculture and the establishment of the Roman Empire, few factual data can be produced that relate to the agriculture/population relationship except that in some regions, like ancient Egypt and Sumer, detailed reconstructions are possible. Even these are subject to interpretation though the archaeological evidence points to the fact that food surpluses were available and that they facilitated trade. In such circumstances, the carrying capacity of the land was not exceeded and the ability to provide food for trade conferred a degree of power. In some instances, such as the annexation of Britain by the Romans in AD 44, this agrarian ability was recognised and appropriated. The ability to produce a food surplus attracted a nation whose eminence derived from a plentiful food supply and whose continued survival as a world power required additional food sources. By modern standards, however, the total population was low, having reached about 200×10^6 by AD 100. By 1500, global population had increased to approximately 400×10^6. The doubling time for global population was, therefore, 1500 years or so for this early period of the current interglacial. Thereafter doubling time decreased to 180 years between 1650 and 1830, to 100 years between 1830 and 1940 and to 45 years between 1930 and 1974 when there were 4×10^9 people. Today there are 5×10^9. It is anticipated that the world's population will reach *c.* 8×10^9 by the year 2020. As will be discussed below, population growth rates are not the same around the globe; this spatial variation gives rise to different stimuli to agricultural change. On a temporal basis the rapid population growth of the seventeenth and eighteenth centuries was mainly in Europe, the former Soviet Union and China. Today, rapid population growth is characteristic of developing nations. In addition, the rapid doubling time of population since *c.* 1600 has increased considerably the pressure on agricultural systems, which until *c.* 1950 was accommodated by an expansion of the area under cultivation. The direct and indirect impacts of such rapid population growth, even acknowledging the fact that the people/resource relationship is complex and varied, also represent an immense strain on Earth surface processes.

Precisely why population increased so strikingly in the sixteenth and seventeenth centuries when compared with the slow increase for the earlier millennia is a matter for speculation. As discussed in Sections 3.2 and 3.3, both innovation in, and the intensification of, agriculture occurred during these periods—especially in Europe and China. What remains speculative, however, is whether the changes in agriculture are the result of, or a stimulus to, population increase. At the same time, the so-called expansion of Europe into the Americas and Africa set in train culturally motivated environmental change on a scale and of a type never hitherto experienced. Although this emigration from Europe began as a quest for mineral resources, notably precious metals, it soon turned into an appropriation of biotic resources, principally agricultural produce, and a means of accommodating Europe's burgeoning population. Indeed, at least some of Europe's rise to pre-eminence as a world power from the fifteenth century must be attributed to its annexation of resources, including food produce, from its colonies. Europe either exported its people or imported the goods. As Table 1.3 shows, substantial changes occurred in global land use after 1700, especially in the New World. Although no such comprehensive data are available for the pre-1700 period there is abundant evidence, mainly from palaeoecology, for widespread landscape change. In Europe and the Near East, for example, more than 50 per cent of natural woodlands had been cleared. Moreover, it is not population increase alone which exerts pressure on the environment; it relates to the resource consumption levels of individuals as reflected in standards of living, including food choice. At different times it is also possible that the role of population increase in relation to food production has changed. For example, the availability of an abundant food source could inherently, or through its provision of a commodity to trade, encourage population growth. Eventually, population would match the food supply and begin to surpass it, thus prompting an expansion or intensification of food production.

In the twenty-first century food production globally will need to increase fourfold to meet the food requirements of a predicted global population of double that of today's 5×10^9 people. Each doubling of the population inevitably increases the pressure on the Earth's ecosystems for two reasons. First, there is a loss of biodiversity, and thus a reduction in the Earth's genetic resource, as ecosystems are transformed into agroecosystems. Secondly, the increasingly intensive use of technology to extract maximum productivity from the agroecosystems has repercussions in terms of air, soil and water quality in the remaining ecosystems. Consequently, their ability to provide services, such as the maintenance of equilibrium in global biogeochemical cycles, is impaired. There are other indirect reverberations. For example, the growth of cities supported by agroecosystems creates an additional range of environmental problems, e.g. waste disposal and air pollution, which in turn impair ecosystems. In the 1990s this pressure is at its greatest. Whilst it is fair to say that the Earth and society have survived population increases in the past, the fact that population doubling time has decreased so rapidly and technology has advanced so rapidly since 1600 means that there has been an ever diminishing time in which ecosystem processes can adjust. It is likely that positive feedback is already occurring in the atmospheric system as a response to perturbations to the global biogeochemical cycle of carbon. Agriculture and agricultural technology and their changes over time have made a substantial contribution to this problem, not least through deforestation and the support of industrialisation.

11.3 TECHNOLOGICAL INNOVATIONS AND POPULATION INCREASE: SPATIAL DIMENSIONS

The changes discussed above in relation to technology and population have not occurred evenly over the globe. This is illustrated by the rise and fall of populations in places such as ancient Egypt and Sumer, and the domination of technological innovation by European nations when compared with the rest of the world in the eighteenth century. Today there is just as much, if not more, spatial heterogeneity. In some parts of the world there is a high level of technology, in others there is very little; population growth is declining in some countries whilst in others it is increasing. Generalisations can be misleading because they tend to mask heterogeneity. However, there are some characteristics that reflect the current global pattern of technology use and population pressure in terms of annual growth rate. These are given in Table 11.2. If the consumption of commercial energy per capita is used as an indicator of technology intensity, it is clear that the high-income nations of the developed world are the most technology intensive. If the rate of use of fertiliser is taken as a measure of agricultural technology, Table 11.2 shows that these nations are also top of the league tables. Where rates of population increase are concerned, however, they are amongst the lowest in the world in the developed nations. Whilst the data for the share of Gross Domestic Product (GDP) are given in absolute values rather than as a percentage of total GDP, all of the developed nations given in Table 11.2 have a value for the share of agriculture in GDP of less than 6 per cent. This implies that the major proportion of GDP comprises industrial goods and services; it also reflects the fact that a relatively small proportion of each country's population is employed in agriculture. Conversely, for the majority of low- and middle-income nations (mostly

Table 11.2 Selected indices to illustrate the spatial distribution of technology use and population pressure (based on World Bank, 1994). Note that data quoted refer to 1992

	Share of agriculture in GDP $\times 10^6$ \$	Fertiliser consumption 1×10^2 gms ha^{-1} of arable land	1992 cereal imports $\times 10^3$ t	Food Aid in cereals $\times 10^3$ t	Rate of population increase 1992–2000	Commercial energy use kg per capita (oil equivalent)
Low income[a]						
Mozambique	n/a	16	1164	591	2.6	32
Nepal	1440	272	15	8	2.4	20
Malawi	473	447	412	321	2.5	40
Bangladesh	8197	1098	1339	1429	1.8	59
Madagascar	925	31	147	41	2.8	38
Burkina Faso	n/a	72	145	n/a	3.0	16
India	69 682	752	3044	299	1.7	235
Nicaragua	562	273	136	128	2.7	253
Pakistan	11 416	889	2044	322	2.7	223
China	137 677	3043	11 661	172	1.0	600
Indonesia	24 279	1093	3178	82	1.4	303
Middle Income[a]						
Philippines	11 380	548	1883	78	2.3	302
Cameroon	2286	26	424	8	3.0	77
Peru	n/a	206	2015	464	1.8	330
Guatemala	2639	759	329	251	2.8	161
Romania	4617	461	1779	375	0.0	1958
Jamaica	177	948	459	181	0.6	1075
Thailand	13 096	365	992	75	1.3	614
Iran	25 711	748	4350	104	2.8	1256
Chile	n/a	706	1095	13	1.3	837
High Income[a]						
New Zealand	n/a	9341	159	0	0.8	4284
Australia	9702	273	33	0	1.2	5263
United Kingdom	15 391	3171	3559	0	0.2	3743
Germany	19 952	2473	3312	0	0.1	4358
United States	n/a	998	3718	0	1.0	7662
Sweden	5139	950	167	0	0.4	5395

[a] Income is based on 1992 Gross National Product per capita. Low income: \$675 or less; middle income \$676–8355; high income \$8356 or more per year

developing nations), agriculture comprises up to 30 per cent GDP and rarely less than *c.* 15 per cent. In both these categories, commercial energy consumption in particular but also fertiliser use are much lower than in high-income nations. There is, therefore, a much heavier reliance on agriculture in developing nations to provide the existing populations with both food and employment than there is in developed nations. Developing nations are also characterised by rapidly increasing populations with doubling times of 30 to 40 years. Most currently import cereals and receive cereals in food aid. Although such data are not given in Table 11.2, all of the developing nations referred to have significant external debts and many rely on the export of agricultural produce to generate foreign income. This renders them particularly vulnerable to changes in world food markets.

The pressures on agriculture in these nations are thus immense, not least because they are exporting their natural resources (i.e. biomass, soil, water, etc.) in order to help sustain developing nations, with concomitant implications for the environment. Local environmental problems become national and/or regional issues and eventually reach global significance. Thus, whilst apparently opposite trends are occurring in agriculture in the two major world sectors of the developed and developing world, i.e. agricultural intensification in the latter and extensification in the former, the two conspire to continue to bring about environmental change at a rate hitherto unprecedented not only in geological history but even in the comparatively short period of culturally directed environmental history. All of the environmental change concerns the harnessing of energy flows (fossil-fuel energy as well as trophic energy), the manipulation of biogeochemical cycles and the acceleration of the rates of Earth surface processes such as soil erosion and desertification.

In a temporal context, the first tool-makers and their successors the agriculturalists harnessed trophic energy flows and in so doing exerted a degree of control on ecosystems, biogeochemical cycles and Earth surface processes. Subsequently, industrial society accelerated the rate of these changes and discovered new ways of exerting control. This has caused the disruption of biogeochemical cycles, especially that of carbon, to alter atmospheric composition. The latter continues, as is reflected in the figures for energy consumption in Table 11.2. The use of fossil-fuel energy in agriculture means that most developed nations produce sufficient food to support their populations, the growth of which is now stabilising (Table 11.2) at near zero. In many of these countries production exceeds demand and policies to reduce production are being implemented. In a spatial but current temporal context, the destruction of ecosystems, and the resulting loss of biodiversity, along with the acceleration of soil erosion, etc., are most intense in the developing nations where the rate of change is apparent even on a decadal basis. Not only is there a problem with feeding rapidly growing populations but also an intensification of pollution, especially atmospheric pollution, as these nations continue to industrialise. Both China and India, for example, are industrialising rapidly. China is one of the fastest developing countries in the world; its GDP per capita grew at an annual growth rate of 7.6 per cent between 1980 and 1992; only the Republic of Korea with a growth rate of 8.5 per cent per capita exceeds it (World Bank, 1994). The latter, with a population of 43.7×10^6, is relatively insignificant when compared with China's 1162.2×10^6. It is difficult to envisage how food production can be increased sufficiently in the next decade or so to accommodate the rates of population increase which characterise most developing nations (see Table 11.2), including that of China which has one of the lowest population growth rates in the developing world. It is difficult to anticipate how such a large and rapid increase in population, and rise in living standards, can be achieved without considerable direct and indirect environmental impact, especially if such nations continue to rely on the export of agricultural produce to generate more than a small proportion of their export income.

11.4 CHALLENGES FOR THE MILLENNIUM

Global agricultural systems face a number of challenges as the millennium approaches if the world's population of *c.* 8×10^9 by the year 2020 is to be fed adequately. These challenges involve adjustments at global, regional (or national) and local levels and within varied political and economic frameworks. At the global level, adjustments to a likely warmer world will be essential.

- As discussed in Section 10.4, predicting how climate will change and what the impact of that change will be on the world's agricultural systems is full of uncertainties. On a regional and local basis, the climatic changes will open up opportunities for some (at local or national levels), whilst for others productivity will decline. Unfortunately, many developing nations are likely to be disadvantaged. In the context of their high rates of population increase, the problem of food production will be compounded.

- Global warming is unlikely to result in an equitable distribution of food. If anything, it may cause greater inequities in food production than occurs at present. Under such circumstances, ways acceptable to food donors and recipients alike must be implemented to minimise, and preferably avoid, problems associated with famine and poor nutrition. This is a difficult issue and one that requires resolution within a global political framework and economic order. The current food aid system (see Table 11.2 for limited data) will be far from adequate. Moreover, it would be preferable if the poor food-producing nations could turn to account other types of resources which could be exchanged for food imports, engender self respect and encourage the development process. It would also avoid an influx of so-called environmental refugees into already heavily populated regions. Taking food to the people is a more appropriate action than vice versa but achieving this in an equitable way is no easy task. Moreover, any nation which does not have the capacity, or near capacity, to feed its people is unlikely to enjoy the same degree of security in a world context than a country that produces more than adequate food.

- At regional or national levels there must be policies to curtail the spread of agriculture into lands currently occupied by natural and semi-natural ecosystems. This will require not only rhetoric but also positive action, including enforced regulation through legislation. Such measures are essential to conserve biodiversity and, through its gene resource, opportunities for the future. The conservation of plant species diversity is also important to maintain the biosphere's ability to absorb carbon dioxide and so limit global warming. These conservation measures require political commitment, a legal infrastructure and active enforcement. The establishment of a sustainable use for natural habitats, e.g. through ecotourism, renewable resource consumption and watershed management, would contribute substantially to conservation whilst providing alternative sources of income to agriculture.

• At national and local levels there is a need to intensify or extensify agriculture. Europe (through the EU's Common Agricultural Policy) and the USA already have set-aside policies to reduce agricultural production. Indirectly, these policies will help curtail some forms of pollution, e.g. cultural eutrophication, and should also encourage conservation by reducing the need to convert remaining natural habitats into agroecosystems. In most developing nations agriculture is intensifying with all the problems that this brings for environmental conservation. Careful planning is essential to minimise environmental disruption and loss of biodiversity, etc. Social issues, such as wealth distribution and land tenure, may need to be addressed as components of agricultural policies.

• At all levels it will be necessary to implement suitable technology. This will require technology transfer from developed to developing countries at affordable costs. It will require training and education at regional and especially local levels if the technology is to bring rewards in terms of improved food productivity without too many adverse environmental impacts. Technology development and employment require investment and planning for the future. It is also imperative that the products of new technologies such as biotechnology (Section 10.1) and genetic engineering (Section 10.2) are adequately tested prior to general use. This requires international co-operation and the realisation of benefits for both the hosts of genetic resources (the richest are the developing countries) and the developers of such resources into valuable products. These are generally transnational companies (TNCs) in developed nations. Such technology will, however, continue to provide benefits if gene banks and genetic resources are preserved (see above in relation to the conservation of biodiversity).

• At all levels agriculture must become sustainable. One way or another the small amount of solar energy that enters agroecosystems (see Section 1.4) must be used at the utmost efficiency. The ways of achieving this efficiency must also be conservational. There is much to be learnt, for example, from the organic farming movement in developed nations (see Section 7.1) as well as agroforestry (Section 7.5) and intercropping (Sections 4.2, 5.1 and 5.4). In the former external energy inputs are kept to a minimum whilst in the latter two the use of soil minerals, water and light are optimised and soil is protected from erosion. In the future, biomass food (and fuel) production could become more important than it is now. The harnessing of micro-organisms, e.g. fungi, algae, bacteria (Section 10.1), that can produce edible biomass (i.e. single-cell protein) from solar energy and/or alternative energy sources on a large scale may provide a viable addition to food produced conventionally.

• At all levels, the involvement of the farmers in decision making as well as policy implementation should be recognised. It is not desirable that food producers should only be the recipients and implementors of policy decided by politicians. Farmers must be widely consulted and be appraised of their vital role in society, especially in developing nations. This policy was included in Agenda 21, a United Nations Programme for sustainable development that was discussed at the United Nations Conference on Environment and Development, in Rio de Janeiro, August 1992.

- At all levels, though scaled, there must be adequate finance for research into, and the development of

  (a) New crops from wild species.
  (b) The collection and conservation of crop genetic resources (there are parallels here with germplasm banks for species in general—see above) to facilitate the improvement of existing crop types.
  (c) The multi-use of crops.
  (d) Large-scale soil and water conservation measures.
  (e) Alternatives to the use of nitrate and phosphate fertilisers.
  (f) Sustainable and environmentally benign energy use and energy production. The latter could involve the use of biomass fuels.
  (g) Environmentally acceptable pest control.
  (h) Efficient recycling of nutrients within agricultural systems.
  (i) The education of individuals in relation to the importance of agriculture to his/her continued well-being and that of their progeny.

All of these challenges are pressing, all are essential for improved agricultural practices and all involve the interplay between environment, agriculture and a range of social factors (Figures 1.1 and 11.1). These factors reflect agriculture's unique position at the interface between culture and environment whilst the varied nature of agriculture on the Earth's surface reflects the astonishing range of human ingenuity that has been brought to bear on the biosphere to create a degree of order out of the complexity and chaos that characterise nature. As a fundamental component in the people/environment relationship, agriculture should serve as a stark reminder that all people, and the cultural diversity they exhibit, are dependent on the efficient harnessing of less than 0.5 per cent of the solar energy that reaches the Earth's surface. This vulnerability or fragility is rarely acknowledged. In the next 30 years, as the world's population almost doubles, its agricultural systems will be severely tested with considerable repercussions for the biosphere in general.

11.5 FURTHER READING

Alexandratos, N. (ed.) (1995) *World Agriculture: Towards 2010. An FAO Study*. John Wiley and Sons, Chichester.

Anderson, J.R. (ed.) (1994) *Agricultural Technology: Policy Issues for the International Community*. CAB International, Wallingford.

Arntzen, C.J. and Ritter, E.M. (eds) (1994) *Encyclopedia of Agricultural Science*, 4 vols. Academic Press, London.

Ausubel, J.H. and Langford, H.D. (eds) (1994) *Technological Trajectories and the Human Environment*. National Academy Press, Washington, DC.

Kates, R.W. (1994) Sustaining Life on Earth. *Scientific American* **271**, 92–99.

Thomas, A. plus 5 others (1994). *Third World Atlas*, 2nd edn. Open University Press, Buckingham.

World Bank (1994) *World Development Report 1994*. Oxford University Press, Oxford.

References

Abbott, R.J. (1994) Ecological risks of transgenic crops. *Trends in Ecology and Evolution* **9**, 280–282.

Adams, W.M. (1990) *Green Development: Environment and Sustainability in the Third World.* Routledge, London.

Adams, W.M., Potkanski, T. and Sutton, J.E.G. (1994) Indigenous farmer-managed irrigation in Sonjo, Tanzania. *The Geographical Journal* **160**, 17–32.

Agrochemical Monitor (1993) Ciba and Mycogen agreement on insect-resistant maize. *Agrochemical Monitor* **97**, 15.

Agrochemical Monitor (1994) Pesticide resistant crops. *Agrochemical Monitor* **102**, 2–12.

Alayev, E.B., Badenkov, Y.P. and Karavaeva, N.A. (1990) The Russian Plain. In B.L. Turner II, W.C. Clark, R.W. Kates, J.F. Richards, J.T. Mathewes and W.B. Maeyer (eds), *The Earth as Transformed by Human Action.* Cambridge University Press, Cambridge and Clark University, Worcester, Massachusetts, pp. 543–560.

Ali, A.M.S. (1987) Intensive paddy agriculture in Shyampur, Bangladesh. In B.L. Turner II and S.B. Brush (eds), *Comparative Farming Systems.* The Guilford Press, New York, pp. 276–312.

Ali-Ibrahim, A. (1991) Excessive use of groundwater resources in Saudi Arabia: impacts and policy options. *Ambio* **22**, 34–37.

Allen, R.C. (1991) The two English agricultural revolutions, 1450–1850. In B.M.S. Campbell and M. Overton (eds), *Land, Labour and Livestock: Historical Studies in European Agricultural Productivity.* Manchester University Press, Manchester, pp. 237–254.

Alström, K. and Åkerman, A.B. (1992) Contemporary soil erosion rates on arable land in southern Sweden. *Geografiska Annaler* **74A**, 101–107.

Altieri, M.A. (1992a) Agroecological foundations of alternative agriculture in California. *Agriculture, Ecosystems and Environment* **39**, 23–53.

Altieri, M.A. (1992b) Sustainable development in Latin America: exploring the possibilities. *Agriculture, Ecosystems and Environment* **39**, 1–21.

Ammerman, A.J. and Cavalli-Sforza, L.L. (1984) *Neolithic Transition and the Genetics of Populations in Europe.* Princeton University Press, Princeton, New Jersey.

Anderson, E. (1956) Man as a maker of new plants and plant communities. In W.A. Thomas Jr. (ed.), *Man's Role in Changing the Face of the Earth.* University of Chicago Press, Chicago, pp. 363–377.

Anderson, M.D. and Lockeretz, W. (1992) Sustainable agriculture research in the ideal and in the field. *Journal of Soil and Water Conservation* **47**, 100–104.

Aneja, R.P. (1990) Dairy development in a changing world: lessons to be learned and principles in dairy development—the experience of India. In *Dairying in a Changing World.* Proceedings of the 23rd International Dairy Congress, Montreal, Quebec, pp. 84–96.

Ansell, D.J. and Tranter, R.B. (1992) *Set-aside: In Theory and Practice.* Department of Agricultural Economics and Management, University of Reading, Miscellaneous Study No. 75 and Centre for Agricultural Strategy, University of Reading, Joint Publication No. 4.

Anthony, D., Telegin, D.Y. and Brown, D. (1991) The origin of horseback riding. *Scientific American* **265**, 44–48.

Arden-Clarke, C. and Evans, R. (1993) Soil erosion and conservation in the United Kingdom.

In D. Pimentel (ed.), *World Soil Erosion and Conservation*. Cambridge University Press, Cambridge, pp. 193–215.
Arnon, I. (1992) *Agriculture in Dry Lands: Principles and Practice*. Elsevier, Amsterdam.
Arulpragasam, P.V. (1992) Disease control in Asia. In K.C. Wilson and M.N. Clifford (eds), *Tea: Cultivation to Consumption*. Chapman and Hall, London, pp. 353–374.
Ash, R.F. (1993) Agricultural policy under the impact of reform. In Y.Y. Kueh and R.F. Ash (eds), *Economic Trends in Chinese Agriculture: The Impact of Post-Mao Reforms*. Clarendon Press, Oxford, pp. 11–45.
Ashley, J. (1993) Oilseeds. In J.R.J. Rowland (ed.), *Dryland Farming in Africa*. Macmillan, London, pp. 240–259.
Askin, D.C. (1990) Pasture establishment. In R.H.M. Langer (ed.), *Pastures: Their Ecology and Management*. Oxford University Press, Auckland, pp. 132–156.
Assadourian, C.S. (1992) The colonial economy: the transfer of the European system of production to New Spain and Peru. *Journal of Latin American Studies* **24**, Supplement, 55–68.
Atkinson, T.C., Briffa, K.R. and Coope, G.R. (1987) Seasonal temperatures in Britain during the past 22,000 years, reconstructed using beetle remains. *Nature* **325**, 587–592.
Austin, R.B., Ford, M.A. and Morgan, C.L. (1989) Genetic improvement in the yield of winter wheat: a further evaluation. *Journal of Agricultural Science (Cambridge)* **112**, 295–301.
Avery, G. (1985) Guaranteed thresholds and the Common Agricultural Policy. *Journal of Agricultural Economics* **36**, 355–364.
Ayres, P. and Paul, N. (1990) Weeding with fungi. *New Scientist* **127** (No. 1732), 36–39.
Bäck, L. (1993) Reindeer management in conflict and co-operation. A geographic land use and simulation study from northernmost Sweden. *Nomadic Peoples* **32**, 65–80.
Bahn, P.G. (1994) Time for a change. *Nature* **367**, 511–512.
Baker, D., Doyle, D. and Lidgate, H. (1991) Grass production. In C. Thomas, A. Reeve and G.E.J. Fisher (eds), *Milk from Grass*, 2nd edn. ICI Agricultural Division, Cleveland, The Scottish Agricultural College, Perth and the Institute of Grassland and Environmental Research, Hurley, Maidenhead, pp. 1–26.
Baker, R.D., Micol, D. and Béranger, C. (1992) Pasture fattening in the humid temperate zone. In R. Jarrige and C. Béranger (eds), *Beef Cattle Production*. Elsevier, Amsterdam, pp. 271–287.
Baltensperger, B.H. (1993) Larger and fewer farms: patterns and causes of farm enlargements on the central Great Plains, 1930–1978. *Journal of Historical Geography* **19**, 299–313.
Barrow, C.J. (1991) *Land Degradation*. Cambridge University Press, Cambridge.
Bar-Yosef, O. and Kislev, M.E. (1989) Early farming communities in the Jordan Valley. In D.R. Harris and G.C. Hillman (eds), *Foraging and Farming. The Evolution of Plant Exploitation*. Unwin Hyman, London, pp. 632–642.
Bater, J.H. (1989) *The Soviet Scene: A Geographical Perspective*. Edward Arnold, London.
Bazzaz, F.A. and Fajer, E.D. (1992) Plant life in a CO_2-rich world. *Scientific American* **266**, 68–74.
Beach, H. (1988) *The Saami of Lapland*. The Minority Rights Group, London.
Beach, H. (1990) Comparative systems of reindeer herding. In J.G. Galaty and D.L. Johnson (eds), *The World of Pastoralism*. The Guildford Press, New York and Belhaven Press, London, pp. 255–298.
Beach, T. (1994) The fate of eroded soil: sediment sinks and sediment budgets of agrarian landscapes in Southern Minnesota, 1851–1988. *Annals of the Association of American Geographers* **84**, 5–28.
Beck, C.I. and Ulrich, T. (1993) Biotechnology in the food industry. *Bio/Technology* **11**, 895–902.
Beckerman, S. (1987) Swidden in Amazonia and the Amazon rim. In B.L. Turner II and S.B. Brush (eds), *Comparative Farming Systems*. The Guildford Press, New York, pp. 55–94.
Bell, M. and Walker, M.J.C. (1992) *Late Quaternary Environmental Change: Physical and Human Perspectives*. Longman, Harlow.
Bennett, D. and Hoffmann, R.S. (1991) Ranching in the New World. In H.J. Viola and C. Margolis (eds), *Seeds of Change*. Smithsonian Institution Press, Washington, pp. 90–111.

Bernus, E. (1990) Dates, dromedaries and drought: diversification in Tuareg pastoral systems. In J.G. Galaty and D.L. Johnson (eds), *The World of Pastoralism*. The Guildford Press, New York and Belhaven Press, London, pp. 149- 176.

Betton, C., Webb, B.W. and Walling, D.E. (1991) Recent trends in NO_2–N concentration and loads in British rivers. *International Association of Hydrological Sciences* **203**, 167–180.

Beukema, H.P. and van der Zaag, D.E. (1990) *Introduction to Potato Production*. Pudoc, Wageningen.

Biagi, P., Cremaschi, M. and Nisbet, R. (1993) Soil exploitation and early agriculture in northern Italy. *The Holocene* **3**, 164–168.

Binns, T. (1994) *Tropical Africa*. Routledge, London.

Birrell, J. (1992) Deer and deer farming in medieval England. *Agricultural History Review* **40**, 112–126.

Biswas, A.K. (1993) Land resources for sustainable agricultural development in Egypt. *Ambio* **22**, 556–560.

Biswas, M.R. (1994) Agriculture and environment: a review, 1972–1992. *Ambio* **23**, 192–197.

Blaikie, P. and Brookfield, H. (1987) *Land Degradation and Society*. Methuen, London.

Blake, F. (1990) Standards: regulating organic food production. In T. Clunies-Ross and T. Weisselberg (eds), *Organic Farming—an Option for the Nineties*. British Organic Farmers and Organic Growers Association, Bristol.

Blanton, R.E., Kowalewski, S.A., Feinman, G.M. and Finsten, L.M. (1993) *Ancient Mesoamerica: A Comparison of Three Regions*, 2nd edn. Cambridge University Press, Cambridge.

Blaschke, P.M., Trustrum, N.A. and DeRose, R.C. (1992) Ecosystem processes and sustainable land use in New Zealand steeplands. *Agriculture, Ecosystems and Environment* **41**, 153–178.

Blumler, M.A. (1992) Independent inventionism and recent genetic evidence on plant domestication. *Economic Botany* **46**, 98–111.

Blumler, M.A. and Byrne, R. (1991) The ecological genetics of domestication and the origins of agriculture. *Current Anthropology* **32**, 23–54.

Boardman, J. (1990) Soil erosion on the South Downs: a review. In J. Boardman, I.D.L. Foster and J.A. Dearing (eds), *Soil Erosion on Agricultural Land*. John Wiley and Sons, Chichester, pp. 87–105.

Boardman, J. (1991) Land use, rainfall and erosion risk on the South Downs. *Soil Use and Management* **7**, 34–38.

Body, R. (1987) *Red or Green Farmers (and the Rest of Us?)*. Broad Leys Publishing, Saffron Walden.

Bonavia, D. and Grobman, A. (1989) Andean maize: its origins and domestication. In D.R. Harris and G.C. Hillman (eds), *Foraging and Farming: The Evolution of Plant Exploitation*. Unwin Hyman, London, pp. 456–470.

Bonny, S. (1993) Is agriculture using more and more energy? A French case study. *Agricultural Systems* **43**, 51–66.

Borget, M. (1992) *Food Legumes*. CTA (Technical Centre for Agricultural and Rural Co-operation), Wageningen, Netherlands and Macmillan, London.

Boserup, E. (1965) *The Conditions of Agricultural Growth: The Economics of Agrarian Change Under Population Pressure*. Aldine, Chicago.

Boucher, S.C. and Powell, J.M. (1994) Gullying and tunnel erosion in Victoria. *Australian Geographical Studies* **32**, 17–26.

Bower, C.A. (1994) Soil and water management for sustainability. In K.L. Campbell, W.D. Graham and A.B. Bottcher (eds), *Environmentally Sound Agriculture*. American Society of Agricultural Engineers, St Joseph, Michigan, pp. 497–498.

Bowring, R. and Kornicki, P. (eds) (1993) *The Cambridge Encyclopedia of Japan*. Cambridge University Press, Cambridge.

Brears, T. and Ryals, J. (1994) Genetic engineering for disease resistance in plants. *Agro-food-Industry Hi-Tech* **5**, 10–13.

Bredero, Th.J. (1991) *Concepts and Guidelines for Crop-Water Management Research: A Case Study for India*. Aspect Publications, Edinburgh.

Brenner, C. (1991) *Biotechnology and Developing Country Agriculture: The Case of Maize*. OECD, Paris.

Briggs, S.A. and staff of the Rachel Carson Council (1992) *Basic Guide to Pesticides: Their Characteristics and Hazards*. Hampshire Publishing Corporation, Washington.

British Organic Farmers in conjunction with the Soil Association (1992) *Organic Farming and the Countryside*. Organic Food and Farming Centre, Bristol.

Brklacich, M., Bryant, C.R. and Smit, B. (1991) Review and appraisal of concept of sustainable food production systems. *Environmental Management* **15**, 1–14.

Broekhaven, G. (1991) Bangladesh. In N.M. Collins, J.A. Sayer and T.C. Whitmore (eds), *The Conservation Atlas of Tropical Forests: Asia and the Pacific*. Macmillan Press, Basingstoke, pp. 92–97.

Brookfield, H., Lian, F.J., Kwai-Sim, L. and Polter, L. (1990) Borneo and the Malay Peninsula. In B.L. Turner II, W.C. Clark, R.W. Kates, J.F. Richards, J.T. Matthews and W.B. Meyer (eds), *The Earth as Transformed by Human Action*. Cambridge University, Worcester, Massachusetts, pp. 495–512.

Browder, J.O. (1988) Public policy and deforestation in the Brazilian Amazon. In R. Repetto and M. Gillis (eds), *Public Policies and the Misuse of Forest Resources*. Cambridge University Press, Cambridge, pp. 247–297.

Brown, A.G. and Barber, K.E. (1985) Late Holocene palaeoecology and sedimentary history of a small lowland catchment in Central England. *Quaternary Research* **24**, 87–102.

Brown, C.G. and Longworth, J.W. (1992) Multilateral assistance and sustainable development: the case of an IFAD project in the pastoral region of China. *World Development* **20**, 1663–1674.

Brown, D. and Meadowcroft, S. (1989) *The Modern Shepherd*. Farming Press, Ipswich.

Brown, E.P. and Nooter, R. (1992) *Successful Small-scale Irrigation in the Sahel*. World Bank, Technical Paper No. 171, The World Bank, Washington, DC.

Bunting, A.H. (1975) Time, phenology and yield of crops. *Weather* **30**, 312–325.

Burrough, P.A. (1986) *Principles of Geographical Information Systems for Land Resources Assessment*. Clarendon Press, Oxford.

Burt, T.P. and Haycock, N.E. (1992) Catchment planning and the nitrate issue: a UK perspective. *Progress in Physical Geography* **16**, 379–404.

Butzer, K.W. (1976) *Early Hydraulic Civilization in Egypt*. University of Chicago Press, Chicago.

Byng, J. (1992) EC organic food standards. *New Farmer and Grower* **3/4**, 18–20.

Byrd, B.F. (1994) From early humans to farmers and herders—recent progress on key transitions in southwest Asia. *Journal of Archaeological Research* **2**, 221–253.

Callaway, B. (1992) A compendium of crop varietal tolerance to weeds. *American Journal of Alternative Agriculture* **7**, 169–180.

Cambrony, H.R. (1992) *Coffee Growing*. The Technical Centre for Agricultural and Rural Co-operation (CTA), Wageningen and Macmillan, Basingstoke.

Camillo, J. and Schiersmann, G. (1992) Beef cattle production in the temperate zone of South America (Argentina and Uruguay). In R. Jarrige and C. Béranger (eds), *Beef Cattle Production*. Elsevier, Amsterdam, pp. 259–269.

Campbell, B.M.S. (1991) Land, labour, livestock, and productivity trends in English seignorial agriculture, 1208–1450. In B.M.S. Campbell and M. Overton (eds), *Land, Labour and Livestock: Historical Studies in European Agricultural Productivity*. Manchester University Press, Manchester, pp. 144–182.

Campbell, B.M.S. and Power, J.P. (1989) Mapping the agricultural geography of medieval England. *Journal of Historical Geography* **15**, 24–39.

Campbell, D.J. (1991) The impact of development upon strategies for coping with drought among the Maasai of Kajiado District, Kenya. In J.C. Stone (ed.), *Pastoral Economies in Africa and Long Term Responses to Drought*. Aberdeen University African Studies Group, Aberdeen, pp. 116–128.

Caporali, F. and Onnis, A. (1992) Validity of rotation as an effective agroecological principle for a sustainable agriculture. *Agriculture, Ecosystems and Environment* **41**, 101–113.

Carey, D.I. (1993) Development based on carrying capacity: a strategy for environmental protection. *Global Environmental Change* **3**, 140–148.

Carne, R.J. (1993) Agroforestry land use: the concept and practice. *Australian Geographical Studies* **31**, 79–90.

Carrillo, J. and Schiersmann, G. (1992) Beef cattle production in the temperate zone of South America (Argentina and Uruguay). In R. Jarrige and C. Béranger (eds), *Beef Cattle Production*. Elsevier, Amsterdam, pp. 259–269.

Carter, G.A. and Miller, R.L. (1994) Early detection of plant stress by digital imaging within narrow stress-sensitive wavebands. *Remote Sensing of Environment* **50**, 295–302.

Cartwright, N., Clark, L. and Bird, P. (1991) The impact of agriculture on water quality. *Outlook on Agriculture* **20**, 145–152.

Carver, A., Dalrymple, M.A., Wright, G., Cottom, D.S., Reeves, D.B., Gibson, Y.H., Keenan, J.L., Barrass, J.D., Scott, A.R., Colman, A. and Garner, I. (1993) Transgenic livestock as bioreactors: stable expression of human alpha-1-antitrypsin by a flock of sheep. *Bio/Technology* **11**, 1263–1270.

Catroux, G. and Amarger, N. (1992) *Rhizobia* as soil inoculants in agriculture. In C. Fry and M. Day (eds), *Release of Genetically Engineered and Other Microorganisms*. Cambridge University Press, Cambridge, pp. 1–13.

Cavalli-Sforza, L.L., Menozzi, P. and Piazza, A. (1993) Demic expansions and human evolution. *Science* **259**, 639–646.

Cavalli-Sforza, L.L., Menozzi, P. and Piazza, A. (1994) *The History and Geography of Human Genes*. Princeton University Press, Princeton, New Jersey.

Chamberlaine, A. (1990) Milk and milk products. In W.J.A. Payne, *An Introduction to Animal Husbandry in the Tropics*. Longman, Harlow, pp. 747–789.

Chang, K.-C. (1986) *The Archaeology of Ancient China*, 4th edn. Yale University Press, New Haven, Connecticut.

Chang, T.T. (1976) The origin, evolution, cultivation, dissemination, and diversification of Asian and African rices. *Euphytics* **25**, 425–441.

Chang, T.T. (1989) Domestication and spread of the cultivated rices. In D.R. Harris and G.C. Hillman (eds), *Foraging and Farming. The Evolution of Plant Exploitation*. Unwin Hyman, London, pp. 408–417.

Chao, K. (1986) *Man and Land in Chinese History*. Stanford University Press, Stanford.

Chapman, J. and Müller, J. (1990) Early farmers in the Mediterranean Basin: the Dalmatian evidence. *Antiquity* **64**, 127–134.

Chapman, R. (1981) The megalithic tombs of Iberia. In J. Evans, B. Cunliffe and C. Renfrew (eds), *Antiquity and Man*. Thames and Hudson, London, pp. 93–106.

Chapman, S.B., Clarke, R.T. and Webb, N.R. (1989) The survey and assessment of heathland in Dorset, England, for conservation. *Biological Conservation* **47**, 137–152.

Charray, J., Humbert, J.M. and Lerif, J. (1992) *Manual of Sheep Production in the Humid Tropics of Africa*. CAB International, Wallingford.

Chase, P.G. (1989) How different was Middle Palaeolithic subsistence? A zooarchaeological perspective on the Middle to Upper Palaeolithic transition. In P. Mellars and C. Stringer (eds), *The Human Revolution: Behaviourial and Biological Perspectives on the Origins of Modern Humans*. Edinburgh University Press, Edinburgh, pp. 321–337.

Chen, Z. and Gu, H. (1993) Plant biotechnology in China. *Science* **262**, 377–378.

Cheng, X., Chunru, H. and Taylor, D. (1992) Sustainable agricultural development in China. *World Development* **20**, 1127–1144.

Childe, V.G. (1936) *Man Makes Himself*. Watts, London.

Chrispeels, M.J. and Sadava, D.E. (1994) *Plants, Genes and Agriculture*. Jones and Bartlett Publishers, Boston.

Christou, P. (1994) Genetic engineering of crop legumes and cereals: current status and recent advances. *Agro-food-Industry Hi-Tech* **5**, 17–27.

Clark, G. (1991) Yields per acre in English agriculture, 1250–1860: evidence from labour inputs. *Economic History Review* **44**, 445–460.

Cleaver, K. (1992) Population, environment and agriculture. In J.A. Sayer, C.S. Harcourt and

N.M. Collins (eds), *The Conservation Atlas of Tropical Forests: Africa.* Macmillan Publishers, Basingstoke, pp. 49–55.
Clutton-Brock, J. (1992) How the wild beasts were tamed. *New Scientist* **133** (No.1808), 41–43.
Cobb, A. (1992) *Herbicides and Plant Physiology*. Chapman and Hall, London.
Coffelt, T.A. (1989) Peanut. In G. Röbbelen, R.K. Downey and A. Ashri (eds), *Oil Crops of the World.* McGraw Hill Publishing Company, New York, pp. 319–338.
Coghlan, A. (1993) Green shoots showing for transgenic rice. *New Scientist* **138** (No. 1877), 20.
Coghlan, A. (1994) Will the scorpion gene run wild? *New Scientist* **142** (No. 1931), 14–15.
Colchester, M. (1993) Guatemala: the clamour for land and the fate of the forests. In M. Colchester, and L. Lohmann (eds) *The Struggle for Land and the Fate of the Forests.* The World Rainforest Movement, Penang, The Ecologist, Sturminister Newton and Zed Books, London, pp. 99–138.
Colinvaux, P. (1993) *Ecology 2.* John Wiley and Sons, New York.
Colwell, R.K. and Coddington, J.A. (1994) Estimating terrestrial biodiversity through extrapolation. *Proceedings of the Royal Society of London*, B **345**, 101–118.
Concar, D. (1994) The organ factory of the future? *New Scientist* **142** (No. 1930), 24–29.
Cooper, A. (1986) Another look at the Great Betrayal. *Agricultural History* **60**, 81–104.
Cory, J.S., Hirst, M.L., Williams, T., Hails, R.S., Goulson, D., Green, B.M., Carty, T.M., Possee, D., Cayley, P.J. and Bishop, D.H.L. (1994) Field trial of a genetically improved baculovirus insecticide. *Nature* **370**, 138–140.
Coste, R. (1992) *Coffee: The Plant and the Product*. Macmillan, Basingstoke.
Cowan, C.W. and Watson, P.J. (1992) Some concluding remarks. In C.W. Cowan and P.J. Watson (eds), *The Origins of Agriculture*. Smithsonian Institution Press, Washington and London, pp. 207–212.
Cox, R. and Collins, M. (1991) Indonesia. In N.M. Collins, J.A. Sayer and T.C. Whitmore (eds), *The Conservation Atlas of Tropical Forests: Asia and the Pacific*. Macmillan Press, Basingstoke, pp. 141–165.
Cramb, R.A. (1988) Shifting cultivation and resource degradation in Sarawak: perception and policies. *Review of Indonesian and Malaysian Affairs* **22**, 115–149.
Cramb, R.A. (1989) The use and productivity of labour in shifting cultivation: An East Malaysian case study. *Agricultural Systems* **29**, 97–115.
Cramb, R.A. (1993) Shifting cultivation and sustainable agriculture in East Malaysia: a longitudinal case study. *Agricultural Systems* **42**, 209–226.
Crawley, M.J., Hails, R.S., Rees, M., Kohn, D. and Buxton, J. (1993) Ecology of transgenic oilseed rape in natural habitats. *Nature* **363**, 620–623.
Cremlyn, R.J. (1991) *Agrochemicals*. John Wiley and Sons, Chichester.
Critchley, W.R.S., Reij, C.P. and Turner, S.D. (1992) *Soil and Water Conservation in Sub-Saharan Africa*. International Fund for Agricultural Development, Rome.
Crosby, A. (1991) Metamorphosis of the Americas. In H.J. Viola and C. Margolis (eds), *Seeds of Change*. Smithsonian Institution Press, Washington, pp. 70–89.
Crosson, P. (1993) Impacts of climatic change on the agriculture and economy of the Missouri, lowa, Nebraska and Kansas (MINK) region. In H.M. Kaiser and T.E. Drennen (eds), *Agricultural Dimensions of Global Climate Change*. St Lucia Press, Delray Beach, Florida, pp. 117–135.
Croston, D. and Pollott, G. (1994) *Planned Sheep Production*. Blackwell, Oxford.
Cudjoe, F. and Rees, P. (1992) How important is organic farming in Great Britain? *Tijdschrift voor Economic en Social Geografie* **83**, 13–24.
Cunningham, E.P. (1993) Annual genetic resources—the perspective for developing countries. In M. Gill, E. Owen, G.E. Pollott and T.J.L. Lawrence (eds), *Animal Production in Developing Countries*. Occasional Publication No. 16, British Society of Animal Production, Edinburgh, pp. 33–35.
Dahl, T.E. (1990) *Wetland Losses in United States 1780s to 1980s*. US Department of the Interior, Fish and Wildlife Service, Washington, DC.
Daly, G.T. (1990) The grasslands of New Zealand. In R.H.M.Langer (ed.), *Pastures: Their Ecology and Management*. Oxford University Press, Auckland, pp. 1–38.
Dansgaard, W.S., Johnsen, S.J., Clausen, H.B., Dahl-Jensen, D., Gundestrup, N.S., Hammer,

C.U., Hridberg, C.S., Steffesnsen, J.P., Sveinbjornsdottir, A.E., Jouzel, J. and Bond, G. (1993) Evidence for general instability of past climate from a 250–Kyr ice-core record. *Nature* **364**, 218–220.

Dao, C., Lifeng, X. and Chaohui, Y. (1991) The rural economy. In X. Guohua and L.J. Peel (eds), *The Agriculture of China*. Oxford University Press, Oxford, pp. 179–234.

Dart, P.J. (1990) Agricultural microbiology: Introduction. In G.J. Persley (ed.), *Agricultural Biotechnology: Opportunities for International Development*. CAB International, Wallingford, pp. 53–77.

Darwin, C. (1859) *The Origin of Species*. Mentor Books, London.

Davies, J.H. (1985) The pyrethroids: an historical introduction. In J.P. Leahy (ed.), *The Pyrethroid Insecticides*. Taylor and Francis, London, pp. 1–41.

Davies, M.S. and Hillman, G.C. (1992) Domestication of cereals. In G.P. Chapman (ed.), *Grass Evolution and Domestication*. Cambridge University Press, Cambridge, pp. 199–224.

Davis, S.J.M. (1987) *The Archaeology of Animals*. B.T. Batsford Ltd, London.

Davis, S.J.M. and Valla, F.R. (1978) Evidence for domestication of the dog 12,000 years ago in the Natufian of Israel. *Nature* **276**, 608–610.

Dazhong, W. (1993) Soil erosion and conservation in China. In D. Pimentel (ed.), *World Soil Erosion and Conservation*. Cambridge University Prcss, Cambridge, pp. 63–85.

Dazhong, W., Yungxin, T., Xunhua, Z. and Yungzhen, H. (1992) Sustainable and productive agricultural development in China. *Agriculture, Ecosystems and Environment* **39**, 55–70.

Dean, W.R.J. and Macdonald, I.A.W. (1994) Historical changes in stocking rates of domestic livestock as a measure of semi-arid and arid rangeland degradation in the Cape Province, South Africa. *Journal of Arid Environments* **26**, 281–298.

Delcourt, H.R. and Delcourt, P.A. (1991) *Quaternary Ecology*. Chapman and Hall, London.

Dempster, J.P. (1987) Effects of pesticides on wildlife and priorities for the future. In K.J. Brent and R.K. Atkin (eds), *Rational Pesticide Use*. Cambridge University Press, Cambridge, pp. 17–25.

Denevan, W.M. (1992) The pristine myth: the landscape of the Americas 1492. *Annals of the Association of American Geographers* **8**, 369–385.

Dennell, R.W. (1992) The origins of crop agriculture in Europe. In C.W. Cowan and P.J. Watson (eds), *The Origins of Agriculture*. Smithsonian Institution Press, Washington and London, pp. 71–100.

Denton, R.I. (1993) Vegetable jute (*Corchorus*) In J.T. Williams (ed.), *Pulses and Vegetables*. Chapman and Hall, London, pp. 167–175.

De Ploey, J. (1986) Soil erosion and possible conservation measures in loess loamy areas. In G. Chischi and R.P.C. Morgan (eds) *Soil Erosion in the European Community*. A.A. Balkema, Rotterdam, pp. 157–163.

DeVay, J.E., El-Zik, K.M., Bourland, F.M., Garber, R.H., Kappleman, A.J., Lyda, S.D., Minton, E.B., Roberts, P.A. and Wallace, T.P. (1989) Strategies and tactics for managing plant pathogens and nematodes. In R.E. Frisbie, K.M. El-Zik and L.T. Wilson (eds), *Integrated Pest Management Systems and Cotton Production*. John Wiley and Sons, New York, pp. 225–266.

Devendra, C. (1990) Goats. In W.J.A. Payne (ed.), *An Introduction to Animal Husbandry in the Tropics*. Longman, Harlow, pp. 505–536.

Devendra, C. (1993) *Sustainable Animal Production from Small farm Systems in South-East Asia*. FAO Animal Production and Health Paper No. 106. FAO, Rome.

Dicks, M.R. (1992) What will be required to guarantee the sustainability of U.S. agriculture in the 21st century? *American Journal Alternative Agriculture* **7**, 190–195.

Dinham, B. (1993) *The Pesticide Hazard*. Zed Books, London.

Dixon, R.K., Brown, S., Houghton, R.A., Solomon, A.M., Trexler, M.C. and Wisniewski, J. (1994) Carbon pools and flux of global forest ecosystems. *Science* **263**, 185–190.

Doebley, J. (1990) Molecular evidence and the evolution of maize. *Economic Botany* **44**, Supplement 3, 6–27.

Doggett, H. (1988) *Sorghum*, 2nd edn. Longman, Harlow.

Donald, C.M. and Hamblin, J. (1976) The biological yield and harvest index of cereals as agronomic and plant breeding criteria. *Advances in Agronomy* **28**, 361–405.

Dong, J.-Z., Yang, M.-Z., Jia, S.-R. and Chua, N.-H. (1991) Transformation of melon (*Cucumis melo* L.) and expression from the cauliflower mosaic virus 355 promoter in transgenic melon plants. *Bio/Technology* **9**, 858–863.

Doolittle, W.E. (1992) Agriculture in North America on the eve of contact: a reassessment. *Annals of the Association of American Geographers* **82**, 386–401.

Döös, B.R. (1994) Environmental degradation, global food production, and risk for large-scale migrations. *Ambio* **23**, 124–130.

Douches, D.S. and Jastrzebski, K., 1993. Potato. In G. Kalloo and B.O. Bergh (eds), *Genetic Improvement of Vegetable Crops*. Pergamon Press, pp. 605–644.

Dove, M.R. (1993a) Smallholder rubber and swidden agriculture in Borneo: a sustainable adaptation to the ecology and economy of the tropical forest. *Economic Botany* **47**, 136–147.

Dove, M.R. (1993b) A revisionist view of tropical deforestation and development. *Environmental Conservation* **20**, 17–56.

Dovrat, A. (1993) *Irrigated Forage Production*. Elsevier, Amsterdam.

Dregne, H.E. (1992) Erosion and soil productivity in Asia. *Journal of Soil and Water Conservation* **47**, 8–13.

Dudley, N. (1990) *Nitrates: The Threat to Food and Water*. Green Print, London.

Dugan, P. (ed.) (1993) *Wetlands in Danger*. Mitchell Beazley, London.

Earle, C. (1992) Into the abyss . . . again: technical change and destructive occupance in the American cotton belt, 1870–1930. In L.M. Dilsaver and C.E. Colton (eds), *The American Environment: Interpretations of Past Geographies*. Rowman and Littlefield Publishers, Lanham, Maryland, pp. 53–88.

Eden, M.J. (1993) Swidden cultivation in forest and savanna in lowland southwest Papua New Guinea. *Human Ecology* **21**, 145–166.

Eden, S.E. (1994) Using sustainable development. The business case. *Global Environmental Change* **4**, 160–167.

Edwards, C.A. (1993) The impact of pesticides on the environment. In D. Pimentel and H. Lehman (eds), *The Pesticide Question: Environment, Economics and Ethics*. Chapman and Hall, New York, pp. 13–46.

Edwards, C.A. and Wali, M.K. (1993) The global need for sustainability in agriculture and natural resources. In C.A. Edwards, M.K. Wali, D.J. Horn and F. Miller (eds), *Agriculture and Environment*. Elsevier, Amsterdam, pp. vii–xxv.

Edwards, K. (1993) Soil erosion and conservation in Australia. In D. Pimentel (ed.), *World Soil Erosion and Conservation*. Cambridge University Press, Cambridge.

Eger, H. (1990) Low-cost soil and water conservation measures for smallholders in the Sudano-Sahelian zone of Burkina Faso. In J. Kotschi (ed.) *Ecofarming Practices for Tropical Smallholdings*. Verlag Josef Margraf, Weikersheim, Germany, pp. 127–158.

Ehrlich, P.R. and Ehrlich, A.H. (1990) *The Population Explosion*. Simon and Schuster, New York.

Ehrlich, P.R. and Wilson, E.O. (1991) Biodiversity studies: science and policy. *Science* **253**, 758–762.

Ellis, S., Taylor, D. and Masood, K.R. (1993) Land degradation in northern Pakistan. *Geography* **78**, 84–87.

El Wakeel, A.S. and Sabah, M.A.A. (1993) Relevance of mobility to rangeland utilization. *Nomadic Peoples* **32**, 33–38.

El-Zik, K.M., Grimes, D.W. and Thaxton, P.M. (1989) Cultural management and pest suppression. In R.E. Frisbie, K.M. El-Zik and L.T. Wilson (eds), *Integrated Pest Management Systems and Cotton Production*. John Wiley and Sons, New York, pp. 11–36.

English, J. (1993) Does population growth inevitably lead to land degradation? In J.P. Srivastava and H. Alderman (eds), *Agriculture and Environmental Challenges*. The World Bank, Washington, DC, pp. 45–58.

Ernst and Young (1993) *Biotech 94: Long Term Value, Short Term Hurdles*. Ernst and Young, New York.

European Communities Commission (1992) *The Mangroves of Africa and Madagascar*. Office for the Official Publications of the European Communities, Luxembourg.

Evans, D.A. (1989) Techniques in plant cell and tissue culture. In S.-D. Kung and C.J. Arntzen (eds), *Plant Biotechnology*. Butterworth, Boston, pp. 53–76.

Evans, L.T. (1993) *Crop Evolution, Adaptation and Yield.* Cambridge University Press, Cambridge.

Evans, R. (1990a) Soil erosion: its impact on the English and Welsh Landscape since woodland clearance. In J. Boardman, I.D.L. Foster and J.A. Dearing (eds), *Soil Erosion on Agricultural Land.* John Wiley and Sons, Chichester, pp. 231–254.

Evans, R. (1990b) Soils at risk of accelerated erosion in England and Wales. *Soil Use and Management* **6**, 125–131.

Evans, R. (1992) Erosion in England and Wales—the present the key to the past. In M. Bell and J. Boardman (eds), *Past and Present Soil Erosion.* Oxbow Monograph No. 22, pp. 53–66.

Evans, R. (1993a) On assessing accelerated erosion of arable land by water. *Soils and Fertilizers* **56**, 1285–1293.

Evans, R. (1993b) Extent, frequency and rates of tilling of arable land in localities in England and Wales. In S. Wicherek (ed.), *Farm Land Erosion in Temperate Plains Environment and Hills.* Elsevier, Amsterdam, pp. 177–190.

Evans, R. (1995) Some methods of directly assessing water erosion of cultivated land—a comparison on measurements made on plots and in fields. *Progress in Physical Geography* **19**, 115–129.

Everhart, M.E. (1991) The preferred grazing system. *Rangelands* **13**, 266–270.

Ewell, P.T. and Merrill-Sands, D. (1987) *Milpa* in Yucatán: a long-fallow maize system and its alternatives in the Maya peasant economy. In B.L. Turner II and S.B. Brush (eds), *Comparative Farming Systems.* The Guildford Press, New York and London, pp. 95–129.

Fearnside, P.M. (1993) Deforestation in Brazilian Amazonia: the effect of population and land tenure. *Ambio* **22**, 537–545.

Feitelson, J.S., Payne, J. and Kim, L. (1992) *Bacillus thuringiensis*: insects and beyond. *Bio/Technology* **10**, 40–43.

Fernández-Baca, S. (1990) Llamoids or New World Camelidae: Llama, alpaca, guanaco and vicuña. In W.J.A. Payne, *An Introduction to Animal Husbandry in the Tropics*. Longman, Harlow, pp. 557–580.

Fick, G.N. (1989) Sunflower. In G. Röbbelen, R.K. Downey and A. Ashri (eds), *Oil Crops of the World.* McGraw-Hill Publishing Company, New York, pp. 301–318.

Fiedel, S.J. (1987) *Prehistory of the Americas.* Cambridge University Press, Cambridge.

Findlay, C., Martin, W. and Watson, A. (1993) *Policy Reform, Economic Growth and China's Agriculture.* Organisation for Economic Co-operation and Development (OECD), Paris.

Finlayson, C.M. (1991) Australasia and Oceania. In I.M. Finlayson and M. Moser (eds), *Wetlands.* Facts on File, Oxford, pp. 179–208.

Finley, E. and Price, R.R. (1994) *International Agriculture.* Delmar Publishers, New York.

Fisher, R. (1993) Biological aspects of the conservation of wetlands. In F.B. Goldsmith and A.Warren (eds), *Conservation in Progress.* John Wiley and Sons, Chichester, pp. 97–113.

Fitzgerald, M.A. (1994) Man with a mission. *New Scientist* **142** (No. 1930), 30–32.

Flannery, K.V. (ed.) (1986) *Guilá Naquitz: Archaic Foraging and Early Agriculture in Oaxaca, Mexico.* Academic Press, New York.

Food and Agriculture Organisation (1988) *An Interim Report on the State of the Forest Resources in the Developing Countries.* Food and Agriculture Organisation, Rome.

Food and Agriculture Organisation (1990) *Production Year Book.* Food and Agriculture Organisation, Rome.

Food and Agriculture Organisation (1991) *Potato Production and Consumption in Developing Countries.* Food and Agriculture Organisation, Rome.

Food and Agriculture Organisation (1992) *Production Yearbook 1991.* Food and Agriculture Organisation, Rome.

Food and Agriculture Organisation (1993) *Production Yearbook 1992.* Food and Agriculture Organisation, Rome.

Food and Agriculture Organisation/United Nations Environment Programme (1981) *Tropical*

Forest Resources Assessment Project (In the Framework of GEMs). Food and Agriculture Organisation, Rome.

Fox, M.W. (1993) *Superpigs and Wondercorn: The Brave New World of Biotechnology and Where it All May Lead.* Lyons and Burford, New York.

Fraley, R.T. (1992) Sustaining the food supply. *Bio/Technology* **10**, 40–43.

Freyssinet, G. and Derose, R.T. (1994) Development of genetically modified crops resistant to herbicides and pests. *Agro-food-Industry Hi-Tech* **5**, 3–7.

Friedel, M.H., Foran, B.D. and Stafford-Smith, D.M. (1990) Where the creeks run dry or ten feet high: pastoral management in arid Australia. *Proceedings of the Ecological Society of Australia* **16**, 185–194.

Frisbie, R.E., Walker, J.K. Jr., El-Zik, K.M. and Wilson, L.T. (1989) Perspective on cotton production and integrated pest management. In R.E. Frisbie, K.M. El-Zik and L.T. Wilson (eds), *Integrated Pest Management Systems and Cotton Production.* John Wiley and Sons, New York, pp. 1–9.

Fujimoto, H., Itoh, K., Yamamoto, M., Kyozuka, J. and Shimamoto, K. (1993) Insect resistant rice generated by introduction of a modified δ-endotoxin gene of *Bacillus thuringiensis. Bio/Technology* **11**, 1151–1155.

Fullen, M.A. and Mitchell, D.J. (1994) Desertification and reclamation in north-central China. *Ambio* **23**, 131–135.

Fullerton, B. and Knowles, R. (1991) *Scandinavia.* Paul Chapman Publishing, London.

Gahukar, R.T. (1993) Food production in Sub-Saharan Africa: major issues and challenges. *Outlook on Agriculture* **22**, 31–38.

Galaty, H.G. and Johnson, D.L. (1990) Introduction: Pastoral systems in global perspective. In J.G. Galaty and D.L. Johnson (eds), *The World of Pastoralism.* The Guildford Press, New York and Belhaven Press, London, pp. 1–31.

Galinat, W.C. (1992) Evolution of corn. *Advances in Agronomy* **47**, 203–231.

Garforth, C. (1993) Karnataka, India: seeing the people for the trees. *Rural Extension Bulletin* **2**, 33–39.

Gasser, C.S. and Fraley, R.T. (1992) Transgenic crops. *Scientific American* **266**, 34–39.

Gatehouse, A.M.R. and Hilder, V.A. (1994) Genetic manipulation of crops for insect resistance. In G. Marshall and D. Walters (eds), *Molecular Biology in Crop Protection.* Chapman and Hall, London.

Gatehouse, A.M.R., Boulter, D. and Hilder, V.A. (1992) Potential of plant-derived genes in the genetic manipulation of crops for insect resistance. In: A.M.R. Gatehouse, V.A. Hilder and D. Boulter (eds), *Plant Genetic Manipulation for Crop Protection.* CAB International, Wallingford, pp. 155–181.

Gatehouse, A.M.R., Shi, Y., Powell, K.S., Brough, C., Hilder, V.A., Hamilton, W.D.O., Newell, C.A., Merryweather, A., Boulter, D. and Gatehouse, J.A. (1993) Approaches to insect resistance using transgenic plants. *Philosophical Transactions of the Royal Society of London*, B **342**, 279–286.

Gatenby, R.M. (1991) *Sheep.* Technical Centre for Agricultural and Rural Co-operation (CTA), Wageningen and Macmillan, Basingstoke.

Ghosh, T.K. (1994) Environmental impact analysis of desertification through remote sensing and land based information system. *Journal of Arid Environments* **25**, 141–150.

Giampietro, M., Cerretelli, G. and Pimentel, D. (1992) Assessment of different agricultural production practices. *Ambio* **21**, 451–459.

Gibbons, A. (1991) Moths take the field against biopesticide. *Science* **354**, 646.

Gilland, B. (1993) Cereals, nitrogen and population: an assessment of global trends. *Endeavour, New Series*, **17**, 84–88.

Gilles, J.L. and Gefu, J. (1990) Nomads, ranchers, and the state: the sociocultural aspects. In J.G. Galaty and D.L. Johnson (eds), *The World of Pastoralism.* The Guildford Press, New York and Belhaven Press, London, pp. 99–118.

Gillespie, A.R., Knudson, D.M. and Geilfus, F. (1993) The structure of four home gardens in the Petén, Guatemala. *Agroforestry Systems* **24**, 157–170.

Glantz, M.H., Rubinstein, A.Z. and Zonn, I. (1993) Tragedy in the Aral Sea basin. *Global Environmental Change* **3**, 174–198.

Gleason, H.A. (1926) The individualistic concept of the plant association. *Bulletin of the Torrey Botanical Club* **53**, 7–26.

Glennie, P. (1991) Measuring Crop yields in early modern England. In B.M.S. Campbell and M. Overton (eds), *Land, Labour and Livestock: Historical Studies in European Agricultural Productivity*. Manchester University Press, Manchester, pp. 255–285.

Gliessman, S.R. (1988) The home garden agroecosystem: A model for developing more sustainable tropical agricultural systems. In P.A. Allen and D. Van Dusen (eds), *Global Perspectives in Agroecology and Sustainable Agriculture*. University of California, Santa Cruz, pp. 445–453.

Gliessman, S.R. (1990) Understanding the basis of sustainability for agriculture in the tropics: Experiences in Latin America. In C.A. Edwards, R. Lal, P. Madden, R.H. Miller, G. House (eds), *Sustainable Agricultural Systems*. Soil and Water Conservation Society, Ankeny, Iowa, pp. 378–390.

Goering, P., Norberg-Hodge, H. and Page, J. (1993) *From the Ground Up: Rethinking Industrial Agriculture*. Zed Books in association with the International Society for Ecology and Agriculture, London.

Goldstein, M.C. and Beall, C.M. (1991) Change and continuity in nomadic pastoralism on the Western Tibetan plateau. *Nomadic Peoples* **28**, 105–122.

Gonsalves, D., Chee, P., Provvidenti, R., Seem, R. and Slightom, J.L. (1992) Comparison of coat protein-meditated and genetically-derived resistance in cucumbers to infection by cucumber mosaic virus under field conditions with natural challenge inoculations by vectors. *Bio/Technology* **10**, 1562–1570.

Gooch, P. (1992) Transhumant pastoralism in Northern India: the Gujar case. *Nomadic Peoples* **30**, 84–96.

Goodman, D. and Redclift, M. (1991) *Rafashioning Nature: Food, Ecology and Culture*. Routledge, London.

Gordon, I.J. and Illius A.W. (1992) Foraging strategy: from monoculture to mosaic. In A.W. Speedy (ed.), *Progress in Sheep and Goat Research*. CAB International, Wallingford, pp. 153–177.

Goudie, A.S. (ed.) (1990) *Techniques for Desert Reclamation*. John Wiley and Sons, Chichester.

Goudie, A.S. (1994) Dryland degradation. In N. Roberts (ed.), *The Changing Global Environment*. Blackwell, Oxford, pp. 351–368.

Goy, P.A., Chasseray, E. and Duesing, J. (1994) Field trials of transgenic plants. *Agro-food-Industry Hi-Tech* **5**, 10–15.

Graham, O.P. (1992) Survey of land degradation in New South Wales, Australia. *Environmental Management* **16**, 205–223.

Grainger, A. (1990) *The Threatening Desert: Controlling Desertification*. Earthscan, London.

Grainger, A. (1992) Characterisation and assessment of desertification processes. In G.P. Chapman (ed.), *Desertified Grasslands: Their Biology and Management*. Academic Press, London, pp. 17–33.

Grant, A. (1988) Animal resources. In G. Astill and A. Grant (eds), *The Countryside of Medieval England*. Blackwell, Oxford, pp. 149–187.

Grayzel, J.A. (1990) Markets and migration: A Fulbe pastoral system in Mali. In J.G. Galaty and D.L. Johnson (eds), *The World of Pastoralism*. The Guildford Press, New York and Belhaven Press, London, pp. 35–67.

Green, G.M. and Sussman, R.W. (1990) Deforestation history of the eastern rain forests of Madagascar from satellite images. *Science* **248**, 212–215.

Green, J.O. (1990) The distribution and management of grasslands in the British Isles. In A.I. Breymeyer (ed.), *Managed Grasslands: Regional Studies*. Elsevier, Amsterdam, pp. 15–35.

Greenhalgh, J. (1992) The principles of animal production, In C.R.W. Spedding (ed.), *Fream's Principles of Food and Agriculture*. Blackwell Scientific Publications, Oxford, pp. 146–192.

Greig, J. (1988) Plant resources. In G. Astill and A. Grant (eds), *The Countryside of Medieval England*. Blackwell, Oxford, pp. 108–127.

Greig, J.R.A. (1991) The British Isles. In W. van Zeist, K. Wasylikowa and K.-E. Behre (eds), *Progress in Old World Palaeoethnobotany*. A.A. Balkema, Rotterdam, pp. 299–334.

Grierson, D. and Schuch, W. (1993) Control of ripening. *Philosophical Transactions of the Royal Society of London*, B **342**, 241–250.

Grigg, D. (1989) *English Agriculture: An Historical Perspective*. Blackwell, Oxford.

Grigg, D. (1992) *The Transformation of Agriculture in the West*. Blackwell, Oxford.

Grigg, D. (1993a) The role of livestock products in world food consumption. *Scottish Geographical Magazine* **109**, 66–74.

Grigg, D. (1993b) International variations in food consumption in the 1980s. *Geography* **78**, 251–266.

Grigg, D. (1993c) *The World Food Problem*, 2nd edn. Blackwell, Oxford.

Grove, J. (1988) *The Little Ice Age*. Methuen, London and New York.

Grun, P. (1990) The evolution of cultivated potatoes. *Economic Botany* **44**, Supplement 3, 39–55.

Guo, J.Y. and Bradshaw, A.D. (1993) The flow of nutrients and energy through a Chinese farming system. *Journal of Applied Ecology* **30**, 86–94.

Gutteridge, R.C. and Shelton, H.M. (1994) The role of forage tree legumes in cropping and grazing systems. In R.C. Gutteridge and H.M. Shelton (eds), *Forage Tree Legumes in Tropical Agriculture*. CAB International, Wallingford, pp. 3–11.

Hadley, P., Summerfield, R.J. and Roberts, E.H. (1982) Effects of temperature and photoperiod on reproductive development of selected grain legume crops. In D.R. Davies and D. Gareth-Jones (eds), *The Physiology, Genetics and Nodulation of Temperate Legumes*. Pitman Books, London, pp. 19–41.

Haferkamp, M.R., Volesky, J.D. Borman, M.M., Heitschmidt, R.K. and Currie, P.O. (1993) Effects of mechanical treatments and climatic factors of the productivity of Northern Great Plains rangelands. *Journal of Range Management* **46**, 346–350.

Hall, A.E. (1990) Physiological ecology of crops in relation to light, water, and temperature. In C.R. Carroll, J.H. Vandermeer and P.M. Rosset (eds), *Agroecology*. McGraw-Hill, New York, pp. 191–234.

Hall, C.A.S. and Hall, M.H.P. (1993) The efficiency of land and energy use in tropical economies and agriculture. *Agriculture, Ecosystems and Environment* **46**, 1–30.

Hall, D. (1988) The late Saxon countryside: villages and their fields. In D. Hooke (ed.), *Anglo-Saxon Settlements*. Blackwell, Oxford, pp. 99–122.

Hall, T. and Croll, B.T. (1993) Treatment processes for nitrate removal from water supplies. In T.P. Burt, A.L. Heathwaite and S.T. Trudgill (eds), *Nitrate: Processes, Patterns and Management*. John Wiley and Sons, Chichester, pp. 869–885.

Hallam, H.E. (1988) Population movements in England, 1086–1350. In H.E. Hallam (ed.), *The Agrarian History of England and Wales, Vol. II 1042–1350*. Cambridge University Press, Cambridge, pp. 508–593.

Hallam, J.S., Edwards, B.J.N., Barnes, B. and Stuart, A.J. (1973) The remains of a Late Glacial elk with associated barbed points from High Furlong, near Blackpool, Lancashire. *Proceedings of the Prehistoric Society* **39**, 100–128.

Hamissa, M.R. and Mahrous, F.N. (1989) Fertilizer use efficiency in rice. In International Rice Research Institute (ed.), *Rice Farming Systems: New Directions*. International Rice Research Institute, Manila, pp. 129–139.

Hannah, L., Lohse, D., Hutchinson, C., Carr, J.L. and Lankerani, A. (1994) A preliminary inventory of human disturbance of world ecosystems. *Ambio* **23**, 246–250.

Harcourt, M. (1992) Madagascar. In J.A. Sayer, C.S. Harcourt and N.M. Collins (eds), *The Conservation Atlas of Tropical Forests: Africa*. Macmillan Publishers, Basingstoke, pp. 221–229.

Harden, C.P. (1993) Land use, soil erosion, and reservoir sedimentation in an Andean drainage basin in Ecuador. *Mountain Research and Development* **13**, 177–184.

Hargrove, T.R. (1990) A grass called Rice. In Speedy, A. (ed.), *Developing World Agriculture*. Grosvenor Press International, London, pp. 38–42.

Harlan, J.R. (1971) Agricultural origins: centers and non-centers. *Science* **174**, 465–473.

Harlan, J.R. (1989) The tropical African cereals. In D.R. Harris and G.C. Hillman (eds), *Foraging and Farming. The Evolution of Plant Exploitation*. Unwin Hyman, London, pp. 335–343.

Harlan, J.R. (1992a) *Crops and Man*. American Society of Agronomy and Crop Science Society of America, Madison, Wisconsin.

Harlan, J.R. (1992b) Origins and processes of domestication. In G.P. Chapman (ed.), *Grass Evolution and Domestication*. Cambridge University Press, Cambridge, pp. 159–175.

Harris, C. (1992) The principles of crop production. In C.R.W. Spedding (ed.), *Fream's Principles of Food and Agriculture*. Blackwell Scientific Publication, Oxford, pp. 94–145.

Hart, J.F. (1991) *The Land that Feeds Us*. W.W. Norton and Company, New York and London.

Hart, R.H., Bissio, J., Samuel, M.J. and Waggoner, J.W., Jr. (1993) Grazing systems, pasture size, and cattle grazing behaviour, distribution and grains. *Journal of Range Management* **46**, 81–87.

Harvey, S. (1988) Domesday England. In H.E. Hallam (ed.), *The Agrarian History of England and Wales, Vol. II 1042–1350*. Cambridge University Press, Cambridge, pp. 45–136.

Hastorf, C.A. (1993) *Agriculture and the Onset of Political Inequality before the Inka*. Cambridge University Press, Cambridge.

Hathaway, J.S. (1993) Alar: the EPA's mismanagement of an agricultural chemical. In D. Pimentel and H. Lehman (eds), *The Pesticide Question: Environment, Economics and Ethics*. Chapman and Hall, New York, pp. 337–343.

Hatje, G. (1989) World importance of oil crops and their products. In G. Röbbelen, R.K. Downey and A. Ashri (eds), *Oil Crops of the World: Their Breeding and Utilization*. McGraw Hill Publishing Company, New York, pp. 1–21.

Hawkes, J.G. (1989) The domestication of roots and tubers in the American tropics. In D.R. Harris and G.C. Hillman (eds), *Foraging and Farming. The Evolution of Plant Exploitation*. Unwin Hyman, London, pp. 481–503.

Hawkes, J.G. (1990) *The Potato: Evolution, Biodiversity and Genetic Resources*. Belhaven Press, London.

Hawkes, J.G. (1991) The evolution of tropical American root and tuber crops with special reference to potatoes. In J.G. Hawkes, R.N. Lester, M. Nee and N. Estrada (eds), *Solonaceae III: Taxonomy, Chemistry, Evolution*. The Royal Botanic Gardens, Kew, for the Linnaean Society of London, pp. 347–356.

Hawkes, N. (1993) The grain of a good idea. *Geographical* **65**, 28–31.

Hawkins, R. (1984) Intercropping maize with sorghum in Central America: a cropping systems case study. *Agricultural Systems* **15**, 1–21.

Haworth, E.Y. (1985) The highly nervous system of the English Lakes: aquatic sensitivity to external changes, as demonstrated by diatoms. *Annual Report of the Freshwater Biological Association* **53**, 60–79.

Haycock, N.E., Pinay, G. and Walker, C. (1993) Nitrogen retention in river corridors: European perspective. *Ambio* **22**, 340–346.

Hearn, A.B. and Fitt, G.P. (1992) Cotton cropping systems. In C.J. Pearson (ed.), *Field Crop Ecosystems*. Elsevier, Amsterdam, pp. 85–142.

Heathcote, R.L. (1987) Land. In G. Davison, J.W. McCarty and A. McLeary (eds), *Australians in 1988*. Fairfax, Syme and Weldon Associates, Sydney, pp. 49–67.

Heathcote, R.L. (ed.) (1988) *The Australian Experience*. Longman Cheshire, Melbourne.

Heathwaite, A.L., Burt, T.P. and Trudgill, S.T. (1993) Overview—the nitrate issue. In A.L. Heathwaite, T.P. Burt and S.T. Trudgill (eds), *Nitrate: Processes, Patterns and Management*. John Wiley and Sons, Chichester, pp. 3–21.

Hedeager, L. (1992) *Iron-Age societies: From Tribe to State in Northern Europe, 500 BC to AD 700*. Blackwell, Oxford.

Hellden, U. (1991) Desertification—time for an assessment? *Ambio* **20**, 372–383.

Hemming, J. (1994) Indians, cattle and settlers: the growth of Roraima. In P.A. Furley (ed.), *The Forest Frontier: Settlement and Change in Brazilian Roraima*. Routledge, London, pp. 39–67.

Henderson-Sellers, A. (1994) Numerical modelling of global climates. In N. Roberts (ed.), *The Changing Global Environment*. Blackwell, Oxford, pp. 99–124.

Henry, D.O. (1989) *From Foraging to Agriculture: The Levant at the End of the Ice Age*. University of Pennsylvania Press, Philadelphia.

Herdt, R.W. (1989) Rice in Egyptian and global agriculture. In International Rice Research Institute (ed.), *Rice Farming Systems: New Directions*. International Rice Research Institute, Manila.

Hernández, M.A., Fasano, J.L. and Bocanegra, E.M. (1992) Overexploitation effects on the aquifer of Mar del Plata (Argentina): marine intrusion and groundwater decline. *Hydrogeology* **3**, 41–47.

Herren, U.J. (1992) Cash from camel milk: the impact of commercial camel milk sales on Garre and Gaaljacel camel pastoralism in Southern Somalia. *Nomadic Peoples* **30**, 97–113.

Hershey, C.H. (1993) Cassava *Manihot esculenta* Grantz. In G. Kalloo and B.O. Bergh (eds), *Genetic Improvement of Vegetable Crops*. Pergamon Press, Oxford, pp. 669–691.

Hesselberg, J. (1993) Food security in Botswana. *Norsk Geografisk Tidsskrift* **47**, 183–195.

Higham, N. (1987) *The Northern Counties to AD 1000*. Longman, London.

Higham, N. (1992) *Rome, Britain and the Anglo-Saxons*. Seaby, London.

Hill, K.K., Jarvis-Egan, N., Halk, E.L., Krahn, K.J., Liao, L.W., Mathewson, R.S., Merlo, D.J., Nelson, S.E., Rashka, K.E. and Loesch-Fries, L.S. (1991) The development of virus-resistant alfalfa, *Medicago sativa* L. *Bio/Technology* **9**, 373–377.

Hill, S.B. and MacRae, R.J. (1992) Organic farming in Canada. *Agriculture, Ecosystems and Environment* **39**, 71–84.

Hillman, G.C. (1989) Late Palaeolithic plant foods from Wadi Kubbaniya in Upper Egypt: dietary diversity, infant weaning, and seasonality in a riverine environment. In D.R. Harris and G.C. Hillman (eds), *Foraging and Farming. The Evolution of Plant Exploitation*. Unwin Hyman, London, pp. 207–239.

Hillman, G.C. and Davies, M.S. (1990) Measured domestication rates in wild wheats and barley under primitive cultivation, and their archaeological implications. *Journal of World Prehistory* **4**, 157–222.

Hillman, G.C., Colledge, S.M. and Harris, D.R. (1989) Plant-food economy during the Epipalaeolithic period at Tell Abu Hureyra, Syria: dietary diversity, seasonality, and models of exploitation. In D.R. Harris and G.C. Hillman (eds), *Foraging and Farming. The Evolution of Plant Domestication*. Unwin Hyman, London, pp. 240–268.

Hobbs, R.J. and Hopkins, A.J.M. (1990) From frontier to fragments: European impact on Australia's vegetation. *Proceedings of the Ecological Society of Australia* **16**, 93–114.

Hoffman, P.T. (1991) Land rents and agricultural productivity: the Paris Basin, 1450–1789. *Journal of Economic History* **51**, 771–805.

Hollis, T. and Bedding J. (1994) Can we stop the wetlands from drying up? *New Scientist* **143** (No. 1932), 30–35.

Holmes, D.L. (1993) Rise of the Nile Delta. *Nature* **363**, 402–403.

Holmes, W. (1977) Choosing between animals. *Philosophical Transactions of the Royal Society of London*, B **281**, 121–137.

Homewood, K.M. (1993) *Livestock Economy and Ecology in El Kala, Algeria: Evaluating Ecological and Economic Costs and Benefits in Pastoralist Systems*. Overseas Development Institute, London, Network Paper No. 35a.

Hopkins, N.S. (1987) Mechanized irrigation in Upper Egypt: the role of technology and the state in agriculture. In B.L. Turner II and S.B. Brush (eds), *Comparative Farming Systems*. The Guilford Press, New York, pp. 223–247.

Horn, D.R. (1988) *Ecological Approach to Pest Management*. Elsevier, Amsterdam.

Horton, D.E. and Anderson, J.L. (1992) Potato production in the context of the world and farm economy. In P.M. Harris (ed.), *The Potato Crop: The Scientific Basis for Improvement*. Chapman and Hall, London, pp. 794–815.

Hoskins, W.G. (1988) *The Making of the English Landscape*. Hodder and Stoughton, London. (Revised 1955 edition with an introduction and commentary by Christopher Taylor).

Houghton, J.T., Jenkins, G.J. and Ephraums, J.J. (eds), (1990) *Climate Change: The IPCC Scientific Assessment*. Cambridge University Press, Cambridge.

Houghton, J.T., Callander, B.A. and Varney, S.K. (eds), (1992) *IPCC Climate Change 1992*. Cambridge University Press, Cambridge.

Hoyle, R. (1993) Herbicide-resistant crops are no conspiracy. *Bio/Technology* **11**, 783–784.

Hu, S.T., Hannaway, D.B. and Youngberg, H.W. (1992) *Forage Resources of China*. Pudoc, Wageningen.

Hughes, J.D. (1992) Sustainable agriculture in ancient Egypt. *Agricultural History* **66**, 12–22.

Hulme, M. (1994) Historic records and recent climatic change. In N. Roberts (ed.), *The Changing Global Environment*. Blackwell, Oxford, pp. 69–98.

Humphries, C.J. and Fisher, C.T. (1994) The loss of Banks's Legacy. *Philosophical Transactions of the Royal Society of London*, B **344**, 3–9.

Huntley, B. and Birks, H.J.B. (1983) *An Atlas of Past and Present Pollen Maps of Europe 0–13,000 Years Ago*. Cambridge University Press, Cambridge.

Hurni, H. (1993) Land degradation, famine and land resource scenarios in Ethiopia. In D. Pimentel (ed.), *World Soil Erosion and Conservation*. Cambridge University Press, Cambridge, pp. 27–61.

IITA (International Institute of Tropical Agriculture) (1992) *Sustainable food production in Sub-Saharan Africa 1. IITA's Contributions*. IITA, Ibadan, Nigeria.

Institute for Remote Sensing Applications (1993) *Annual Report 1992*. Joint Research Centre, Commission of the European Communities.

International Federation of Organic Agriculture Movements (IFOAM) (1990) Standards for Organic Agriculture. IFOAM, Thors-Theley, Germany.

Isager, S. and Skydsgaard, J.E. (1992) *Ancient Greek Agriculture*. Routledge, London and New York.

Ivens, G.W. (1993) *The UK Pesticide Guide*, 6th edn. CAB International, Wallingford and the British Crop Protection Council, Farnham.

Ives, J.D. and Messerli, B. (1989) *The Himalayan Dilemma*. Routledge, London.

Jabbar, M.A. (1993) Evolving crop livestock farming systems in the humid zone of West Africa: potential and research needs. *Outlook on Agriculture* **22**, 13–21.

Jagtap, T., Chavan, V.S. and Untawale, A.G. (1993) Mangrove ecosystems of India: a need for protection. *Ambio* **22**, 252–254.

Jamal, A. and Huntsinger, L. (1993) Deterioration of a sustainable agro-silvo-pastoral system in the Sudan: the gum gardens of Kardofan. *Agroforestry Systems* **23**, 23–38.

Jarrige, R. and Auriol, P. (1992) An outline of world beef production. In R. Jarrige and C. Béranger (eds), *Beef Cattle Production*. Elsevier, Amsterdam, pp. 3–27.

Jarvis, P.J. (1993) Environmental changes. In R.W. Furness and J.J.D. Greenwood (eds), *Birds as Monitors of Environmental Change*. Chapman and Hall, London, pp. 42–85.

Jensen, M. (1993) Soil conditions, vegetation structure and biomass of a Javanese homegarden. *Agroforestry Systems* **24**, 171–186.

Jin, Y., Xiong, Y. and Ervin, R.T. (1990) Energy efficiency of grassland animal production in Northwest China. *Agriculture, Ecosystems and Environment* **31**, 63–76.

John, M.E. (1994) Re-engineering cotton fibre. *Chemistry and Industry* **17**, 676–679.

Johnes, P.J. and Burt, T.P. (1993) Nitrate in surface waters. In T.P. Burt, A.L. Heathwaite and S.T. Trudgill (eds), *Nitrate: Processes, Patterns and Management*. John Wiley and Sons, Chichester, pp. 269–317.

Johns, A.D. (1988) Economic development and wildlife conservation in Brazilian Amazonia. *Ambio* **17**, 302–306.

Johnson, K.S. (1992) Exploitation of the Tertiary–Quaternary Ogallala aquifer in the high plains of Texas, Oklahoma and New Mexico, southwestern USA. *Hydrogeology* **3**, 249–264.

Jones, A., Roberts, D.L. and Slingo, A. (1994) A climate model study of indirect radiative forcing by anthropogenic sulphate aerosols. *Nature* **370**, 450–453.

Jones, D.W. and O'Neill, R.V. (1993) Human-environmental influences and interactions in shifting agriculture when farmers form expectations rationally. *Environment and Planning, A* **25**, 121–136.

Jones, J.S. (1991) Farming is in the blood. *Nature* **351**, 97–98.

Jones, M. (1988) *The Sami of Lapland*. Minority Rights Group, London.

Jones, P.D. (1993) Is climatic change occurring? Evidence from the instrumental record. In H.M. Kaiser and T.E. Drennan (eds), *Agricultural Dimensions of Global Climatic Change*. St Lucie Press, Delray Beach, Florida, pp. 27–44.

Jordan, C.F. (ed.) (1989) *An Amazonian Rainforest: The Structure and Function of a Nutrient*

Stressed Ecosystem and the Impact of Slash-and-Burn Agriculture. UNESCO, Paris and the Parthenon Publishing Group, Carnforth.

Kaichen, D. (1991) The historical and social background. In X. Guohua and L.J. Peel (eds), *The Agriculture of China*. Oxford University Press, Oxford, pp. 42–72.

Kang, B.T. (1993) Alley cropping: past achievements and future directions. *Agroforestry Systems* **23**, 141–155.

Kass, D.C.L., Foletti, C., Szott, L.T., Landaverde, R. and Nolasco, R. (1993) Traditional fallow systems of the Americas. *Agroforestry Systems* **23**, 207–218.

Kathuri, L., Polastro, E.T. and Mellor, N. (1992) Biotechnology in an uncommon market. *Bio/Technology* **10**, 1545–1547.

Keatinge, R.W. (1988) Preface. In R.W. Keatinge (ed.), *Peruvian Prehistory*. Cambridge University Press, Cambridge, pp. xiii–xvii.

Kebin, Z. and Kaiguo, Z. (1989) Afforestation for sand fixation in China. *Journal of Arid Environments* **16**, 3–10.

Keller, M., Jacob, D.J., Wofsy, S.C. and Harriss, R.C. (1991) Effects of tropical deforestation on global and regional atmospheric chemistry. *Climatic Change* **19**, 139–158.

Kelly, R. and Marshal, T. (1993) Sheep and wool industries need to improve their performance. *Journal of Agriculture, Western Australia* **34**, 9–15.

Kendall, H.W. and Pimentel, D. (1994) Constraints on the expansion of the global food supply. *Ambio* **23**, 198–205.

Kessler, J.-J. (1993) Agroforestry and sustainable land-use in semi-arid Africa. *Zeitschrift für Wirtschaftsgeographie* **37**, 68–77.

Khair, K., Haddad, F. and Fattouh, S. (1992) The effects of overexploitation on coastal aquifers in Lebanon, with special reference to saline intrusion. *Hydrogeology* **3**, 349–362.

Khan, F.K. (1991) *A Geography of Pakistan*. Oxford University Press, Oxford.

Khan, T.A. and Liang, T. (1989) Mapping pesticide contamination potential. *Environmental Management* **13**, 233–242.

Khoshoo, T.N. and Tejwani, K.G. (1993) Soil erosion and conservation in India (status and policies). In D. Pimentel (ed.), *World soil Erosion and Conservation*. Cambridge University Press, Cambridge, pp. 110–145.

Kidd, G. and Dvorak, J. (1994) Agracetus' cotton patent draws opposition. *Bio/Technology* **12**, 659.

King, F.H. (1911) *Farmers for Forty Centuries: Permanent Agriculture in China, Korea and Japan*. Madison Womens Institute, Wisconsin.

Kirkbride, M.P. and Reeves, A.D. (1993) Soil erosion caused by low-intensity rainfall in Angus, Scotland. *Applied Geography* **13**, 299–311.

Kishokumar, M., Peng, K.S., Barlow, H., Pong, T.Y., Chiew, T.H., bin Kavanaghm, M., bin Ariffin, I., Kumari, K., Rahim, A.A., Parish, D., Chung, C.S., Ng, F. and Tuan, M.S. (1991) Peninsular Malaysia. In N.M. Collins, J.A. Sayer and T.C. Whitmore (eds), *The Conservation Atlas of Tropical Forests: Asia and the Pacific*. Macmillan Press, Basingstoke, pp. 183–191.

Kislev, M.E., Bar-Yosef, O. and Gropher, A. (1986) Early Neolithic domesticated and wild barley from the Netiv Hagdud region of the Jordan Valley. *Israel Journal of Botany* **35**, 197–201.

Kislev, M.E., Nadel, D. and Carmi, I. (1992) Epipalaeolithic (19,000 BP) cereal and fruit diet at Ohalo II, Sea of Galilee, Israel. *Review of Palaeobotany and Palynology* **73**, 161–166.

Klein, R.G. (1989) Biological and behaviourial perspectives on modern human origins in southern Africa. In P. Mellars and C. Stringer (eds), *The Human Revolution: Behaviourial and Biological Perspectives on the Origins of Modern Humans*. Edinburgh University Press, Edinburgh, pp. 529–546.

Klein, R.G. (1992) The archaeology of modern human origins. *Evolutionary Anthropology* **1**, 5–14.

Koepf, H.H., Petterson, B.D. and Schaumann, W. (1976) *Bio-dynamic Agriculture: An Introduction*. Anthroposophic Press, Spring Valley, New York.

Koizumi, H., Usami, Y. and Satoh, M. (1990) Annual net primary production and efficiency of solar energy utilization in three double-cropping agro-ecosystems in Japan. *Agriculture, Ecosystems and Environment* **32**, 241–255.

Kolata, A. (1991) The technology and organization of agriculture production in the Tiwanaku state. *Latin American Antiquity* **2**, 99–125.

Kovda, V.A. (1980) *Land Aridization and Drought Control.* Westview Press, Boulder, Colorado.

Koziel, M.G., Beland, G.L., Bowman, C., Carozzi, N.B., Crenshaw, R., Crossland, L., Dawson, J., Desai, N., Hill, M., Kadwell, S., Launis, K., Lewis, K., Maddox, D., McPherson, K., Meghji, M.R., Merlin, E., Rhodes, R., Warren, G.W., Wright, M. and Evola, S.V. (1993) Field performance of elite transgenic maize plants expressing an insecticidal protein derived from *Bacillus thuringiensis. Bio/Technology* **11**, 194–200.

Kronvang, B., Aertebjerg, G., Grant, R., Kristensen, P., Hormand, M. and Kirkegaard, J. (1993) Nationwide monitoring of nutrients and their ecological effects: state of the Danish aquatic environment. *Ambio* **22**, 176–187.

Kruip, A.M. and Spaan, W.J.M. (1992a) *In vitro* embryo production and manipulation. In The Biopol Team, *Biotechnological Innovations in Animal Productivity*. Butterworth-Heinemann, Oxford, pp. 37–54.

Kruip, A.M. and Spaan, W.J.M. (1992b) Gene transfer to a whole animal: the production of transgenic (livestock) animals. In The Biopol Team, *Biotechnogical Innovations in Animal Productivity*. Butterworth-Heinemann, Oxford, pp. 67–77.

Kukal, S.S., Sur, H.S. and Gill, S.S. (1991) Factors responsible for soil erosion hazard in submontane Punjab, India. *Soil Use and Management* **7**, 38–44.

Kumar, B.M., George, S.J. and Chinnamani, S. (1993) Diversity, structure and standing stock of wood in the homegardens of Kerala in Peninsular India. *Agroforestry Systems* **25**, 243–262.

Kunstadter, P. (1987) Swiddeners in transition: Lua' farmers in Northern Thailand. In B.L. Turner II and S.B. Brush (eds), *Comparative Farming Systems*. The Guilford Press, New York, pp. 130–155.

Kuznar, L.A. (1991) Transhumant goat pastoralism in the high sierra of the South Central Andes: Human responses to environmental and social uncertainty. *Nomadic Peoples* **28**, 93–104.

Ladizinsky, G. (1987) Pulse domestication before cultivation. *Economic Botany* **41**, 60–65.

Lagudah, E.S. and Appels, R. (1992) Wheat as a model system. In G.P. Chapman (ed.), *Grass Evolution and Domestication*, Cambridge University Press, Cambridge, pp. 225–265.

Lal, R. (1987) Managing the soils of sub-Saharan Africa. *Science* **236**, 1069–1076.

Lal, R. (1993) Soil erosion and conservation in West Africa. In D. Pimentel (ed.), *World Soil Erosion and Conservation*. Cambridge University Press, Cambridge, pp. 7–25.

Lal, R. (1994) Soil erosion by wind and water: problems and prospects. In R. Lal (ed.), *Soil Erosion Research Methods*, 2nd edn. Soil and Water Conservation Society, Ankeny and St Lucie Press, Delray Beach, Florida, pp. 1–9.

Lal, R. (1994) Sustainable land use systems and soil resilience. In D.J. Greenland and I. Szabolcs (eds), *Soil Resilience and Sustainable Land Use*. CAB International, Wallingford, pp. 41–67.

Lal, R. and Elliot, W. (1994) Erodibility and erosivity. In R. Lal (ed.), *Soil Erosion Research Methods*, 2nd edn. Soil and Water Conservation Society, Ankeny and St Lucie Press, Delray Beach, Florida, pp. 181–208.

Lampkin, N. (1990) *Organic Farming*. Farming Press Books, Ipswich.

Langer, R.M.H. and Hill, G.D. (1991) *Agricultural Plants*, 2nd edn. Cambridge University Press, Cambridge.

Lanly, J.P., Singh, K.D. and Janz, K. (1991) FAO's 1990 reassessment of tropical forest cover. *Nature and Resources* **27**, 21–26.

Laquihon, W.A. and Pagbilao, M.V. (1994) Sloping Agricultural Land Technology (SALT) in the Philippines. In R.C. Guttridge and H.M. Shelton (eds), *Forage Tree Legumes in Tropical Agriculture*. CAB International, Wallingford, pp. 366–373.

Larson, J.S. (1991) North America. In M. Finlayson and M. Moser (eds), *Wetlands*. Facts on File, Oxford, pp. 57–84.

Lawrence, G. and Vanclay, F. (1992) Agricultural production and environmental degradation in the Murray-Darling Basin. In G. Lawrence, F. Vanclay and B. Furze (eds), *Agriculture, Environment and Society: Contemporary Issues for Australia*. Macmillan, Melbourne, pp. 33–59.

Layton, R., Foley, R. and Williams, E. (1991) The transition between hunting and gathering and the specialised husbandry of resources. *Current Anthropology* **32**, 255–274.

Le Houérou, H.N. (1989) *The Grazing Land Ecosystems of the African Sahel*. Springer-Verlag, Berlin.

Le Houérou, H.N. (1990) Agroforestry and sylvopastoralism to combat land degradation in the Mediterranean Basin: old approaches to new problems. *Agriculture, Ecosystems and Environment* **33**, 99–109.

Leaf, M.J. (1987) Intensification in peasant farming: Punjab in the Green Revolution. In B.L. Turner II and S.B. Brush (eds), *Comparative Farming Systems*. The Guilford Press, New York, pp. 248–295.

Leeming, F. (1994) Necessity, policy and opportunity in the Chinese countryside. In D. Dwyer (ed.), *China: The Next Decades*. Longman, Harlow, pp. 77–94.

Lehman, H. (1993) Values, ethics and the use of synthetic pesticides in agriculture. In D. Pimentel and H. Lehman (eds), *The Pesticide Question: Environment, Economics and Ethics*. Chapman and Hall, New York, pp. 347–379.

Leishy, D.J. and Van Beek, N. (1992) Baculoviruses: possible alternatives to chemical insecticides. *Chemistry and Industry* (6th April), 250–254.

Leonard, C.S. (1989) The distribution of land and agricultural output in non-blackearth Russia on the eve of emancipation (Maloga uezd). In G. Grantham and C.S. Leonard (eds), *Agrarian Organization in the Century of Industrialization: Europe, Russia and North America. Research in Economic History*. Supplement 5, pp. 353–368.

Leonard, R.A., Krusel, W.G. and Still, D.A. (1987) GLEAMS: Groundwater Loading Effects of Agricultural Management Systems. *Transactions of the American Society of Agricultural Engineers* **30**, 1403–1418.

Leonard, R.A., Truman, C.C., Krusel, W.G. and Davis, F.M. (1992) Pesticide runoff simulations: long term annual means versus event extremes? *Weed Technology* **6**, 725–730.

Lepart, J. and Debussche, M. (1992) Human impact on landscape patterning: Mediterranean examples. In A.J. Hansen and F.D. Castri (eds), *Landscape Boundaries*. Springer-Verlag, Berlin, pp. 76–106.

Lewin, R. (1993) *Human Evolution: An Illustrated Introduction*, 3rd edn. Blackwell Scientific Publications, Oxford.

Liebman, M. (1992) Research and extension efforts for improving agricultural sustainability in the north central and northeastern United States. *Agriculture, Ecosystems and Environment* **39**, 101–122.

Lillesand, T.M. and Kiefer, R.W. (1994) *Remote Sensing and Image Interpretation*, 2nd edn. John Wiley and Sons, Chichester.

Lindow, S.E. (1990) Use of genetically altered bacteria to achieve plant frost control. In J.P. Nakas and C. Hagerdorn (eds), *Biotechnology of Plant–Microbe Interactions*. McGraw-Hill, New York, pp. 145–187.

Lindskog, P. and Tengberg, A. (1994) Land degradation, natural resources and local knowledge in the Sahel zone of Burkina Faso. *GeoJournal* **33**, 365–375.

Loftus, R.T., MacHugh, D.E., Bradley, D.G., Sharp, P.M. and Cunningham, P. (1994) Evidence for two independent domestications of cattle. *Proceedings of the National Academy of Sciences USA* **91**, 2757–2761.

Logan, T.J. (1993) Agricultural best management practices for water pollution control: current issues. *Agriculture, Ecosystems and Environment* **46**, 223–231.

Logeman, J. and Schell, J. (1993) The impact of biotechnology on plant breeding, or how to combine increases in agricultural productivity with an improved protection of the environment. In I. Chet (ed.), *Biotechnology in Plant Disease Control*. Wiley-Liss, New York, pp. 1–14.

Loker, W.M. (1993) Where's the beef?: Incorporating cattle into sustainable agroforestry systems in the Amazon Basin. *Agroforestry Systems* **25**, 227–241.

Loomis, R.S. and Connor, D.J. (1991) *Crop Ecology: Productivity and Management in Agricultural Systems*. Cambridge University Press, Cambridge.

Loy, T.H., Spriggs, M. and Wickler, S. (1992) Direct evidence for human use of plants 28,000 years ago: starch residues on stone artifacts from the northern Solomon Islands. *Antiquity*, **66**, 898–912.

MacDicken, K.G. and Vergara, N.T. (1990) Introduction to agroforestry. In K.G. MacDicken and N.T. Vergara (eds), *Agroforestry: Classification and Management*. John Wiley and Sons, New York, pp. 1–130.

MacKenzie, D. (1994) Battle for the world's seed banks. *New Scientist* **143** (No. 1932), 4.

MacNeish, R.S. (1992) *The Origins of Agriculture and Settled Life*. University of Oklahoma Press, Norman and London.

Madeley, J. (1993) Raising rice in the savannas. *New Scientist* **138** (No. 1878), 36–39.

Maghimbi, S. (1991) The riverside Masai: cattle economy, drought and settlement pattern in the Pangani River valley. In J.C. Stone (ed.), *Pastoral Economies in Africa and Long Term Responses to Drought*. Aberdeen University African Studies Group, Amsterdam, pp. 62–68.

Mannion, A.M. (1986) Energy flow in ecosystems. In R.D. Thompson, A.M. Mannion, M. Parry and J.R.G. Townshend, *Processes in Physical Geography*. Longman, London, pp. 262–274.

Mannion, A.M. (1991) *Global Environmental Change*. Longman, Harlow.

Mannion, A.M. (1992a) Acidification and eutrophication. In A.M. Mannion and S.R. Bowlby (eds), *Environmental Issues in the 1990s*. John Wiley and Sons, Chichester, pp. 177–195.

Mannion, A.M. (1992b) Sustainable development and biotechnology. *Environmental Conservation* **19**, 297–306.

Mannion, A.M. (1993a) Biotechnology: its place in geography. *GeoJournal* **31**, 347–354.

Mannion, A.M. (1993b) Biotechnology and global change. *Global Environmental Change* **3**, 320–329.

Mannion, A.M. (1994) *Biodiversity and Industry*. Department of Geography, University of Reading, Geographical Papers No. 115.

Mannion, A.M. (1995a) Pesticides, Insecticides. In D.E. Alexander and R.W. Fairbridge (eds), *Encyclopedia of Environmental Science*. Chapman and Hall, New York. (In Press).

Mannion, A.M. (1995b) Fungi, Fungicides. In D.E. Alexander and R.W. Fairbridge (eds), *Encyclopedia of Environmental Science*. Chapman and Hall, New York. (In Press).

Mannion, A.M. (1995c) Agriculture, environment and biotechnology. *Agriculture, Ecosystems and Environment* **53**, 31–45.

Mannion, A.M. (1995d) Biotechnology and environmental quality. *Progress in Physical Geography* **19**, 192–215.

Mannion, A.M. and Bowlby, S.R. (1992) Introduction. In A.M. Mannion and S.R. Bowlby (eds), *Environmental Issues in the 1990s*. John Wiley and Sons, Chichester, pp. 1–20.

Martin, P.S. (1984) Prehistoric overkill: the global model. In P.S. Martin and R.G. Klein (eds), *Quaternary Extinctions, a Prehistoric Revolution*. University of Arizona Press, Tucson, pp. 354–403.

Mason, D.C., O'Conaill, M.A. and Bell, S.B.M. (1994) Handling four-dimensional data in environmental GIS. *International Journal of Geographical Information Systems* **8**, 191–215.

Mather, J.R. and Sdasyuk, G.V. (eds), (1991) *Global Change: Geographical Approaches*. University of Arizona Press, Tucson, Arizona.

Matthewman, R.W. (1993) *Dairying*. The Macmillan Press, Basingstoke.

Maule, J. (1992) Productivity of tropical cattle: a re-appraisal. *Outlook on Agriculture* **21**, 47–55.

May, R.M. (1989) How many species? In L. Friday and R. Laskey (eds), *The Fragile Environment*. Cambridge University Press, Cambridge, pp. 24–39.

May, R.M. (1994) Conceptual aspects of the quantification of the extent of biological diversity. *Philosophical Transactions of the Royal Society of London*, B **345**, 13–20.

Mayne, S., Reeve, A. and Hutchinson, M. (1991) Grazing. In C. Thomas, A. Reeve and G.E.J. Fisher (eds), *Milk from Grass*, 2nd edn. ICI Agricultural Division, Billingham, Scottish Agricultural College, Perth and the Institute of Grassland and Environmental Research, Hurley, Maidenhead, pp. 53–71.

McChesney, I.G., Sharp, B. and Hayward, J.A. (1981) Energy in New Zealand agriculture: current use and future trends. *Energy in Agriculture* **1**, 141–153.

McCloskey, M. (1993) Note on the fragmentation of primary rainforest. *Ambio* **22**, 250–251.

McCorriston, J. and Hole, F.A. (1991) The ecology of seasonal stress and the origins of agriculture in the Near East. *American Anthropologist* **93**, 46–69.

McCown, R.L. and Williams, J. (1991) The water environment and implications for productivity. In P.A. Werner (ed.), *Savanna Ecology and Management: Australian Perspectives and Intercontinental Comparisons*. Blackwell, Oxford, pp. 169–176.

McCown, R.L., Keating, B.A., Probert, M.E. and Jones, R.K. (1992) Strategies for sustainable crop production in semi-arid Africa. *Outlook on Agriculture* **21**, 21–31.

McNeill, W.H. (1991) American food crops in the old World. In H.J. Viola and C. Margolis (eds), *Seeds of Change*. Smithsonian Institution Press, Washington, pp. 43–59.

Meadows, M.P. (1992) Environmental release of *Bacillus thuringiensis*. In J.C. Fry and M.J. Day (eds), *Release of Genetically Engineered and Other Organisms*. Cambridge University Press, Cambridge, pp. 120–134.

Meeks, Y.J. and Dean, J.D. (1990) Evaluating ground-water vulnerability to pesticides. *Journal of Water Resources Planning and Management* **116**, 693–707.

Meissner, F. with Morrison N. (1991) *Seeds of Change: Stories of IDB Innovation in Latin America*. Inter-American Development Bank, Distributed by the Johns Hopkins University Press, Washington, DC.

Mellanby, K. (1992) *The DDT Story*. The British Crop Protection Council, Farnham.

Mellars, P.A. (1973) The character of the Middle–Upper Palaeolithic transition in south-west France. In C. Renfrew (ed.), *The Explanation of Culture Change*. Duckworth, London, pp. 255–276.

Mepham, T.B. (1993) Approaches to the ethical evaluation of animal biotechnologies. *Animal Production* **57**, 353–359.

Merchant, J.W. (1994) GIS-based groundwater pollution hazard assessment: a critical review of the DRASTIC model. *Photogrammetric Engineering and Remote Sensing* **60**, 1117–1127.

Mestel, R. (1994) Rich pickings for cotton's pioneers. *New Scientist* **141** (No. 1913), 13–14.

Meyer, W.B. and Turner, B.L. II. (1992) Human population growth and global land-use/cover change. *Annual Review of Ecology and Systematics* **23**, 39–61.

Michalk, D.L., Fu, N.-P. and Zhu, C.-M. (1993a) Improvement of dry tropical rangelands in Hainan Island, China: 1. Evaluation of pasture legumes. *Journal of Range Management* **46**, 331–339.

Michalk, D.L., Fu, N.-P. and Zhu, C.-M. (1993b) Improvement of dry tropical rangelands in Hainan Island, China: 2. Evaluation of pasture grasses. *Journal of Range Management* **46**, 339–345.

Micklin, P.P. (1988) Dessication of the Aral Sea: a water management disaster in the Soviet Union. *Science* **241**, 1170–1176.

Micklin, P.P. (1992) The Aral Crisis: introduction to the special issue. *Post-Soviet Geography* **33**, 275.

Miller, E. (1991) People. In E. Miller (ed.), *The Agrarian History of England and Wales, Vol. III 1348–1500*. Cambridge University Press, Cambridge, pp. 1–8.

Miller, G.D. and Anderson, J.C. (1994) Nitrogen management, irrigation method, and nitrate leaching. In K.L. Campbell, W.D. Graham and A.B. Bottcher (eds), *Environmentally Sound Agriculture*. American Society of Agricultural Engineers, St Joseph, Michigan, pp. 113–119.

Miller, N.F. (1991) The Near East. In W. van Zeist, K. Wasylikowa and K.-E. Behre (eds), *Progress in Old World Palaeoethnobotany*. A.A. Balkema, Rotterdam, pp. 133–160.

Miller, S.K. (1994) Genetic first upsets food lobby. *New Scientist* **142** (No. 1927), 6.

Minc, L.D. and Vandermeer, J.H. (1990) The origin and spread of agriculture. In C.R. Carroll, J.H. Vandermeer and P. Rosset (eds), *Agroecology*. McGraw-Hill Publishing Company, New York, pp. 65–111.

Ministry of Agriculture, Fisheries and Food. (1987) *Farm Extensification Scheme: A Consultative Document*. MAFF, London.

Ministry of Agriculture, Fisheries and Food (MAFF) (1990) *Agriculture in the United Kingdom, 1989*. Her Majesty's Stationery Office, London.

Minnis, P.E. (1992) Earliest plant cultivation in the desert borderlands of North America. In C.W. Cowan and P.J. Watson (eds), *The Origins of Agriculture*. Smithsonian Institution Press, Washington and London, pp. 121–141.

Mintz, S.W. (1991) Pleasure, profit and satiation. In H.J. Viola and C. Margolis (eds), *Seeds of Change*. Smithsonian Institution Press, pp. 112–129.

Mitsch, W.J. and Gosselink, J.G. (1993) *Wetlands*, 2nd edn. Van Nostrand Reinhold, New York.

Moffat, A.S. (1992) Plant biotechnology explored in Indianapolis. *Science* **255**, 25.

Moffat, A.S. (1993) New chemicals seek to outwit insect pests. *Science* **261**, 550–551.

Molinillo, M.F. (1993) Is traditional pastoralism the cause of erosive processes in mountain environments? The case of the Cumbres Calchaquies in Argentina. *Mountain Research and Development* **13**, 189–202.

Molleson, T. (1994) The eloquent bones of Aby Hureyra. *Scientific American* **271**, 60–65.

Monsalve, J. (1985) A pollen core from the Hacienda Lusitania. *Pro Calima* **4**, 40–44.

Montserrat, P. and Fillat, F. (1990) The systems of grassland management in Spain. In A.I. Breymeyer (ed.), *Managed Grasslands: Regional Studies*. Elsevier, Amsterdam, pp. 37–70.

Moog, F.A. (1985) Forages in integrated food cropping systems. In G.J. Blair, D.A. Ivory and T.R. Evans (eds), *Forages in Southeast Asian and South Pacific Agriculture*. ACIAR Proceedings Series No. 12, Canberra, pp. 152–156.

Moore, A.M.T. and Hillman, G.C. (1992) The Pleistocene to Holocene transition and human economy in southwest Asia: the impact of the Younger Dryas. *American Antiquity* **57**, 482–494.

Moran, J.M., Morgan, M.D. and Wiersma, J.H. (1986) *Introduction to Environmental Science*, 2nd edn. W.H. Freeman, New York.

Morgan, R.P.C. (1985) Assessment of soil erosion risk in England and Wales. *Soil Use and Management* **1**, 127–131.

Morgan, R.P.C. (1986) *Soil Erosion and Conservation*. Longman, London.

Morgan, W.B. and Solarz, J.A. (1994) Agricultural crisis in sub-Saharan Africa: development constraints and policy problems. *The Geographical Journal* **160**, 57–73.

Mortimore, M. (1989) *Adapting to Drought: Farmers, Famines and Desertification in West Africa*. Cambridge University Press, Cambridge.

Moss, B. (1988) *Ecology of Freshwaters: Man and Medium*, 2nd edn. Blackwell, Oxford.

Mount, T.D. (1994) Climate change and agriculture: a perspective on priorities for economic policy. *Climatic Change* **27**, 121–138.

Munro, J.M. (1987) *Cotton*, 2nd edn. Longman, Harlow.

Muraleedharan, N. (1992) Pest control in Asia. In K.C. Willson and M.N. Clifford (eds), *Tea: Cultivation to Consumption*. Chapman and Hall, London, pp. 375–412.

Murata, N., Ishizaki-Nishizawa, O., Higashi, S., Hayashi, H., Tasaka, K. and Nishida, I. (1992) Genetically engineered alteration in the chilling sensitivity of plants. *Nature* **356**, 710–713.

Murdy, C.N. (1990) Prehispanic agriculture and its effects on the valley of Guaternala. *Frost and Conservation History* **34**, 179–190.

Murray, J.D. and Oberbauer, A.M. (1992) Growth hormone manipulation and growth promotants in sheep. In A.W. Speedy (ed.), *Progress in Sheep and Goat Research*. CAB International, Wallingford, pp. 217–234.

Myers, N. (1982) Depletion of tropical moist forests: a comparative review of rates and causes in the three main regions. *Acta Amazonia* **12**, 745–758.

Myers, N. (1988) Threatened biotas: 'hot spots' in tropical forests. *The Environmentalist* **8**, 187–208.

Myers, N. (1990) The biodiversity challenge: expanded hot-spots analysis. *The Environmentalist* **10**, 243–256.

Myers, N. (ed.) (1993a) *The Gaia Atlas of Planet Management*, 2nd edn. Gaia Books, London.

Myers, N. (1993b) Population, environment and development. *Environment Conservation* **20**, 205–216.

Myers, N. (1993c) Tropical forests: the main deforestation fronts. *Environmental Conservation* **20**, 9–16.

Mytton, L.R. and Skøt, L. (1993) Breeding for improved symbiotic nitrogen fixation. In M.D. Hayward, N.O. Bosemark, I. Romagosa and M. Cerezo (eds), *Plant Breeding. Principles and Prospects*. Chapman and Hall, London, pp. 451–472.

Nadel, D. and Hershkovitz, I. (1991) New subsistence data and human remains from the earliest Levantine Epipalaeolithic. *Current Anthropology* **32**, 631–635.

Naeem, S., Thompson, L.J., Lawler, S.P., Lawton, J.H. and Woodfin, R.M. (1994) Declining biodiversity can alter the performance of ecosystems. *Nature* **368**, 734–736.

Nahlik, A.J. de (1992) *Management of Deer and Their Habitat*. Wilson Hunt, Gillingham, Dorset.

Nair, P.K.R. (1985) Classification of agroforestry systems. *Agroforestry Systems* **3**, 97–128.

Nair, P.K.R. (1990) Classification of agroforestry systems. In K.G. MacDicken and N.T. Vergara (eds) *Agroforestry: Classification and Management*. John Wiley and Sons, New York, pp. 31–57.

Nair, P.K.R. (1991) State-of-the-art of agroforestry systems. *Forest Ecology and Management* **45**, 5–29.

Nair, P.K.R. (1993) *An Introduction to Agroforestry*. Kluwer Academic Publishers, Dordrecht, in co-operation with the International Centre for Research in Agroforestry (ICRAF), Nairobi.

Napier, T.L. (1990) The evolution of US soil-conservation policy: from voluntary adoption to coercion. In J. Boardman, I.D.L. Foster and J.A. Dearing (eds), *Soil Erosion on Agricultural Land*. John Wiley and Sons, Chichester, pp. 627–644.

National Research Council (1993) *Vetiver Grass: A Thin Green Line Against Erosion*. National Academy Press, Washington, DC.

Nations, J. and Komer, D. (1987) Rainforests and the hamburger society. *The Ecologist* **17**, 4–5.

Neher, D. (1992) Ecological sustainability in agricultural systems: definition and measurement. *Journal of Sustainable Agriculture* **2**, 51–61.

Newman, E.I. (1993) *Applied Ecology*. Blackwell, Oxford.

Nilsson, A. (1992) *Greenhouse Earth*. John Wiley and Sons, Chichester.

Nisanka, S.K. and Misra, M.K. (1990) Ecological study of an Indian village ecosystem: energetics. *Biomass* **23**, 165–178.

Njoroge, J.M. and Kimemia, J.K. (1993) Current intercropping observations and future trends in Arabica coffee, Kenya. *Outlook on Agriculture* **22**, 43–49.

Norris, P.E. and Shabman, L.A. (1992) Economic and environmental considerations for nitrogen management in the mid-Atlantic coastal plain. *American Journal of Alternative Agriculture* **7**, 148–160.

Novoa, C. and Wheeler, J.C. (1984) Llama and alpaca. In I.L. Mason (ed.), *Evolution of Domesticated Animals*, Longman, London, pp. 116–128.

O'Brien, S.J. and Knight, J.A. (1987) The future of the giant panda. *Nature* **325**, 758–759.

Odemerho, F.O. and Avwunudiogba, A. (1993) The effects of changing cassava management practices on soil loss: a Nigerian example. *The Geographical Journal* **159**, 63–69.

Odend'hal, S. (1993) Intermediary agricultural energetics: a case study of solar energy linkage with Chinese working cattle. *Agriculture, Ecosystems and Environment* **43**, 217–233.

Odum, E.P. (1975) *Ecology*. Rinehart and Winston, New York.

Offer, A. (1989) *The First World War: An Agrarian Interpretation*. Clarendon Press, Oxford.

O'Hara, S.L., Street-Perrott, F.A. and Burt, T.P. (1993) Accelerated soil erosion around a Mexican highland lake caused by prehispanic agriculture. *Nature* **362**, 48–51.

Oka, H.I. (1988) *Origin of Cultivated Rice*. Elsevier/Japan Scientific Society, Amsterdam.

O'Kelly, M.J. (ed.) (1982) *Newgrange*. Thames and Hudson, London.

Okigbo, B.N. (1990) Sustainable agricultural systems in tropical Africa. In C.A. Edwards, R. Lal, P. Madden, R.H. Miller and G. House (eds) *Sustainable Agricultural Systems*. Soil and Water Conservation Society, Ankeny, lowa, pp. 323–352.

Okigbo, B.N. (1991) *Development of Sustainable Agricultural Production Systems in Africa*. International Institute of Tropical Agriculture (IITA), Ibadan, Nigeria.

Ong, C.K. and Black, C.R. (1994) Complementarity of resource use in intercropping and agroforestry systems. In J.L. Monteith, R.K. Scott and M.H. Unsworth (eds), *Resource Capture by Crops*. Nottingham University Press, Nottingham, pp. 255–278

Opdam, P., van Apeldoorn, R., Schotman, A. and Kalkhoven, J. (1993) Population response to landscape fragmentation. In C.C. Vos and P. Opdam (eds), *Landscape Ecology of a Stressed Environment*. Chapman and Hall, London, pp. 147–171.

O'Riordan, T. (1991) The new environmentalism and sustainable development. *The Science of the Total Environment* **108**, 5–15.

O'Riordan, T. and Bentham, G. (1993) The politics of nitrate in the UK. In T.P. Burt, A.L. Heathwaite and S.T. Trudgill (eds), *Nitrate: Processes, Patterns and Management*. John Wiley and Sons, Chichester, pp. 403–416.

Ortloff, C.R. and Kolata, A.L. (1993) Climate and collapse: agro-ecological perspectives on the decline of the Tiwanaku State. *Journal of Archaeological Science* **20**, 195–221.

Osmond, C.B., Winter, K. and Ziegler, H. (1982) Functional significance of different pathways of CO_2 fixation in photosynthesis. In O.L. Lange, P.S. Nobel, C.B. Osmond and H. Ziegler (eds), *Physiological Plant Ecology II*. Springer-Verlag, Heidelberg, pp. 589–613.

Otzen, U. (1992) Stabilisation of agricultural resources: preconditions for sustainable development. *Quarterly Journal of International Agriculture* **31**, 132–148.

Otzen, U. (1993) Reflections on the principles of sustainable agricultural development. *Environmental Conservation* **20**, 310–316.

Overton, M. (1991) The determinants of crop yields in early modern England. In B.M.S. Campbell and M. Overton (eds), *Land, Labour and Livestock: Historical Studies in European Agricultural Productivity*. Manchester University Press, Manchester, pp. 284–322.

Owens, F.N. and Geay, Y. (1992) Nutrition of growing and finishing cattle. In R. Jarrige and C. Béranger (eds), *Beef Cattle Production*. Elsevier, Amsterdam, pp. 225–243.

Oxford Hammond Atlas of the World (1993) Oxford University Press.

Padoch, C., Chota, J., de Jong, W. and Unruh, J. (1985) Amazonian agroforestry: A market orientated system in Peru. *Agroforestry Systems* **3**, 47–58.

Pagot, J. (1992) *Animal Production in the Tropics*. The Macmillan Press, Basingstoke.

Pallot, J. (1991) The countryside under Gorbachev. In M.J. Bradshaw (ed.), *The Soviet Union: A New Regional Geography*. Belhaven Press, London.

Papendick, R.I. (1994) Maintaining soil physical conditions. In D.J. Greenland and I. Szabolcs (eds), *Soil Resistance and Sustainable Land Use*. CAB International, Wallingford, pp. 215–234.

Parry, M.L. (1990) *Climate Change and World Agriculture*. Earthscan, London.

Parry, M. and Rosenzweig, C. (1993) The potential effects of climatic change on world food supply. In D. Atkinson (ed.), *Global Climate Change: Its Implications for Crop Protection*. British Crop Protection Council Monograph No. 56. BCPC, Farnham, pp. 33–55.

Parry, M.L. and Swaminathan, M.S. (1992) Effects of climate change on food production. In I.M. Mintzer (ed.), *Confronting Climate Change*. Cambridge University Press, Cambridge, pp. 113–125.

Parry, M.L., Carter, T.R. and Konijn, N.T. (eds), (1988) *The Impact of Climatic Variations on Agriculture, Volumes I and II*. Kluwer, Dordrecht.

Parsons, A.J. (1994) Exploiting resource capture—grassland. In J.L. Monteith, R.K. Scott and M.H. Unsworth (eds), *Resource Capture by Crops*. Nottingham University Press, Nottingham, pp. 315–349.

Paul, C.L. (1990) *Sorghum Agronomy*. ICRISAT (International Crops Research Institute for the Semi-Arid Tropics), Patancheru, India.

Payne, W.J.A. (1990) *An Introduction to Animal Husbandry in the Tropics*. Longman, Harlow.

Payton, R.W., Christiansson, C., Shishira, E.K., Yanda, P. and Eriksson, M.G. (1992) Landform, soils and erosion in the north-eastern Irangi Hills, Kondoa, Tanzania. *Geografiska Annaler*, A 7, 65–79.

Pearce, F. (1992) *The Dammed*. Bodley Head, London.

Pearce, F. (1994a) High and dry in Aswan. *New Scientist* **142** (No. 1924), 28–32.

Pearce, F. (1994b) Neighbours sign deal to save Aral Sea. *New Scientist* **141** (No. 1909), 10.

Pearsall, D.M. (1992) The origins of plant cultivation in South America. In C.W. Cowan and

P.J. Watson (eds), *The Origins of Agriculture*. Smithsonian Institution Press, Washington and London, pp. 173–205.

Pen, J., Verwoerd, T.C., van Paridon, P.A., Bendeker, R.F., van den Elzen, J.M., Geerse, L., van den Kilis, J.D., Versteegh, H.A.J., van Ooyan, A.J.J. and Hockema, A. (1993) Phytase-containing transgenic seeds as a novel feed additive for improved phosphorus utilization. *Bio/Technology* **11**, 811–814.

Penner, J.E. (1994) Atmospheric chemistry and air quality. In W.B. Meyer and B.L. Turner II (eds) *Changes in Land Use and Land Cover: A Global Perspective*. Cambridge University Press, Cambridge, pp. 175–209.

Perez, R.S. (1990) Intensive livestock production system based on local resources in Cuba. In A. Speedy (ed.), *Developing World Agriculture*. Grosvenor Press International, London.

Perkins, J. and Holochuk, N. (1993) Pesticides: historical changes demand ethical choices. In D. Pimentel and H. Lehman (eds), *The Pesticide Question: Environment, Economics and Ethics*. Chapman and Hall, New York, pp. 370–417.

Perlas, N. (1994) *Overcoming Illusions about Biotechnology*. Third World Network, Penang.

Perry, T.W. (1992) Feedlot fattening in North America. In R. Jarrige and C. Béranger (eds), *Beef Cattle Production*. Elsevier, Amsterdam, pp. 289–305.

Persley, G.J., Giddings, L.V. and Juma, C. (1992) *Biosafety: The Safe Application of Biotechnology in Agriculture and the Environment*. International Service for National Agricultural Research, The Hague.

Pickett, A.A. (1993) *Hybrid Wheat: Results and Problems*. Paul Parey Scientific Publishers, Berlin.

Pickup, G. and Chewings, V.H. (1994) A grazing gradient approach to land degradation assessment in arid areas from remotely-sensed data. *International Journal of Remote Sensing* **15**, 597–617.

Pickup, G., Chewings, V.H. and Nelson, D.J. (1993) Estimating changes in vegetation cover over time in arid rangelands using Landsat MSS data. *Remote Sensing of the Environment* **43**, 243–263.

Pigram, J.J. (1986) *Issues in the Management of Australia's Water Resources*. Longman Cheshire, Longman.

Pimentel, D. (ed.) (1993) *World Soil Erosion and Conservation*. Cambridge University Press, Cambridge.

Pimentel, D., Allen, J., Beers, A., Guinard, L., Linder, R., McLaughlin, P., Meer, B., Musonda, D., Perdue, D., Poisson, S., Siebert, S., Stoner, K., Salazar, R. and Hawkins, A. (1987) World agriculture and soil erosion. *Bioscience* **37**, 277–283.

Pimentel, D., McLaughlin, L., Zepp, A., Lakitan, B., Kraus, T., Kleinman, P., Vancini, F., Roach, W.J., Graap, E., Keeton, W.S. and Selig, G. (1991) Environmental and economic impacts of reducing US agricultural pesticide use. In D. Pimentel (ed.), *Handbook on Pest Management in Agriculture*. LRC Press, Boca Raton, pp. 679–718.

Pimentel, D., Acquay, H., Biltonen, M., Rice, P., Silva, M., Nelson, J., Lipner, V., Giordando, S., Horowitz, A. and D'Amore, M. (1993a) Assessment of environmental and economic impacts of pesticide use. In D. Pimentel and H. Lehman (eds), *The Pesticide Question: Environment, Economics and Ethics*. Chapman and Hall, New York, pp. 47–84.

Pimentel, D., Allen, J., Beers, A., Guinard, L., Hawkins, A., Linder, R., McLaughlin, P., Meer, B., Musonda, D., Perdue, D., Poisson, S., Salazar, R., Siebert, S. and Stoner, K. (1993b) Soil erosion and agricultural productivity. In D. Pimentel (ed.), *World Soil Erosion and Conservation*. Cambridge University Press, Cambridge, pp. 277–292.

Pineda, R.F. (1988) The late Preceramic and Initial period. In R.W. Keatinge (ed.), *Peruvian Prehistory*, Cambridge University Press, Cambridge, pp. 67–96.

Pinedo-Vasques, M., Zavin, D. and Jipp, P. (1992) Economic returns from forest conversion in the Peruvian Amazon. *Ecological Economics* **6**, 163–173.

Piperno, D.R. (1989) Non-affluent foragers: resource availability, seasonal shortages, and the emergence of agriculture in Panamanian tropical forests. In D.R. Harris and G.C. Hillman (eds), *Foraging and Farming: The Evolution of Plant Exploitation*. Unwin Hyman, London, pp. 538–554.

Piperno, D.R. (1990) Aboriginal agriculture and land usage in the Amazon Basin, Ecuador. *Journal of Archaeological Science* **17**, 665–557.
Plant, R.E. and Stone, N.D. (1991) *Knowledge-Based Systems in Agriculture*. McGraw-Hill, New York.
Pleydell-Bouverie, J. (1994) Cotton without chemicals. *New Scientist* **143** (No. 1944), 25–29.
Pockney, B.P. (1993) *Agriculture in the New Russian Federation*. Agra Europe Special Report No. 69, Tunbridge Wells.
Poesen, J. (1993) Gully typology and gully control measures in the European loess belt. In S. Wicherek (ed.), *Farm Land Erosion: In Temperate Plains Environment and Hills*. Elsevier, Amsterdam, pp. 221–239.
Pollott, G.E. (ed.) (1992) *Milk and Meat from Forage Crops*. British Grassland Society, Hurley, Maidenhead.
Postel, S. (1990) Saving water for agriculture. In L.R. Brown plus 12 others, *State of the World 1990*. W.W. Norton and Company, New York, pp. 39–58.
Postel, S. (1992) *Last Oasis: Facing Water Scarcity*. W.W. Norton and Company, New York.
Postel, S. (1993) Water and agriculture. In Gleick, P.H. (ed.), *Water in Crisis: A Guide to the World's Fresh Water Resources*. Oxford University Press, Oxford, pp. 56–59.
Poswal, M.A.T. (1993) Evaluation of local and introduced cotton (*Gossypium hirsutum* L.) cultivars in the Nigerian savanna. *Tropical Agriculture* (Trinidad) **70**, 208–213.
Potter, C. (1988) Making waves: farm extensification and conservation. *Ecos* **9**, 32–37.
Potter, T.W. (1987) *Roman Italy*. British Museum Publications, London.
Potts, G.R. (1986) *The Partridge: Pesticides, Predation and Conservation*. Collins, London.
Power, J.P. and Campbell, B.M.S. (1992) Cluster analysis and the classification of medieval demesne-farming systems. *Transactions of the Institute of British Geographers New Series* **17**, 227–245.
Primrose, S.B. (1991) *Molecular Biotechnology*, 2nd edn. Blackwell, Oxford.
Qazi, M.H., Mohamahd, A.S., Rama, M.A. and Khaliq, P. (1992) Current status and future prospects of biotechnology research in Pakistan. In J.P. Moss (ed.), *Biotechnology and Crop Improvement in Asia*. ICRISAT: Andhra Pradesh, India, pp. 53–59.
Quick, G.R. (1989) Oilseeds as energy crops. In G. Röbbelen, R.K. Downey and A. Ashri (eds), *Oil Crops of the World*. McGraw-Hill Publishing Company, New York, pp. 118–131.
Quilter, J., Bernadino, O.E., Pearsall, D.M., Sandweiss, D.H., Jones, J.G. and Wing, E.S. (1991) Subsistence economy at El Paraíso, an Early Peruvian site. *Science* **251**, 277–283.
Quine, T.A. and Walling, D.E. (1991) Rates of soil erosion on arable fields in Britain: quantitative data from caesium-137 measurements. *Soil Use and Management* **7**, 169–177.
Rackham, O. (1986) *The History of the Countryside*. Dent, London.
Ramachandran, S. (1992) Agricultural research under conditions of biotic and abiotic stress and scarce resources. In J.P. Moss (ed.), *Biotechnology and Crop Improvement in Asia*. ICRISAT, Andhra Pradesh, India, pp. 9–15.
Ramakrishnan, P.S. (1990) Agricultural systems of the Northeastern hill region of India. In S.R. Gliessman (ed.), *Agroecology: Researching the Ecological Basis for Sustainable Agriculture*. Springer-Verlag, New York, pp. 251–274.
Ramakrishnan, P.S. (1992) *Shifting Agriculture and Sustainable Development*. UNESCO, Paris and Parthenon Publishing Group, Carnforth, Lancashire, UK.
Ranasinghe, D.M.S.H.K. and Newman, S.M. (1993) Agroforestry research and practice in Sri Lanka. *Agroforestry Systems* **22**, 119–130.
Rao, A.R. and Singh, I.J. (1977) Choice of foods to shorten food chains in India. In W. Lockeretz (ed.), *Agriculture and Energy*. Academic Press, New York, pp. 581–595.
Rao, P.P.N. and Mohankumar, A. (1994) Cropland inventory in the command area of Krishnarajasagar project using satellite data. *International Journal of Remote Sensing* **15**, 1295–1305.
Rapp, A. (1987) Desertification. In K.J. Gregory and D.E. Walling (eds), *Human Activity and Environmental Processes*. John Wiley and Sons, Chichester, pp. 425–443.
Rappaport, R. (1992) *Controlling Crop Pests and Diseases*. Macmillan, New York.
Ray, J.G., Travleev, A.P. and Belova, N.A. (1993) The extinction of some perennial grass vegetation and the degradation of chernozem due to anthropozoogenial factors, in some

steppes of Ukrainian S.S.R. In S. Wicherek (ed.), *Farm Land Erosion: In Temperate Plains Environment and Hills*. Elsevier, Amsterdam, pp. 241–251.

Redclift, M. (1992) The meaning of sustainable development. *Geoforum* **23**, 395–403.

Redclift, M. (1994) Reflections on the 'sustainable development' debate. *International Journal of Sustainable Development and World Ecology* **1**, 3–21.

Redman, M. (1991) *The Organic Fact File: A Guide to the Production and Marketing of Organic Produce in the UK*. British Organic Farmers and Organic Growers Association in conjunction with Safeway plc.

Reenberg, A. (1994) Land-use dynamics in the Sahelian zone in eastern Niger—monitoring change in cultivation strategies in drought prone areas. *Journal of Arid Environments* **27**, 179–192.

Rees, J. (1987) Agriculture and horticulture. In J. Wacher (ed.), *The Roman World*, Vol. II. Routledge and Kegan Paul, London, pp. 481–503.

Reganold, J.P., Palmer, A.S., Lockhart, J.C. and Macgregor, A.N. (1993) Soil quality and financial performance of biodynamic and conventional farms in New Zealand. *Science* **260**, 344–349.

Reid, W.V. (1992) How many species will there be? In T.C. Whitmore and J.A. Sayer (eds), *Tropical Deforestation and Species Extinction*. Chapman and Hall, London, pp. 55–73.

Reid, W.V., Laird, S.A., Meyer, C.A., Gámez, R., Sittenfield, A., Janzen, D.H., Crollin, M.A. and Juma, C. (1993) *Biodiversity Prospecting: Using Genetic Resources for Sustainable Development*. World Resources Institute, USA; Instituto Nacional de Biodiversidad (INbio), Costa Rica; Rainforest Alliance, USA and African Centre for Technology Studies (ACTS), Kenya.

Reilly, J. (1994) Crops and climate change. *Nature* **367**, 118–119.

Reiners, W.A. (1973) Terrestrial detritus and the carbon cycle. In G.M. Woodwell and E.V. Pecan (eds), *Carbon and the Biosphere*. National Technical Information Service, Springfield, Virginia, pp. 303–327.

Reinken, G., Hartfiel, W. and Körner, E. (1990) *Deer Farming: A Practical Guide to German Techniques*. Farming Press, Ipswich.

Ren, C. (1991) The components of agriculture. In X. Guohua and L.J. Peel (eds), *The Agriculture of China*. Oxford University Press, Oxford, pp. 73–107.

Renfrew, C. (1987) *Archaeology and Language: The Puzzle of Indo-European Origins*. Jonathan Cape, London.

Repetto, R. and Gillis, M. (eds) (1988) *Public Policies and the Misuse of Forest Resources*. Cambridge University Press, Cambridge.

Reynolds, S.G. (1988) Pastures and cattle under coconuts. *FAO Plant Production and Protection* Paper No. 91, pp. 321.

Rhoades, J.D. (1990) Soil salinity—causes and controls. In A.S. Goudie (ed.), *Techniques for Desert Reclamation*. John Wiley and Sons, Chichester, pp. 109–134.

Riasanovsky, N.V. (1993) *A History of Russia*, 5th edn. Oxford University Press, Oxford.

Rice, R.A. and Vandermeer, J. (1990) Climate and the geography of agriculture. In C.R. Carroll, J.H. Vandermeer and P.M. Rosset (eds), *Agroecology*. McGraw-Hill, New York, pp. 21–63.

Richards, J.F. (1990) Land transformation. In B.L. Turner II, W.C. Clark, R.W. Kates, J.F. Richards, J.T. Mathews and W.B. Meyer (eds) *The Earth as Transformed by Human Action*. Cambridge University Press, Cambridge, pp. 163–178.

Riebsame, W.E. (1990) The United States Great Plains. In B.L. Turner II, W.C. Clark, R.W. Kates, J.F. Richards, J.T. Mathews and W.B. Meyer (eds) *The Earth as Transformed by Human Action*. Cambridge University Press with Clark University, Cambridge, pp. 561–575.

Rijal, K., Bansal, N.K. and Grover, P.D. (1993) Energy and subsistence: Nepalese agriculture. *Bioresource Technology* **37**, 61–69.

Roberts, J.M. (1992) *History of the World*. Helicon Publishing, London.

Robinette, A. (1991) Land management applications of GIS in the state of Minnesota. In D.J. Maguire, M.F. Goodchild and D.W. Rhind (eds), *Geographical Information Systems: Principles and Applications. Volume 2: Applications*. Longman, Harlow, pp. 275–283.

Robinson, C. (1993) Biotechnology in Cuba: tackling Third-World problems with front-line technology. *Trends in Biotechnology* **11**, 80–84.

Robison, D.M. and McKean, S.J. (1992) *Shifting Cultivation and Alternatives. An Annotated Bibliography, 1972–1989*. CAB International, Wallingford.

Roca, W.M. (1989) Cassava production and utilization problems and their biotechnological solutions. In A. Sasson and V. Costarini (eds), *Plant Biotechnologies for Developing Countries*. Technical Centre for Agriculture and Rural Co-operation and the Food and Agriculture Organisation of the United Nations, Rome, pp. 213–219.

Rodwell, J.S. (ed.) (1992) *British Plant Communities, Volume 3. Grasslands and Montane Communities*. Cambridge University Press, Cambridge.

Rogers, P. (1994) Hydrology and water quality. In W.B. Meyer and B.L. Turner II (eds), *Changes in Land Use and Land Cover: A Global Perspective*. Cambridge University Press, Cambridge, pp. 231–257.

Roosevelt, A.C. (1984) Population, health and the evolution of subsistence: conclusions from the conference. In M.N. Cohen and G.A. Armelagos (eds), *Paleopathology and the Origins of Agriculture*. Academic Press, Orlando, pp. 559–583.

Rosenberg, R., Elmgren, R., Fleischer, S., Jonsson, P., Persson, G. and Dahlin, H. (1990) Marine eutrophication: case studies in Sweden. *Ambio* **19**, 102–108.

Rosenzweig, C. and Parry, M.L. (1993) Potential impacts of climate change on world food supply: a summary of a recent international study. In H.M. Kaiser and T.E. Drennen (eds), *Agricultural Dimensions of Global Climate Change*. St Lucie Press, Delray Beach, Florida, pp. 87–116.

Rosenzweig, C. and Parry, M.L. (1994) Potential impact of climate change on world food supply. *Nature* **367**, 133–138.

Rotmans, J., Hulme, M. and Downing, T.E. (1994) Climate change implications for Europe: an application of the ESCAPE model. *Global Environmental Change* **4**, 97–124.

Rowntree, J. (1993) Marketing channels and price determination for agricultural commodities. In G.M. Craig (ed.), *The Agriculture of Egypt*. Oxford University Press, Oxford, pp. 420–444.

Rudel, T.K., with Horowitz, B. (1993) *Tropical Deforestation: small Farmers and Land Clearing in the Ecuadorian Amazon*. Columbia University Press, New York.

Ruiz, M.E. (1990) Milk production systems in Latin America: constraints and potentials. In *Dairying in a Changing World*. Proceedings of the 23rd International Dairy Congress, Montreal, Quebec, pp. 171–187.

Runge, C.F. (1992) A policy perspective on the sustainability of production environments: towards a land theory of value. *Quarterly Journal of International Agriculture* **31**, 149–161.

Russel, A.J.F. (1994) Fibre production from South American camelids. *Journal of Arid Environments* **26**, 33–37.

Ruthenberg, H. (1980) *Farming Systems in the Tropics*, 3rd edn. Oxford University Press, Oxford.

Ruttan, V.W. (1991) Constraints on sustainable growth in agricultural production: into the 21st century. *Outlook on Agriculture* **20**, 225–234.

Ruvinsky, A., Nezavitin, A. and Ruvinskaya, L. (1992) The Siberian agroindustrial complex: problems and opportunities. *Outlook on Agriculture* **21**, 137–140.

Sachs, W. (ed.) (1993) *Global Ecology: A New Arena of Political Conflict*. Zed Books, London.

Salamini, F. and Motto, M. (1993) The role of gene technology in plant breeding. In M.D. Hayward, N.O. Bosemark, I. Romagosa and M. Cerezo (eds), *Plant Breeding. Principles and Prospects*. Chapman and Hall, London, pp. 138–159.

Salati, E., Dourojeanni, M.J., Novaes, F.C., de Oliviera, A.E., Perritt, R.W., Schubart, H.O.R. and Umana, J.C. (1990) Amazonia. In B.L. Turner II, W.C. Clark, R.W. Kates, J.F. Richards, J.T. Mathews and W.B. Meyer (eds), *The Earth as Transformed by Human Action*. Cambridge University Press, Cambridge and Clark University, Worcester, Massachusetts, pp. 479–493.

Sallares, R. (1991) *The Ecology of the Ancient Greek World*. Duckworth, London.

Samad, M., Merrey, D., Vermillion, D., Fuchs-Carsch, M., Mohtadullah, K. and Lenton, R.

(1992) Irrigation management strategies for improving the performance of irrigated agriculture. *Outlook on Agriculture* **21**, 279–286.

Sarmiento, L., Monasterio, M. and Montilla, M. (1993) Ecological bases, sustainability, and current trends in traditional agriculture in the Venezuelan high Andes. *Mountain Research and Development* **13**, 167–176.

Sasson, A. (1990) *Feeding Tomorrow's World.* UNESCO, Paris.

Sattler, F. and Wistinghausen, E. von (1992) *Bio-Dynamic Farming Practice.* (Translated from the German by A.R. Meuss and first published in 1989 by Ewgen Ulmer GmbH and Co, Stuttgart, Germany). Bio-Dynamic Agricultural Association, Stourbridge, West Midlands.

Sauer, C.O. (1952) *Agricultural Origins and Dispersals.* American Geographical Society, Bowman Memorial Lecture 5, series 2.

Sayer, J.A. (1992) A future for Africa's tropical forests. In J.A. Sayer, C.S. Harcourt and N.M. Collins (eds), *The Conservation Atlas of Tropical Forests: Africa.* Macmillan Publishers, Basingstoke, pp. 81–93.

Sayer, J.A and Whitmore, T.C. (1991) Tropical moist forests: destruction and species extinction. *Biological Conservation* **55**, 199–214.

Scanlan, J.C., Prinsley, R., Pigot, J.P., Wakefield, S., Van der Somman, F., Duncan, F., Stadler, T., McLellan, R and Farago, A. (1992) Retention of native woody vegetation on farms in Australia: management considerations, planning guidelines and information gaps. *Agroforestry Systems* **20**, 141–166.

Schaller, N. (1993) The concept of agricultural sustainability. *Agriculture, Ecosystems and Environment* **46**, 89–97.

Schreier, H., Brown, S., Schmidt, M., Shah, P., Shrestha, B., Nakarmi, G., Subba, K. and Wymann, S. (1994) Gaining forests but losing ground: a GIS evaluation of a Himalayan watershed. *Environmental Management* **18**, 139–150.

Schulman, B.J. (1991) *From Cotton Belt to Sunbelt.* Oxford University Press, New York.

Schusky, E.L. (1989) *Culture and Agriculture.* Bergin and Garvey Publishers, New York.

Scoones, I. (1992) Coping with drought: Responses of herders and livestock in contrasting savanna environments in Southern Zimbabwe. *Human Ecology* **20**, 293–314.

Scott, C.A. (1994) Facing environmental degradation in the Aravalli Hills, India. In A.C. Millington and K. Pye (eds), *Environmental Change in Drylands: Biogeographical and Geomorphological Perspectives.* John Wiley and Sons, Chichester, pp. 413–426.

Scott, G.A.J. (1987) Shifting cultivation where land is limited. In C.F. Jordan (ed.), *Amazonian Rain Forests.* Springer-Verlag, New York, pp. 34–45.

Scott, R.K., Jaggard, K.W. and Sylvester-Bradley, R. (1994) Resource capture by arable crops. In J.L. Monteith, R.K. Scott and M.H. Unsworth (eds), *Resource Capture by Crops.* Nottingham University Press, Nottingham, pp. 279–302.

Selincourt, K. de (1993) Europe's home-grown fuel. *New Scientist* **140** (No. 1895), 22–24.

Serageldin, I. (1993) Agriculture and environmentally sustainable development. In J.P. Srivastava and H. Alderman (eds), *Agriculture and Environmental Challenges.* The World Bank, Washington, DC, pp. 5–16.

Shakya, J.D. (1985) *The Production of Potatoes from True Seed by Transplanting and Sowing.* Unpublished PhD Thesis, University of Reading.

Sharma, A., Martin, M.J., Okabe, J.F., Trugho, R.A., Dhanjal, N.K., Logan, J.S. and Kumar, R. (1994) An isologous porcine promotor permits high level expression of human haemoglobin in transgenic swine. *Bio/Technology* **12**, 55–59.

Shengxiu, L. and Ling, X. (1992) Distribution and management of drylands in the People's Republic of China. *Advances in Soil Science* **18**, 147–302.

Shepherd, G. (1993) Meeting the need for trees: land rights, planting choice and the protection of the environment. *Rural Extension Bulletin* **2**, 10–17.

Shipman, P., Bosler, W. and Davis, K.L. (1981) Butchering of giant geladas at an Acheulian site. *Currently Anthropology* **22**, 257–268.

Shiva, V. (1993) *Monocultures of the Mind.* Zed Books, London and Third World Network, Penang.

Silva, J.F. and Moreno, A. (1993) Land use in Venezuela. In M.D. Young and O.T. Solbrig (eds), *The World's Savannas.* UNESCO, Paris and Parthenon Press, New York, pp. 239–257.

Simmons, I.G. (1989) *Changing the Face of the Earth: Culture, Environment, History*. Basil Blackwell, Oxford.

Slade, C.F.R. (1990) The UK sheep industry: current position, economics and emerging trends. In C.F.R. Slade and T.L.J. Lawrence (eds), *New Developments in Sheep Production*. British Society of Animal Production, Occasional Publication No. 14, pp. 7–12.

Slater, K. (1991) *The Principles of Dairy Farming*. Farming Press, Ipswich.

Smiet, A.C. (1992) Forest ecology on Java: human impact and vegetation of montane forest. *Journal of Tropical Ecology* **8**, 129–152.

Smika, D.E. (1992) Cereal systems of the North American Central Great Plains. In C.J. Pearson (ed.), *Field Crop Ecosystems*. Elsevier, Amsterdam, pp. 401–412.

Smil, V. (1987) *Energy, Food and Environment*. Clarendon Press, Oxford.

Smil, V. (1991) *General Energetics: Energy in the Biosphere and Civilization*. John Wiley and Sons, New York.

Smil, V. (1993a) *Global Ecology: Environmental Change and Social Flexibility*. Routledge, London.

Smil, V. (1993b) *China's Environmental Crisis: An Inquiry into the Limits of National Development*. M.E. Sharpe, Armonk, New York.

Smith, A.B. (1992) Origins and spread of pastoralism in Africa. *Nomadic Peoples* **32**, 91–105.

Smith, B.D. (1989) Origins of Agriculture in Eastern North America. *Science* **246**, 1566–1571.

Smith, B.D. (1992) Prehistoric plant husbandry in Eastern North America. In C.W. Cowan and P.J. Watson (eds), *The Origins of Agriculture*. Smithsonian Institution Press, Washington and London, pp. 101–119.

Smith, F.D.M., May, R.M., Pellew, R., Johnson, T.H. and Walter, K.R. (1993a) Estimating extinction rates. *Nature* **364**, 494–496.

Smith, F.D.M., May, R.M., Pellew, R., Johnson, T.H. and Walter, K.R. (1993b) How much do we know about the current extinction rate? *Trends in Ecology and Evolution* **8**, 375–378.

Soil Association (1992) *Standards for Organic Food and Farming*. The Soil Association Marketing Company, Bristol.

Sokal, R.R., Oden N.L. and Wilson, C. (1991) Genetic evidence for the spread of agriculture in Europe by demic diffusion. *Nature* **351**, 143–145.

Somerville, C.R. (1993) Production of industrial material in trangenic plants. *Philosophical Transactions of the Royal Society of London*, B **342**, 251–257.

Sonnenfeld, D.A. (1992) Mexico's 'Green Revolution', 1940–1980: towards an environmental history. *Environmental History Review* **16**, 29–52.

Soulé, J.D. and Piper, J.K. (1992) *Farming in Nature's Image: An Ecological Approach to Agriculture*. Island Press, Washington, DC.

Southerton, N.W. and Rands, M.R.W. (1987) The environmental interest of field margins to game and other wildlife: a Game Conservancy view. In J.M. Way and P.W. Greig-Smith (eds), *Field Margins*. British Crop Protection Council, Thornton Heath, pp. 67–75.

Southgate, D. and Basterrechea, M. (1992) Population growth, public policy and resource degradation: The case of Guatemala. *Ambio* **21**, 460–464.

Spalding, R.F. and Exner, M.E. (1993) Occurrence of nitrate in groundwater—a review. *Journal of Environmental Quality* **22**, 392–402.

Späth, H.-J.W. (1987) Dryland wheat farming on the Central Great Plains: Sedgwick County, Northeast Colorado. In B.L. Turner II and S.B. Brush (eds), *Comparative Farming Systems*. The Guilford Press, New York, pp. 313–344.

Spelman, C.A. (1994) *Non-food Uses of Agricultural Raw Materials: Economics, Biotechnology and Politics*. CAB International, Wallingford.

Squires, V.R. and Vera, R.R. (1992) Commercial beef ranching in tropical and semi-arid zones. In R. Jarrige and C. Béranger (eds), *Beef Cattle Production*. Elsevier, Amsterdam, pp. 437–454.

Sridhar, V.N., Dadhwal, V.K., Chaudhari, K.N., Sharma, R., Bairagi, G.D. and Sharma, A.K. (1994) Wheat production forecasting for a predominantly unirrigated region in Madhya Pradesh (India). *International Journal of Remote Sensing* **15**, 1307–1316.

Stadel, C. (1986) Del Valle Al Monte: altitudinal patterns of agricultural activities in the Patate-Pelileo area of Ecuador. *Mountain Research and Development* **6**, 53–64.

Stafford Smith, D.M. and Foran, B.D. (1991) 22/RANGEPACK: the philosophy underlying the development of a microcomputer-based decision support system for pastoral land management. In P.A. Werner (ed.), *Savanna Ecology and Management: Australian Perspectives and Intercontinental Comparisons*. Blackwell, Oxford, pp. 197–202.

Ståhl, M. (1993) Land degradation in East Africa. *Ambio* **22**, 505–508.

Stanhill, G. (1981) The Egyptian agro-ecosystem at the end of the 18th century—an analysis based on the 'Description de L'Egypt'. *Agro-Ecosystems* **6**, 305–314.

Stanley, D.J. and Warne, A.G. (1993) Sea level and initiation of predynastic culture in the Nile delta. *Nature* **363**, 435–438.

Stark, B.A. and Wilkinson, J.M. (eds) (1992) *Whole-crop Cereals*. Chalcombe Publications, Kingston, nr Canterbury, UK.

Sterling, W.L., El-Zik, K.M. and Wilson, L.T. (1989) Biological control of pest populations. In R.E. Frisbie, K.M. El-Zik and L.T. Wilson (eds), *Integrated Pest Management Systems and Cotton Production*. John Wiley and Sons, New York, pp. 155–189.

Steven, M.D. and Clark, J.A. (eds) (1990) *Applications of Remote Sensing in Agriculture*. Butterworths, London.

Steyhaert, L.T. and Goodchild, M.F. (1994) Integrating geographic information systems and environmental simulation models: a status review. In W.K. Michener, J.W. Brunt and S.G. Stafford (eds), *Environmental Information Management and Analysis*. Taylor and Francis, London, pp. 333–355.

Stocking, M.A. (1994) Assessing vegetative cover and management effects. In R. Lal (ed.), *Soil Erosion Research Methods*, 2nd edn. Soil and Water Conservation Society, Ankeny and St Lucie Press, Delray Beach, Florida, pp. 211–232.

Stone, N.D., Rummel, D.R., Carroll, S., Makela, M.E. and Frisbie, R.E. (1990) Simulation of bull weevil (Coleoptera: Curculionidae) spring emergence and overwintering survival in the Texas rolling plains. *Environmental Entomology* **19**, 91–98.

Stoskopf, N.C. with Tomes, D.T. and Christie, B.R. (1993) *Plant Breeding: Theory and Practice*. Westview Press, Boulder, Colorado.

Strobel, G.A. (1992) Biological control of weeds. *Scientific American* **265**, 50–60.

Suarez, D. (1992) Perspective on irrigation management and salinity. *Outlook on Agriculture* **21**, 287–291.

Summy, K.R. and King, E.G. (1992) Cultural control of cotton insect pests in the United States. *Crop Protection* **11**, 307–319.

Swift, M.J. (1994) Maintaining the biological status of soil: a key to sustainable land management. In D.J. Greenland and I. Szabolcs (eds), *Soil Resilience and Sustainable Land Use*. CAB International, Wallingford, pp. 235–247.

Tacher, G. and Jahnke, H.E. (1992) Beef production in tropical Africa. In R. Jarrige and C. Béranger (eds), *Beef Cattle Production*. Elsevier, Amsterdam, pp. 419–436.

Taconet, O., Benallegue, M., Vidal-Madjar, D., Prevot, L., Dechambre, M. and Normad, M. (1994) Estimation of soil and crop parameters for wheat from airborne radar backscattering data in C and X bands. *Remote Sensing of the Environment* **50**, 287–294.

Takeo, T. (1992) Green and semi-fermented teas. In K.C. Willson and M.N. Clifford (eds), *Tea: Cultivation to Consumption*. Chapman and Hall, London, pp. 413–457.

Tao, S. (1991a) Crop production on Chinese farms. *Outlook on Agriculture* **20**, 25–29.

Tao, S (1991b) China's animal husbandry: the past ten years. *Outlook on Agriculture* **20**, 191–194.

Taylor, A.E.B., O'Callaghan, P.W. and Probert, S.D. (1993) Energy audit of an English Farm. *Applied Energy* **44**, 315–335.

Tegart, W.J.McG., Sheldon, G.W. and Griffiths, D.C. (eds) (1990) *Climate Change: The IPCC Impacts Assessment*. Australian Government Publishing Service, Canberra.

Thiele, G. (1993) The dynamics of farm development in the Amazon: The *Barbecho* crisis model. *Agricultural Systems* **42**, 179–197.

Thirsk, J. (1987) *England's Agricultural Regions and Agrarian History, 1500–1750*. Macmillan, London.

Thomas, C., Reeve, A. and Fisher, G.E.J. (eds) (1991) *Milk from Grass*, 2nd edn. ICI

Agricultural Division, Cleveland, The Scottish Agricultural College, Perth and the Institute of Grassland and Environmental Research, Hurley, Maidenhead.

Thomas, D.S.G. and Middleton, N.J. (1993) Salinization: new perspectives on a major desertification issue. *Journal of Arid Environments* **24**, 95–105.

Thomas, D.S.G. and Middleton, N.J. (1994) *Desertification: Exploding the Myth*. John Wiley and Sons, Chichester.

Thomas, J.A. (1991) Rare species conservation: case studies of European butterflies. In I. Spellerberg, B. Goldsmith and M.G. Morris (eds), *The Scientific Management of Temperate Communities for Conservation*. Blackwell, Oxford, pp. 149–197.

Thomas, J.A. and Morris, M.G. (1994) Patterns, mechanisms and rates of extinction among invertebrates in the United Kingdom. *Philosophical Transactions of the Royal Society of London*, B **344**, 47–54.

Thompson, K.F. and Poppi, D.P. (1990) Livestock production from pasture. In R.H.M. Langer (ed.), *Pastures: Their Ecology and Management*. Oxford University Press, Auckland, pp. 263–283.

Thompson, L. and Moseley-Thompson, E. (1989) One-half millennia of tropical climate variability as recorded in the stratigraphy of the Quelccaya ice cap, Peru. *Geophysical Union Monograph* 55.

Thompson, P.B. (1993) Genetically modified animals: ethical issues. *Journal of Animal Science* **71**, Supplement 3, 51–56.

Thwaites, T. (1993) Modified clover could create super sheep. *New Scientist* **138** (No. 1867), 19.

Tiffen, M., Mortimore, M. and Gichuki, F. (1994) *More People, Less Erosion: Environmental Recovery in Kenya*. John Wiley and Sons, Chichester.

Tilman, D., May, R.M., Lehman, C.L. and Nowak, M.A. (1994) Habitat destruction and the extinction debt. *Nature* **371**, 65–66.

Tivy, J. (1990) *Agricultural Ecology*. Longman, Harlow.

Todd, M. (1987) *The South-West to AD 1000*. Longman, London.

Tolba, M.K. (1992) *Saving Our Planet*. Chapman and Hall, London.

Tolba, M.K., El-Kholy, O.A., El-Hinnawi, E., Holdgate, M.W. McMichael, D.F. and Munn, R.E. (eds) (1992) *The World Environment 1972–1974*. Chapman and Hall, London.

Treacher, T.T. (1990) Grazing management and supplementation for the lowland sheep crop. In C.F.R. Slade and T.L.J. Lawrence (eds), *New Developments in Sheep Production*. British Society of Animal Production, Occasional Publication No. 14, pp. 45–54.

Trimble, S.W. (1992) The Alcovy River swamps: the result of culturally accelerated sedimentation. In L.M. Dilsaver and C.E. Colten (eds), *The American Environment: Interpretations of Past Geographies*. Rowman and Littlefield Publishers, Lanham, Maryland, pp. 21–32.

Truve, E., Aaspôllu, A., Honkanen, J., Puska, K., Mehto, M., Hassï, A., Teeri, T.H., Kelve, M., Seppänen, P. and Saarma, M. (1993) Transgenic potato plants expressing mammalian 2′-5′ oligoadenylate synthetase are protected from potato virus X infection under field conditions. *Bio/Technology* **11**, 1048–1052.

Tsatsarelis, C.A. (1993) Energy inputs and outputs for soft winter wheat production in Greece. *Agriculture, Ecosystems and Environment* **43**, 109–118.

Tuan, F.C. (1993) The livestock sector. In Y.Y. Kueh and R.F. Ash (eds), *Economic Trends in Chinese Agriculture: The Impact of Post-Mao Reforms*. Clarendon Press, Oxford, pp. 203–228.

Tucker, C.J., Dregne, H.E. and Newcomb, W.W. (1991) Expansion and contraction of the Sahara Desert from 1980 to 1990. *Science* **253**, 299–301.

Turner II, B.L. and Brush, S.B. (1987) Introduction to Part III. In B.L. Turner and S.B. Brush (eds), *Comparative Farming Systems*. The Guilford Press, New York, pp. 191–194.

Turner II, B.L. and Meyer, W.B. (1994) Global land-use and land-cover change: an overview. In W.B. Meyer and B.L. Turner II (eds), *Changes in Land Use and Land Cover*. Cambridge University Press, Cambridge, pp. 3–10.

Turner II, B.L., Meyer, W.B. and Skole, D.L. (1994) Global land-use/land-cover change: towards an integrated study. *Ambio* **23**, 91–95.

United Nations (1994) *The State of World Population*. United Nations, New York.

United Nations Environment Programme (1991) *Status of Desertification and the Implementation of the United Nations Plan of Action to Combat Desertification*. Report of the executive director to the governing council, third special session, Nairobi.

United Nations Environment Programme (1992) *World Atlas of Desertification*. Edward Arnold, Sevenoaks.

United States Department of Agriculture (USDA) (1980) *Report and Recommendations on Organic Farming*. United States Department of Agriculture, US Government Printing Office, Washington, DC.

Utting, P. (1993) *Trees, People and Power. Social Dimensions of Deforestation and Forest Protection in Central America*. Earthscan, London.

Vanclay, J.K. (1993) Saving the tropical forest: needs and prognosis. *Ambio* **22**, 225–231.

Vandaele, K. (1993) Assessment of factors affecting ephemeral gully erosion in cultivated catchments of the Belgian Loam belt. In S. Wicherek (ed.), *Farmland Erosion: In Temperate Plains, Environment and Hills*. Elsevier, Amsterdam, pp. 125–136.

Van den Berg, J.C.T. (1990) *Strategy for Dairy Development in the Tropics and Subtropics*. Pudoc, Wageningen.

Van den Elzen, P.J.M., Jongedyk, E., Melchers, L.S. and Cornelissen, B.J.C. (1993) Virus and fungal resistance: from laboratory to field. *Philosophical Transactions of the Royal Society of London*, B **342**, 271–278.

Van de Meeberg, R. (1992) The world trade in tea. In K.C. Willson and M.N. Clifford (eds), *Tea: Cultivation to Consumption*. Chapman and Hall, London, pp. 649–687.

Vandermeer, J. (1989) *The Ecology of Intercropping*. Cambridge University Press, Cambridge.

Van der Merwe, N.J. (1982) Carbon isotopes, photosynthesis and archaeology. *American Scientist* **70**, 596–606.

Van Duivenbooden, N. (1993) Grazing as a tool for rangeland management in semiarid regions: a case study in the north-western coastal zone of Egypt. *Agriculture, Ecosystems and Environment* **43**, 309–324.

Van Keulen, H. and Breman, H. (1990) Agricultural development in the West African Sahelian region: a cure against land hunger. *Agriculture, Ecosystems and Environment* **32**, 177–197.

Van Tuijl, W. (1993) *Improving Water Use in Agriculture: Experiences in the Middle East and North Africa*. World Bank Technical Paper No. 201, The World Bank, Washington, DC.

Van Vliet, L.J.P., Kline, R. and Hall, J.W. (1993) Effects of three tillage treatments on seasonal runoff and soil loss in the Peace River region. *Canadian Journal of Soil Science* **73**, 469–480.

Van Zeist, W. and Bakker-Heeres, J.A.H. (1979) Some economic and ecological aspects of the plant husbandry of Tell Aswad. *Paléorient* **5**, 161–169.

Vasil, V., Castillo, A.M., Fromm, M.E. and Vasil, I.K. (1992) Herbicide resistant fertile transgenic wheat plants obtained by microprojectile bombardment of regenerable embryonic callus. *Bio/Technology* **10**, 667–674.

Vavilov, N.I. (1992) *Origin and Geography of Cultivated Plants*. Cambridge University Press, Cambridge. (Translation of Vavilov's papers by D. Löve.)

Villachica, H., Silva, J.E., Peres, J.R. and de Rocha, C.M. (1990) Sustainable agricultural systems in the humid tropics of South America. In C.A. Edwards, R. Lal, P. Madden, R.H. Miller and G. House (eds), *Sustainable Agricultural Systems*. Soil and Water Conservation Society, Ankeny, Iowa, pp. 391–437.

Vogel, H. (1992) Effects of conservation tillage on sheet erosion from sandy soils at two experimental sites in Zimbabwe. *Applied Geography* **12**, 229–242.

Vos, C.C. and Zonneveld, J.I.S. (1993) Patterns and processes in a landscape under stress: the study area. In C.C. Vos and P. Opdam (eds), *Landscape Ecology of a Stressed Environment*. Chapman and Hall, London, pp. 1–27.

Vossen, P. (1992) Forecasting national crop yields of EC countries: the approach developed by the agriculture project. In F. Toselli and J. Meyer-Roux (eds), *Conference on the Application of Remote Sensing to Agricultural Statistics*. Joint Research Centre, Ispra, Commission of the European Communities, pp. 159–176.

Wacher, J. (1987) *The Roman Empire*. Dent, London.

Wade, G., Mueller, R., Cook, P. and Doralswamy, P. (1994) AVHRR map products for crop production assessment: a geographic information systems approach. *Photogrammetric Engineering and Remote Sensing* **60**, 1145–1150.

Wainwright, M. (1992) *An Introduction to Fungal Biotechnology*. John Wiley and Sons, Chichester.

Walling, D.E. and Quine, T.A. (1990) Use of caesium-137 to investigate patterns and rates of soil erosion on arable fields. In J. Boardman, I.D.L. Foster and J.A. Dearing (eds), *Soil Erosion on Agricultural Land*. John Wiley and Sons, Chichester, pp. 33–53.

Walton, J.R. (1990) Agriculture and rural society 1730–1914. In R.A. Dodgshon and R.A. Butlin (eds), *An Historical Geography of England and Wales*, 2nd edn. Academic Press, London, pp. 323–350.

Ward, P.N. (1993) Systems of agricultural production in the Delta. In G.M. Craig (ed.), *The Agriculture of Egypt*, Oxford University Press, Oxford, pp. 229–264.

Watson, D.J. (1971) Size, structure and activity of the productive system of crops. In P.F. Wareing and J.P. Cooper (eds), *Potential Crop Production*. Heinemann, London, pp. 76–88.

Watson, J.S. (1990) Steady improvements in cotton production. In A. Speedy (ed.), *Developing World Agriculture*. Grosvenor Press International, London, pp. 57–60.

Weatherston, J. (1992) Historical introduction. In K.C. Willson and M.N. Clifford (eds), *Tea: Cultivation to Consumption*. Chapman and Hall, London, pp. 1–23.

Webb, N.R. (1989) Studies on the invertebrate fauna of fragmented heathland in Dorset, UK, and the implications for conservation. *Biological Conservation* **17**, 153–165.

Webb, N.R. and Thomas, J.R. (1994) Conserving insect habitats in heathland biotopes: a question of scale. In P.J. Edwards, R.M. May and N.R. Webb (eds), *Large-scale Ecology and Conservation Biology*. Blackwell, Oxford, pp. 129–151.

Webster, G.M. and Povey, G.M. (1990) Nutrition of the finishing lamb. In C.F.R. Slade and T.L.J. Lawrence (eds), *New Developments in Sheep Production*. British Society of Animal Production, Occasional Publication No. 14, pp. 71–82.

Weightman, B. (1989) *Agriculture in Vanuatu*. British Friends of Vanuatu, Cheam, Surrey.

Weischet, W. and Caviedes, C.N. (1993) *The Persisting Ecological Constraints of Tropical Agriculture*. Longman, Harlow.

Weiss, H., Courty, M.-A. Wetterstrom, W., Guichard, F., Senior, L., Meadow, R. and Curnow, A. (1994). The genesis and collapse of third millennium North Mesopotamian civilisation. *Science* **261**, 995–1004.

Wen, D. and Pimentel, D. (1992) Ecological resource management to achieve a productive, sustainable agricultural system in northeast China. *Agriculture, Ecosystems and Environment* **4**, 215–230.

Wendorf, F., Schild, R., Close, A.E., Hillman, G.C., Gautier, A., van Neer, W., Donahue, D.J., Jull, A.J.T. and Linick, T.W. (1988) New radiocarbon dates and Late Palaeolithic diet at Wadi Kubbaniya, Egypt. *Antiquity* **62**, 279–283.

Wendorf, F., Close, A.E., Schild, R., Wasylikowa, K., Housley, R.A., Harlan, J.R. and Krolik, H. (1992) Saharan exploitation of plants 8000 years BP. *Nature* **359**, 721–724.

West, B. and Zhou, B.-X. (1988) Did chickens go north? New evidence for domestication. *Journal of Archaeological Science*, **15**, 515–533.

West, N.E. (1993) Biodiversity of rangelands. *Journal of Range Management* **46**, 2–13.

Whitmore, T.C. and Sayer, J.A. (1992) Deforestation and species extinction in tropical moist forests. In T.C. Whitmore and J.A. Sayer (eds), *Tropical Deforestation and Species Extinction*. Chapman and Hall, London, pp. 1–14.

Whitmore, T.M. and Turner, B.L. II (1992) Landscapes of cultivation in mesoamerica on the eve of the conquest. *Annals of the Association of American Geographers* **82**, 402–425.

Whittaker, R.H. and Likens, G.E. (1973) Carbon in the biota. In G.M. Woodwell and E.V. Pecan (eds), *Carbon and the Biosphere*. National Technical Information Service, Springfield, Virginia, pp. 281–302.

Whitten, A.J. (1991) Agricultural settlement schemes. In N.M. Collins, J.A. Sayer and T.C. Whitmore (eds), *The Conservation Atlas of Tropical Forests: Asia and the Pacific*. Macmillan Press, Basingstoke, pp. 36–42.

Whitten, M.J. and Oakeshott, J.G. (1990) Biocontrol of insects and weeds. In G.J. Persley (ed.), *Agricultural Biotechnology*. CAB International, Wallingford, pp. 123–142.

Wibberley, E.J. (1989) *Cereal Husbandry*. Farming Press, Ipswich, UK.
Wigley, T.M.L. (1991) Could reducing fossil-fuel emissions cause global warming? *Nature* **349**, 503–506.
Wildin, J.H. (1994) Beef production from Broadacre Leucaena in Central Queensland. In R.C. Gutteridge and H.M. Shelton (eds), *Forage Tree Legumes in Tropical Agriculture*. CAB International, Wallingford, pp. 352–356.
Wilkes, G. (1989) Maize: domestication, racial evolution and spread. In D.R. Harris and G.C. Hillman (eds), *Foraging and Farming. The Evolution of Plant Domestication*. Unwin Hyman, London, pp. 440–455.
Williams, M. (1989a) Deforestation: past and present. *Progress in Human Geography* **13**, 176–208.
Williams, M. (1989b) *Americans and Their Forests*. Cambridge University Press, Cambridge.
Williams, M. (1990) Forests. In B.L. Turner II, W.C. Clarke, R.W. Kates, J.F. Richards, J.T. Mathewes and W.B. Meyer (eds), *The Earth As Transformed by Human Action*. Cambridge University Press, Cambridge and Clark University, Worcester, Massachusetts, pp. 179–201.
Willis, L.M., Forrest, S.B., Nissen, J.A., Hiscock, J.G. and Kirby, P.V. (1994) Analysis of on-farm best management practices in the Everglades Agricultural Area. In K.L. Campbell, W.D. Graham and A.B. Bottcher (eds), *Environmentally Sound Agriculture*. American Society of Agricultural Engineers, St Joseph, Michigan, pp. 93–99.
Willson, K.C. (1992a) Field Operations 1. In K.C. Willson and M.N. Clifford (eds), *Tea: Cultivation to Consumption*. Chapman and Hall, London, pp. 201–226.
Willson, K.C. (1992b) Field Operations 2. In K.C. Willson and M.N. Clifford (eds), *Tea: Cultivation to Consumption*. Chapman and Hall, London, pp. 227–267.
Wilson, A.D. (1990) The effect of grazing on Australian ecosystems. *Proceedings of the Ecological Society of Australia* **16**, 235–244.
Wilson, E.O. (ed.) (1988) *Biodiversity*. National Academy Press, Washington, DC.
Wilson, E.O. (1992) *The Diversity of Life*. Belknap Press, Cambridge, Massachusetts, USA.
Wilson, P., Clark, R., McAdam, J.H. and Cooper, E.A. (1993) Soil erosion in the Falkland Islands: an assessment. *Applied Geography* **13**, 329–352.
Wing, E.S. and Wheeler, J.C. (eds) (1988) *Economic Prehistory of the Central Andes*. British Archaeology Reports International Series No. 427.
Wisherek, S. (1993) The soil asset: preservation of a natural resource. In S. Wicherek (ed.), *Farm Land Erosion in Temperate Plains Environment and Hills*. Elsevier, Amsterdam, pp. 1–29.
Witherick, M. and Carr, M. (1993) *The Changing Face of Japan: A Geographical Perspective*. Hodder and Stoughton, Sevenoaks.
Witney, B.D. and McRae, D.C. (1992) Mechanisation of crop production and handling operations. In P.M. Harris (ed.), *The Potato Crop: The Scientific Basis for Improvement*. Chapman and Hall, London, pp. 570–607.
Witte, J. (1993) Zaire: landlessness and deforestation. In M. Colchester and L. Lohmann (eds), *The Struggle for Land and the Fate of the Forests*. The World Rainforest Movement, Penang, The Ecologist, Sturminister Newton and Zed Books, London, pp. 179–197.
Wolfe, D.W. and Erickson, J.D. (1993) Carbon dioxide effects on plants: uncertainties and implications for modelling crop response to climate change. In H.M. Kaiser and T.E. Drennen (eds), *Agricultural Dimensions of Global Climate Change*. St Lucie Press, Delray Beach, Florida, pp. 153–178.
Wood Mackenzie Consultants Ltd. (1994) Agrochemical products part 1: the key agrochemical product groups. Agrochemical Service, products Section Update. Wood Mackenzie Consultants Ltd., London, p. 1.
World Bank (1994) *World Development Report 1994*. Oxford University Press, Oxford.
World Commission on Environment and Development (1987) *Our Common Future*. Oxford University Press, Oxford.
World Conservation Monitoring Centre (1992) *Global Biodiversity: Status of the Earth's Living Resources*. Chapman and Hall, London.
World Resources Institute (1992) *World Resources 1992–93*. Oxford University Press, Oxford and New York.

Wright, K.I. (1994) Ground-stone tools and hunter-gatherer subsistence in southwest Asia: implications for the transition to farming. *American Antiquity* **59**, 238–263.

Wrigley, E.A and Schofield R.S. (1988) *The Population History of England 1541–1871: A Reconstruction*, 2nd edn. Cambridge University Press, Cambridge.

Yan, W. (1991) China's earliest rice agriculture remains. *Bulletin of the Indo-Pacific Prehistory Association*, **10**, 118–126.

Yelling, D. (1990) Agriculture 1500–1730. In R.A. Dodgshon and R.A. Butlin (eds), *An Historical Geography of England and Wales*, 2nd edn. Academic Press, London, pp. 181–198.

Yilma, T. (1993) Transfer of technologies in molecular biology to developing countries. Recombinant vaccines and rapid diagnostic kits for diseases in the developing world. In G.T. Tzozotos (ed.), *Biotechnology R and D Trends*. The New York Academy of Sciences, New York, pp. 22–31.

Zaimeche, S.E. (1994) The consequences of rapid deforestation: a North African example. *Ambio* **23**, 136–140.

Zalom, E. and Fry, W. (1992) *Food Crop Pests and the Environment: The Needs and Potential for Biologically Intensive Integrated Pest Management*. American Phytopathology Society, St Paul, Minneapolis.

Zhang, X., Quine, T.A., Walling, D.E. and Zhou, L. (1994) Application of the caesium-137 technique in a study of soil erosion on gully slopes in a yuan area of the loess plateau near Xifeng, Gansu Province, China. *Geografiska Annaler* **76A**, 103–120.

Zhao, S. (1994) *Geography of China*. John Wiley and Sons, New York.

Zhaohua, Z., Mantang, C., Shiji, W. and Youxu, J. (eds) (1991) *Agroforestry Systems in China*. Chinese Academy of Forestry, Beijing and the International Development Research Centre, Ottawa.

Zhenda, Z., Xizhang, W., Wiei, W., Guoding, K., Che, Z., Fafeng, Y. and Tao, W. (1992) China: desertification mapping and desert reclamation. In UNEP (ed.) *World Atlas of Desertification*. Edward Arnold, London, pp. 46–49.

Zimmerer, K.S. (1993) Agricultural biodiversity and peasant rights to subsistence in the central Andes during Inca rule. *Journal of Historical Geography* **19**, 15–32.

Zohary, D. (1989) Domestication of southwest Asian neolithic crop assemblage of cereals, pulses and flax: the evidence from the living. In D.R. Harris and G.C. Hillman (eds), *Foraging and Farming: The Evolution of Plant Exploitation*. Unwin Hyman, London, pp. 358–373.

Zohary, D. and Hopf, M. (1993) *Domestication of Plants in the Old World*. 2nd edn. Oxford University Press, Oxford.

Index